2023 5th Iranian International Conference on Microelectronics (IICM 2023)

Tehran, Iran
25 – 26 October 2023

IEEE Catalog Number: CFP23BX3-POD
ISBN: 979-8-3503-6020-2

**Copyright © 2023 by the Institute of Electrical and Electronics Engineers, Inc.
All Rights Reserved**

Copyright and Reprint Permissions: Abstracting is permitted with credit to the source. Libraries are permitted to photocopy beyond the limit of U.S. copyright law for private use of patrons those articles in this volume that carry a code at the bottom of the first page, provided the per-copy fee indicated in the code is paid through Copyright Clearance Center, 222 Rosewood Drive, Danvers, MA 01923.

For other copying, reprint or republication permission, write to IEEE Copyrights Manager, IEEE Service Center, 445 Hoes Lane, Piscataway, NJ 08854. All rights reserved.

*** This is a print representation of what appears in the IEEE Digital Library. Some format issues inherent in the e-media version may also appear in this print version.

IEEE Catalog Number: CFP23BX3-POD
ISBN (Print-On-Demand):   979-8-3503-6020-2
ISBN (Online):   979-8-3503-6019-6

Additional Copies of This Publication Are Available From:

Curran Associates, Inc
57 Morehouse Lane
Red Hook, NY 12571 USA
Phone: (845) 758-0400
Fax: (845) 758-2633
E-mail: curran@proceedings.com
Web: www.proceedings.com

2023 5th Iranian International Conference on Microelectronics (IICM 2023)

Tehran, Iran
25 – 26 October 2023

IEEE Catalog Number: CFP23BX3-POD
ISBN: 979-8-3503-6020-2

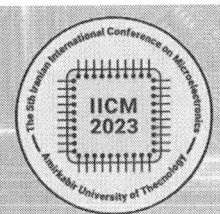

The Fifth Iranian International Conference on Microelectronics

iicm2023.aut.ac.ir

Amirkabir University of Technology
(Tehran Polytechnic)

TABLE OF CONTENTS

PAPERS

SESSION 1 .... Analog and Mixed-Signal Integrated Circuits and Systems

1. A Bootstrapped Switch Based Efficient CMOS Full-Wave Active Rectifier for Biomedical Implants 1

2. A High-Precision Low-Dropout Regulator With High Current Efficiency and Slew-Rate Enhancement 6

3. Double-OTA External Capless Low-power LDO Regulator with Enhanced PSR and Transient Response 11

4. A Low-Power Bandgap Voltage Reference Circuit With Ultra-Low Temperature Coefficient 16

5. A Nanowatt Low Voltage Subthreshold CMOS Voltage Reference Based On 2-T 21

6. A Curvature Compensated CMOS Bandgap Voltage Reference With 6.8 ppm/°C Temperature Coefficient and Low Quiescent Current 26

SESSION 2 .... Optoelectronic Circuits and Devices (1)

7. Two-Wavelength Quantum Dot Mid-Infrared Photodetectors Using Solution Process Method 31

8. Role of Doping Concentration of n- and p-Strip Regions on Optoelectronical Characterization in IBC-SHJ Solar Cell 36

9. Broadened Graded Asymmetric Waveguide Structure for Low Divergence 915nm Diode Laser 41

10. Base Transit Time Investigation of InP/InGaAs HBT Optoelectronic Mixer Using Different Base Doping Profiles 46

SESSION 3 .... Radio Frequency Integrated Circuits and Systems (1)

11. Current-Mode Wideband Frontends With Linearity Enhancement for 5G Receivers 51

12. Design of a Calibration Circuit for Adaptive Phase-Locked Loop in the 5GHz Range Using CMOS 180nm Technology 56

13. A 0.9-8 GHz Highly Linear SAW-Less Direct-Conversion Receiver Front-End for 5G Communication Standard 62

14. A New Design for 1.75 to 2.55 GHz GaN Power Amplifier with More Than 40 dBm Output Power and 12 dB Maximum Gain 67

15. Design and Performance Analysis of a Diplexer for Simultaneous GSM and Bluetooth Communication 72

SESSION 4 .... Digital Integrated Circuits and Systems

16. High-Level Synthesis-Based Approach for CNN Acceleration on FPGA 77

17. Neural Networks & Logistic Regression for FPGA Hardware Trojan Detection 82

18. Ultra Low Power SRAM-PUF for IoT Devices Based on CNTFETs 86

19. A Novel Approach for Offline and Online Application-Dependent testing of FPGA interconnects 91

SESSION 5 .... Biomedical Circuits, Systems, and Devices

20. An Integrated Wearable Bio-Impedance Spectroscopy System for Remote Monitoring Heart Failure in 65nm CMOS Technology 97

21. 2D Axisymmetric Modeling of Circular PCB Coils and Solenoids in COMSOL Multiphysics 102

22. An Asynchronous Strategy for Efficient Audio Processing for Better Perception in Cochlear Implants Based on Peak and Trough Detection 107

23. Numerical Analysis of Studying the Importance of Choosing the Right Image Reconstruction Algorithms in Tomography's Accuracy and Processing Time 112

24. Design of Electrical Stimulation Circuit in 180 nm/1.8 V Standard CMOS Process 117

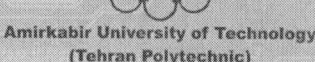

The Fifth Iranian International Conference on Microelectronics

iicm2023.aut.ac.ir

Amirkabir University of Technology
(Tehran Polytechnic)

TABLE OF CONTENTS

SESSION 6 .... **Modeling and Simulation of Circuits and Devices**

25   Thorough Analysis of mm-Wave Broadband Planar and Vertical Transitions for Loss Reduction of Interconnects in Multilayer PCBs 122

26   Impact of Geometrical and Process Design Parameters on the Performance of Schottky Barrier Reconfigurable Field Effect Transistor 128

27   Modeling GaN-HEMT Electrostatic Band Diagram under full depletion approximation 134

28   Magnetic Properties of Permalloy (Co-Ni-Fe) Electroplated Film on Graphene-Oxide (GO) Thin Film Based on Copper Substrate 139

29   Analysis of Electrostatic Interaction Between a Charge Trap and a Quantum Dot Based Single Electron Transistor 145

SESSION 7 .... **Radio Frequency Integrated Circuits and Systems (2)**

30   A Low-Power Differential Ring VCO Using an Active Inductor for Wireless Applications 150

31   A Low-Noise Amplifier with Bandwidth Extension and Noise Cancellation for 5G Receivers 155

32   A Low-Power Inductor-Less Linear Wideband CMOS Balun-LNA Using Current Reuse and Linearity Techniques 160

33   Optimization of 6.5 GHz CMOS Low Noise Amplifier Applying Multi-objective Firefly Algorithm 165

34   Design of a High-Efficiency Deep Bias Class-AB Power Amplifier With 70% PAE at P1dB 172

SESSION 8 .... **MEMS and NEMS Devices and Sensors**

35   Piezoresistive Pressure Sensor Based on Flexo Photosensitive Resin Plate 177

36   A MEMS Resonant Pressure Sensor Based on 2D Graphene Material 181

37   ZnO-Based Surface Acoustic Wave Droplet Sensor 186

38   Design and Fabrication of Carbon Nanoparticles-Based Sensor by Arc Discharge Method 190

39   Synthesis of TiNb2O7 by Mechanical Alloying and Subsequent Heat Treatment as an Anode Material for Li-ion Batteries 195

SESSION 9 .... **Data Converters**

40   A Nonlinear, Low-Power, VCO-Based ADC for Neural Recording Applications 199

41   Design of 1-1-1 Cascaded Discrete-Time Delta-Sigma Modulator based on Tracking Quantizer 204

42   Optimizing High Dynamic Range Current Measurement Circuit for IoT Applications 211

SESSION 10 .... **Optoelectronic Circuits and Devices (2)**

43   Few-Layered Phosphorene Synthesis by CVD Approach as an Anode for Sodium-Ion Battery 216

44   First Principles Study of Optical and Electrical Properties for Mixed-halide 2D BA2PbBr4-xClx (x=0, 2, and 4) as an Active Layer of Perovskite Light Emitting Diode 219

45   Light Trapping in InAsSb-Based Barrier Photodetectors for Enhanced Mid-Wave Infrared Bio-Medical Sensing: A Study on Jurkat Biomarker Detection 222

46   Possible Teleportation of Quantum States using Squeezed Sources and Photonic Integrated Circuits 227

KEYNOTE SPEAKERS **233**

INDEX TO AUTHORS **235**

SCIENTIFIC COMMITTEE **237**

ORGANIZING COMMITTEE **239**

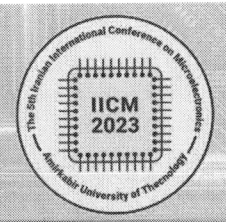

The Fifth Iranian International Conference on Microelectronics

iicm2023.aut.ac.ir

Amirkabir University of Technology
(Tehran Polytechnic)

COVER

CONFERENCE NAME

2023 5th Iranian International Conference on Microelectronics (IICM)

PERIOD

5

LOCATION

**Amirkabir University of Technology (Tehran Polytechnic)
Tehran, Iran**

Date

October 25-26

Amirkabir University of Technology

انجمن میکروالکترونیک ایران

The 5th Iranian International Conference on

Microelectronics

October 25-26, 2023, Amirkabir University of Technology (Tehran Polytechnic)

Organizing Committee:

General Chair
Prof. Ahad Tavakoli

Chair
Prof. Mohammad Yavari

Technical Program Chair
Prof. Hassan Kaatuzian

Industrial Relations
Prof. Majid Shalchian

International Relations
Prof. Samad Sheikhaei

Workshops and Tutorials
Prof. Mohsen Moezzi

Informatics
Prof. Amir Jahanshahi

Publications Chair
Mr. Amir Kashi

Communications and Information
Prof. Sirous Toofan

Finance
Mr. Abbas Khalili

Executive Secretariat
Mr. Farshad Gozalpour
Mr. Ali Sajadi
Mr. Rasoul Fathipour
Mr. Soroush Hashemibani
Ms. Mahboubeh Liaghati Rad

Conference Topics:

- Analog and Mixed-Signal Integrated Circuits and Systems
- Data Converters
- Radio Frequency Integrated Circuits and Systems
- Digital Integrated Circuits and Systems
- Microwave and mm-Wave Circuits and Devices
- Power and Energy Management Circuits and Devices
- Biomedical Circuits, Systems and Devices
- Optoelectronic Circuits and Devices
- Quantum Circuits and Devices
- Displays and Imaging Circuits and Systems
- MEMS and NEMS Devices and Sensors
- Device Fabrication and Packaging Technologies
- Reliability of Circuits and Devices
- Modelling and Simulation of Circuits and Devices
- Emerging Electronic Devices and Materials
- Discrete Electronics Circuits and Systems
- Microelectronic Business and Economy

Important Dates:

- **Paper Submission Deadline: August 13, 2023**
- **Notification of Paper Acceptance: September 03, 2023**
- **Final Manuscript Submission: September 22, 2023**

- As previous IICM conferences, upon approval of the conference by IEEE, English accepted papers will be submitted to IEEE for inclusion in the IEEE Xplore digital library.
- The international authors may present their paper on site as well as online based on their convenience.
- The extended version of selected papers can be submitted to AUT Journal of Electrical Engineering for rapid decision made only by one reviewer.

Secretariat Address:

Department of Electrical Engineering, Abu Reihan Building, Amirkabir University of Technology (Tehran Polytechnic), No. 350, Hafez Ave, Valiasr Square, Tehran, Iran 15875-4413

Website: iicm2023.aut.ac.ir

Email: iicm2023@aut.ac.ir

Tel: +98-21-64543307

Fax: +98-21-66406469

A Bootstrapped Switch Based Efficient CMOS Full-Wave Active Rectifier for Biomedical Implants

Mahmood Alibakhshi
Khajeh Nasir Toosi University of
Technology
Tehran, Iran
mahmood.alibakhshi@.kntu.ac.ir

Farshad Gozalpour
Amirkabir University of Technology
Tehran, Iran
f.gozalpour@aut.ac.ir

Yarallah Koolivand
Khajeh Nasir Toosi University of
Technology
Tehran, Iran
y.koolivand@kntu.ac.ir

Abstract—In implantable medical devices (IMDs), CMOS active rectifiers convert the received AC signal to DC voltage with a high efficiency as possible. One of the main challenges in increasing the power conversion efficiency (PCE) is the reverse current caused by off-delay of the comparators and buffers in the rectifier. In this paper, a new technique has been proposed for the rectifier, which significantly reduces the reverse current without using any compensation loop. Instead of very large PMOS switches, NMOS bootstrapped switches are employed, which reduces the peak value of reverse current by 71% compared to the conventional rectifiers, for almost equal power delivered to the load (PDL). The proposed 6.78 MHz rectifier is implemented with TSMC 180-nm CMOS technology 3.3 V transistors, while the load resistor and off-chip capacitor are 500 Ω and 20 nF, respectively. The simulation results show that for 3 V input voltage amplitude, the PCE of the proposed rectifier is 87%, about 15% more respect to conventional design. Also, the voltage conversion ratio (VCR) of the proposed and conventional rectifiers are 91.6% and 87.5%, respectively.

Keywords—CMOS full-wave active rectifier, reverse current, power conversion efficiency (PCE), bootstrapped switch.

I. INTRODUCTION

Wireless power transmission (WPT) eliminates the need for batteries in implantable medical devices (IMDs) such as cardiac pacemakers [1], cochlear implants [2], and retinal prosthesis [3]. It removes the wiring through the skin, and as a result, prevents the risk of infection and chemical leakage of the batteries. WPT has significant benefits for patient safety and comfortability, potentially extending the life-time of the implanted device as well as reducing the need for further surgery. Inductive links [4] are one of the most widely used and efficient methods in wireless power transmission for IMDs.

Fig. 1 shows an inductive power transmission system, where an active rectifier converts the received AC signal in the secondary coil into a DC voltage in order to provide the required power of biomedical implants. Improving the power conversion efficiency (PCE) of the rectifier is essential to increase the overall efficiency of the transmission system. Fig. 2(a) and Fig. 2(b) show two different implementations of CMOS full-wave active rectifier. In these rectifiers, the

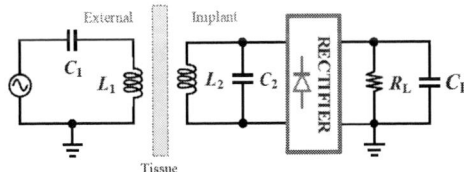

Fig. 1. Wireless inductive power transmission system.

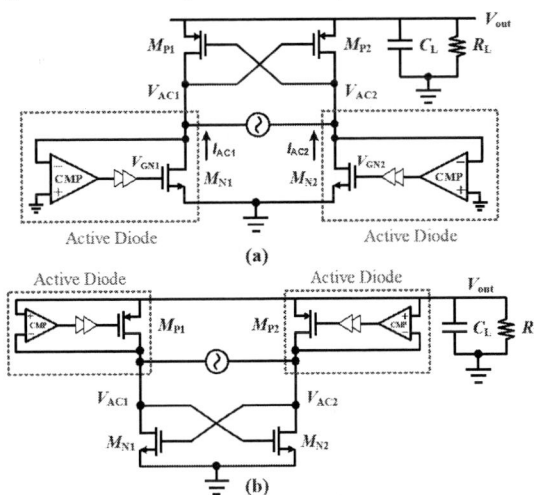

Fig. 2. CMOS full-wave active rectifier with (a) cross-coupled PMOS transistors and comparator-controlled NMOS switches, (b) cross-coupled NMOS transistors and comparator-controlled PMOS switches.

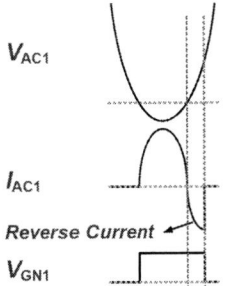

Fig. 3. Reverse current phenomenon in active rectifiers.

979-8-3503-6020-2/23 $31.00 © 2023 IEEE

forward voltage drop is converted to the voltage drop between drain and source nodes of the switches, and as a result, by choosing large *W/L* for switches, it is possible to achieve a high voltage conversion ratio (VCR) for the rectifier. In these rectifiers, switches M_{N1} and M_{P2} are turned on in one cycle, and switches M_{N2} and M_{P1} are turned on in the next cycle. A major challenge in these rectifiers is related to the reverse current (illustrated in Fig. 3), which discharges the load capacitor and reduces the PCE. The reverse current occurs due to the delay of comparator and driving buffer in turning off the NMOS switches M_{N1} and M_{N2} (Fig. 2(a)) or PMOS switches M_{P1} and M_{P2} (Fig. 2(b)).

Different methods have been proposed for delay compensation of the comparators. In these methods an artificial offset voltage is created on the comparator input, which causes the comparator to turns off the rectifier switch earlier and prevents the reverse current. Some of these methods are unbalanced biasing [5], unbalanced sizing [6], offset current injection [7, 8], comparator with voltage mode switched offset [9], and hybrid modes [10]. In unbalanced biasing, two different biasing current sources are employed inside the comparator and an offset voltage is imposed to the comparator [5]. In unbalance sizing, by using switches, the sizing of the transistors in comparator is changed when they are turned on and off, and this causes different currents and imposes an offset voltage [6]. In offset current injection, which is called current mode (CM) switched also, offset currents proportional to the delay of comparator are injected into the branches of the comparator, which create an offset voltage [7, 8]. In voltage mode (VM) switched offset, the offset voltage is directly created on the comparator input without additional current consumption [9]. Finally, the hybrid mode (HM) is the combination of voltage mode and current mode switched offsets [10].

In this paper, a new method is proposed for rectification, in which NMOS bootstrapped switches are used instead of large PMOS switches. Replacing the large PMOS switches with smaller NMOS bootstrapped switches, reduces the parasitic capacitance and the delay time for turning off the switches, leading to lower reverse current than conventional rectifiers, without using any off-delay compensation feedback loops.

The rest of paper is organized as follows. In section II, the proposed rectifier and its working principals is presented. In section III, the simulation results of the proposed rectifier are presented. Finally, the conclusion of paper is given in section IV.

II. PROPOSED RECTIFIER

In conventional rectifiers, two PMOS transistors are used as the switches, which have a very large area (about four times of an NMOS transistor to show an equal ON resistance [7, 10, 11]). Therefore, the reverse current caused by the delay of comparator and the buffer will be very high. In previous works [5-10], complex compensation loops have been employed to eliminate the reverse current. In this paper, in order to reduce the reverse current without using any compensation loops, NMOS bootstrapped switches are used instead of very large PMOS switches. Since NMOS switches have higher mobility than PMOS witches, smaller width can be considered for switches, and as a result, the reverse current of the proposed rectifier will be reduced without using any compensation loops.

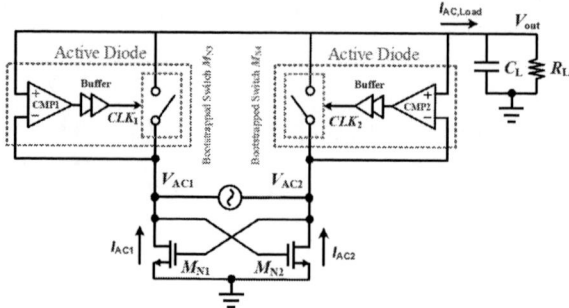

Fig. 4. Proposed bootstrapped switch based CMOS full-wave active rectifier.

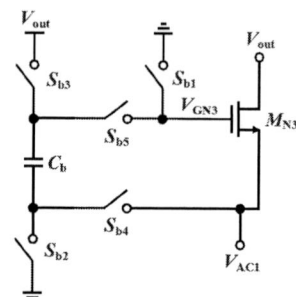

Fig. 5. Structure of the bootstrapped switch M_{N3}.

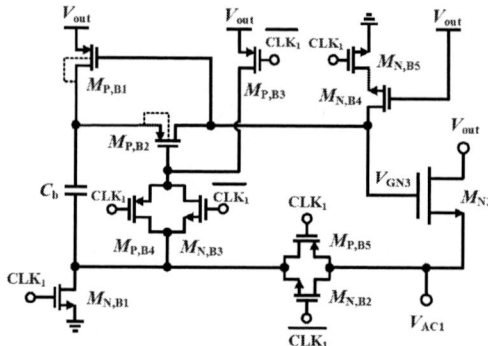

Fig. 6. Transistor-level implementation of the bootstrapped switch M_{N3}.

Fig. 4 shows the structure of the proposed rectifier. In the proposed rectifier, unlike the conventional rectifiers that use the comparators and buffers to switch the large PMOS switches (Fig. 2(b)), they are employed to provide the required clock signals for the bootstrapped NMOS switches M_{N3} and M_{N4} in Fig. 4. According to Fig. 4, the proposed rectifier works as follow: when V_{AC1} is higher than V_{out}, M_{N2} turns on and the output of comparator #1 (CMP1) becomes zero. With inverting the output of the CMP1 twice by the buffer, the required clock signals CLK_1 and $\overline{CLK_1}$ (shown later in Fig. 6) for the bootstrapped NMOS switch M_{N3} are provided. As a result, when the both NMOS switch M_{N2} and bootstrapped NMOS switch M_{N3} are turned on, the received AC voltage is connected to the output node V_{out}. In the next half cycle, when V_{AC2} is higher than V_{out}, M_{N1} turns on, and also, the comparator #2 (CMP2) provides the clock signals CLK_2 and $\overline{CLK_2}$ needed for bootstrapped NMOS switch M_{N4}, and as a result, the received AC voltage is connected to the output node. It is worth mentioning that the parasitic body-drain PN junctions of bootstrapped switches M_{N3} and M_{N4} have the main rule in rectifier start-up process. During the start-up ($V_{AC} > V_{out}$), parasitic PN junctions are forward-

The 5th Iranian International Conference on Microelectronics (IICM2023)

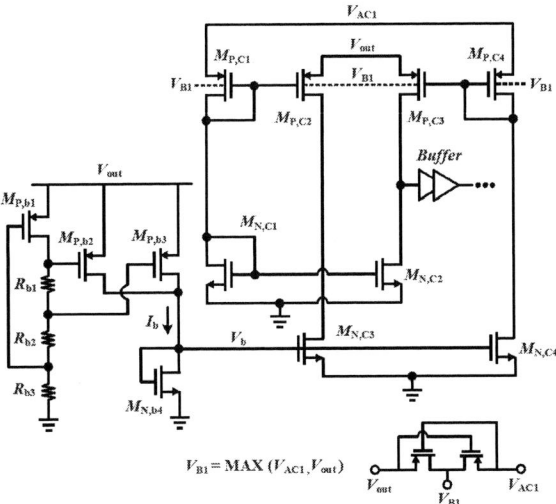

Fig. 7. Structure of the employed comparator in the proposed rectifier.

biased. When the output capacitor C_L is charged up to a minimum required supply voltage of the comparators, the bootstrapped switches M_{N3} and M_{N4} are then activated and the rectifier will operate as designed. In the following, the different parts of proposed rectifier are explained.

A. Bootstrapped Switch

Fig. 5 shows the structure of the bootstrapped switch M_{N3}. In the first phase, the switches S_{b4} and S_{b5} are turned off to disconnect the M_{N3} from the capacitor C_b, and the switches S_{b2} and S_{b3} are turned on to charge the capacitor C_b. Also, to turn off M_{N3}, the switch S_{b1} is turned on. In the next phase, to turn on the M_{N3}, the switches S_{b1}, S_{b2}, and S_{b3} are turned off and the switches S_{b4} and S_{b5} are turned on, so that the capacitor C_b is connected between the gate and source of M_{N3}. The transistor level implementation of the bootstrapped switch is illustrated in Fig. 6. The switch S_{b1} is implemented by transistors $M_{N,B4}$ and $M_{N,B5}$. Switches S_{b2}, S_{b3}, and S_{b5} are implemented by transistors $M_{N,B1}$, $M_{P,B1}$, and $M_{P,B2}$, respectively. Also, the switch S_{b4} is implemented by a transmission gate (transistors $M_{N,B2}$ and $M_{P,B5}$). When S_{b5} is turned off, the transistor $M_{P,B3}$ connects its gate to the highest voltage, and when S_{b5} is turned on, transistors $M_{N,B3}$ and $M_{P,B4}$ connect its gate to a voltage that is lower than its source voltage [12-14].

B. Comparator

The comparator in the proposed rectifier should compare between the output DC voltage V_{out} and the input AC voltage V_{AC}, and then provides the required clock signals for the bootstrapped switch. Shown in Fig. 7 in the designed comparator, PMOS transistors do the comparison work. When V_{AC1} is greater than V_{out}, transistors $M_{P,C1}$ and $M_{P,C4}$ are on, and $M_{N,C2}$, which is connected to $M_{N,C1}$ as a current mirror, is on and drops the comparator output to zero. When V_{AC1} is less than V_{out}, transistor $M_{P,C3}$ is on and connects the comparator output node to V_{out} [6, 15]. It should be mentioned that the bulks of the PMOS transistors are connected to V_{B1}, which is the maximum of V_{out} and V_{AC} through the dynamic body biasing circuit. Also, a peaking current source [11] is employed to generate the required bias voltage V_b in comparator.

TABLE I.   Component Values of the Proposed Rectifier.

Parameter	Value	Parameter	Value
$(W/L)_{N1}$	1800 μm / 0.35 μm	$(W/L)_{N2}$	1800 μm / 0.35 μm
$(W/L)_{N3}$	600 μm / 0.35 μm	$(W/L)_{N4}$	600 μm / 0.35 μm
$(W/L)_{P,C1}$	16 μm / 0.5 μm	$(W/L)_{P,C2}$	16 μm / 0.5 μm
$(W/L)_{P,C3}$	16 μm / 0.5 μm	$(W/L)_{P,C4}$	16 μm / 0.5 μm
$(W/L)_{N,C1}$	8 μm / 0.35 μm	$(W/L)_{N,C2}$	8 μm / 0.35 μm
$(W/L)_{N,C3}$	16 μm / 0.5 μm	$(W/L)_{N,C4}$	16 μm / 0.5 μm
$(W/L)_{N,B1}$	8 μm / 0.35 μm	$(W/L)_{N,B2}$	4 μm / 0.35 μm
$(W/L)_{N,B3}$	4 μm / 0.35 μm	$(W/L)_{N,B4}$	4 μm / 0.35 μm
$(W/L)_{N,B5}$	4 μm / 0.35 μm	$(W/L)_{P,B1}$	16 μm / 0.5 μm
$(W/L)_{P,B2}$	8 μm / 0.35 μm	$(W/L)_{P,B3}$	8 μm / 0.35 μm
$(W/L)_{P,B4}$	8 μm / 0.35 μm	$(W/L)_{P,B5}$	8 μm / 0.35 μm
$(W/L)_{P,b1}$	2 μm / 1 μm	$(W/L)_{P,b2}$	24 μm / 1 μm
$(W/L)_{P,b3}$	6 μm / 1 μm	$(W/L)_{N,b4}$	3 μm / 1 μm
C_b	10 pF	C_L	20 nF
R_{b1}	57 kΩ	R_{b2}	57 kΩ
R_{b3}	240 kΩ		

III. SIMULATION RESULTS

In order to evaluate the presented idea, the rectifier has been implemented with TSMC 180-nm CMOS technology 3.3 V transistors. The device parameters have been summarized in Table I. It is worth mentioning that in these simulations, the values of the load resistance R_L and the load capacitance C_L are considered to be 500 Ω and 20 nF, respectively. Also, input AC signal amplitude and the frequency are 3 V and 6.78 MHz, respectively.

Fig. 8(a) shows the transient input and output voltages of the rectifier. In addition, I_{AC1} and I_{AC2} illustrated in Fig. 8(b). It can be seen that after around 2.5 μs, the operation of the rectifier is settled and the voltage V_{out} is reached

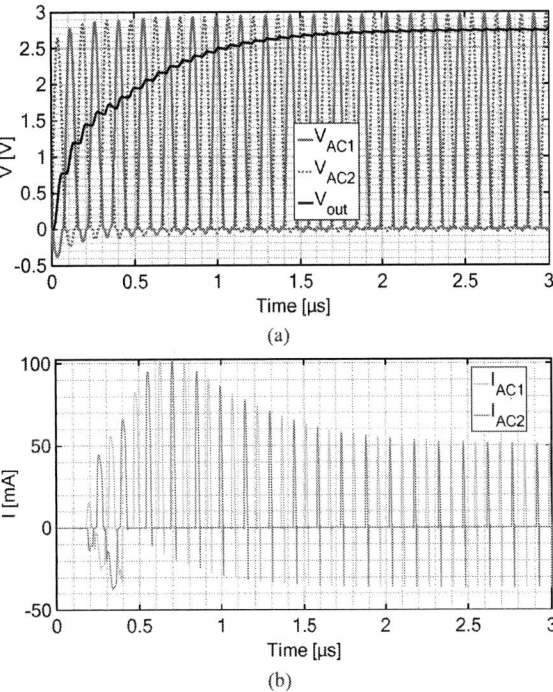

Fig. 8. Start-up simulation of the proposed rectifier with 500 Ω loading and 3V input voltage V_{AC} amplitude. (a) V_{AC1}, V_{AC2}, V_{out}. (b) I_{AC1}, I_{AC2}.

The 5ᵗʰ Iranian International Conference on Microelectronics (IICM2023)

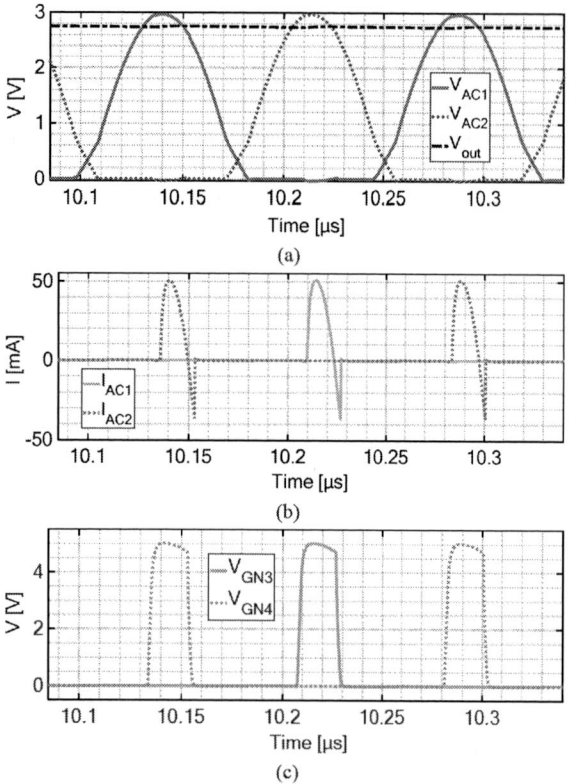

Fig. 9. Steady state simulation of the proposed rectifier with 500 Ω loading and 3V input voltage V_{AC} amplitude. (a) V_{AC1}, V_{AC2}, V_{out}. (b) I_{AC1}, I_{AC2}. (c) V_{GN3}, V_{GN4}.

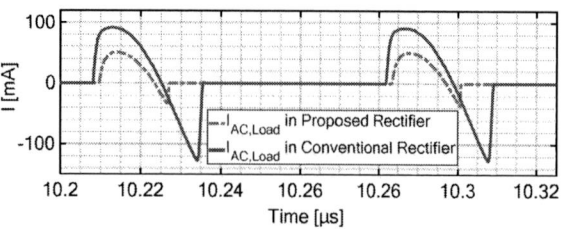

Fig. 10. Reverse current comparison for the proposed and conventional rectifiers with 500 Ω loading and 3V input voltage V_{AC} amplitude.

$\left(|V_{AC}| - |V_{ds,MN1}| - |V_{ds,MN4}| \right)$ in the steady state, where $|V_{ds,MN1}|$ and $|V_{ds,MN4}|$ are the drain-source voltage of the switches M_{N1} and M_{N4}, respectively. The steady state waveforms of V_{AC1}, V_{AC2}, V_{out}, I_{AC1}, I_{AC2}, V_{GN3}, and V_{GN4} are illustrated in Fig. 9, where V_{GN3} and V_{GN4} are the gate voltage of bootstrapped switches M_{N3} and M_{N4} (Fig. 4 and Fig. 6), respectively.

Fig. 10 shows the $I_{AC,Load}$ (shown in Fig. 4) of the proposed and conventional rectifiers with 500 Ω loading and input voltage V_{AC} amplitude of 3 V. The peak values of the reverse current in the proposed and conventional rectifiers are 36.7 mA and 127.8 mA, respectively. So, without using any off-delay compensation loop, reverse current of the proposed rectifier is 71% lower than that of conventional rectifier.

Fig. 11 shows the Monte Carlo simulation of $I_{AC,Load}$ in the proposed rectifier with 500 Ω loading and input voltage V_{AC} amplitude of 3 V. The maximum reverse current peak in this simulation is 40.5 mA.

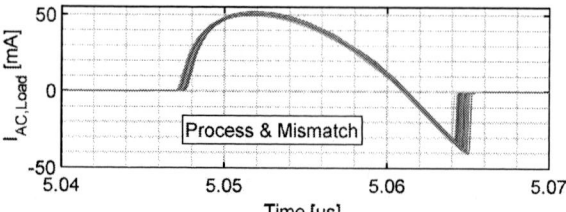

Fig. 11. Monte Carlo simulation of $I_{AC,Load}$ in proposed rectifier with 500 Ω loading and input voltage V_{AC} amplitude of 3 V.

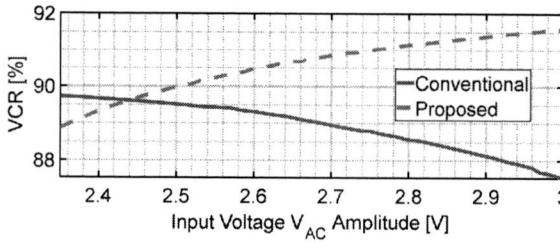

(a)

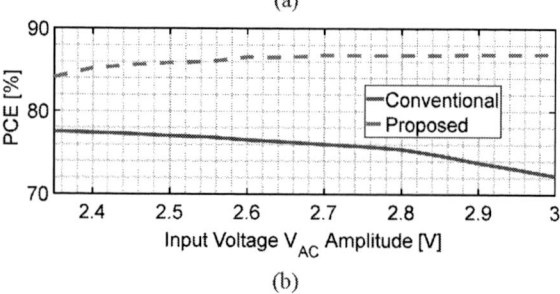

(b)

Fig. 12. Simulated (a) VCR and (b) PCE of the proposed and conventional rectifiers with 500 Ω loading.

Fig. 12 shows the simulated VCR and PCE of the proposed and conventional rectifiers with 500 Ω load, while the input voltage V_{AC} amplitude is swept from 2.35 V to 3 V. Due to the less reverse current, the proposed rectifier shows 87% power conversion efficiency, about 15% more respect to the conventional design with 3 V input amplitude.

The summary of presented work is reported in Table II to provide a comparison with other state-of-the-art counterparts. It can be seen that in most of previous works, the complex feedback loops are employed to compensate the On/Off delay of comparators and buffers. In our proposed work, which is implemented without delay compensation feedback loops, the obtained PCE is significant and comparable to the previous works.

I. CONCLUSION

In this paper, a new CMOS full-wave active rectifier has been proposed which employs NMOS bootstrapped switches instead of very large PMOS switches. Since NMOS transistors have higher mobility than PMOS transistors, smaller width can be considered for NMOS switches, and as a result, the reverse current of the proposed rectifier will be reduced significantly without using any off-delay compensation loop. The peak value of the reverse current in the proposed rectifier decreases by 71% compared to the conventional one. As a result, about 15% improvement in PCE is obtained with 3 V input amplitude.

979-8-3503-6020-2/23 $31.00 © 2023 IEEE

The 5th Iranian International Conference on Microelectronics (IICM2023)

TABLE II. COMPARISON WITH PREVIOUS PUBLISHED WORKS.

	Process [μm]	Area [mm²]	Frequency [MHz]	Output Capacitor	Input Range [V]	$P_{L,Max}$ [mW]	VCR [%]	PCE [%]	Delay Compensation
This Work [#]	**0.18**	**N/A**	**6.78**	**20 nF off-chip**	**2.35-3**	**15.07**	**88.9-91.55** ($R_L = 500\ \Omega$)	**84-86.8** ($R_L = 500\ \Omega$)	**No Compensation**
TCASII'20 [16] [##]	0.13	0.166 *	40.68	500 pF off-chip	1-1.5	9	91-97 ($R_L = 500\ \Omega$)	85.3-93.2 ($R_L = 500\ \Omega$) 92.13-94.2 ($R_L = 500\ \Omega$) [#]	Adaptive On/Off
SSCL'20 [17] [##]	0.18	0.203 **	13.56	10 nF off-chip	1.5-2	54	91-92.5 ($R_L = 500\ \Omega$)	87.1-90.7 ($R_L = 500\ \Omega$)	Adaptive On/Off
TCASII'22 [18] [##]	0.18	0.79 *	40.68	1.53 nF on-chip	1.2-1.8	11.7	N/A	87.3-93.3 ($R_L = 500\ \Omega$)	Adaptive On/Off
TCASII'12 [19] [##]	0.18	0.009 **	13.56	10 μF off-chip	0.9-2	3.2	82-89 ($R_L = 1\ k\Omega$)	81.9 ($R_L = 1\ k\Omega$) [$]	No Compensation
JSSC'16 [8] [##]	65 nm	1.44 *	13.56	4.8 nF on-chip	1.3-2.5	248.1	94.8-97.7 ($R_L = 500\ \Omega$) 91.7-95.2 ($R_L = 100\ \Omega$)	88.5-91 ($R_L = 500\ \Omega$) 91.3-94.6 ($R_L = 100\ \Omega$)	Adaptive On/Off
ACCESS'22 [10] [#]	0.18	0.519 *	6.78	100 nF off-chip	1.6-3	45	92.7-97.5 ($R_L = 500\ \Omega$) 90.6-96.7 ($R_L = 200\ \Omega$)	91.2-92.8 ($R_L = 500\ \Omega$) 94.6-95.4 ($R_L = 200\ \Omega$)	Adaptive On/Off
TCASI'20 [20] [##]	0.18	1.58 *	40.68	2 nF on-chip	2.5-4	56.6	73.2-84.1 ($R_L = 500\ \Omega$)	70.7-80.9 ($R_L = 500\ \Omega$)	Adaptive On/Off
ACCESS'18 [21] [##]	0.13	1.24 *	13.56	1 nF off-chip	2.14-3.6	89	86-96.4 [¥]	88-91.9 [¥]	No Compensation

#. Simulation result, ##. Measurement result, *. Including pads, **. Excluding pads, ¥. $R_L = 100\ \Omega - 1\ k\Omega$ and $V_{AC} = 3.6$ V, $. Maximum PCE.

REFERENCES

[1] U. Anwar, O. A. Ajijola, K. Shivkumar, and D. Marković, "Towards a leadless wirelessly controlled intravenous cardiac pacemaker," *IEEE Trans Biomedical Engineering*, vol. 69, no. 10, pp. 3074-3086, 2022.

[2] F. Gozalpour and M. Yavari, "An Improved FSK-Modulated Class-E Power and Data Transmitter for Biomedical Implants," *AEU-International Journal of Electronics and Communications*, p. 154786, 2023.

[3] A. Akinin et al., "An optically addressed nanowire-based retinal prosthesis with wireless stimulation waveform control and charge telemetering," *IEEE Journal of Solid-State Circuits*, vol. 56, no. 11, pp. 3263-3273, 2021.

[4] M. J. Karimi, A. Schmid, and C. Dehollain, "Wireless power and data transmission for implanted devices via inductive links: a systematic review," *IEEE Sensors Journal*, vol. 21, no. 6, pp. 7145-7161, 2021.

[5] S. Guo and H. Lee, "An efficiency-enhanced CMOS rectifier with unbalanced-biased comparators for transcutaneous-powered high-current implants," *IEEE Journal of Solid-State Circuits*, vol. 44, no. 6, pp. 1796-1804, 2009.

[6] X. Li, C.-Y. Tsui, and W.-H. Ki, "A 13.56 MHz wireless power transfer system with reconfigurable resonant regulating rectifier and wireless power control for implantable medical devices," *IEEE Journal of Solid-State Circuits*, vol. 50, no. 4, pp. 978-989, 2015.

[7] L. Cheng, W.-H. Ki, Y. Lu, and T.-S. Yim, "Adaptive on/off delay-compensated active rectifiers for wireless power transfer systems," *IEEE Journal of Solid-State Circuits*, vol. 51, no. 3, pp. 712-723, 2016.

[8] C. Huang, T. Kawajiri, and H. Ishikuro, "A near-optimum 13.56 MHz CMOS active rectifier with circuit-delay real-time calibrations for high-current biomedical implants," *IEEE Journal of Solid-State Circuits*, vol. 51, no. 8, pp. 1797-1809, 2016.

[9] K. Noh, J. Amanor-Boadu, M. Zhang, and E. Sánchez-Sinencio, "A 13.56-MHz CMOS active rectifier with a voltage mode switched-offset comparator for implantable medical devices," *IEEE Trans Very Large Scale Integration (VLSI) Systems*, vol. 26, no. 10, pp. 2050-2060, 2018.

[10] K. S. Zheng, X. Liu, X. Wang, Q. Su, and Y. Liu, "A 6.78 MHz CMOS Active Rectifier With Hybrid Mode Delay Compensation for Wireless Power Transfer Systems," *IEEE Access*, vol. 10, pp. 46176-46186, 2022.

[11] Y. Lu and W.-H. Ki, "A 13.56 MHz CMOS active rectifier with switched-offset and compensated biasing for biomedical wireless power transfer systems," *IEEE Trans Biomedical Circuits and Systems*, vol. 8, no. 3, pp. 334-344, 2013.

[12] B. Razavi, "The bootstrapped switch [a circuit for all seasons]," *IEEE Solid-State Circuits Magazine*, vol. 7, no. 3, pp. 12-15, 2015.

[13] B. Razavi, "The design of a bootstrapped sampling circuit [the analog mind]," *IEEE Solid-State Circuits Magazine*, vol. 13, no. 1, pp. 7-12, 2021.

[14] P. Pouya, A. Ghasemi, and H. Aminzadeh, "A low-voltage high-speed high-linearity MOSFET-only analog bootstrapped switch for sample-and-hold circuits," in *2015 2nd International Conference on Knowledge-based engineering and Innovation (KBEI)*, 2015: IEEE, pp. 418-421.

[15] R. Erfani, F. Marefat, S. Nag, and P. Mohseni, "A 1–10-MHz frequency-aware CMOS active rectifier with dual-loop adaptive delay compensation and > 230-mW output power for capacitively powered biomedical implants," *IEEE Journal of Solid-State Circuits*, vol. 55, no. 3, pp. 756-766, 2019.

[16] S. Pal, and W.-H. Ki, "40.68 MHz digital on-off delay-compensated active rectifier for WPT of biomedical applications," *IEEE Trans Circuits and Systems II: Express Briefs*, vol. 67, no. 12, pp. 3307-3311, 2020.

[17] Y. Ma, K. Cui, Z. Ye, Y. Sun, and X. Fan, "A 13.56-MHz active rectifier with SAR-assisted coarse-fine adaptive digital delay compensation for biomedical implantable devices," *IEEE Solid-State Circuits Letters*, vol. 3, pp. 122-125, 2020.

[18] Q. Duan, C. Chen, X. Han, and L. Cheng, "A 40.68-MHz Active Rectifier Using an Inverter-Based Conduction-Time Generator for Wirelessly Powered Implantable Medical Devices," *IEEE Trans Circuits and Systems II: Express Briefs*, vol. 69, no. 11, pp. 4334-4338, 2022.

[19] H.-K. Cha, W.-T. Park, and M. Je, "A CMOS rectifier with a cross-coupled latched comparator for wireless power transfer in biomedical applications," *IEEE Trans Circuits and Systems II: Express Briefs*, vol. 59, no. 7, pp. 409-413, 2012.

[20] L. Cheng, X. Ge, L. Hu, Y. Yao, W.-H. Ki, and C.-Y. Tsui, "A 40.68-MHz active rectifier with hybrid adaptive on/off delay-compensation scheme for biomedical implantable devices," *IEEE Trans Circuits and Systems I: Regular Papers*, vol. 67, no. 2, pp. 516-525, 2020.

[21] H.-C. Cheng, C.-S. A. Gong, and S.-K. Kao, "A 13.56 MHz CMOS high-efficiency active rectifier with dynamically controllable comparator for biomedical wireless power transfer systems," *IEEE Access*, vol. 6, pp. 49979-49989, 2018.

The 5th Iranian International Conference on Microelectronics (IICM2023)

25 – 26 October 2023

A High-Precision Low-Dropout Regulator With High Current Efficiency and Slew-Rate Enhancement

Yeganeh Moradzadeh Rezaei
Department of Electrical and Computer Engineering
Urmia University
Urmia, Iran
st_y.moradzadeh@urmia.ac.ir

Mortaza Mojarad
Department of Electrical and Computer Engineering
Urmia University
Urmia, Iran
m.mojarad@urmia.ac.ir

Abstract— **A fast-settling Low-Dropout (LDO) regulator has been presented in this paper. A new frequency compensation scheme has been proposed to ensure stability for all loading conditions without using external compensation capacitors. The new LDO regulator has been designed and simulated in a standard 0.18 μm CMOS process. The simulation results show that the LDO achieves 4.42 MHz and 3.6 MHz unity-gain frequency for 0 mA and 350 mA of load current, respectively, while the minimum phase margin is almost 60 degrees. From the results of the simulations, it can be seen that the worst-case settling time for the load step current from 0 mA to 350 mA is 5.2 μs, which is reduced to 1.5 μs after using a newly proposed slew-rate enhancement circuit. Also, the settling time for the load step current from 350 mA to 0 mA is almost 3 μs. The LDO consumes a quiescent current of 16.5 μA and the power supply rejection is -50 dB at 20 kHz frequency offset.**

Keywords— **Low Dropout, Settling time, LDO Regulator, Power supply rejection, Stability**

I. INTRODUCTION

On-chip and current efficient low dropout regulators are commonly used in portable electronic devices, such as Internet of Things (IoT) and wearable health care equipment [1]. The LDO regulator is inherently unstable due to the common source structure of the pass transistor. Moreover, due to the changes in the load current, the locations of the poles change which causes stability problems. Therefore, it is necessary to design a compensation system for the LDO regulator. Conventional LDO regulators use a large off-chip capacitor to produce a dominant pole to ensure stability. The drawback of this method is the increase in cost and the significant decrease in bandwidth, which slows down the response of the regulator to load and line variations. Another important factor in LDO regulator design is the settling or the recovery time which should be always taken into account. The smaller the settling time, the faster the LDO responds to variations in the load current and the input supply voltage.

Recently, various techniques such as Nested Miller Compensation, active feedback frequency compensation and damping factor control have been introduced to stabilize the LDO regulator without using an external capacitor [2]-[5]. However, achieving high stability for a wide range of the load currents without the use of an external capacitor is still

one of the fundamental challenges. In [6], Nested Miller Compensation technique is used to ensure stability and increase the phase margin. However, the LDO suffers from small loop-gain, low accuracy, and the output voltage changes dramatically with load current variations. In [7], to increase the response speed, the bandwidth has been increased. However, the power consumption is significantly elevated which makes it inappropriate for low-power applications.

In this paper, a low-quiescent-current low-dropout regulator has been proposed. A new compensation scheme has been proposed which facilitates increasing DC gain and bandwidth simultaneously while guaranteeing stability for a large range of load currents. Moreover, a new auxiliary circuit to improve the transient response has been proposed which effectively reduces the settling time while consuming almost zero DC power at the steady state. Also, the newly proposed compensation scheme does not require the conventional off-chip capacitor and does not lead to bandwidth shrinkage resulting in a significantly superior small-signal settling compared to state-of-the-art.

II. STRUCTURE AND CIRCUIT IMPLEMENTATION

The block diagram of the LDO with the proposed compensation scheme has been shown in Fig. 1. The error amplifier consists of three low-gain stages A_1, A_2 and A_3 and a high-gain stage A_4.

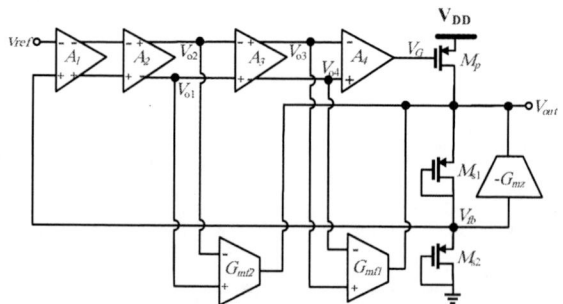

Fig. 1   The block diagram of the LDO with the proposed compensation scheme

979-8-3503-6020-2/23 $31.00 © 2023 IEEE

$\left(\frac{W}{L}\right)$	value
$\left(\frac{W}{L}\right)_{M_P}$	$\left(\frac{100}{0.18}\right)$
$\left(\frac{W}{L}\right)_{1,3,5}$	$\left(\frac{5}{0.18}\right)$
$\left(\frac{W}{L}\right)_{2,4,6,7,8,9}$	$\left(\frac{1.5}{0.18}\right)$
$\left(\frac{W}{L}\right)_{M_{S1},M_{S2}}$	$\left(\frac{0.25}{10}\right)$
$\left(\frac{W}{L}\right)_{17}$	$\left(\frac{1}{0.18}\right)$
$\left(\frac{W}{L}\right)_{18-22}$	$\left(\frac{0.5}{0.18}\right)$
$\left(\frac{W}{L}\right)_{23,24}$	$\left(\frac{3}{0.25}\right)$
$\left(\frac{W}{L}\right)_{25,30}$	$\left(\frac{55}{0.18}\right)$
$\left(\frac{W}{L}\right)_{26-29}$	$\left(\frac{3}{5}\right)$
$\left(\frac{W}{L}\right)_{31}$	$\left(\frac{5}{0.75}\right)$
$\left(\frac{W}{L}\right)_{32-35}$	$\left(\frac{2}{0.18}\right)$
$\left(\frac{W}{L}\right)_{36,38}$	$\left(\frac{0.5}{0.18}\right)$
$\left(\frac{W}{L}\right)_{37}$	$\left(\frac{1}{0.18}\right)$

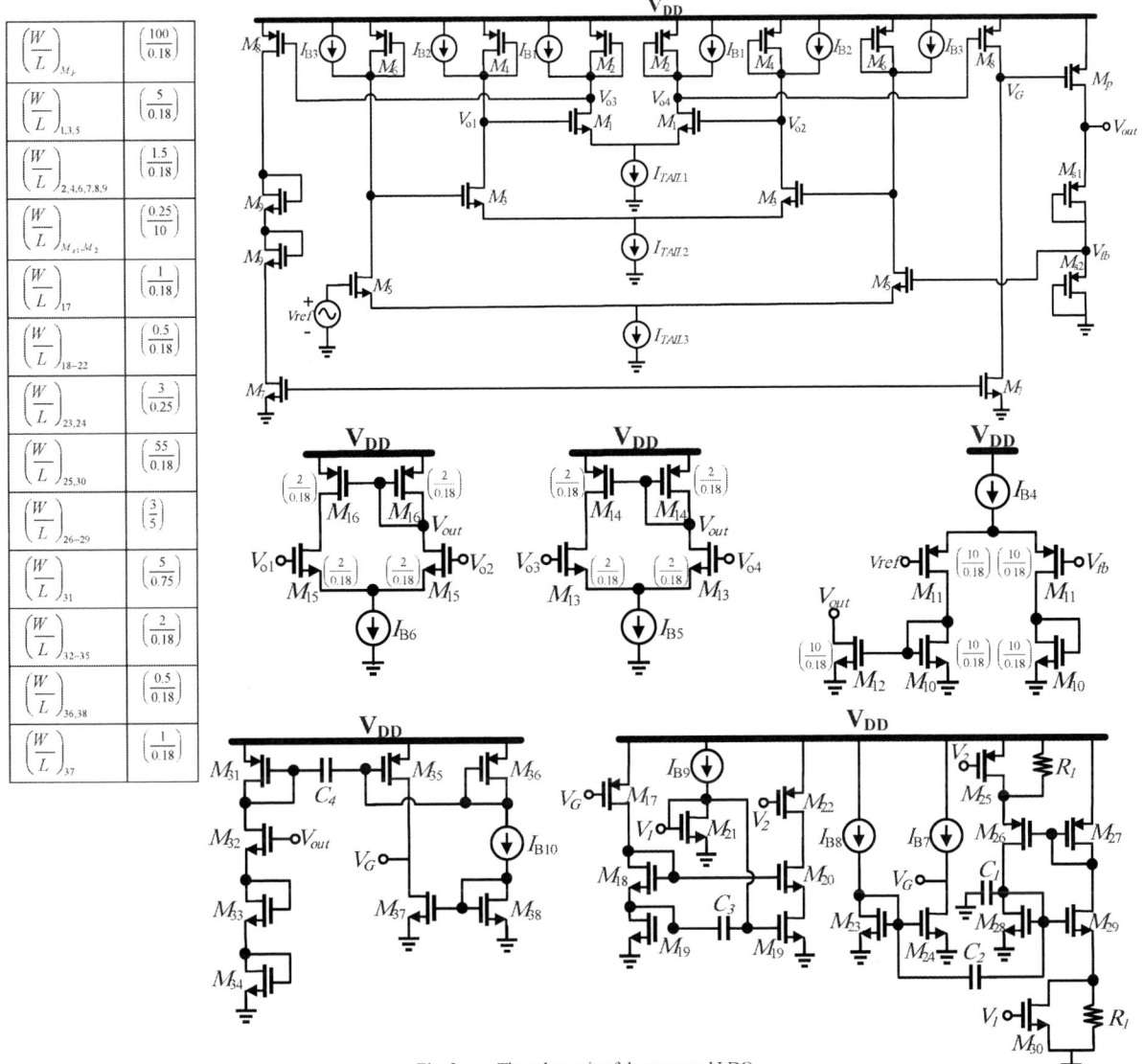

Fig. 2   The schematic of the proposed LDO

Also, the presented LDO structure is comprised of a power transistor, M_P, the transistors M_{S1} and M_{S2} to sample the output voltage, and three transconductance amplifiers, G_{mf1} and G_{mf2}, and G_{mz}. Herein, the frequency response has been analyzed using the small-signal model for the amplifiers and transconductors. There are three high-frequency poles in the transfer function of the LDO, which are associated with the output nodes of the three low-gain stages, A_1, A_2, and A_3. The pole frequencies at the output nodes of the first (A_1), second (A_2), and third (A_3) stages are denoted by ω_{p1}, ω_{p2}, and ω_{p3}, respectively. The frequency of these poles have been calculated as follows:

$$\omega_{p1} = \frac{1}{\frac{1}{g_{m6}} \times C_{par1}} \tag{1}$$

$$\omega_{p2} = \frac{1}{\frac{1}{g_{m4}} \times C_{par2}} \tag{2}$$

$$\omega_{p3} = \frac{1}{\frac{1}{g_{m2}} \times C_{par3}} \tag{3}$$

where considering the schematic in Fig. 2, g_{m2}, g_{m4}, and g_{m6} are the transconductances of M_2, M_4, and M_6, respectively. Moreover, the parasitic capacitances C_{par1}, C_{par2}, and C_{par3} can be approximately described by

$$C_{par1} \approx C_{gs3} + C_{db5} + C_{gs6} + C_{db6} \tag{4}$$

$$C_{par2} \approx C_{gs1} + C_{db3} + C_{gs4} + C_{db4} \tag{5}$$

$$C_{par3} \approx C_{gs8} + C_{db1} + C_{gs2} + C_{db2} \tag{6}$$

Since to implement the first three stages, transistor with small aspect ratios are used, the parasitic capacitance C_{par1}, C_{par2}, and C_{par3} are small. The transconductances g_{m2}, g_{m4} and g_{m6} can be adjusted by using the tail currents I_{TAIL1}, I_{TAIL2}, and I_{TAIL3}, and the biasing currents I_{B1}, I_{B2}, and I_{B3}. As an example, according to the schematic depicted in Fig. 2, larger I_{TAIL1} and smaller I_{B1} results in larger g_{m2} and this means that from (3), ω_{p3} is pushed to higher frequencies and have negligible effect on the phase margin. On the other hand, the gain of each stage of the error amplifier shown in Fig. 1 can be given by

$$A_1 = \frac{g_{m5}}{g_{m6}}, A_2 = \frac{g_{m3}}{g_{m4}}, A_3 = \frac{g_{m1}}{g_{m2}}, A_4 = g_{m8}\left(r_{o8} \| r_{o7}\right). \quad (7)$$

From (1) to (3) and (7), it can be deducted that for adequately large g_{m2}, g_{m4}, and g_{m6}, these poles are located at higher frequencies and have marginal impact on the phase margin. This implies that A_1, A_2, and A_3 are low-gain amplifiers that because are connected in a cascade configuration boost the loop-gain significantly leading to higher precision and improved power supply rejection (PSR) for the LDO.

The non-dominant pole is associated with the output node of the LDO and can be described by:

$$\omega_{p,non-dom} = \frac{C_{gs} + g_{mp} R_{eq} C_{gd}}{R_{eq} C_{gs} (C_L + C_{gd})} \quad (8)$$

where g_{mp}, C_{gs} and C_{gd} represent the transconductance, gate-source parasitic capacitance, and gate-drain parasitic capacitance of the power transistor M_P, respectively. The dominant pole is generated at the gate of the power transistor M_P and has been obtained as:

$$\omega_{p,dom} = \frac{1}{\left(r_{o8} \| r_{o7}\right)(C_{gs} + g_{mp} R_{eq} C_{gd})} \quad (9)$$

It is interesting to note that in the proposed frequency compensation scheme, no Miller capacitor to perform the pole-splitting task has been utilized. However, from (9) it can be concluded that a pole-splitting has been carried out for the parasitic gate-drain capacitance of M_P, and as a result a dominant pole has been generated which improves the phase margin. Therefore, since no other Miller capacitance has been used, the dominant pole is not further pushed towards the origin and the bandwidth is not much reduced. Consequently, the phase margin will not be sufficient to guarantee stability for all loading conditions. In the proposed compensation topology, two active feed-forward paths have been devised as shown in Fig. 1. These feed-forward transconductors, G_{mf1} and G_{mf2}, give rise to a left-half-plane (LHP) zero which has been calculated as follows:

$$\omega_z = \frac{A_3 g_{m4} g_{mp}}{(A_3 G_{mf2} + G_{mf1})(C_{gs} + C_{gd})} \quad (10)$$

By equating this zero to the non-dominant pole $\omega_{p,non-dom}$ in (8), a design constraint can be obtained as follows:

$$A_3 G_{mf2} + G_{mf1} = \frac{A_3 g_{m4} g_{mp} R_{eq} C_L}{C_{gs} + g_{mp} R_{eq} C_{gd}} \quad (11)$$

By satisfying (11) a single-pole property will be achieved for frequencies below the unity-gain frequency and sufficient

phase margin and stability will be ensured. Here, the power transistor has been implemented as a common source for two reason: first, to have a high gain which increases the accuracy of the LDO and load and line regulations and to enhance the PSR. Second, the dropout voltage becomes smaller and the efficiency increases. The transconductance amplifier G_{mz}, consists of transistor M_{10}-M_{12} and is proposed to boost the stability of the LDO. It reduces the output resistance of the LDO and pushes the non-dominant pole of the output node to higher frequencies, improving the phase margin and increasing the stability of the LDO. Transconductance amplifiers G_{mf1} and G_{mf2} are feedforward paths and includes transistors M_{13}-M_{16}.

In order to minimize the quiescent current of M_P, transistors M_{s1} and M_{s2} are very small. Therefore, the transconductance of M_{s1} and M_{s2} become small, and as a result, their equivalent resistances ($1/g_{ms1}$ and $1/g_{ms2}$) are large. By increasing the equivalent resistance, the quiescent current decreases. However, it causes the non-dominant pole at the output of the LDO regulator to move towards the origin and reduces the phase margin. Herein, to mitigate this issue, the transconductance G_{mz} has been added as shown in Fig. 3. Based on the calculation in this section, that G_{mz} reduces the output equivalent resistance.

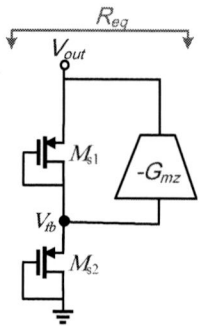

Fig. 3   Reduction of the ac equivalent resistance of R_{eq}.

The equivalent resistance R_{eq} according to Fig. 3 can be obtained as:

$$R_{eq} = \frac{\dfrac{1}{g_{ms1}} + \dfrac{1}{g_{ms2}}}{1 + G_{mz}\dfrac{1}{g_{ms2}}} \quad (12)$$

For $G_{mz}/g_{ms2} \gg 1$ becomes much smaller than $1/g_{ms1}+1/g_{ms2}$, leading to better phase margin by pushing the $\omega_{p,non-dom}$ to higher frequencies. When the load current changes from 0 mA to 350 mA and an undershoot occurs at the output node, transistor M_{17}, which samples the output current at a scale of 1/250000, senses the variation in load current. At this moment, by shorting the capacitor C_3, the current copied from the output is copied by transistor M_{19}. As a consequence, the current of transistor M_{22} increases and the gate voltage of M_{22} is pulled down and V_2 decreases. On the other hand, with the increase in the drain current of M_{19}, the current of M_{21} also increases and as a result, the voltage V_1 increases. Since the transistor M_{25} and M_{30} are biased in the triode region, by increasing V_1 and decreasing V_2, their on-resistance decreases and the current of both branches increases. By shorting capacitor C_2, the current is copied by transistors M_{23} and M_{24}. Since the drain of M_{24} is connected to the gate of M_P, this extra current causes the gate capacitor to discharge faster which increases the settling speed.

Transistor M_{32} is biased in the triode region. When the output voltage is set at 1.6 Volts and the LDO is in the steady state mode, the current of transistors M_{31}-M_{34} is in the range of a few nano-Ampers and does not consume any power. When an overshoot occurs at the output node, the gate-source voltage of M_{32} increases. Accordingly, its on-resistance is lowered and the current of transistors M_{32}-M_{34} increases sharply. Therefore, capacitor C_4 is short-circuited and the current is copied by current mirror (M_{35} and M_{36}) and is injected to the gate of M_p. As a result, the large gate capacitor is charged with more current and the gate voltage of M_P increases at a faster pace. In this way, the output voltage is reduced and the output settles quickly. The significant advantage of slew-rate enhancement circuits is that after the LDO reaches steady state, they are turned off and do not consume any DC power.

III. SIMULATION RESULTS

The proposed LDO voltage regulator has been implemented in a standard 0.18 μm CMOS process. The frequency response simulation results for no load (I_{LOAD}=0mA) and heavy load (I_{LOAD}=350mA) conditions are shown in Fig.4. For I_{LOAD}=0mA the unity gain frequency is equal to 4.42 MHz while the phase margin is equal to 60 degrees. For the heavy load condition, the unity gain frequency and phase margin are equal to 3.6 MHz and 62 degrees, respectively.

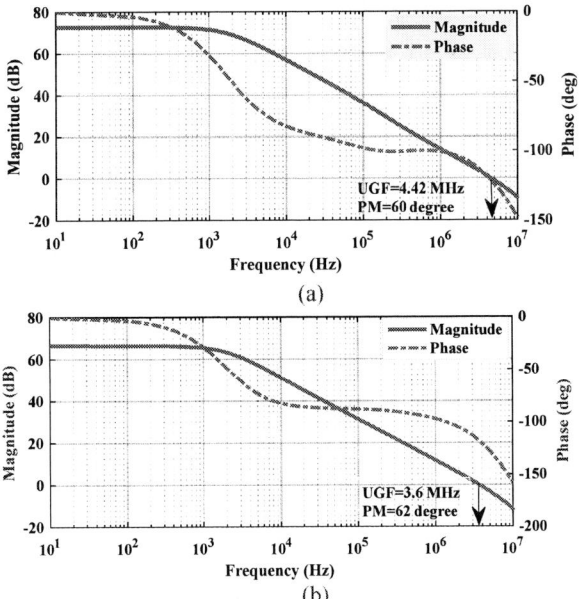

Fig. 4   Frequency response of the proposed LDO for a) no load and b) heavy load conditions

The simulated transient responses for different load steps are shown in Fig.5 and Fig. 6. For the load step of 0 mA to 350 mA without using the proposed slew rate enhancement circuits, the recovery time is almost 5.2μs, and when the slew rate enhancement circuits are included, the recovery time is roughly equal to 1.5μs, as shown in Fig. 5. Fig. 6 shows the result for the 350 mA to 0 mA load step and the settling time is roughly equal to 3μs.

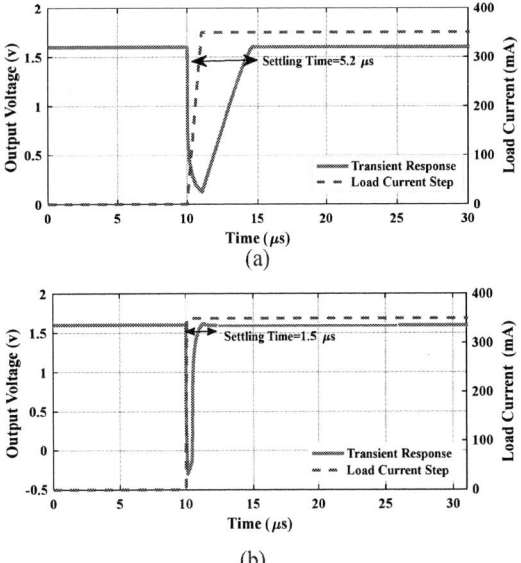

Fig .5   a) Transient response for load step of 0-350 mA without proposed slew rate enhancement circuit, b) Transient response for load step of 0-350 mA with proposed slew rate enhancement circuit,

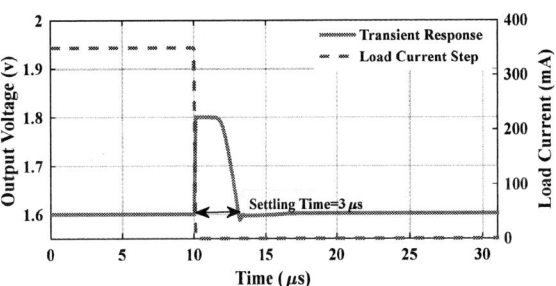

Fig .6   Transient response for load step of 350-0 mA

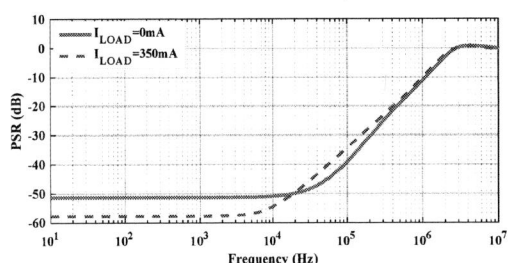

Fig.7   PSR simulation for I_{load}=0,350 mA

Table. 1   The simulation results for different corners

	TT @ 27		FF @ -40		SS @ 85	
	0 mA	350mA	0 mA	350mA	0 mA	350mA
Gain(dB)	70	66.4	62	61.2	71	59.1
UGF(MHz)	4.42	3.6	1.4	4.9	2.8	1.72
PM(deg.)	60	62	67	55	52	66.5

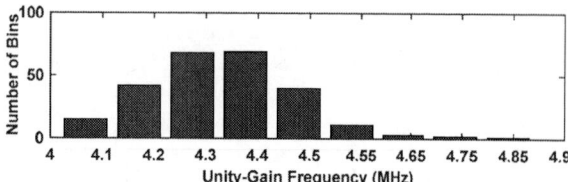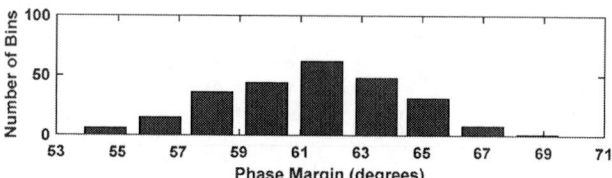

Fig.8 Monte Carlo simulation for the unity-gain frequency and phase margin for no-load condition

Table. 2 Comparison to Previous Art

Parameter	Unit	This work[*]	[8]	[9][*]	[10][*]	[11][*]	[12][*]	[13][*]	[14][*]
Technology	nm	180	65	65	180	55	350	90	180
Input Voltage (V_{in})	V	1.8-3	1.2	1.2	1.35	0.66-1	1.8-3	0.9-1.2	2
Quiescent Current (I_Q)	μA	16.5	297.5	60	15.27	30	25	1.74	1.99
Max. Load Current (I_{Load})	mA	350	25	10	1	10	100	100	10
Settling Time (T_{settle})	μs	3	1.5	65	63.3	1.67	2.5	1.2	1.35
Load Regulation	mV/mA	0.0018	0.042	-	6.2	0.6	-	0.006	0.02
Line Regulation	mV/V	2.42	3.8	-	116.8	12.5	-	0.4	3
PSR	dB	-40dB @100 kHz	-52dB @1MHz	-21dB @1M	-	-	83@10Hz	-51@ 1 kHz	- 66dB@100 Hz
FOM[**]	ns	0.141	17.85	390	966.591	5.01	0.625	20.9	0.26

[*]Simulation results

[**]FOM=$T_{settle} \times I_Q / I_{Load,max}$

The simulation results for the PSR have also been shown in Fig. 7. The simulation results for different corners with temperature variations spanning from -40°C to 85°C are reported in Table. 1. The Monte Carlo simulation results are also shown in Fig. 8.

The performance of the proposed LDO has been summarized in Table. 2 and a comparison has been conducted to some of the prior works. As it is obvious, both line and load regulations have been improved due to the gain-boosting and the new frequency compensation scheme. Regarding the overall performance of the LDO and the figure of merit (FoM), the proposed LDO outperforms the previously reported regulators.

IV. CONCLUSION

In this paper, a new capacitor-less and low-power low-dropout regulator with fast settling, high accuracy, and high PSR at a wide frequency range has been proposed. In order to ensure the stability, a novel frequency compensation technique has been proposed which guarantees stability for the load current from 0 to 350 mA. Moreover, a new slew rate enhancement technique has been proposed which helps the LDO achieve small settling time. The proposed LDO consumes low quiescent current and the comparison of the LDO to the state-of-the-art demonstrates a substantial improvement in the performance of the LDO.

REFERENCES

[1] M. Boanloo, M., Yavari, M. A Low-Power High-Gain Low-Dropout Regulator for Implantable Biomedical Applications. *Circuits Syst Signal Process* 40, pp. 1041-1060 (2021).

[2] J. Guo and K. N. Leung, "A 6-μ W Chip-Area-Efficient Output-Capacitor-less LDO in 90-nm CMOS Technology," in *IEEE Journal of Solid-State Circuits*, vol. 45, no. 9, pp. 1896-1905, Sept. 2010.

[3] G. A. Rincon-Mora and P. E. Allen, "A low-voltage, low quiescent current, low drop-out regulator," in *IEEE Journal of Solid-State Circuits*, vol. 33, no. 1, pp. 36-44, Jan. 1998.

[4] Ka Nang Leung and P. K. T. Mok, "A capacitor-free CMOS low-dropout regulator with damping-factor-control frequency compensation," in *IEEE Journal of Solid-State Circuits*, vol. 38, no. 10, pp. 1691-1702, Oct. 2003.

[5] S. Chong and P. K. Chan, "A 0.9-/spl mu/A Quiescent Current Output-Capacitor-less LDO Regulator With Adaptive Power Transistors in 65-nm CMOS," in *IEEE Transactions on Circuits and Systems I: Regular Papers*, vol. 60, no. 4, pp. 1072-1081, April 2013.

[6] Huang WJ, Liu SI. Capacitor-free low-dropout regulators using nested Miller compensation with active resistor and 1-bit programmable capacitor array. IET Circuits, Devices and Systems 2008; 2(3):306-316.

[7] P. Hazucha, T. Karnik, B. A. Bloechel, C. Parsons, D. Finan and S. Borkar, "Area-efficient linear regulator with ultra-fast load regulation," in *IEEE Journal of Solid-State Circuits*, vol. 40, no. 4, pp. 933-940, April 2005.

[8] Y. Lim, J. Lee, S. Park, Y.jo, J. Choi. "An external capacitor -less low-dropout regulator with high PSR at all frequencies from 10 kHz to 1 GHz using an adaptive supply-ripple cancellation technique." *IEEE Journal of Solid-State Circuits* 53.9 (2018): 2675-2685.

[9] S. Gweon, J. Lee, K. Kim and H .J. Yoo, "93.8% Current Efficiency and 0.672 ns Transient Response Reconfigurable LDO for Wireless Sensor Network Systems," *IEEE Int. Symp. Circ. Syst. (ISCAS)*, 2019, pp. 1-5.

[10] C .H. Hsieh, C. Y. Du and S. Y. Lee, "Power management with energy harvesting from a headphone jack," *2014 IEEE Int. Symp. Circ. Syst. (ISCAS)*, Melbourne, VIC, Australia, 2014, pp. 1989-1992.

[11] N.Utomo, B. C. T. Teo, X. Y. Lim, V. Navaneethan, Z. Liu, C. B. Tan, Y. D. B. Seah, Y. H. Y. Lam, L. Siek. "Low Voltage Low Power Output Programmable OCL-LDO with Embedded Voltage Reference," *2021 IEEE International Symposium on Circuits and Systems (ISCAS)*, Daegu, Korea, 2021, pp. 1-5.

[12] M. Mojarad and M. Yavari, "A novel frequency compensation scheme for on-chip low-dropout voltage regulators," *IEEE International Conference on Electronics, Circuits, and Systems*, 2011, pp. 318-321.

[13] M. Moradian Boanloo and M. Yavari, "A push-pull FVF based LDO voltage regulator with slew rate enhancement at the gate of power transistor", Microelectronics Journal, vol. 122, April 2022.

[14] V. B. Khezerlu, M. Yavari and M. Mojarad, "A Low-Power High-Precision Low-Dropout Regulator For Biomedical Implants," *30th International Conference on Electrical Engineering (ICEE)*, Tehran, Iran, Islamic Republic of, 2022, pp. 857-863.

The 5th Iranian International Conference on Microelectronics (IICM2023)

25 – 26 October 2023

Double-OTA External Capless Low-power LDO Regulator with Enhanced PSR and Transient Response

Mohammad Ahmadi
IC Design Research Lab,
Electrical Engineering Department,
Shahrood University of Technology,
Shahrood, Iran
mohammad.ahmadi@shahroodut.ac.ir

Emad Ebrahimi
IC Design Research Lab,
Electrical Engineering Department,
Shahrood University of Technology,
Shahrood, Iran
eebrahimi@shahroodut.ac.ir

Abstract—Low drop-out (LDO) regulators play a vital role in power management devices, serving as a critical component. This research paper introduces an innovative LDO regulator that offers enhanced load transient response and power supply rejection (PSR). In the proposed LDO, by using two OTAs as error amplifiers in a complementary scheme and transferring effectively the over/undershoot of the output voltage to their Tail transistors, the bias current of each OTA is modulated dynamically to improve the slew rate and load transient response. Besides, creating a zero by a low pass filter in the voltage supply terminal of the first error amplifier is caused to reduce the effect of the pole which is located in the gate of the pass transistor, which results in the improvement of the output PSR. These zeroes made the proposed LDO stable without any external capacitor with a phase margin of 60 degrees. The proposed design with an input voltage of 1.8 to 2.4 volts has a PSR of -69 dB at a 1 kHz frequency. In addition, the LDO quiescent current is as low as 600 nA, while delivers maximum current of 50 mA to the load.

Keywords—External capacitor-less LDO, fast transient, low drop-out (LDO) regulator, power supply rejection (PSR).

I. INTRODUCTION

The adoption of portable and wearable biomedical devices powered by batteries has revolutionized medical care, and LDO regulators have emerged as an essential component within this ecosystem. By providing stable voltage regulation, minimizing voltage drop, and optimizing power utilization, LDO regulators have played a pivotal role in making medical care more accessible, convenient, and efficient. Moreover, these circuits need to exhibit a quick response to load transients, ensuring minimal output voltage fluctuations during the transition between sleep and active states or vice versa [1]. Furthermore, these equipment incorporate circuits that are susceptible to noise, including analog-to-digital converters (ADCs), where the consecutive switching of digital circuits and noise can generate voltage ripples in the power supply. Given the sensitivity of the analog circuits, it is crucial for the LDO employed in such equipment to possess a high Power Supply Rejection (PSR) capability.

Typically, the power supply rejection (PSR) of a low dropout regulator (LDO) is characterized across three

frequency ranges: 1- The DC PSR covers frequencies in the range of a few Hz. 2- The mid-frequency PSR pertains to frequencies around 10 kHz. 3- The high-frequency PSR applies to frequencies exceeding 10 kHz. This paper centers on enhancing PSR at mid-frequency and high-frequency, which is commonly used in biomedical equipment [2]. According to Fig. 1, various paths exist through which power supply ripples are transmitted to the LDO's output, and each of these paths exerts a distinct influence on the output Power Supply Rejection (PSR). The output PSR in a conventional LDO is influenced by the ripples that propagate through four main paths, i.e. the ripples through the (1) reference voltage, (2) error amplifier, (3) drain-source conductance of the pass transistor and (4) the parasitic capacitance which is located in the pass transistor gate which is caused by the high dimensions of the pass transistor [3]. The most critical path that leads to the significant degradation of PSR in a conventional LDO is the parasitic capacitance situated in the gate of the pass transistor. This parasitic capacitance arises due to the large dimensions of the pass transistor [3]. Therefore it is considered as the primary contributor to the destruction of PSR in the system.

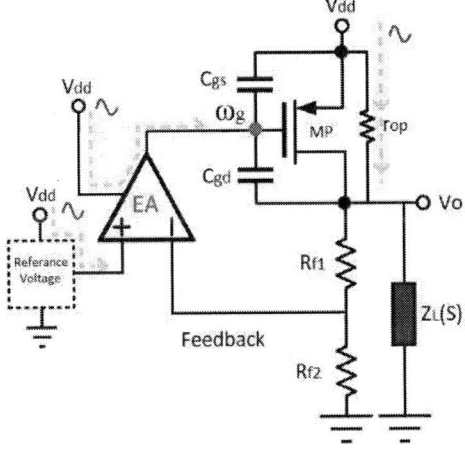

Fig. 1 . Schematic of a conventional LDO.

979-8-3503-6020-2/23 $31.00 © 2023 IEEE

The 5th Iranian International Conference on Microelectronics (IICM2023)

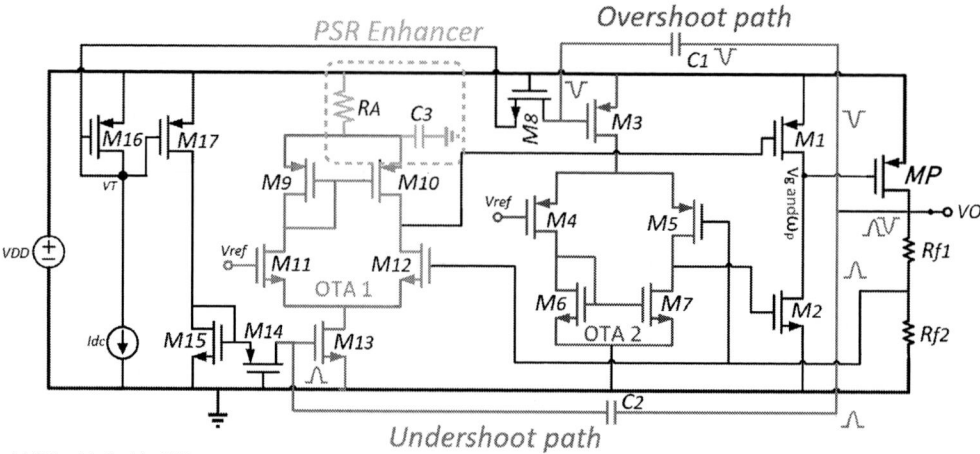

Fig. 2. Proposed LDO with double OTA.

One of the most important parameters that can play a major role in improving of PSR is the LDO's loop gain. Clearly, when the frequency of the pole at the gate of the pass transistor is increased, it leads to a reduction in the loop gain, which results in severe PSR degradation. Also, this pole reduces the feedback loop's speed, which results in a rise in the level of Over/Undershoot voltage when a load current step is present. Several methods have been suggested to minimize the effects of the pole situated at the gate of the pass transistor. One method involves parallelizing the negative capacitor with the pass transistor's gate [4]. However, these techniques often require additional complex auxiliary circuits, resulting in higher power consumption. Another approach suggested in [5] and [6], is the utilization of an NMOS transistor, instead of PMOS, as a pass transistor. However, this approach is not suitable due to the use of a charge pump circuit that results in higher power consumption and output PSR degradation at clock frequency of the charge pump.

Recently, several techniques have been presented to improve load transient response and PSR of the LDO regulators. Most of these techniques draw a lot of current from the supply to move the non-dominant poles to high frequencies, which increases the power consumption of the circuit. On the other hand, some other methods such as [7], [8] use large external capacitors to filter the PSR and improve the load transient response at the output node of LDO. However, the use of off-chip capacitor increase the PCB's area and the material's price.

The rest of the paper is structured as follows. The proposed LDO and its analysis is presented in Section II. summary of the proposed LDO's performance and comparisons with previous works is given in Section III. Finally, Section IV concludes this paper.

II. PROPOSED LDO AND ITS ANALYSIS

Fig. 2 shows the proposed LDO regulator. The proposed LDO regulator comprised of a PMOS pass transistor (Mp), two OTAs as error amplifiers, a feedback network (R_{f1} and R_{f2}), and a Bandgap reference voltage. The feedback resistors sample the output voltage which is compared with Vref by the OTAs. The OTAs generate the appropriate error signal

and apply it to the pass transistor in a negative feedback loop to regulate the output voltage. Apparently the output regulated voltage is determined by the resistive feedback factor (β) and the reference voltage (Vref). The main novelty of the proposed circuit is using two OTAs in a complementary scheme and an R-C network to improve transient response as well as PSR.

According to Fig. 2, two simple OTAs are used as error amplifiers. The output of the first OTA with NMOS input transistors is connected to the gate of transistor M_1, and the output of the second OTA with PMOS input transistors is connected to M_2. Also, in both OTAs, the inverting input is connected to the feedback network and the non-inverting input is connected to the reference voltage. Capacitor C_3 and resistor R_A together form a low-pass filter. As known, to increase the speed of the circuit, the bias current should be maximized and the parasitic capacitance should be minimized. Since the proposed LDO should be used in battery-powered applications that power saving is very important, maximizing bias current is not possible for increasing the life time of battery. Therefore, to save power, the OTAs are biased in subthreshold region and to increase the speed of the loop, the parasitic capacitances are minimized by minimizing the transistors' dimensions. In order to improve the loop speed more a dynamic biasing scheme is also proposed.

A. PSR Enhancement Circuit

One of the parameters that improves the Low-frequency PSR is the loop gain of the LDO. Since the error amplifiers employed in the suggested LDO operate with biasing in the subthreshold region to decrease power consumption and quiescent current, 100 dB loop gain is achieved in low frequencies. Considering that the pole which is located in the gate of the pass transistor causes PSR destruction at high frequencies, a zero is needed to decrease the effect of the pole which is located in the gate of the pass transistor. According to [9] in OTAs with NMOS input transistors, the supply ripples are almost transferred to the output. Therefore, according to Fig. 2, if a low-pass filter is placed on the supply voltage of OTA$_1$, it can create a zero from the input to the gate of the pass transistor and reduce the pole's effect which is located in the gate of the pass transistor. Note that since the OTA is biased at very low bias current, the voltage drop on

979-8-3503-6020-2/23 $31.00 © 2023 IEEE

The 5th Iranian International Conference on Microelectronics (IICM2023)

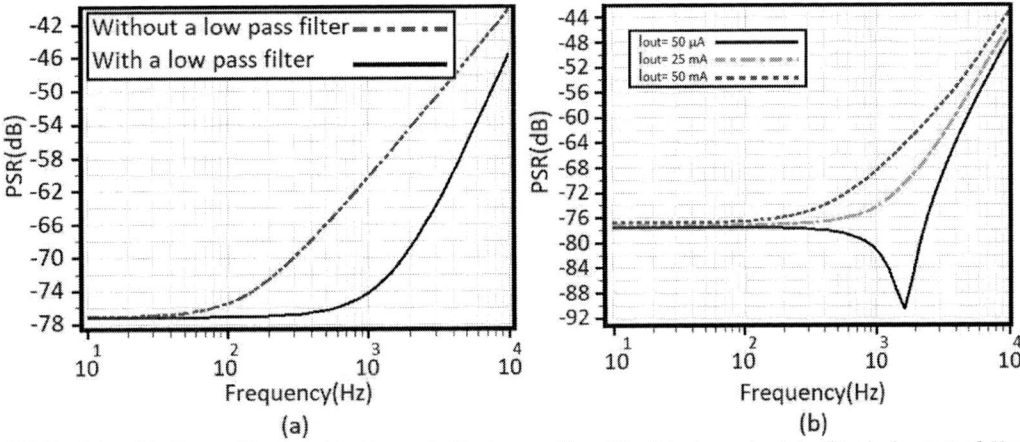

Fig. 3. PSR Simulation of the Proposed Regulator (a) without and with a low pass filter at 25 mA load current and also (b) at load currents of 50mA, 25mA and 50 μA

R_A resistor is negligible. The followings represents the transfer function and zero created by the low-pass filter.

$$\frac{V_g}{V_{dd}} = \frac{R_A C_3 S}{R_A C_3 S + 1} g_{m1} (r_{ds1} \| r_{ds2})$$ (1)

In Eq 1, C_3, R_A are the capacitor and the resistor of the low-pass filter respectively. Also g_{m1} and V_g are the transconductance of the transistor M_1 and gate voltage of pass transistor, and r_{ds1} and r_{ds2} are the drain-source resistance of the transistor M_1 and M_2, respectively.

According to (1) the created zero is at origin and the pole is at $-1/R_A C_3 S$. Considering that C_3 is in the pf range and R_A resistance is in KΩ range, it can be said that the pole is situated in the high-frequency domain, and has no effect on circuit performance at medium and low frequencies. The high dimensions of the pass transistor increase the parasitic capacitances and creates a low frequency pole in the gate of this transistor, which can deteriorate the output PSR in the Mid frequencies. So the created zero reduces the effect of this pole to a suitable extent and improve Mid frequency PSR. Fig. 3(a) demonstrates the simulation results of output PSR with and without a low-pass filter, which apparently the PSR improvement in Mid frequencies is observed. Fig. 3(b) also depicts the output PSR of the LDO under various load currents of 50μA, 25mA and 50 mA, that accordingly shows the effectiveness of proposed technique at all load currents.

B. Over/Undershoot Voltage Improvement and Stability Analysis.

One of the issues that is important in LDOs is the load transient response. Generally, if the feedback loop's speed in the LDO is not high enough, the sudden changes in the load current will cause an over/undershoot voltage in the LDO output. One of the things that slows down the speed of the loop is the pole which is located in the pass transistor's gate. Because of the substantial dimensions of the pass transistor and Miller effect, this pole is not far enough from the origin and can affect the loop speed. The value of this pole is as follows:

$$\omega_p = -\frac{1}{(C_{gsp} + \underbrace{(1 + A_p)C_{gdp}}_{C \ miller})(r_{ds1} \| r_{ds2})}$$ (2)

where C_{gsp} and C_{gdp} are the gate-source and gate-drain internal capacitance of the pass transistor (M_p) and the voltage gain of the pass transistor is Ap. According to the Fig. 2 the use of two OTAs in a complementary way increases the feedback loop speed in the presence of output current changes, which in turn reduces the amount of over/undershoot voltage in the circuit. Since over/undershoot voltage is considered as large signals [10], to examine the performance of the suggested LDO under the influence of over/undershoot voltage, large signal models should be used. According to Fig. 2, When the load current undergoes sudden variations, causing an increase in the output voltage (i.e. overshoot occurs), it causes an increase in the gate-source voltage of transistor M_{12} in OTA_1 and then is transferred to the gate of M_1 with a phase difference of 180 degree. Finally, this change reaches to the gate of the pass transistor (Mp) with another 180-degree phase difference. Since in the MOSFET transistor, a phase difference of 180 degrees exists between the gate and drain voltages, the transferred voltage changes to the gate of Mp is inverted in the output and in this way the overshoot is suppressed. Besides, in order to increase the loop's speed as much as possible, the output overshoot voltage is directly transferred to the gate of the tail transistor of OTA_1 through the capacitor C_2. Actually, this provides a dynamic biasing for OTA_1 that instantly increases the tail current in the presence of overshoot, increases the speed of the OTA_1 and a better suppression of overshoot in LDO output is achieved.

In the same way, when the output voltage decreases due to sudden changes in the load current (i.e. undershoot occurs), it causes an increase in the gate-source voltage of transistor M_5 in OTA_2 and then it is transmitted to the gate of transistor M_2 with a 180-degree phase difference. Finally, this change reaches to the the pass transistor's gate (Mp) with another 180-degree phase difference and suppress the undershoot at output node. In order to increase the loop's speed as much as possible, the output undershoot voltage is straightly transferred to the tail transistor's gate of OTA_2 through the capacitor C_1, and this undershoot voltage instantly increases the tail current which results in an increase in the speed of the OTA$_2$ and a better suppression of undershoot in LDO output.

Briefly, in the proposed LDO each OTA is biased in subthreshold to decrease the quiescent current. In order to increase the loop speed for fast suppression of over/undershoot dynamic biasing is exploited. Using two OTAs in a complementary scheme guaranty the effective suppression of both overshoot and undershoot simultaneously. Fig. 4 shows the proposed LDO's load transient response, based on which the maximum amount of over/undershoot in the current range of 500µA to 50 mA is equal to 164 mV.

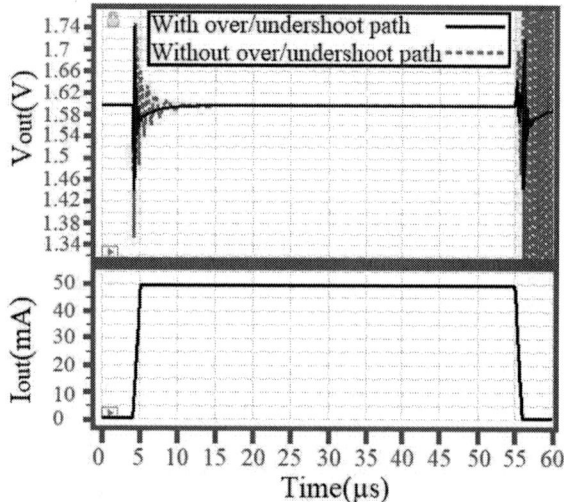

Fig. 4. Proposed LDO's load transient response without over/undershoot path and with over/undershoot path at 1 µs load current edge

Furthermore, using two nested capacitor from the output to the gate of Tail transistor in each OTA, a zero has been created that enables the proposed LDO to be stable without an off-chip capacitor. Fig. 5. illustrates the open-loop frequency response of the proposed design, from which it is determined that the phase margin of the proposed Low Dropout Regulator (LDO) is 47 degrees.

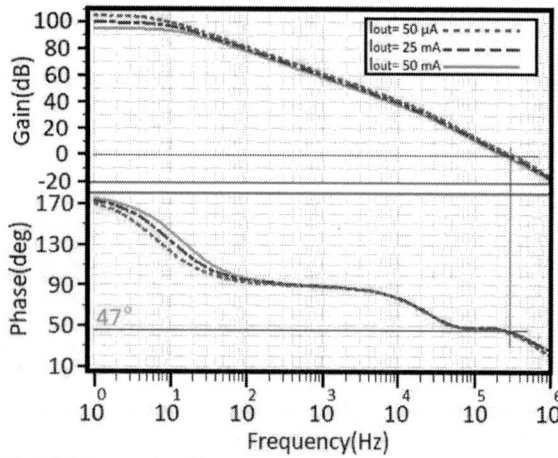

Fig. 5. LDO's open-loop frequency response

III. PVT ANALYSIA AND PERFORMANCE SUMMARY FOR THE PROPOSED LDO AND PRIOR WORKS

To examine the effect of process, voltage, and temperature (PVT) on the LDO performance, The PVT simulation results on the proposed LDO on TT, SS and FF corners in the temperature range of -20 to 85 degrees are presented in Table I, based on which the phase margin and PSR of the circuit in the worst case are equal to 42 degrees and -60 dB, respectively. Also, the amount of over/undershoot and line regulation and load regulation of the circuit in the worst case is only 162 mV, 488 µV/V and 1.58 µV/mA respectively. In addition Fig. 6 illustrates the outcomes of Monte Carlo simulations concerning power supply rejection (PSR) at a 1 kHz frequency and phase margin under a load current of 50 µA. A comparative analysis of the performance between the suggested low-dropout regulator (LDO) and previous research is presented in Table II. To ensure a fair evaluation, a widely recognized figure of merit (FOM) [11], expressed as (3), is employed for the comparison.

$$FOM = K \left(\frac{\Delta V_{out} I_Q}{\Delta I_L} \right)$$

$$K = \frac{\Delta t \text{ used in the measurment}}{The \text{ smallest } \Delta t \text{ among designs for comparison}}$$

(3)

Table. I. Process, Voltage, and Temperature simulation results of the proposed LDO

Corners	Temp (°C)	Over/Undershoot Voltage (mV)	Load Regulation (µV/mA)	Load Current	Line Regulation (µV/V)	Phase margin (degree)	PSR(dB) @1kHz
TT	-20	118	0.8	50µA	15	42	-80
				50mA	21	46.3	-73
	+85	166	2.4	50µA	140	44	-74
				50mA	130	52	-64
FF	-20	148	1.48	50µA	90	48	-77
				50mA	79	49	-70
	+85	114	7	50µA	488	57	-60
				50mA	445	64	-66
SS	-20	119	0.72	50µA	11	42	-80
				50mA	9	46	-76
	+85	139	1.58	50µA	50	43	-72
				50mA	49	50	-60

Tabel II. Summary of the performance of the proposed LDO in comparison to previous works.

Parameters	This work	[12]	[13]	[14]	[15]
Year	2023	2022	2021	2023	2020
Technology (nm)	180	90	90	28	130
Off chip capacitor	No	No	No	Yes	No
I$_{Load(max)}$ (mA)	50	100	40	10	100
I$_{Load(min)}$ (µA)	50	40	30	0	100
V$_{out}$ (V)	1.6	0.75	0.75	0.4V	0.4-1.1
V$_{Drop-out}$ (mV)	200	150	150	200	100
I$_Q$ (µA)	0.6	1.74	1.83	31.4	21
C$_{On chip(pf)}$	4	1.1	0.49	-	11.28
C$_{Load}$	0-20pf	0-100pf	0-100pf	1µf	50pf
Line-reg (µV/V)	125.9	400	1000	9700	24800
Load-reg (µV/mA)	4.6	6	36	800	150
PSR (dB)	-69@1kHz -43@10kHz	-50@1kHz -30@10kHz	-43@1kHz -43@10kHz	NA@1kHz -36@10kHz	-30@1kHz -30@10kHz
Δ I$_{LOAD}$(mA)	49.5	100	49.97	9.9	100
Δ V$_{out}$ (mV/µs)	162/1	350/1.2	440/0.2	10/0.1	89/0.3
FOM (µV)	19.6	73.1	32.2	31.7	56

The 5th Iranian International Conference on Microelectronics (IICM2023)

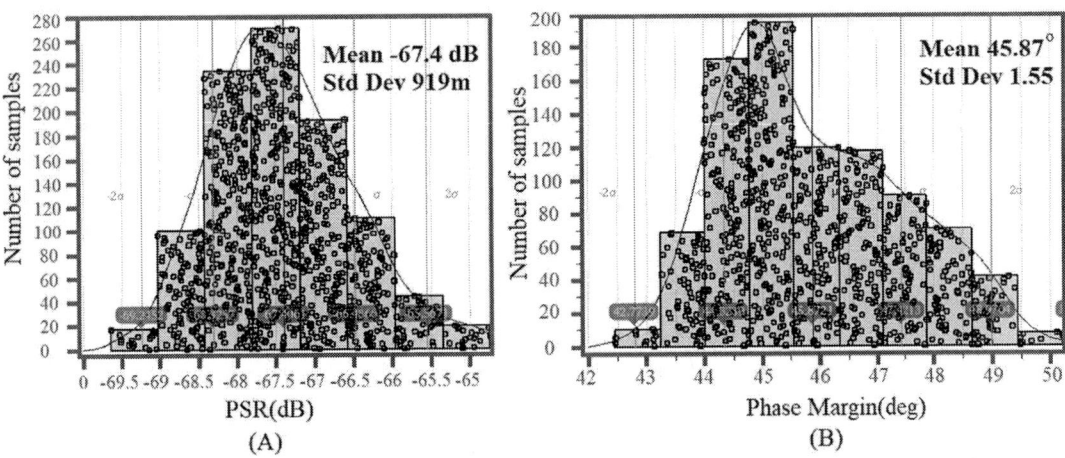

Fig. 6. Monte Carlo simulation outcomes of A, power supply rejection (PSR) at a frequency of 1 kHz, B, phase margin under a load current of 50 μA

IV. CONCLUSION

Low dropout regulators (LDOs) play a crucial role in analog circuits. These circuits require a reliable output PSR and transient response, as they are sensitive to supply ripples and sudden changes in the voltage of the output. Considering that the proposed LDO design is intended for battery-powered circuits, it incorporates a novel approach to dynamically increase the bias current of the error amplifiers. This design achieves a low quiescent current of as low as 600nA and a maximum output over/undershoot of 162mV in the presence of full load step current, making it suitable for low-power circuits. In addition, by creating a zero from the power supply to the gate of the pass transistor by using a low-pass filter, the proposed LDO has an impressive output PSR of -69 dB and -43 dB at frequencies of 1 kHz and 10 kHz, respectively. Furthermore, using two nested capacitor from the output to the gate of Tail transistor in each OTA, a zero has been created that enables the proposed LDO to reach a phase margin of 47 degrees without requiring any external capacitor. So, while the PSR and transient response of the proposed LDO are enhanced, the LDO is stabilized without any external capacitor.

REFERENCES

[1] Y. Huang, Y. Lu, F. Maloberti, and R. P. Martins, 'Nano-ampere low-dropout regulator designs for IoT devices', IEEE Trans. Circuits Syst. I Regul. Pap., vol. 65, no. 11, pp. 4017–4026, Nov. 2018.

[2] S. Heng and C.-K. Pham, 'A low-power high-PSRR low-dropout regulator with bulk-gate controlled circuit', IEEE Trans. Circuits Syst. II Express Briefs, vol. 57, no. 4, pp. 245–249, Apr. 2010.

[3] C.-J. Park, M. Onabajo, and J. Silva-Martinez, 'External capacitor-less low drop-out regulator with 25 dB superior power supply rejection in the 0.4–4 MHz range', IEEE J. Solid-State Circuits, vol. 49, no. 2, pp. 486–501, Feb. 2014.

[4] S. Joo and S. Kim, 'Output-capacitor-free LDO design methodologies for high EMI immunity', IEEE Trans. Electromagn. Compat., vol. 60, no. 2, pp. 497–506, Apr. 2018.

[5] C. Desai, D. Mandal, B. Bakkaloglu, and S. Kiaei, 'A 1.66 mV FOM output cap-less LDO with current-reused dynamic biasing and 20 ns settling time', IEEE Solid-state Circuits Lett., vol. 1, no. 2, pp. 50–53, Feb. 2018.

[6] D. Mandal, C. Desai, B. Bakkaloglu, and S. Kiaei, 'Adaptively biased output cap-less NMOS LDO with 19 ns settling time', IEEE Trans. Circuits Syst. II Express Briefs, vol. 66, no. 2, pp. 167–171, Feb. 2019.

[7] J. S. Kim, K. Javed, K. H. Min, and J. Roh, 'A 13.5-nA quiescent current LDO with adaptive ultra-low-power mode for low-power IoT applications', IEEE Trans. Circuits Syst. II Express Briefs, pp. 1–1, 2023.

[8] I. Jeon, T. Guo, and J. Roh, '300 mA LDO using 0.94 μA IQ with an additional feedback path for buffer turn-off under light-load conditions', IEEE Access, vol. 9, pp. 51784–51792, 2021.

[9] M. Ahmadi and E. Ebrahimi, 'A low power external capacitor - less low drop - out regulator with low over/undershoot voltage and high PSR', Int. J. Circuit Theory Appl., Jun. 2023.

[10] R. Magod, B. Bakkaloglu, and S. Manandhar, 'A 1.24 μA quiescent current NMOS low dropout regulator with integrated low-power oscillator-driven charge-pump and switched-capacitor pole tracking compensation', IEEE J. Solid-State Circuits, vol. 53, no. 8, pp. 2356–2367, Aug. 2018.

[11] P. Manikandan and B. Bindu, 'High-PSR capacitorless LDO with adaptive circuit for varying loads', J. Circuits Syst. Comput., vol. 29, no. 11, p. 2050178, Sep. 2020.

[12] Boanloo, Mehdi Moradian, and Mohammad Yavari. "A push-pull FVF based LDO voltage regulator with slew rate enhancement at the gate of power transistor." Microelectronics Journal 122 (2022): 105389.

[13] Moradian Boanloo, Mehdi, and Mohammad Yavari. "A low-power high-gain low-dropout regulator for implantable biomedical applications." Circuits, Systems, and Signal Processing 40 (2021): 1041-1060.

[14] Kim, Jung Sik, Khurram Javed, Kyoung Hyun Min, and Jeongjin Roh. "A 13.5-nA Quiescent Current LDO with Adaptive Ultra-Low-Power Mode for Low-Power IoT Applications." IEEE Transactions on Circuits and Systems II: Express Briefs (2023).

[15] C. Yang, K. Ye, and M. Tan, 'A 0.5-V capless LDO with 30-dB PSRR at 10-kHz using a lightweight local generated supply', IEEE Trans. Circuits Syst. II Express Briefs, vol. 67, no. 10, pp. 1785–1789, Oct. 2020.

The 5th Iranian International Conference on Microelectronics (IICM2023)

25 – 26 October 2023

A Low-Power Bandgap Voltage Reference Circuit With Ultra-Low Temperature Coefficient

Elaheh Pakravan
Department of Electrical and Computer Engineering
Urmia University
Urmia, Iran
st_e.pakravan@urmia.ac.ir

Mortaza Mojarad
Department of Electrical and Computer Engineering
Urmia University
Urmia, Iran
m.mojarad@urmia.ac.ir

Behboud Mashoufi
Department of Electrical and Computer Engineering
Urmia University
Urmia, Iran
b.mashouf@urmia.ac.ir

Abstract—**A highly accurate bandgap reference with ultra-low sensitivity to temperature variations has been proposed. In the presented work, new techniques and circuits have been proposed to provide voltages with negative and positive temperature coefficients. Moreover, a novel fully MOS CTAT voltage generator without using BJT transistors has been proposed. This voltage and PTAT voltages with different sensitivities to temperature are finally summed up leading to the generation of a reference with a very low temperature dependency. The proposed bandgap reference has been simulated in a 0.18μm CMOS process using ADS. The temperature coefficient is almost 5 ppm/°C, over -40°C to 100°C temperature variation, consuming almost 2μA quiescent current.**

Keywords—bandgap reference; CTAT voltage; PTAT voltage; temperature coefficient; curvature compensation

I. INTRODUCTION

High-precision bandgap reference circuits are important building blocks for analog and mixed-signal systems and circuits such as voltage regulators, amplifiers, data converters, and phase-locked loops. Voltage references generate stable and temperature-independent DC voltages, which is critical for all electronic systems to operate efficiently. There are numerous factors that affect the accuracy of a bandgap reference (BGR). Moreover, different design parameters determine the overall performance of a BGR, such as temperature sensitivity, power consumption, resiliency to process variations, and minimum supply voltage. In other words, minimizing the temperature and supply dependency and also coming up with a low-power circuit are the main challenges in BGR design.

In conventional bandgap reference structures, temperature insensitivity is achieved by the summation of a Complementary-To-Absolute-Temperature (CTAT) voltage and a Proportional-To-Absolute-Temperature (PTAT) voltage, using appropriate coefficients. In Fig. 1, this concept has been illustrated. This structure includes a CTAT voltage generator which is often created by using a p-n junction. The voltage across a p-n junction offers a positive temperature coefficient. In recent years bipolar junction transistors are often utilized instead. However, BJT transistors that are fabricated in the CMOS process have low β coefficients resulting in considerable base current giving rise to offset voltages degrading the accuracy of the reference. In order to solve this problem, in [1] voltage buffers with small output impedances and small offset voltages are placed in the base terminals of all of the BJTs. The drawback of this method is the need for highly accurate voltage buffers which leads to high power dissipation. Another challenge that was mentioned earlier, is supply dependency. In [2], a BGR is implemented which operates with a 1-V power supply, but it is fabricated in BiCMOS technology which is cost-inefficient and has a high power consumption. In [3], an ultra-low-power BGR has been presented, nonetheless, it is of high sensitivity to temperature variations. Some recent works exploit second-order curvature compensation techniques for acquiring BGRs with high precision [4]. However, they require high-voltage power supplies which makes them inappropriate for low-voltage applications. In [5], by using a chopping technique the offset is transferred to high frequencies and is filtered as an offset mitigation technique. Also, employing a single room-temperature trimming using switched-capacitor circuits to remove the PTAT errors is discussed. Although it boosts the accuracy of the BGR, due to the inclusion of digital systems, the proposed BGR is of high complexity and considerable power consumption.

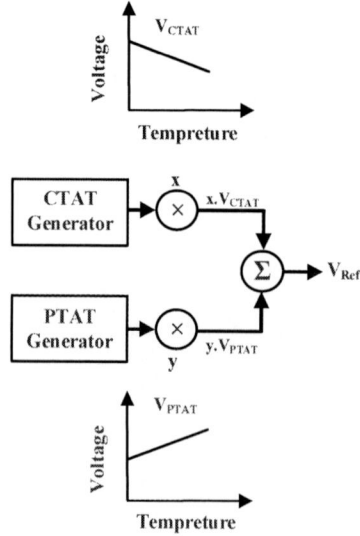

Fig. 1. The basic concept of a conventional BGR structure

979-8-3503-6020-2/23 $31.00 © 2023 IEEE

The BGR in [6] achieves a nano-watts range of power dissipation with enhanced PSR, in exchange for low precision. In traditional BGRs, CTAT voltages are generated using the base-emitter voltage (V_{BE}) of the BJT transistors. These BGRs are highly accurate. However, they consume a substantial amount of quiescent current. In order to reduce cost and save power, producing CTAT voltages by using MOSFET transistors in the sub-threshold region is an interesting alternative [7].

In this paper, an ultra-low-power highly accurate bandgap voltage reference has been proposed. In this work, no BJT transistors have been used and it only exploits MOS transistors. This paper is organized as follows. In Section II, the circuit implementation of the proposed BGR is presented. In Section III, the simulation results are given, and Section IV concludes the paper.

II. THE BASIC STRUCTURE AND IMPLEMENTATION

A. Characteristics of gate-source voltage of MOS transistor

The major advantage of exploiting MOS transistors in the subthreshold region is very low quiescent current consumption. More importantly, the I/V characteristic of a MOS transistor biased in the subthreshold region is similar to that of a BJT and low-β BJTs can be replaced by MOS transistors. Therefore, gate-source voltage of a subthreshold MOS transistor can be considered as a CTAT voltage generator and can be considered as an alternative to the base-emitter voltage of a BJT. The drain current and the gate-source voltage of a MOS transistor in the subthreshold region can be described as follows [8]:

$$I_D = KI_0 \exp(\frac{V_{GS}}{\eta V_T}) \tag{1}$$

$$V_{GS} = \eta V_T \ln(\frac{I_D}{KI_0}) \tag{2}$$

where

$$I_0 = \mu Cox(1-\eta)V_T^2 \tag{3}$$

where $K=(W/L)$, C_{ox} is the gate–oxide capacitance per unit area, and $\eta > 1$ is a non-ideality factor.

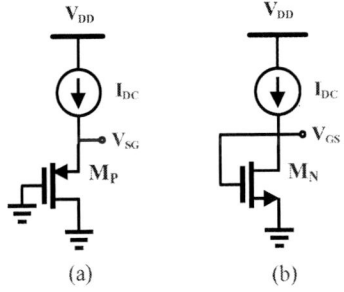

Fig. 2. (a) PMOS and (b) NMOS CTAT voltage generators.

Based on the above, the simplest CTAT voltage generators using PMOS and NMOS transistors are depicted in Fig. 2.

These circuits have been simulated in a 0.18µm CMOS technology and the gate-source voltages are shown in Fig. 3.

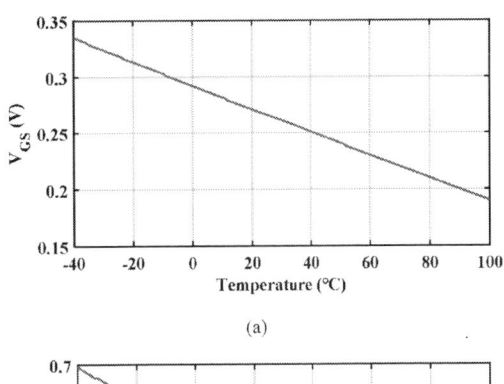

(a)

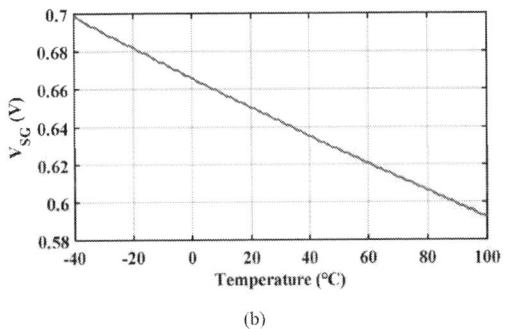

(b)

Fig 3. Simulation result for CTAT voltage generation circuits.

Based on the simulation results shown in Fig. 3, the gate-source voltage of a MOS transistor that is biased by a DC constant current has a CTAT characteristic.

B. PTAT Voltage Generator

Previous PTAT generators often make use of the difference between the base-emitter voltages (ΔV_{BE}) of two BJT transistors. This approach is on the other hand not suitable for low-cost applications. In Fig. 4, a circuit to generate PTAT voltage exploiting MOS transistors has been depicted. This circuit is made up of a source–coupled pair and a current mirror. Also, all of the transistors are biased in the subthreshold region [9].

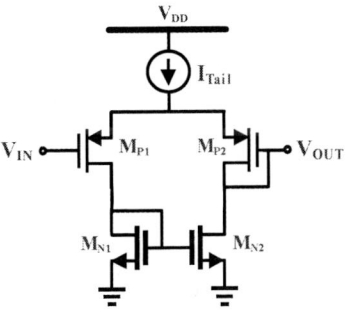

Fig. 4. The CMOS PTAT Voltage generator [9].

In Fig. 4, I_{Tail} is the constant bias current and the output voltage V_{OUT} can be obtained as follows:

$$V_{OUT} = V_{IN} + V_{SG,M_{P1}} - V_{SG,M_{P2}} \tag{4}$$

From (4) and (2) V_{OUT} can be given by

$$V_{OUT} = V_{IN} + \eta V_T \ln\left(\frac{I_{DP1}K_{P2}}{I_{DP2}K_{P1}}\right)$$

$$= V_{IN} + \eta V_T \ln\left(\frac{(\frac{W}{L})_{N1}(\frac{W}{L})_{P2}}{(\frac{W}{L})_{N2}(\frac{W}{L})_{P1}}\right) \tag{5}$$

According to (5), since V_T is proportional to absolute temperature, V_{OUT} is a PTAT voltage.

III. THE PROPOSED BGR

In this paper, in order to enhance the accuracy of the bandgap reference, a new topology has been proposed. In the previous PTAT supplies, temperature sensitivity is small. On the other hand, the sensitivity of the CTAT voltage produced by a BJT transistor is high. As a result, numerous PTAT voltage generators should be cascaded to nullify the CTAT voltage to obtain temperature insensitivity. This approach leads to more complexity, cost, and power consumption. Herein, in order to curtail this issue, a highly sensitive PTAT voltage generator has been proposed.

A. CTAT Voltage Generator

In this work, a fully MOS voltage generator with negative temperature coefficient has been proposed. The presented CTAT voltage circuit is shown in Fig. 5, which is comprised of transistors M_1-M_5. All of the transistors are biased in the subthreshold region, and consequently the power consumption is very low.

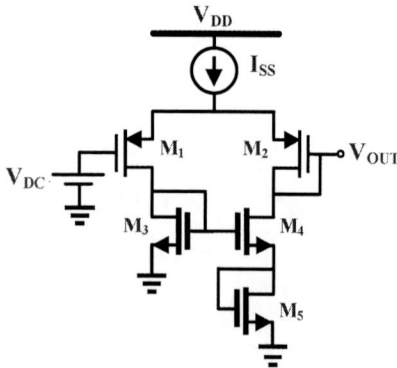

Fig. 5. Proposed CTAT Voltage generator.

The output voltage V_{OUT} in Fig. 5 can be described by

$$V_{OUT} = V_{DC} + V_{SG1} - V_{SG2} \tag{6}$$

By substituting (2) in (6), the output voltage can be calculated as

$$V_{OUT} = V_{DC} + \eta V_T \ln\left(\frac{I_{D1}K_2}{I_{D2}K_1}\right) \tag{7}$$

Applying Kirchhoff's voltage law yields

$$V_{GS3} = V_{GS4} + V_{GS5} \tag{8}$$

From (2) and (8)

$$\eta V_T \ln\left(\frac{I_{D3}}{K_3 I_0}\right) = \eta V_T \ln\left(\frac{I_{D4}}{K_4 I_0}\right) + \eta V_T \ln\left(\frac{I_{D5}}{K_5 I_0}\right)$$

$$\eta V_T \ln\left(\frac{I_{D3}}{K_3 I_0}\right) = \eta V_T \ln\left(\frac{I_{D4}^2}{K_4 K_5 I_0^2}\right). \tag{9}$$

From (9)

$$\frac{I_{D3}}{K_3 I_0} = \frac{I_{D4}^2}{K_4 K_5 I_0^2} \tag{10}$$

and from (10)

$$I_{D4} = \sqrt{\frac{K_4 K_5}{K_3} I_0 I_{D3}}. \tag{11}$$

By inspecting the circuit in Fig. 5, $I_{D3}=I_{SS} - I_{D4}$, and from (11)

$$I_{D4} = \sqrt{\frac{K_4 K_5}{K_3} I_0 (I_{SS} - I_{D4})} \tag{12}$$

From (12) and by assuming that I_0 is very small and for the sake of improving the power efficiency, the drain current of M_4 can be obtained as follows:

$$I_{D4} \approx \frac{\sqrt{4\alpha I_{ss}}}{2} \tag{13}$$

where

$$\alpha = \frac{K_4 K_5}{K_3} I_0 \tag{14}$$

According to (14) and (3), it can be concluded that the coefficient α is a second-order function of temperature. Therefore, from (13), as temperature is raised, the drain current I_{D4} and therefore I_{D2} increases considerably and since $I_{D1}=I_{SS} - I_{D2}$, I_{D1} decreases. From (7), it is apparent that the increase in I_{D2} and the decrease in I_{D1} results in substantial decline in the output voltage, hence V_{OUT} has a CTAT characteristic without using any BJT transistor. The DC voltage in Fig. 5 is realized by using the source-gate voltage of a PMOS transistor as illustrated in Fig. 2(a).

B. The Proposed PTAT Voltage Generator

In Fig. 6, the proposed PTAT voltage generator with improved temperature sensitivity has been proposed. The circuit is composed of transistors M_6-M_9. The main distinction between this circuit and the conventional PTAT circuit in Fig. 4, is the addition of a DC current I_x. Similar to the prior calculations, the output voltage can be given by

$$V_{OUT} = V_{IN} + \eta V_T \ln\left(\frac{I_{D6} K_7}{I_{D7} K_6}\right)$$
$$= V_{IN} + \eta V_T \ln\left(\frac{K_7 (I_{D8} + I_x)}{a K_6 I_{D8}}\right) \tag{15}$$

which can be simplified into

$$V_{OUT} = V_{IN} + \eta V_T \ln\left(\frac{K_7}{a K_6}\left(1 + \frac{I_x}{I_{D8}}\right)\right) \tag{16}$$

It is obvious that the new circuit will be more sensitive to temperature variations, if I_x is of a positive temperature coefficient.

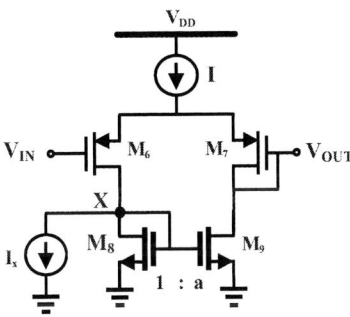

Fig. 6. Improved PTAT Voltage Generator.

In order to generate the PTAT current I_x, a new circuit has been proposed in this paper. The complete schematic of the newly proposed PTAT voltage generator is depicted in Fig. 7. The PTAT current generation is carried out by transistors M_{13}-M_{18} and the amplifier A_1. By assuming V_{OUT} as a PTAT voltage and due to the high gain of the amplifier A_1, the inverting input terminal of the amplifier becomes a PTAT voltage. Each of the transistors M_{18} act as a resistor with a very high resistance, R_{18}.

Therefore, the PTAT current I_{PTAT} in Fig. 7 is equal to V_{OUT}/R_{18} which is of a very small value. This current is then copied by two current mirrors and $I_x = I_{PTAT}$ is formed.

C. Amplifier Implementation

The amplifier A_1, used in the PTAT voltage circuit of Fig. 7, has been realized by a single-ended two-stage Miller compensated amplifier shown in Fig. 8. The first stage is formed by transistors M_{19}-M_{22} and the second stage is composed of transistors M_{23} and M_{24}. The frequency compensation has been performed by the capacitor C_C.

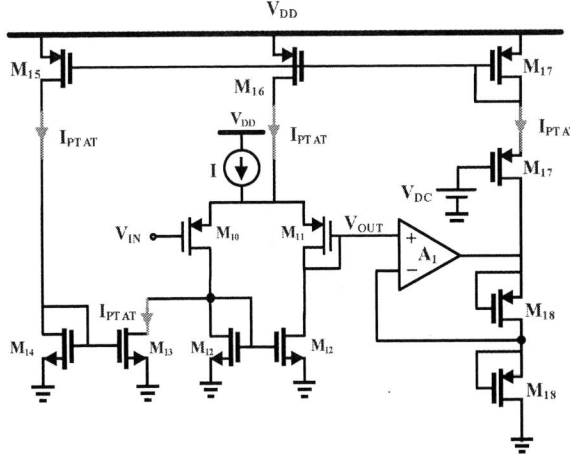

Fig. 7. Proposed PTAT Current Generator.

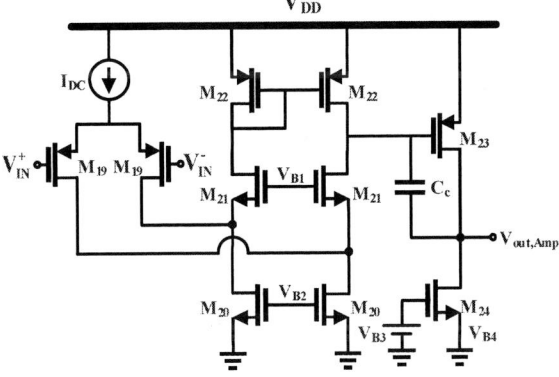

Fig. 8. Schematic of the amplifier.

IV. SIMULATION RESULTS

The proposed bandgap reference which is comprised of the new CTAT voltage generator cascaded by 2 newly presented PTAT voltage generators, and 4 conventional PTAT circuits in Fig. 4, has been simulated in a standard 0.18μm CMOS process using ADS. The reference voltage against temperature variations has been depicted in Fig. 9. It is deduced that the maximum voltage deviation over the temperature span from -40°C to 100°C is almost equal to 0.85 mV.

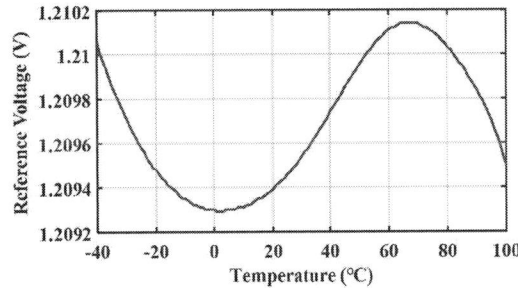

Fig. 9   The reference voltage against temperature variations

979-8-3503-6020-2/23 $31.00 © 2023 IEEE

The 5th Iranian International Conference on Microelectronics (IICM2023)

Table 1   Comparison to Previously Published Works.

	[1]	[2]	[3]	[4]	[5]	[6]	[10]	This Work
CMOS Technology	0.35 μm CMOS	0.5 μm BiCMOS	0.18 μm CMOS	0.13 μm CMOS	0.18 μm CMOS	0.18 μm CMOS	0.13 μm CMOS	0.18 μm CMOS
Supply Voltage (V)	1.4	1	1.8	3.3	1.8	1.8	-	1.8
V_{REF} (V)	0.858	0.19	1.09 –0.548	1.16	1.1419	1.17	1.252	1.2
Temperature range (°C)	-20 to 100	-40 to 125	-40 to 120	-40 to 150	-40 to 125	-40 to 140	-20-110	-40 to 100
Temperature Coefficient (ppm/°C)	12.4	11	147	5.78 – 13.5	3.2 – 5.5	41	68	5.01
Current Consumption (μA)	162	-	0.1	-	-	0.192	-	2.07
Power Consumption (μW)	-	20	-	120	17	-	0.00826	3.726

The other interesting feature of the proposed BGR is low quiescent current consumption. As shown in Fig. 10, the total DC current consumption over the entire temperature span is almost 2μA. Moreover, the current consumption against the input supply variations has been depicted in Fig. 11. The output reference voltage against the supply variation is also shown in Fig. 12 which proves that the reference voltage produced by the new BGR has also a low dependency on the input power supply. The PSR and the simulated temperature coefficient for different corners have been reported in Fig. 13.

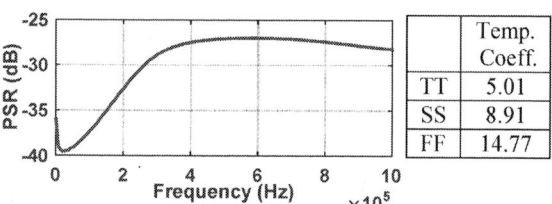

	Temp. Coeff.
TT	5.01
SS	8.91
FF	14.77

Fig. 13 Simulated PSR and the temperature coefficient for different corners

V. CONCLUSION

In this paper, a new bandgap voltage reference has been proposed which achieves very low temperature sensitivity and consumes low quiescent current. In order to accomplish this, new techniques and circuits have been proposed to generate voltages with both negative and positive temperature coefficients. The simulation results and the comparison to the previous works demonstrate the efficacy of the adopted methods.

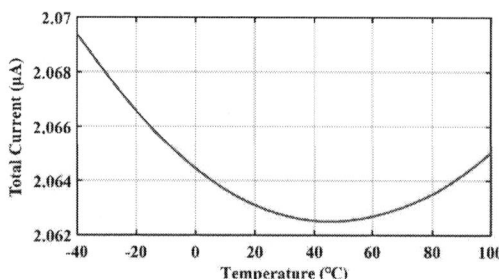

Fig. 10   Current consumption versus temperature

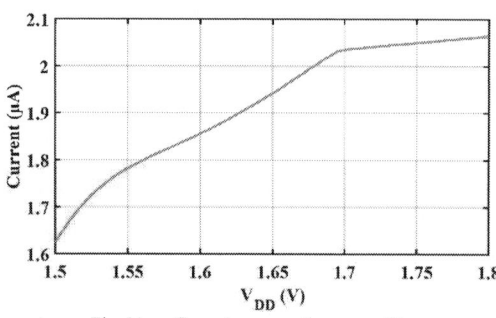

Fig. 11   Current consumption versus V_{DD}

Fig. 12   Reference voltage versus power supply voltage

REFERENCES

[1] R. T. Perry, S. H. Lewis, A. P. Brokaw, and T. R. Viswanathan, "A 1.4-V Supply CMOS Fractional Bandgap Reference" *IEEE J. Solid-State Circuits*, vol. 42, no. 10, Oct. 2007.

[2] K. Sanborn, D. Ma, and V. Ivanov, "A Sub-1-V Low-Noise Bandgap Voltage Reference," *IEEE J. Solid-State Circuits*, vol. 42, no. 11, Nov. 2007.

[3] Y. Osaki, T. Hirose, N. Kuroki, and M. Numa, "1.2-V Supply, 100-nW, 1.09-V Bandgap and 0.7-V Supply, 52.5-nW, 0.55-V Subbandgap Reference Circuits for Nanowatt CMOS LSIs" *IEEE J. Solid-State Circuits*, vol. 48, no. 6, Jun. 2013.

[4] K. Chen, L. Petruzzi, R. Hulfachor, and M. Onabajo, "A 1.16-V 5.8-to-13.5-ppm/°C Curvature-Compensated CMOS Bandgap Reference Circuit With a Shared Offset-Cancellation Method for Internal Amplifiers" *IEEE J. Solid-State Circuits*, vol. 56, no. 1, Jan. 2021.

[5] J.-H. Boo, et al, "A Single-Trim Switched Capacitor CMOS Bandgap Reference With a 3σ Inaccuracy of +0.02%, −0.12% for Battery-Monitoring Applications" *IEEE J. Solid-State Circuits*, vol. 56, no. 4, Apr. 2021.

[6] M. Kim and S. Cho, "A Single BJT Bandgap Reference With Frequency Compensation Exploiting Mirror Pole" *IEEE J. Solid-State Circuits*, vol. 56, no. 10, Dec. 2020.

[7] C. M. Andreou, S. Koudounas, and J. Georgiou, "A Novel Wide-Temperature-Range, 3.9 ppm/°C CMOS Bandgap Reference Circuit," *IEEE J. Solid-State Circuits*, vol. 47, no. 2, pp. 574-581, Feb. 2012.

[8] B. Razavi, Design of Analog CMOS Integrated Circuits, McGraw-Hill, Second edition, 2016.

[9] T. Hirose, K. Ueno, N. Kuroki, and M. Numa, "A CMOS bandgap and sub-bandgap voltage reference circuits for nanowatt power LSIs," *IEEE Asian Solid-State Circuits Conference*, Beijing, China, pp. 1-4, 2010.

[10] S. Wang and P. K. T. Mok, "An 8-nW Resistor-Less Bandgap Reference Based on a Single-Branch Floating PTAT Voltage," *IEEE Solid-State Circuits Letters*, vol. 3, pp. 74-77, 2020.

The 5th Iranian International Conference on Microelectronics (IICM 2023)

October 25-26, 2023

A Nanowatt Low Voltage Subthreshold CMOS Voltage Reference Based On 2-T

Nima Dehghan and Mohammad Yavari

Integrated Circuits Design Laboratory, Department of Electrical Engineering, Amirkabir University of Technology (Tehran Polytechnic), P.O. 15875-4413, Tehran 15914, Iran.

Emails: nimadehghan@aut.ac.ir, myavari@aut.ac.ir

Abstract— **This article presents the design and implementation of a nanowatt CMOS voltage reference circuit capable of operating with a low supply voltage and functioning over a wide temperature range for IoT and biomedical applications. The core of the proposed voltage reference circuit is based on a two-transistor structure. By utilizing the PMOS leakage current biasing technique, the circuit achieves ultra-low voltage operation. This method does not rely on on-chip resistors or operational amplifiers (opamps), thereby reducing complexity and improving reliability. The design is implemented using 0.18-µm CMOS technology. The circuit consumes approximately 69 nA from a minimum supply voltage of 0.9 V at room temperature. The untrimmed output voltage is 560 mV, exhibiting an average temperature coefficient (TC) of 43.4 ppm/°C over the temperature range of -40°C to 85°C. The line-voltage sensitivity (LS) is measured 0.303%/V when the input supply voltage varies from 0.9 V to 1.8 V. Finally the DC power supply rejection ratio (PSRR) at 100 Hz is -81.5 dB.**

Keywords— *nanowatt, CMOS voltage reference circuit, IoT applications, PMOS leakage current, low voltage, reliability.*

I. INTRODUCTION

In many electrical circuits, ranging from amplifiers to analog-to-digital converters (ADCs), digital-to-analog converters (DACs) and phase-locked loop (PLL) circuits, require stable and constant voltage and current reference sources. The term constant refers to the fact that the reference voltage or current should be independent of the power supply and insensitive to temperature variations. The supply voltage of a circuit can vary within an allowable range. However, this voltage often contains a significant amount of noise due to current variations in different circuit components caused by switching or voltage range fluctuations, among other factors. When a constant reference source is required, it is necessary to design the reference source in such a way that it remains independent of the circuit's power supply and unaffected by its variations. This independent reference source is referred to as a source independent of the power supply.

Various methodologies for designing CMOS voltage references (CVR) have been explored in the literature [1]–[16]. The widely adopted approach uses a bandgap voltage reference (BGR) employing parasitic BJTs (bipolar junction transistors). BGR using BJT is a commonly used type of voltage reference that generates a reference voltage with minimal variation due to process, voltage, and temperature (PVT) effects. However, BGRs are not well-suited for applications requiring low supply voltage and low power consumption [1, 2, 10, 11]. This technique combines two voltages with opposite temperature characteristics, a complementary-to-absolute-temperature (CTAT) voltage and a proportional-to-absolute-temperature (PTAT) voltage, to generate a temperature-independent output voltage. Alternatively, some designs utilize the combination of PTAT and CTAT currents, rather than voltages, to achieve a temperature-independent output voltage [3]–[6]. Replacing bipolar transistors with diode-connected MOS transistors biased in the subthreshold region is a suitable alternative in other methods [7]–[9]. The obtained voltage in this case is much lower than the emitter-base voltage of bipolar transistors, allowing them to be used in low-voltage applications. A two-transistor (2-T) subthreshold CVR, as described in [9], obtains the reference voltage by comparing the threshold voltages of two distinct transistor types, resulting in picowatt power consumption and a minimum supply voltage of 0.5 V. Nonetheless, the autonomy of the reference voltage concerning the supply voltage may be compromised by the channel-length modulation effect on the upper transistor linked to VDD in this 2-T structure circuit.

This paper proposes an ultra-low nanowatt voltage reference with sufficiently high line sensitivity (LS) using only seven transistors. In order to protect the delicate core from V_{DD} and to enable circuit operation at high temperatures, a temperature-dependent cascode current generator has been introduced. The suggested system exhibits exquisite performance metrics over the supply voltage range of 0.9 V-1.8 V in a 180-nm prototype.

The structure of the paper is organized as follows: In Section II, the CMOS voltage reference architecture and operational principles are introduced. Section III outlines simulation results, including a comparison with previous work, and finally, Section IV concludes the paper.

II. PROPOSED CMOS VOLTAGE REFERENCE

A. Circuit Description

The conventional structure of a subthreshold voltage reference circuit based on 2-T is illustrated in Fig. 1(a). In this configuration, a native transistor with a threshold voltage close to zero (NVT) is employed as the bias of M_{10}. Since transistor M_{11} is an NVT transistor and its gate is connected to ground, it operates in the subthreshold region,

979-8-3503-6020-2/23 $31.00 © 2023 IEEE

producing a very small current that results in generating a relatively low voltage. To generate higher voltages, the structure in Fig. 1(b) can be utilized. In this arrangement, instead of an nMOS, a pMOS with a moderate threshold voltage (MVT) is used as the bias. Similar to the previous circuit, this structure also encounters issues such as low line sensitivity, PSRR, and other challenges which will be discussed in the following section. The proposed subthreshold CVR circuit based on 2-T is shown in Fig. 1(c). In this arrangement, an attempt has been made to enhance the performance metrics by introducing a few transistors to structure (b). To fulfill the criteria of limited space, reduced voltage, and minimal power usage in internet of things (IoT) applications, all components in this design function in the subthreshold zone. The efficiency of the suggested circuit is additionally improved by the exclusion of operational amplifiers and passive elements like resistors and capacitors.

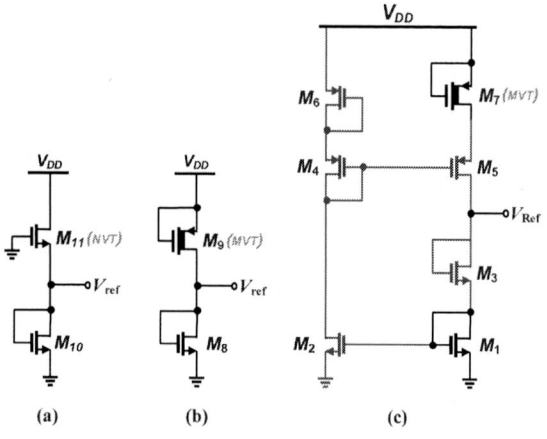

Fig. 1: Schematic of (a) Conventional 2-T CVR used in [9] and (b) 2-T CVR using MVT pMOS and (c) Proposed CVR based on 2-T.

When the voltage across the drain and source of a transistor is sufficiently higher than the thermal voltage (V_T), the subthreshold current can be defined as follows:

$$I_{sub} = \mu C_{ox} \frac{W}{L} (\zeta - 1) V_T^2 \exp\left(\frac{V_{GS} - V_{TH}}{\zeta V_T}\right) \quad (1)$$

where μ is the mobility and C_{ox} is gate oxide capacitance per unit area. W and L represent the width and length of the transistor, respectively. ζ is subthreshold slope factor ($\zeta = 1 + C_d / C_{ox}$ where C_d is depletion capacitance per unit area), V_T is thermal voltage (kT/q), V_{GS} is gate-source voltage and V_{TH} is transistor threshold voltage. In the proposed configuration, by considering the similarity between their depletion capacitance (C_d) and gate oxide capacitance (C_{ox}), their subthreshold slope factors can be assumed to be equal. Furthermore, by equalizing the currents flowing through M_1 and M_7, we can write:

$$\mu_n C_{ox} \left(\frac{W}{L}\right)_1 \exp\left(\frac{V_{REF}/2 - V_{TH,1}}{\zeta V_T}\right) =$$
$$\mu_p C_{ox} \left(\frac{W}{L}\right)_7 \exp\left(\frac{0 - |V_{TH,7}|}{\zeta V_T}\right) \quad (2)$$

Then, the output reference voltage can be expressed as:

$$V_{Ref} = 2 \times \left[|V_{TH,7}| - V_{TH,1} + \zeta V_T \ln\left(\frac{\mu_p}{\mu_n} \frac{\left(\frac{W}{L}\right)_7}{\left(\frac{W}{L}\right)_1}\right) \right] \quad (3)$$

The threshold voltage difference between nMOS and pMOS transistors can be expressed below [12].

$$V_{TH,1} - |V_{TH,7}| = \Delta V_{TH}(T_0) + (\alpha_1 - \alpha_7)T_0 - (\alpha_1 - \alpha_7)T \quad (4)$$

Here, $\Delta V_{TH}(T_0)$ represents the variation in threshold voltage between M_1 and M_7 under normal room temperature conditions. α_1 and α_7 represent the temperature factors associated with $V_{TH,1}$ and, $|V_{TH,7}|$ respectively. Since α_7 is greater than α_1, the disparity in the threshold becomes directly PTAT. It is crucial to ensure that W_7 is more petite than W_1 to achieve a CTAT in the last term of equation (3). Consequently, After compensation, the zero-temperature coefficient (ZTC) V_{Ref} will be:

$$V_{Ref} = 2 \times \left[|V_{TH,7}(T_0)| - V_{TH,1}(T_0) + (\alpha_7 - \alpha_1)T_0 \right] \quad (5)$$

B. Line sensitivity and PSRR

As illustrated in Fig. 1(b), the bias transistor M_9 operates in the subthreshold region with its gate and source linked to VDD, serving as a biasing component. Consequently, during startup, as the input voltage increases, the current also increases, causing the output reference voltage to reach its final value. It is worth noting that The drain terminal of transistor M_9 is likely connected to a constant voltage (V_{Ref}), while its source terminal is connected to V_{DD}. Thus, the ripple present in the power supply is directly reflected in the drain-source voltage of transistor M_9; even if the channel length and drain-source voltage are maximized, the sensitivity and Power Supply Rejection Ratio (PSRR) are low due to the channel length modulation effect. Hence, the objective of transistor M_5 is to decouple the output from the drain of transistor M_9, where the drain-source voltage of transistor M_9 is maintained independently of the power supply by configuring the bias network in such a way that variations can be transferred to the drain. The biasing of transistors M_6 and M_7 is achieved by mirroring of the primary circuit branch current. By appropriately sizing M_6 and M_7 transistors, the input voltage variations that directly appearing at the output in the two-transistor structure can be significantly reduced by adding the left-side path. As a result, this approach increases PSRR and improves LS. By considering the exponential term of the subthreshold current effect, the output voltage reference can be redefined as follows:

979-8-3503-6020-2/23 $31.00 © 2023 IEEE

$$V_{Ref} = 2 \times \left[\mid V_{TH,7} \mid - V_{TH,1} + \zeta V_T \ln \left(\frac{\mu_p \frac{W}{L})_7}{\mu_n \frac{W}{L})_1} \right) \right. \tag{6}$$
$$\left. + \zeta V_T \ln \left(1 - \exp(\frac{V_{SG5} - V_{SG6} - V_{SG7}}{V_T}) \right) \right]$$

Unlike the 2-T structure, where the expression V_{DD} was present internally, in this configuration, the output reference voltage related to the gate-to-source voltage difference, leading to an enhancement in line sensitivity.

Considering the small-signal equivalent circuit of the 2-T structure, which is essentially a resistive divider, the PSRR can be expressed as follows:

$$PSRR \Big|_{2-T} = \frac{1/g_{m8}}{1/g_{m8} + r_{ds,9}} = \frac{1}{1 + g_{mn}r_{dsp}} \tag{7}$$

g_{mn} represents the transconductance of nMOS transistors and r_{dsp} denotes the output resistance of pMOS transistors. Fig. 2 depicts the proposed AC equivalent model for calculating the PSRR of the proposed structure which PSRR can be evaluated as below:

$$\frac{V_{Ref}}{V_{dd}} = \left(\frac{g_{m1} + g_{m3}}{g_{m3}} \right) \left(\frac{1}{1 + g_{m5}r_{ds5}} \right) \left(\frac{g_{m4} + g_{m6}}{g_{m4} + g_{m6} + g_{m4}g_{m6}r_{ds2}} \right)$$
$$\times \left(\frac{g_{m1}g_{m4}g_{m6}(1 + g_{m5}r_{ds5})}{g_{m1}g_{m4}g_{m6}(1 + g_{m5}r_{ds5}) - g_{m5}g_{m2}(g_{m4} + g_{m6})} \right) \tag{8}$$

If we assume $I_2 / I_1 = \beta$ hence, $g_{m2,4,6} = \beta (g_{m1,3,5,7}) = g_{mn}$ and $r_{ds2,4,6} = (1/\beta) r_{ds1,3,5,7} = (1/\beta) r_{dsp}$. ultimately, PSRR can be reformulated as follows:

$$PSRR \Big|_{Proposed} = (1 + \frac{1}{\beta}) \times \frac{1}{1/\beta g_{mn}r_{dsn}} \times \frac{2}{2 + g_{mn}r_{dsn}} \times \frac{g_{mn}r_{dsn}}{g_m r_{dsn} - 2}$$
$$\approx \frac{2\beta}{g_{mn}r_{dsp} \times g_{mn}r_{dsn}} \times (1 + \frac{1}{\beta}) \tag{9}$$

By comparing equations (7) and (9), the improvement in PSRR can be determined.

$$PSRR \Big|_{Proposed} = \frac{2(\beta + 1)}{g_{mn}r_{dsn}} PSRR \Big|_{2-T} \tag{10}$$

The enhancement will be significant depending on β and $g_{mn}r_{dsn}$. Scaling down increases the PSRR compared to the 2-T structure. However, increasing β leads to higher power consumption. Therefore, a careful selection of β and $g_{mn}r_{dsn}$ can achieve both high PSRR and low power dissipation.

III. SIMULATION RESULTS

The 0.18-μm CMOS process is employed to simulate the voltage reference circuit being discussed, and the specific device dimensions can be found in Table I. The size ratios are fine-tuned to optimize the line sensitivity and temperature coefficient of the V_{Ref} output. TC is a significant

parameter of a voltage reference circuit. It evaluates the level of independence of the output reference voltage concerning temperature variations. Fig. 3 shows the variations of the output reference voltage concerning temperature at different corner cases. By examining the extent of voltage fluctuations within the temperature spectrum, one can ascertain TC using the subsequent formula:

$$TC = \frac{V_{Ref,max} - V_{Ref,min}}{V_{Ref,average}} \times \frac{1}{T_{max} - T_{min}} \tag{11}$$

Where $V_{Ref,max}$ represents the maximum value of the reference voltage, $V_{Ref,min}$ denotes the minimum value of the reference voltage. and $V_{Ref,average}$ is the average output reference voltage within the temperature range from T_{min} to T_{max}. Considering that the temperature coefficient is on a micro-scale, it is typically multiplied by one million and reported in parts per million (ppm). Fig. 4 shows the output reference voltage versus temperature with different power supply which demonstrates that the proposed voltage reference circuit is independent of the input voltage.

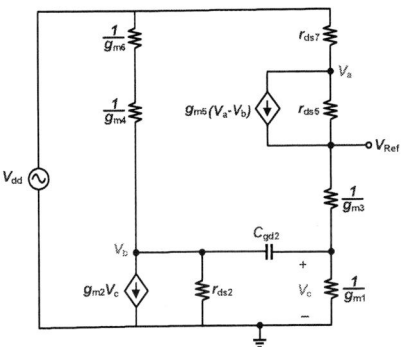

Fig. 2: AC equivalent circuit diagram exploited for PSRR calculation.

TABLE I: TRANSISTOR SIZES OF THE PROPOSED CIRCUIT.

Transistors	W [μm]	L [μm]	M
$M_{1,2,3}$	9	18	9
M_4	14	2.2	1
M_5	70	2.2	1
M_6	0.22	5	1
M_7	45	5	2

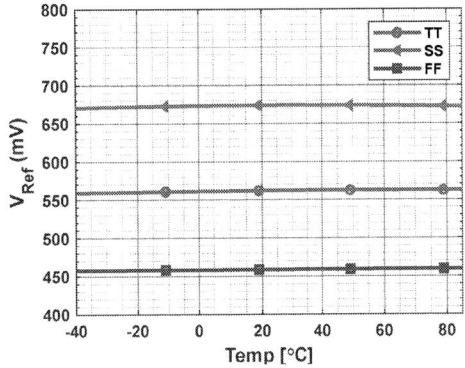

Fig. 3: V_{Ref} vs. temperature with different corner cases @ 1.8 V supply.

The 5th Iranian International Conference on Microelectronics (IICM 2023)

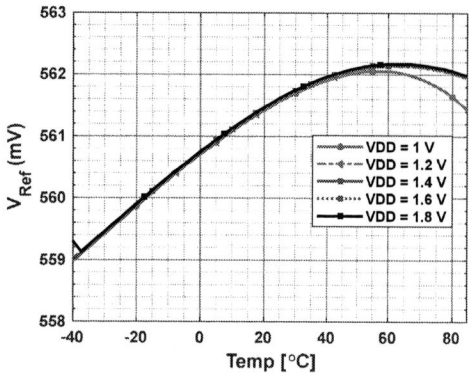

Fig. 4: V_{Ref} vs. temperature with different input supply voltage.

The LS parameter can be calculated based on Fig. 5 which compares the dependency of V_{Ref} on V_{DD} for the 2-T and modified structure. The LS parameter actually evaluates the independence level of V_{Ref} with to supply voltage and can be defined as follows:

$$LS = \frac{V_{Ref,max} - V_{Ref,min}}{V_{Ref,average}} \times \frac{1}{V_{DD,max} - V_{DD,min}} \quad (12)$$

Where $V_{DD,max}$ and $V_{DD,min}$ are the maximum and minimum values of the input supply voltage, respectively. The minimum value of supply voltage in this structure is 0.9 volt.

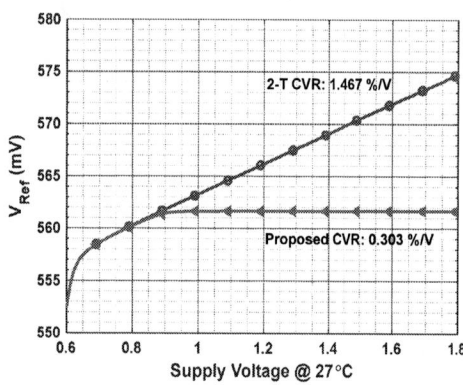

Fig. 5: V_{Ref} vs. supply voltage of the proposed CVR compared with 2-T.

In Fig. 6 the PSRR is graphed against frequency under room temperature conditions, considering various corner cases and a 1.8 V input supply voltage. The chart reveals that the PSRR exhibits strong attenuation at low frequencies, approximately -85 dB for frequencies up to 100 Hz. Notably, at higher frequencies, the PSRR performance gradually diminishes, reaching approximately -50 dB at 1 MHz. Fig. 7 illustrates the startup waveform of the output voltage reference when the input voltage changes from 0 to 0.9 volts. Based on the depicted waveform, the startup time of the voltage reference circuit is approximately 6 ms. Fig. 8 shows the noise spectrum of V_{Ref} with a 1.8 V supply at room temperature, where no load capacitor is present. The output noise level is measured at 0.56 μV/ √ Hz at 1 Hz, and the integrated noise over the frequency range of 0.1 Hz to 10

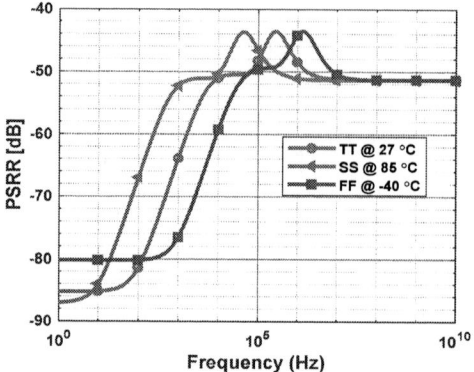

Fig. 6: Assessed the PSRR of the CVR under various corner cases @ 1.8 V supply.

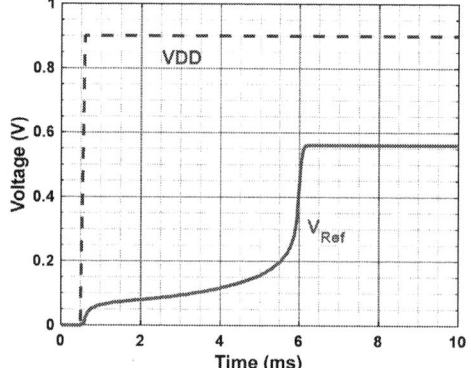

Fig. 7: Start-up response of V_{Ref} in room temperature at @ 0.9 V supply.

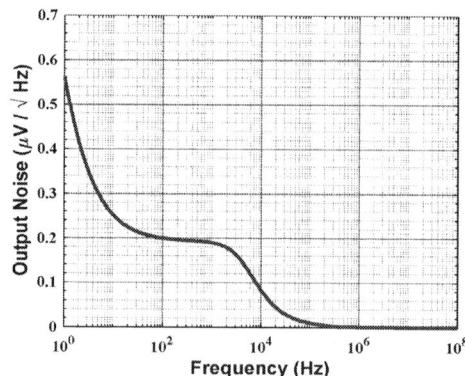

Fig. 8: Output noise for TT corner @ 1.8 V supply.

Hz amounts to 3.366 μV.

Table II provides a summary and compares the performance of the proposed design with suggested designs in recent years. In brief, the temperature coefficient and line sensitivity are enhanced. The improvement of PSRR was achieved by incorporating of transistor M_5 into the circuit, and finally, this method exhibits exceptionally low noise. The proposed circuit is more suitable compared to other references in terms of TC and PSRR. Furthermore, [13, 14, 15,17] have an appropriate LS, but relatively low PSRR, and the proposed circuit also has a similar LS to [16].

979-8-3503-6020-2/23 $31.00 © 2023 IEEE   24

The 5th Iranian International Conference on Microelectronics (IICM 2023)

TABLE II: PERFORMANCE SUMMARY AND COMPARISON WITH OTHER WORKS.

Reference	[9]	[13]	[14]	[15]	[16]	[17]*	This work*
Technology (nm)	130	180	180	180	180	180	180
Minimum Input Voltage (V)	0.5	0.4	1.5	0.9	0.5	0.6	0.9
Reference Voltage (mV)	174.9	151	985	261	288	0.3078	560
Temp. Range (°C)	-20~80	-40~125	-40~85	-40~130	-10~100	-20~80	-40~85
LS (% / V)	0.033	0.0154	0.003	0.01	0.23	0.02	0.303
TC (ppm / °C)	231	89.83	60.86	62	90	24.8	43.4
PSRR [dB]	-53 @ 100 Hz	-47 @ 1 kHz	-93 @ 10 Hz	-73.5 @ 100 Hz	-45 @ 100 Hz	-54 @ 100 Hz	-81.5 @ 100 Hz
Power (nW)	0.0022	1	63	1.8	0.5	0.0214	61.59
No. Transistors	2	6	36	12	5	6	7

*SIMULATION RESULTS

IV. CONCLUSION

A highly efficient CMOS voltage reference with 0.56 V output voltage has been introduced, utilizing a 0.18-μm CMOS process. The proposed structure introduces improvements in certain evaluation parameters by incorporating a number of transistors into the two-transistor (2-T) configuration. The proposed subthreshold CMOS voltage reference achieves a remarkable temperature coefficient (TC) across a wide temperature range. Simulation results confirmed the high line sensitivity (LS), power supply rejection ratio (PSRR), and low output noise.

REFERENCES

[1] A. P. Brokaw, "A simple three-terminal IC bandgap reference," *IEEE J. Solid-State Circuits*, vol. SC-9, pp. 388–393, Dec. 1974.

[2] B.-S. Song and P. R. Gray, "A precision curvature-compensated CMOS bandgap reference," *IEEE J. Solid-State Circuits*, vol. SC-18, pp. 634–643, Dec. 1983.

[3] H. Banba, H. Shiga, A.Umezawa, T.Miyaba, T. Tanzawa, S. Atsumi, and K. Sakui, "A CMOS bandgap reference circuit with sub-1V operation," *IEEE J. Solid-State Circuits*, vol. 34, no. 5, pp. 670–674, May 1999.

[4] K. N. Leung and P. K. T. Mok, "A Sub 1 V 15 ppm C CMOS bandgap voltage reference without requiring low threshold voltage device," *IEEE J. Solid-State Circuits*, vol. 37, no. 4, pp. 526–530, Apr. 2002.

[5] A. Boni, "Op-amps and startup circuits for CMOS bandgap references with near 1 V supply," *IEEE J. Solid-State Circuits*, vol. 37, no. 10, pp. 1339–1343, Oct. 2002.

[6] J. Doyle, Y. J. Lee,Y.-B. Kim,H.Wilsch, and F. Lombardi, "A CMOS sub-bandgap reference circuit with 1 V power supply voltage," *IEEE J. Solid-State Circuits*, vol. 39, no. 1, pp. 252–255, Jan. 2004.

[7] P. Kinget, C. Vezyrtzis, E. Chiang, B. Hung, and T. L. Li, "Voltage references for ultra-low supply voltages," in Proc. *IEEE Custom Integr. Circuits Conf*, 2008, pp. 715–720.

[8] L. Magnelli, F. Crupi, P. Corsonello, C. Pace, and G. Iannaccone, "A 2.6 nW, 0.45 V temperature-compensated subthreshold CMOS voltage reference," *IEEE J. Solid-State Circuits*, vol. 46, no. 2, pp. 465–474, Feb. 2011.

[9] M. Seok, G. Kim, D. Blaauw, and D. Sylvester, "A portable 2-transistor picowatt temperature-compensated voltage reference operating at 0.5 V," *IEEE J. Solid-State Circuits*, vol. 47, no. 10, pp. 2534–2545, Oct. 2012.

[10] B. Ma and F. Yu, "A Novel 1.2–V 4.5-ppm/°C Curvature-Compensated CMOS Bandgap Reference," *IEEE Trans. Circuits and Sys. I: Reg. Papers*, vol. 61, no. 4, pp. 1026-1035, April 2014.

[11] H. Chen, C. Lee, S. Jheng, W. Chen and B. Lee, "A Sub-1 ppm/°C Precision Bandgap Reference With Adjusted-Temperature-Curvature Compensation," *IEEE Trans. Circuits and Sys. I: Reg. Papers*, vol. 64, no. 6, pp. 1308-1317, June 2017.

[12] Wang and C. Zhan, "A 0.7-V 28-nW CMOS Subthreshold Voltage and Current Reference in One Simple Circuit," *IEEE Trans. Circuits Syst. I, Reg. Papers*, vol. 66, no. 9, pp. 3457-3466, Sept. 2019.

[13] J. Lin, L. Wang, C. Zhan and Y. Lu, "A 1-nW Ultra-Low Voltage Subthreshold CMOS Voltage Reference With 0.0154%/V Line Sensitivity, " *IEEE Transactions on Circuits and Systems II: Express Briefs*, vol. 66, no. 10, pp. 1653-1657, Oct. 2019.

[14] Y. Chen and J. Guo, "A 42nA IQ, 1.5–6V VIN, Self-Regulated CMOS Voltage Reference With –93dB PSR at 10 Hz for Energy Harvesting Systems," *IEEE Trans. Circuits Syst. II, Exp. Briefs*, pp. 1-5, Early Access, Feb. 2021.

[15] C. -Z. Shao, S. -C. Kuo and Y. -T. Liao, "A 1.8-nW, −73.5-dB PSRR, 0.2-ms Startup Time, CMOS Voltage Reference With Self-Biased Feedback and Capacitively Coupled Schemes, " *IEEE Journal of Solid-State Circuits*, vol. 56, no. 6, pp. 1795-1804, June 2021.

[16] J. Wang, X. Sun and L. Cheng, "A Picowatt CMOS Voltage Reference Operating at 0.5-V Power Supply With Process and Temperature Compensation for Low-Power IoT Systems, " *IEEE Transactions on Circuits and Systems II: Express Briefs*, vol. 70, no. 4, pp. 1336-1340, April 2023.

[17] L. Colbach, T. Jang, and Y. Ji, "A 21.4 pW Subthreshold Voltage Reference with 0.020 %/V Line Sensitivity Using DIBL Compensation," *Sensors*, vol. 23, no. 4, p. 1862, Feb. 2023.

The 5th Iranian International Conference on Microelectronics (IICM2023)

25 – 26 October 2023

A Curvature Compensated CMOS Bandgap Voltage Reference With 6.8 ppm/°C Temperature Coefficient and Low Quiescent Current

Elaheh Pakravan
Department of Electrical and Computer Engineering
Urmia University
Urmia, Iran
st_e.pakravan@urmia.ac.ir

Mortaza Mojarad
Department of Electrical and Computer Engineering
Urmia University
Urmia, Iran
m.mojarad@urmia.ac.ir

Behboud Mashoufi
Department of Electrical and Computer Engineering
Urmia University
Urmia, Iran
b.mashouf@urmia.ac.ir

Abstract—**This paper presents a fully MOS current-mode curvature compensated bandgap voltage reference. A new method has been proposed for Complementary-To-Absolute-Temperature voltage generation. In addition, temperature compensation with curvature correction has been carried out. The bandgap reference circuit has been simulated in the 0.18 μm CMOS process. The simulation results demonstrate that the new BGR produces a 1.18 V reference voltage with ultra-low current consumption of 2.19 μA. The temperature coefficient is almost equal to 6.8 ppm/°C and the PSR is -68 dB at 1 MHz offset.**

Keywords— bandgap reference; curvature compensation; temperature coefficient; power consumption; sensitivity;

I. INTRODUCTION

In recent years, Bandgap Reference (BGR) circuits with high accuracy have become one of the major building blocks in today's analog and digital circuits such as mixed-signal systems, amplifiers, data converters, DRAMs, and flash memory controller circuits. The function of a BGR is to produce a stable voltage that is of high resiliency to temperature and supply variations. As a result, BGR is expected to provide a reference voltage with very low temperature and supply dependency.

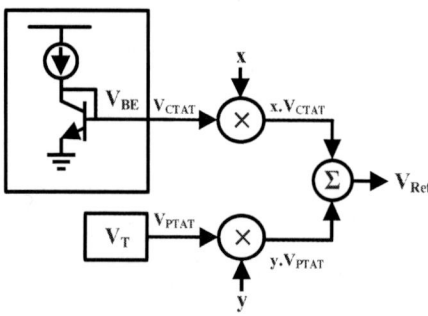

Fig. 1. Conventional BGR Block Diagram.

Fig. 1 depicts a conventional BGR operation concept. The BGR circuit consists of a Complementary-To-Absolute-Temperature (CTAT) voltage and a Proportional-To-Absolute-Temperature (PTAT) voltage, which are added to

each other with proper weights to generate a DC voltage with zero temperature coefficient (ZTC).

The classic method for generating the CTAT voltage is to bias a BJT transistor in the active region and the base-emitter voltage (V_{BE}) of the transistor will exhibit a negative temperature coefficient and is of CTAT characteristic. On the other hand, generating PTAT voltage requires two active BJTs, and the difference between their base-emitter voltages (ΔV_{BE}) has a positive temperature coefficient, hence a PTAT voltage can be developed. A first-order -temperature compensated BGR is formed by summing the weighted CTAT and PTAT voltages. However, since the base-emitter voltage of a BJT is a nonlinear function of temperature and ΔV_{BE} has a linear relationship with temperature, to improve BGR's accuracy, some of the previous works have utilized second-order temperature compensation [1], [2]. This method nevertheless leads to elevated power consumption.

One of the design challenges of BGRs is to improve the precision while consuming low quiescent current. In other words, in modern BGRs, there is a trade-off between the power consumption and sensitivity of the reference voltage to temperature and the supply voltage. For instance, in [3] and [4], the power consumption of the presented BGRs is in the nano-watts range. However, these circuits suffer from considerably high temperature coefficients. Another crucial factor in determining the performance of a BGR is the systematic offset which is due to the high base currents of the BJTs. The considerable base current is due to exploiting low-β BJTs fabricated in the CMOS process. In [5], β-compensation has been carried out by placing a unity-gain buffer at the base terminal of each BJT which leads to more power dissipation and higher complexity and cost.

In this paper, a new bandgap reference circuit has been proposed which is comprised of only MOS transistors. By using novel techniques, high precision and reduced dependency to temperature have been achieved while consuming a very low quiescent current. The MOS transistors have been biased in the subthreshold region because the I/V characteristic of a subthreshold NMOS or PMOS transistors is quite similar to that of a BJT. Hence, the power consumption problem is solved by utilizing the gate-source voltage of the MOS transistor instead of the base-emitter voltage of a BJT transistor. This paper is

979-8-3503-6020-2/23 $31.00 © 2023 IEEE

organized as follows. In Section II, the circuit implementation is discussed. In Section III, the simulation results are given and Section IV concludes the paper.

II. THE PROPOSED BGR

In recent years, the efforts to design a low-power, highly accurate bandgap references with low sensitivity to temperature is increased. The major challenge is the temperature dependency compensation consuming low quiescent current. In this work, the structure proposed for temperature sensitivity compensation is illustrated in Fig. 2. This structure presents a solution by using a curvature compensation. By summing few weighted CTAT and PTAT currents with temperature sensitivity compensation for each of them, an almost zero temperature coefficient current is achieved. Finally, by conveying these currents into a resistor, a ZTC voltage has been produced.

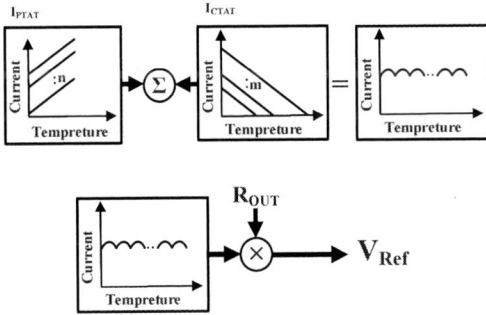

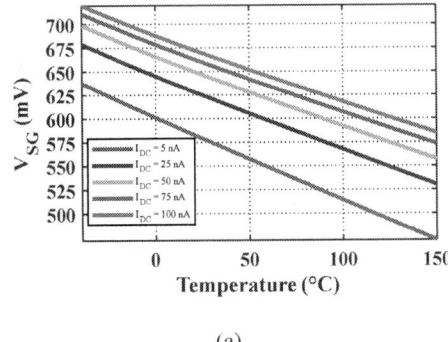

Fig. 2. The Proposed BGR.

A. CTAT Voltage Generator

As mentioned in the previous section, in conventional BGRs to generate the CTAT voltage the V_{BE} of a BJT transistor has been exploited. However, in order to reduce the power consumption, in this paper, MOS transistors in the subthreshold region which have source-gate voltages with inherent CTAT characteristics have been used [6]. To further investigate the behavior of the MOS transistor's gate-source voltage in the subthreshold region that is biased by a constant current, simulations using a standard CMOS process have been conducted. The simulation results are depicted in Fig. 3.

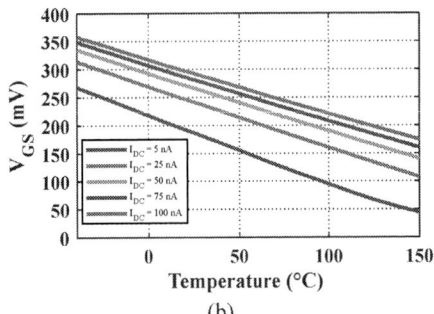

(b)

Fig. 3. (a) PMOS source-gate voltage, and (b) NMOS gate-source voltage CTAT behavior

B. PTAT Voltage Generator

In most previous BGRs, for generating the PTAT voltage the difference between two base-emitter voltages of BJT transistors have been used. However, due to the importance of power consumption, to achieve a PTAT voltage, the difference between two gate-source voltages of two MOS transistors in the subthreshold region has been utilized. In this article, in accordance with [7], a PTAT voltage has been generated from the ΔV_{GS} of MOS transistors, and it has been converted to current.

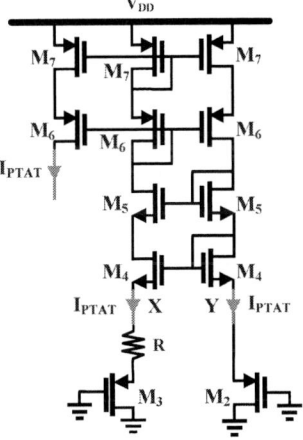

Fig. 4. The PTAT Current Generator.

In Fig. 4, the transistors M_2 to M_7, and the resistor R, form a PTAT voltage and a PTAT current generator (by utilizing the ΔV_{GS}). The transistors M_4 to M_7 force the voltages of the nodes X and Y to be equal and the current mirror formed by transistors M_7 causes the drain currents of M_2 and M_3 to be equal. Therefore, the gate-source voltages of the transistors M_4 become equal which implies that the nodes of X and Y one of the same potential. By inspecting the circuit in Fig. 4, the PTAT current, I_{PTAT} can be obtained as follows:

$$I_{PTAT} = \frac{V_{GS2} - V_{GS3}}{R} \qquad (1)$$

Transistors M_2 and M_3 are biased in the subthreshold region and the drain current of the transistors can be expressed as [7]:

979-8-3503-6020-2/23 $31.00 © 2023 IEEE   27

The 5th Iranian International Conference on Microelectronics (IICM2023)

$$I_D = K I_0 \exp(\frac{V_{GS}}{\eta V_T})$$
$$I_0 = \mu_n Cox (\eta - 1) V_T^2 \qquad (2)$$

where K is the aspect ratio (W/L) of the transistor, μ_n is the carrier mobility, C_{ox} is the gate-oxide capacitance, η is the subthreshold slope factor, V_{GS} is the gate-source voltage, and V_T is the thermal voltage. Based on (2), the gate-source voltage of a MOS transistor can be obtained as follows:

$$V_{GS} = \eta V_T \ln(\frac{I_D}{K I_0}) \qquad (3)$$

$$V_{GS2} - V_{GS3} = \eta V_T \ln(n) \qquad (4)$$

From (1) and (4), I_{PTAT} can be given by

$$I_{PTAT} = \frac{\eta V_T \ln(n)}{R} \qquad (5)$$

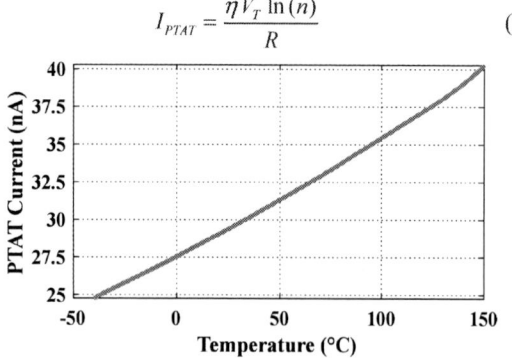

Fig. 5. Generated PTAT Current Simulation Result.

As Fig. 5 illustrates, the generated current has PTAT behavior with temperature variations.

C. Proposed CTAT Current Generator Circuit

In order to produce the CTAT current, a new CTAT current generator is proposed which has been depicted in Fig. 6.

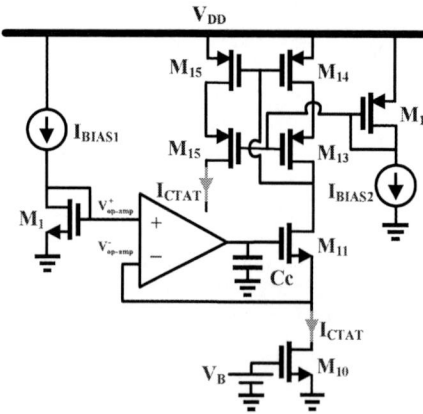

Fig. 6. CTAT current generator circuit.

As mentioned earlier, the source-gate voltage of a MOS transistor has a negative temperature coefficient. In the presented circuit shown in Fig. 6, the gate-source voltage of the transistor M_1 is a CTAT voltage, and the op-amp in the circuit has a high voltage gain. As a result, the voltage of its positive terminal is equal to the voltage of its negative terminal ($V_{op\_amp}^- = V_{op\_amp}^+$). The biasing voltage of V_B is adjusted in a way so that the transistor M_{10} is biased in the triode region. As a result, the transistor M_{10} acts an active resistor, and its on-resistance and the current I_{CTAT} can be obtained as follows:

$$R_{ON,10} = \frac{1}{\mu_n Cox (W/L)_{10} (V_{GS} - V_{TH})}$$

$$I_{CTAT} = \frac{V_{DS,10}}{R_{ON,10}} \qquad (6)$$

Since the drain-source voltage of M_{10} is equal to the potential of the input terminals of the op-amp, the CTAT current can be described by

$$I_{CTAT} = \frac{V_{Gs1}}{R_{ON,10}} \qquad (7)$$

The produced CTAT current is subsequently mirrored by transistors M_{12}-M_{15}. Although this method proves to be effective, it has a major drawback that causes designers to prefer BJTs. From the simulation results reported in Fig. 3, it can be deduced that the sensitivity of the gate-source voltage of a MOS transistor to temperature variations is high and the generated CTAT voltage is substantially dependent to temperature. In this work, biasing current I_{BIAS1} in Fig. 6, is replaced by a PTAT current produced by the circuit shown in Fig. 4. Therefore, the gate-source voltage of the transistor M_1 in Fig. 6, becomes less sensitive to temperature. The complete circuit of the CTAT current generator is shown in Fig. 9.

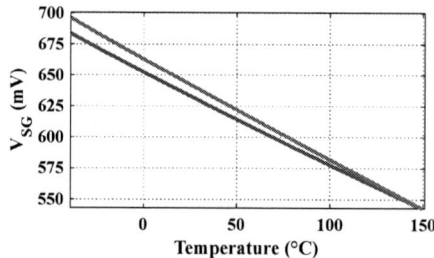

Fig. 7. Comparison simulation result of $V_{GS, M1}$, once biased I_{PTAT}, and once other constant current, versus temperature variations.

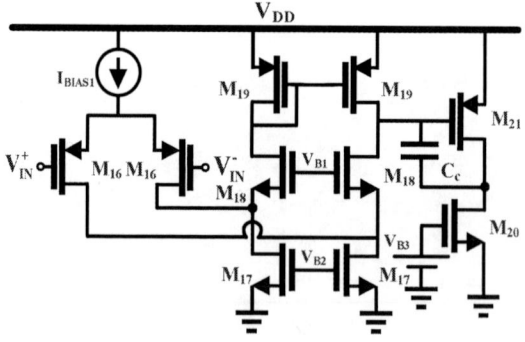

Fig. 8. Op_amp schemematic

According to [7], in Fig. 8, a two-stage high-gain amplifier is utilized to implement the op-amp shown in Fig. 6. This structure is formed by a folded-cascode differential amplifier including transistors M_{16} to M_{19}, and the second stage consists of transistors M_{20} to M_{21}. The capacitor C_C is devised in the circuit to perform the frequency compensation for acquiring adequate phase margin and stability.

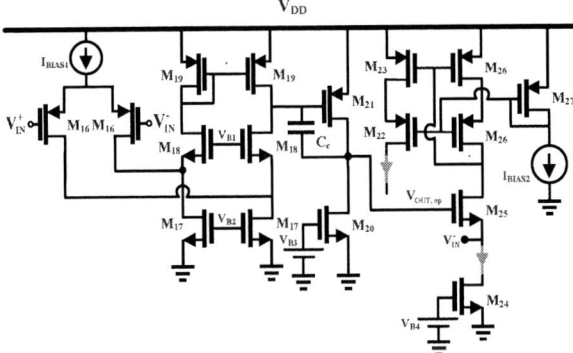

Fig . 9. The CTAT current generator circuit.

III. SIMULATION RESULTS

The presented bandgap reference in Fig. 15, has been simulated by using ADS in a standard 0.18μm CMOS process. The reference voltage variations versus the temperature has been investigated in Fig. 10, for a temperature span from -40°C to 150 °C with the output voltage of 1.18 V and the results prove the temperature coefficient is 6.803 ppm/°C. Another important factor in the design of a BGR is the total current consumption. The result for the quiescent current consumption has been depicted in Fig. 11. The reference voltage against supply variations has been reported in Fig. 12 and Fig. 13 gives the current consumption versus the supply voltage variations. The result for the PSR simulation has also been shown in Fig. 14 with the total current consumption of 2.19 μA. The simulation results for the temperature coefficient against different process corners are presented in Table 1. Monte Carlo simulation result has been shown in Fig. 16. The performance of the newly presented BGR has been reported and compared to the state-of-the-art in Table 2. It can be concluded that the BGR reported here, outperforms all other prior works in terms of temperature coefficient. It is interesting to note that under a specific circumstance, the temperature coefficient in [1] is 5.78 ppm/°C. However, it consumes enormous power compared to the proposed circuit in this paper.

IV. CONCLUSION

In this paper, a novel current-mode bandgap voltage reference is introduced. In this proposed BGR, besides decreasing the voltage reference sensitivity to temperature, ultra-low power consumption has been achieved. In this work, in order to fulfill this primary goal, new realizations for voltages with both negative and positive temperature coefficients has been proposed. The simulation results and comparison of this work to previous works shows the superiority of the presented work to previously published BGRs in terms of accuracy and PSR.

Table 1   Simulation results for different corners

	TT	SS	FF	SF	FS
Temperature Coefficient (ppm/°C)	6.803	41.195	72.738	16.757	6.874

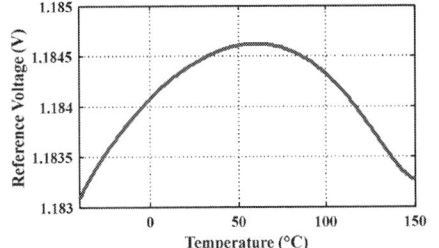

Fig. 10. The Reference voltage against Temperature variations.

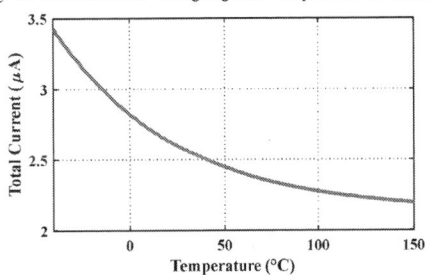

Fig. 11. The current consumption against Temperature variations.

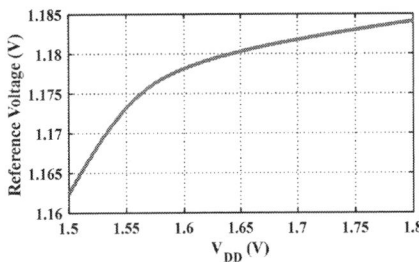

Fig. 12. The reference voltage versus Voltage Supply variations.

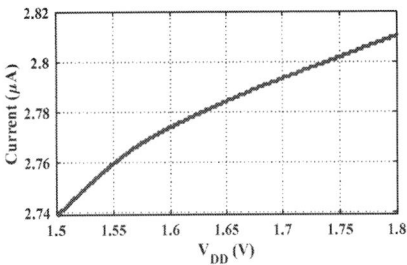

Fig. 13. The Current consumption versus Voltage Supply variations.

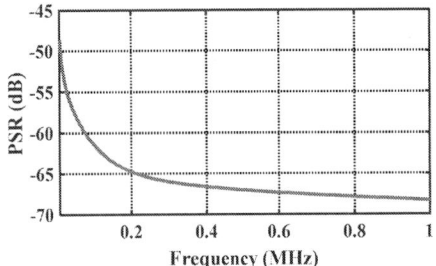

Fig. 14. PSR of the complete Bandgap Reference.

The 5th Iranian International Conference on Microelectronics (IICM2023)

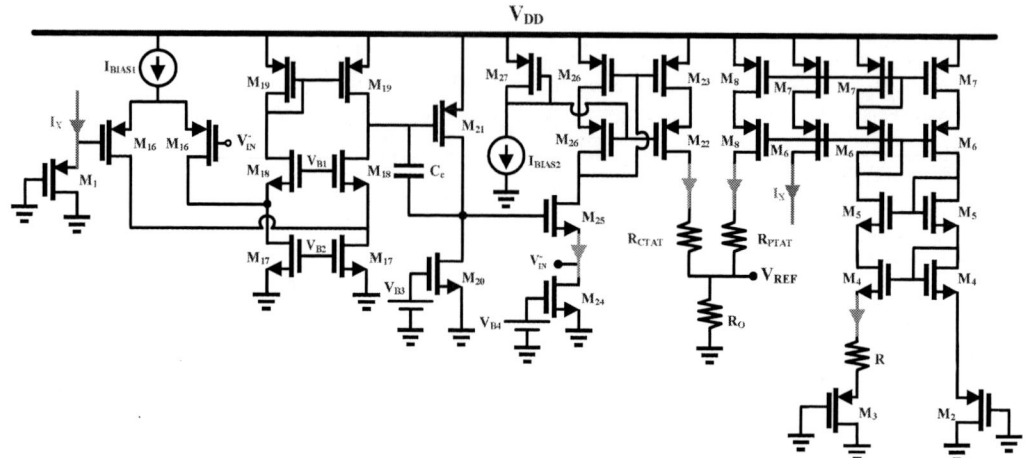

Fig. 15. The proposed Bandgap Reference

Table 2 Comparison to Previously Published Works

	[1]	[2]	[3]	[4]	[5]	[8]	[9]	This Work
CMOS Technology	0.13 µm CMOS	0.5 µm BiCMOS	0.18 µm CMOS	0.35 µm CMOS	0.35 µm CMOS	0.18 µm CMOS	0.18 µm CMOS	0.18 µm CMOS
Supply Voltage (V)	3.3	1	1.8	1.1-3.3/1.3-3.3	1.4	1.3-1.8	1.8	1.8
V_{REF} (V)	1.16	0.19	0.548/1.09	1.18/0.553	0.858	1.17	1.1419	1.18
Temperature range (°C)	-40 - 150	-40 - 125	-40 - 120	-20 - 80	-20 - 100	-40 - 140	-40 - 125	-40 - 150
Temperature Coefficient (ppm/°C)	5.78 – 13.5	11	114/147	394/215	12.4	26.3 Avg. 12.2 Best 41 Worst	3.2-505 4.3 (Avg.)	6.803
Current Consumption (µA)	-	-	0.0525/0.1	-	162	-	17	2.19
Power Consumption (µW)	120	20	-	0.11/0.108	-	0.192	-	3.942
PSR (dB @ Hz)	-	-	-	-	-	-52/ -44 @100/1M	-	-48.69/-61.5/-68.26 @10K/100K/1000K
PSRR (dB @ Hz)	-82/-30 @10/100K	-	-56/-62 @100	-	-	-	76 @ DC	-

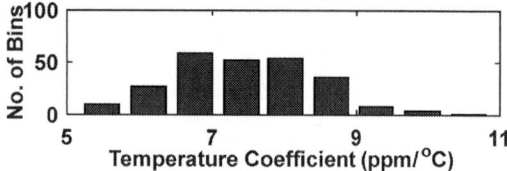

Fig. 16 Monte Carlo simulation results

REFERENCES

[1] K. Chen, L. Petruzzi, R. Hulfachor, and M. Onabajo, "A 1.16-V 5.8-to-13.5-ppm/°C Curvature-Compensated CMOS Bandgap Reference Circuit With a Shared Offset-Cancellation Method for Internal Amplifiers" *IEEE J. Solid-State Circuits*, vol. 56, no. 1, Jan. 2021.

[2] K. Sanborn, D. Ma, and V. Ivanov, "A Sub-1-V Low-Noise Bandgap Voltage Reference," *IEEE J. Solid-State Circuits*, vol. 42, no. 11, Nov. 2007.

[3] Y. Osaki, T. Hirose, N. Kuroki, and M. Numa, "1.2-V Supply, 100-nW, 1.09-V Bandgap and 0.7-V Supply, 52.5-nW, 0.55-V Subbandgap Reference Circuits for Nanowatt CMOS LSIs" *IEEE J. Solid-State Circuits*, vol. 48, no. 6, Jun. 2013.

[4] T. Hirose, K. Ueno, N. Kuroki and M. Numa, "A CMOS bandgap and sub-bandgap voltage reference circuits for nanowatt power LSIs," *2010 IEEE Asian Solid-State Circuits Conference*, Beijing, China, 2010.

[5] R. T. Perry, S. H. Lewis, A. P. Brokaw, and T. R. Viswanathan, "A 1.4-V Supply CMOS Fractional Bandgap Reference" *IEEE J. Solid-State Circuits*, vol. 42, no. 10, Oct. 2007.

[6] G. Giustolisi, G. Palumbo, M. Criscione and F. Cutri, "A low-voltage low-power voltage reference based on subthreshold MOSFETs," *IEEE J. Solid-State Circuits*, vol. 38, no. 1, pp. 151-154, Jan. 2003.

[7] B. Razavi, Design of Analog CMOS Integrated Circuits, McGraw-Hill, Second edition, 2016.

[8] M. Kim and S. Cho, "A Single BJT Bandgap Reference With Frequency Compensation Exploiting Mirror Pole" *IEEE J. Solid-State Circuits*, vol. 56, no. 10, Dec. 2020.

[9] J.-H. Boo, et al, "A Single-Trim Switched Capacitor CMOS Bandgap Reference With a 3σ Inaccuracy of +0.02%, −0.12% for Battery-Monitoring Applications" *IEEE J. Solid-State Circuits*, vol. 56, no. 4, Apr. 2021.

The 5th Iranian International Conference on Microelectronics (IICM2023)

25 – 26 October 2023

Two-wavelength Quantum Dot Mid-Infrared Photodetectors Using Solution Process Method

Hannaneh Dortaj
Faculty of Electrical & Computer Engineering
University of Tabriz
Tabriz, Iran
h_dortaj@tabrizu.ac.ir

Samiye Matloub
Faculty of Electrical & Computer Engineering
University of Tabriz
Tabriz, Iran
matloub@tabrizu.ac.ir

Abstract—This novel report is a way to achieve a mid-infrared photodetector based on quantum dots, which is designed and simulated to be fabricated by the solution process method, which provides low cost and ease of fabrication. To design the proposed structure, the absorber layer is composed of two different sizes of quantum dots, in which, thanks to the intersubband transitions of carriers, two separate wavelengths of the mid-infrared spectrum are detected. Then, with the help of resonant tunneling effect, the excited carriers flow towards the interdigitated contacts and the output photocurrent is generated. For theoretical modeling, the Coupled rate equations and the Schrödinger-Poisson equations are calculated. Simulation results indicate that the peak responsivities are about 7.5(A/W) and 8(A/W), and the specific detectivities D* are about 5×10^{11}(cm.Hz$^{1/2}$W^{-1}) and 6×10^{11}(cm.Hz$^{1/2}$W^{-1}) for the wavelengths of 3μm and 5μm, respectively.

Keywords—Photodetector, Mid-infrared, Quantum dot, Solution process, Detectivity

I. INTRODUCTION

Multi-wavelength infrared photodetectors have many practical applications in night vision cameras, land-mine detection [1], remote thermometers [2], space situational awareness [3], geology, oceanography, military, identification of muzzle flashes from firearms [4], alarm systems in firefighting [5], material classification, and medical diagnosis.

A critical restriction of multi-wavelength photodetectors is the failure to choose peak wavelength reactions without the utilize of multi-channel electrical contacts and optical channels. Including detector components, such as optical channels, decreases the radiation transmission of the detector device and causes the detector device, bulky, complex and costly [6], [7]. Besides, the utilize of numerous electrical contacts on the detector requires complex handling methods. Quantum dot infrared photodetectors (QDIPs) were favored over customary photodetectors since they are able of detecting different wavelengths by modifying the size of QDs and applying an balanced bias voltage to them. QDIP has many advantages, such as the ability to provide high resolution images in the MWIR and LWIR ranges [8], long carrier lifetimes [9], sensitivity to normal infrared radiation, low dark current, high operating temperature, high sensitivity, and detectivity [10]–[12].

Several studies have attempted to improve the performance of QDIP using novel heterostructured systems, such as AlGaAs [13] and InGaAs [14], [15] barriers, sub-monolayer QDs (ML) [16], evolution of QD AlAs/InAs/GaAs

superlattices [17], inter-band transitions of InGaAs/GaAs-based QDs [6], [18], dot-in-well structures (DWELL) [19], dot in double well (DDWELL) [20], [21] formulation, tunneling QDIP structure [22], enhanced containment barrier (CE) [23] and others, but the majority of them rely heavily on the barrier strategy. Furthermore, QDIPs provide exceptional performance and simultaneously function at specific broadband wavelengths, which cannot be done in this manner.

The extensive use of multi-wavelength photodetectors is further limited by the expensive, complex, and , and high defect structures [24]. Following the progress of colloidal quantum dots (CQDs) over the past few decades, high-performance optoelectronic devices, including photodetectors, have been developed covering important atmospheric windows from the short-wave infrared (1.5 - 2.5 μm) to mid-infrared (3 - 5 μm) range [25]. Moreover, CQDs, due to their solution processability, provide many benefits such as lower dark current, low-temperature operating, low cost, excellent photoelectric properties, large-area fabrication, direct coating on silicon electronics for imaging [26]–[29], etc.

In this study, an innovative design for two-wavelength colloidal QDIP is reported that can be fabricated by the solution process technology method, which is an easy, low-cost and available method. An important point is that in all the simulations, some non-idealities caused by the synthesis limitations are taken into account so that the introduced theoretical model is closer to the practical example that can be fabricated in future. To design the proposed structure, the absorber layer is composed of two different sizes of CdSe/ZnS core/shell QDs, in which, thanks to the intersubband transitions of carriers, two separate wavelengths

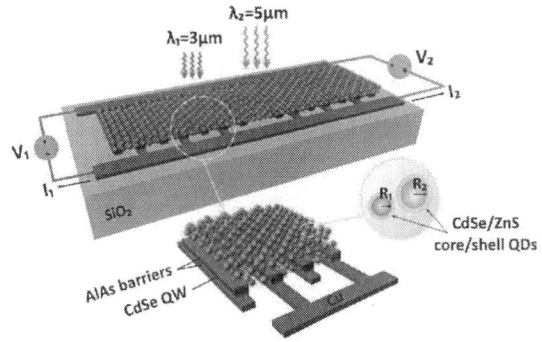

Fig. 1. Schematic view of the proposed two-wavelength photodetector.

979-8-3503-6020-2/23 $31.00 © 2023 IEEE

The 5th Iranian International Conference on Microelectronics (IICM2023)

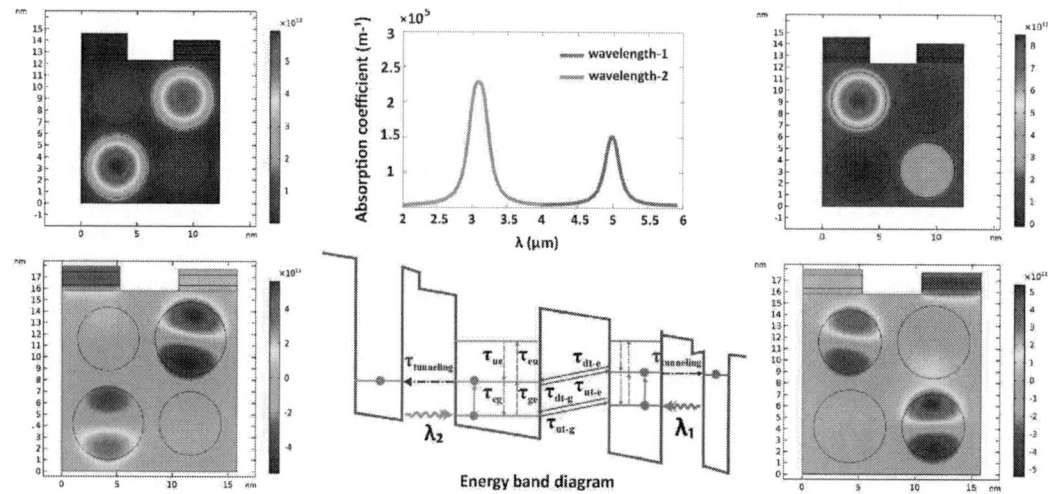

Fig. 2. Modal analysis of designed two-wavelength photodetector by use of FEM software. Absorption coefficient, wave functions of GS and ES of conduction band, and the energy band diagram calculated from solving Schrödinger equation.

(3, 5 μm) of the mid-infrared spectrum are detected. Then, with the help of resonant tunneling effect (RTE), the excited carriers flow towards the interdigitated contacts (IDC) and the output photocurrent is generated.

The simulation results show that the peak responsivities are about 7.5(A/W) and 8(A/W), and the specific detectivities D* are about 5×10^{11}(cm.Hz$^{1/2}$W^{-1}) and 6×10^{11}(cm.Hz$^{1/2}$W^{-1}) for the wavelengths of 3μm and 5μm, respectively. Because of its great detecting capacity, this multi-wavelength QDIP can be employed in a variety of applications, including multi-wavelength infrared camera, spectroscopy, night vision technology, and broadband spectrum analyzer.

II. STRUCTURE DESCRIPTION

In this study, two-wavelength MIR photoconductor composed of an array of two different sizes of CdSe/ZnS core shell QDs which is spin coated on the interdigitated contacts (IDCs) as an absorber layer. Also, a quantum well (QW) structure composed of CdSe as the well and AlAs as barriers are exploited to be deposited on IDCs to extract excited carries from QDs through resonance tunneling. The schematic view of the introduced structure is depicted in Fig. 1, which shows that two various wavelengths of MIR light are simultaneously applied on the top surface of the device. The common contact of both power supplies for the output electrical circuits is connected to one side of each IDCs, as shown in Fig. 1. For two different sizes of QDs that produce two distinct electrical currents in isolated circuits, the other side of IDCs is split into two sections. Regarding the geometrical aspects, to design a two-wavelength QDIP, the radiuses of CdSe QDs are selected R₁=2nm for wavelength-1 (blue QDs) and R₂=2.7nm for wavelength-2 (orange QDs). The well width for QWs related to small QDs is equal to 0.8 nm and for big QDs is 1.3 nm.

III. THEORETICAL MODELING

This section has been discussed in two subsections: First, 3D Schrodinger-Poisson equations have been self-consistently calculated for the introduced model's modal analysis. Second, the Rung-Kutta method is used to compute the coupled rate equations in order to examine the optical behavior of QDIP.

A. Modal analysis of proposed structure

Energy band diagram and the solved wave functions by use of the FEA (finite element analysis) software at the resonant tunneling level for wavelength-1&2 are illustrated in Fig. 2. The energy band diagram in Fig. 2 also shows the time constants related to the relaxation, recombination, and tunneling processes of electrons for both QDs' sizes.

The theoretical model used for the description of energy levels in the cross section of a QD consists of a nonlinear Poisson equation for the electrostatic potential φ coupled with an eigenvalue problem for Schrodinger's equation [30],

$$-\frac{\hbar}{2}\nabla\left(\frac{1}{m^*}\nabla\psi_n\right)+(V_h-q\phi)\psi_n=E_n\psi_n \qquad (1)$$

$$\nabla\left[\varepsilon\nabla\varphi\right]=-\rho\left[\varphi\right] \qquad (2)$$

Where $\hbar$ is the reduced Planck's constant, m^* the effective electron mass, V_h the heterojunction step potential, q is the unit electric charge, ψ_n the wave function belonging to energy level E_n, ε is the dielectric constant, and ρ is the total charge density.

It is obvious that in the photoconductor devices, photocurrent generates through optical absorption process. In proposed quantum-based structure, since there are quantized energy levels, so intersubband transitions of carries play a key role in this process. As a result, by absorbing incident IR light, electrons of ground state (GS) will transfer to excited state (ES) of conduction band. In the specific bias voltage in which the energy levels of GS of QW structures and ES of QDs are in the same level, after being released from the Schottky barrier between QW energy contacts and IDCs by these excited electrons' tunneling to QW structures, the output electrical circuit can then simultaneously operate with two isolated photocurrents, one for each wavelength of input light.

TABLE I. Material parameters exploited in modal analysis [31], [32].

Description	Symbol	Materials	
		CdSe	ZnS
Bandgap Energy	E_g [eV]	1.68	3.715

979-8-3503-6020-2/23 $31.00 © 2023 IEEE

Electron affinity	χ [eV]	4.95	3.8
Effective mass of the electron	m_e^*	$0.13m_0$	$0.22m_0$
Effective mass of the hole	m_h^*	$0.3m_0$	$1.76m_0$

B. The developed coupled rate equations

To analyze the optical behavior and proficiency of the proposed device, developed coupled rate equations have been solved through (3)-(8).

$$\frac{dn_{g_{n,1}}(t)}{dt} = -G(t) - \frac{n_{g_{n,1}}(t)(1-f_{g_{n,2}}(t))}{\tau_{ut-g}} + \frac{n_{g_{n,2}}(t)(1-f_{g_{n,1}}(t))}{\tau_{dt-g}}$$
$$+ \frac{n_{e_{n,1}}(t)(1-f_{g_{n,1}}(t))}{\tau_{eg}} - \frac{n_{g_{n,1}}(t)(1-f_{e_{n,1}}(t))}{\tau_{ge}} - \frac{n_{g_{n,1}}(t)}{\tau_{gr}} \quad (3)$$

$$\frac{dn_{g_{n,2}}(t)}{dt} = -G(t) + \frac{n_{g_{n,1}}(t)(1-f_{g_{n,2}}(t))}{\tau_{ut-g}} - \frac{n_{g_{n,2}}(t)(1-f_{g_{n,1}}(t))}{\tau_{dt-g}}$$
$$+ \frac{n_{e_{n,2}}(t)(1-f_{g_{n,2}}(t))}{\tau_{eg}} - \frac{n_{g_{n,2}}(t)(1-f_{e_{n,2}}(t))}{\tau_{ge}} - \frac{n_{g_{n,2}}(t)}{\tau_{gr}} \quad (4)$$

$$\frac{dn_{e_{n,1}}(t)}{dt} = +G(t) - \frac{n_{e_{n,1}}(t)(1-f_{e_{n,2}}(t))}{\tau_{ut-e}} + \frac{n_{e_{n,2}}(t)(1-f_{e_{n,1}}(t))}{\tau_{dt-e}}$$
$$- \frac{n_{e_{n,1}}(t)(1-f_{g_{n,1}}(t))}{\tau_{eg}} + \frac{n_{e_{n,1}}(t)(1-f_{e_{n,1}}(t))}{\tau_{ge}} - \frac{n_{e_{n,1}}(t)}{\tau_{er}}$$
$$- \frac{n_{e_{n,1}}(t)(1-f_{u_{n,1}}(t))}{\tau_{eu}} + \frac{n_{u_{n,1}}(t)(1-f_{e_{n,1}}(t))}{\tau_{ue}}$$
$$- n_{e_{n,1}}(t)(1-f_{c_{n,1}}(t))W_{tun,1} \quad (5)$$

$$\frac{dn_{e_{n,2}}(t)}{dt} = +G(t) + \frac{n_{e_{n,1}}(t)(1-f_{e_{n,2}}(t))}{\tau_{ut-e}} - \frac{n_{e_{n,2}}(t)(1-f_{e_{n,1}}(t))}{\tau_{dt-e}}$$
$$- \frac{n_{e_{n,2}}(t)(1-f_{g_{n,2}}(t))}{\tau_{eg}} + \frac{n_{e_{n,2}}(t)(1-f_{e_{n,2}}(t))}{\tau_{ge}} - \frac{n_{e_{n,2}}(t)}{\tau_{er}}$$
$$- \frac{n_{e_{n,2}}(t)(1-f_{u_{n,2}}(t))}{\tau_{eu}} + \frac{n_{u_{n,2}}(t)(1-f_{e_{n,2}}(t))}{\tau_{ue}}$$
$$- n_{e_{n,2}}(t)(1-f_{c_{n,2}}(t))W_{tun,2} \quad (6)$$

$$\frac{dn_{c_{n,i}}(t)}{dt} = n_{e_{n,i}}(t)(1-f_{c_{n,i}}(t))W_{tun,i} - \frac{n_{c_{n,i}}(t)}{\tau_{cr}} \quad (7)$$

$$\frac{dn_{u_{n,i}}(t)}{dt} = \frac{n_{e_{n,i}}(t)(1-f_{u_{n,i}}(t))}{\tau_{eu}} - \frac{n_{u_{n,i}}(t)(1-f_{e_{n,i}}(t))}{\tau_{ue}} \quad (8)$$

Here G is the optical generation attained by:

$$G = \frac{P_{in}}{h\nu}(1-e^{-\alpha L}) \quad (9)$$

where P_{in} is the input IR optical power, $h\nu$ is a photon energy, α is described as the linear intersubband absorption coefficient, and L is the thickness of the absorber layer. absorption coefficient (α) is obtained through (10) [33]:

$$\alpha_{m,n,i}^i = \frac{1}{cn_r\varepsilon_0\hbar}\frac{E_n^i}{V_{QD,i}}|<\Psi_{g,i}|\hat{e}\hat{r}|\Psi_{e,i}>|^2 B_{m,n}(E_{n,i}-E_{m,i})G_n(E_{n,i}) \quad (10)$$

Where the index of i is equal to 1 for wavelength-1 and 2 for wavelength-2, c is the free space light speed, ε_0 is the free space permittivity, n_r is the refractive index of QDs, e is the electron charge, and $V_{QD,i}$ is the volume of the corresponding QD. The term $|<\Psi_{g,i}|\hat{e}\hat{r}|\Psi_{e,i}>|$ expresses the intersubband transition dipole moment. The homogenous broadening (HB) [34] and the inhomogeneous broadening (IHB) [34], [35] arising from fabrication limitations and ambient conditions are modeled by Lorentzian ($B_{m,n}$) and the Gaussian (G_n) functions, respectively that are obtained by:

$$E_n^{1(2)} = E_0^{1(2)} - (M+1-n)\Delta E \qquad n = 1,...,2M+1 \quad (11)$$

$$B_{m,n}(E_m^{1(2)}-E_n^{1(2)}) = \frac{1}{\pi}\frac{\Gamma_{HB}/2}{\left(E_m^{1(2)}-E_n^{1(2)}\right)^2 + (\Gamma_{HB}/2)^2} \qquad m,n=1 \quad (12)$$

$$G_n^{1(2)}(E_0^{1(2)}) = \frac{1}{\sqrt{2\pi}\xi_0}\exp\left[\frac{-\left(E_n^{1(2)}-E_0^{1(2)}\right)^2}{2\xi_0^2}\right] \quad (13)$$

In coupled rate equations, where $n_{g_{n,i}}$, $n_{e_{n,i}}$, $n_{c_{n,i}}$, $n_{u_{n,i}}$, $f_{g_{n,i}}$, $f_{e_{n,i}}$, $f_{c_{n,i}}$, and $f_{c_{n,i}}$ are illustrated as the quantity of electrons and the related carrier occupation probabilities in the GS, the ES, the CS (the ground states of QW energy contacts), and the US (upper states), respectively. Density of CdSe QDs per volume is 3.5×10^{16} cm^{-3}.

In the proposed photodetector, excited electrons tunnel from the absorber layer of QDs to QW energy contacts, producing photocurrent. To obtain the tunneling rate W_{tun}, the transmission coefficient T_{tun} has been calculated through (14) [37].

$$T_{tun,i}(E,V) = \frac{(\Gamma/2)^2}{(E-(E_r-(eV_i/2)))^2 + (\Gamma/2)^2} \quad (14)$$

$$W_{tun,i} = \frac{1}{2d}\sqrt{\frac{3KT}{m_e^*}}T_{tun,i} \quad (15)$$

Here E_r is the energy of the resonant level relative to the bottom of the well at its center, Γ is the resonance width considered 5meV, d is the diameter of QDs, K is the Boltzmann constant, T is the temperature.

IV. RESULTS AND DISCUSSION

In this section, for evaluating the performance of the modeled two-wavelength QDIP, the photodetector's figures of merit, including dark current, photocurrent, responsivity, and detectivity, have been studied.

The number of electrons of each energy level have been computed through (3)-(8). As a result, the amount of electrons tunneling to the IDCs determine the output photocurrent:

$$I_i = qW_{tun,i}n_{c_{n,i}}(1-f_{g_{n,i}}) \quad (16)$$

The responsivity and the specific detectivity are attained by [38]:

$$R_i = \frac{I_i}{P_{in}} \quad (17)$$

$$D_i^* = \frac{R_i\sqrt{A_{eff}\Delta f}}{i_n} \qquad i_n = (2qI_{dark,i}\Delta f)^{1/2} \quad (18)$$

TABLE II. Comparison of the responsivity and detectivity of the proposed device with competing samples.

Material (at 300°K)	Operating wavelengths	Responsivity (A/W)	Detectivity (cm.Hz$^{1/2}$W^{-1})
PbS QDs/ZnO heterostructures [39]	532 nm	2.73	2.39×10^{12}
	808 nm	0.42	3.65×10^{11}
HgTe CQDs [40]	2.5 µm	0.05	6×10^{10}
	4 µm		6.5×10^{8}
In$_{0.53}$Ga$_{0.47}$As sub-detector	1640 nm	0.57	2.63×10^{11}
In$_{0.53}$Ga$_{0.47}$As/GaAs$_{0.5}$Sb$_{0.5}$ sub-detector [41]	2 µm	0.22	1.96×10^{9}
CdSe/ZnS CQDs (this work)	3 µm	7.5	5×10^{11}
	5 µm	8	6×10^{11}

where R is the responsivity, A_{eff} is the detector's photosensitive active area, Δf is the noise bandwidth, i_n is the shot noise, and I_{dark} is the dark current. In all simulations, the quantum efficiency is assumed 100%.

Figures of merit calculated by (16)-(18) are depicted in Fig. 3. In Fig. 3(a), responsivities and detectivities as a function of applied light energy have been illustrated for both wavelengths of 3µm and 5µm. In Fig. 3(b), the responsivities and detectivities versus applied bias voltage of V$_1$ have been depicted for wavelength-1. In Fig. 3(c), the responsivities and detectivities versus applied bias voltage of V$_2$ have been depicted for wavelength-2. In all Fig. 3(a,b,c) green lines is related to responsivities and red lines is related to detectivities. It can be conclude that, by applying two laser beams with two different wavelengths of MIR range we can have two distinct photocurrents in separate electrical circuits simultaneously. Some non-idealities arising from fabrication limitations are included in all simulations and also in the rate equations, so the results of responsivities and detetivies demonstrated in Fig. 3 will be close to practical samples. A comparison of this work with other similar studies has been reported in Table II.

V. CONCLUSION

In this research, novel design of two-wavelength QDIP with distinct output currents based on solution-processed CdSe/ZnS core/shell QDs has been presented. The reported model is simulated and designed to detect tunable wavelengths of the MIR light spectra simultaneously by absorbing the incoming light through intersubband transitions. For extracting the stimulated carriers from the absorber layer to the IDCs, QW structures are exploited to choose desired wavelengths based on the specified applied bias voltage through tunneling effect. For modal analysis and evaluation of optical behavior, Schrodinger–Poisson equations and the coupled rate equations have been calculated. Based on the simulation results, the peak responsivities are about 7.5(A/W) and 8(A/W), and the specific detectivities D* are about 5×10^{11}(cm.Hz$^{1/2}$W^{-1}) and 6×10^{11}(cm.Hz$^{1/2}$W^{-1}) for the wavelengths of 3µm and 5µm, respectively.

REFERENCES

[1] A. Goldberg, P. N. Uppal, and M. Winn, "Detection of buried land mines using a dual-band LWIR/LWIR QWIP focal plane array," *Infrared Phys. Technol.*, vol. 44, no. 5–6, pp. 427–437,

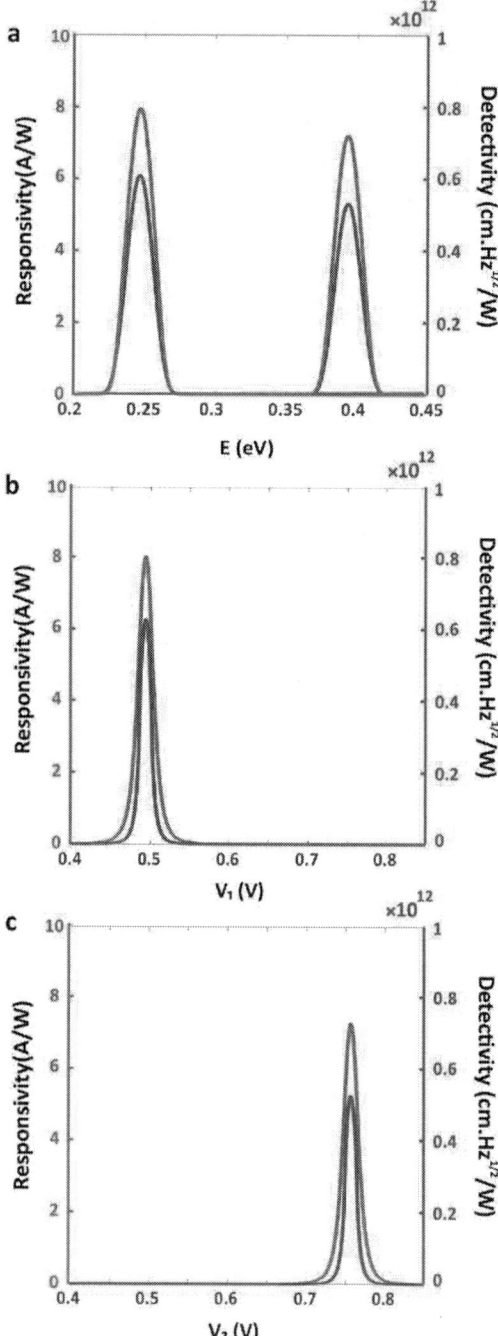

2003.

Fig. 3. Figures of merit for proposed two-wavelength photodetector. Fig. 3(a), responsivities and detectivities as a function of applied light energy for both wavelengths of 3µm and 5µm. Fig. 3(b), the responsivities and detectivities versus applied bias voltage of V$_1$ wavelength-1. Fig. 3(c), the responsivities and detectivities versus applied bias voltage of V$_2$ for wavelength-2.

[2] C. Liang *et al.*, "Chinese military evaluation of a portable near-infrared detector of traumatic intracranial hematomas," *Mil. Med.*, vol. 183, no. 7–8, pp. e318–e323, 2018.

[3] P. M. Alsing, D. A. Cardimona, D. H. Huang, T. Apostolova, W. R. Glass, and C. D. Castillo, "Advanced space-based detector research at the Air Force Research Laboratory," *Infrared Phys.*

[4] D. B. Law, E. M. Carapezza, and C. J. Csanadi, "DARPA countersniper program: phase 1 Acoustic Systems Demonstration results," *SPIE Proc. 2938*, vol. 288, 1997.

[5] X. Yang, L. Tang, H. Wang, and X. He, "Early detection of forest fire based on unmaned aerial vehicle platform," in *2019 IEEE International Conference on Signal, Information and Data Processing (ICSIDP)*, 2019, pp. 1–4.

[6] G. Ariyawansa, A. G. U. Perera, X. H. Su, S. Chakrabarti, and P. Bhattacharya, "Multi-color tunneling quantum dot infrared photodetectors operating at room temperature," *Infrared Phys. Technol.*, vol. 50, no. 2–3, pp. 156–161, 2007.

[7] A. G. U. Perera, G. Ariyawansa, G. Huang, and P. Bhattacharya, "Quantum dot nanostructures for multi-band infrared detection," *Infrared Phys. Technol.*, vol. 52, no. 6, pp. 252–256, 2009.

[8] A. Karim and J. Y. Andersson, "Infrared detectors: Advances, challenges and new technologies," in *IOP Conference Series: Materials Science and Engineering*, 2013, vol. 51, no. 1, p. 12001.

[9] S. Adhikary, Y. Aytac, S. Meesala, S. Wolde, A. G. Unil Perera, and S. Chakrabarti, "A multicolor, broadband (5–20 μm), quaternary-capped InAs/GaAs quantum dot infrared photodetector," *Appl. Phys. Lett.*, vol. 101, no. 26, p. 261114, 2012.

[10] P. Bhattacharya, X. H. Su, S. Chakrabarti, G. Ariyawansa, and A. G. U. Perera, "Characteristics of a tunneling quantum-dot infrared photodetector operating at room temperature," *Appl. Phys. Lett.*, vol. 86, no. 19, p. 191106, 2005.

[11] D. Z. Ting *et al.*, "Development of quantum well, quantum dot, and type II superlattice infrared photodetectors," *J. Appl. Remote Sens.*, vol. 8, no. 1, p. 84998, 2014.

[12] X. Wang, Z. Cheng, K. Xu, H. K. Tsang, and J.-B. Xu, "High-responsivity graphene/silicon-heterostructure waveguide photodetectors," *Nat. Photonics*, vol. 7, no. 11, pp. 888–891, 2013.

[13] B. Lai, Y. Jin, L. Liu, Y. Yu, and S. Yu, "Impact of Double Al 0.1 Ga 0.9 As Barrier on nin InAs/GaAs Quantum Dot Infrared Photodetectors," in *2019 Asia Communications and Photonics Conference (ACP)*, 2019, pp. 1–3.

[14] T. Murata, S. Asahi, S. Sanguinetti, and T. Kita, "Infrared photodetector sensitized by InAs quantum dots embedded near an Al0. 3Ga0. 7As/GaAs heterointerface," *Sci. Rep.*, vol. 10, no. 1, pp. 1–11, 2020.

[15] B. Zhang, H. D. Lu, W. G. Ning, and F. M. Guo, "Photocurrent optimize of InAs/GaAs pip quantum dots infrared photodetectors," in *2016 IEEE 11th Annual International Conference on Nano/Micro Engineered and Molecular Systems (NEMS)*, 2016, pp. 148–151.

[16] Y. Kim, J. O. Kim, S. J. Lee, and S. K. Noh, "Submonolayer quantum dots for optoelectronic devices," *J. Korean Phys. Soc.*, vol. 73, no. 6, pp. 833–840, 2018.

[17] M. Razeghi, A. Dehzangi, and J. Li, "Multi-band SWIR-MWIR-LWIR Type-II superlattice based infrared photodetector," *Results Opt.*, vol. 2, p. 100054, 2021.

[18] S. Chakrabarti, X. H. Su, P. Bhattacharya, G. Ariyawansa, and A. G. U. Perera, "Characteristics of a multicolor InGaAs-GaAs quantum-dot infrared photodetector," *IEEE Photonics Technol. Lett.*, vol. 17, no. 1, pp. 178–180, 2004.

[19] W. Chen *et al.*, "Demonstration of InAs/InGaAs/GaAs quantum dots-in-a-well mid-wave infrared photodetectors grown on silicon substrate," *J. Light. Technol.*, vol. 36, no. 13, pp. 2572–2581, 2018.

[20] S. Sengupta and S. Chakrabarti, "Optical and Spectral Characterization of Sub-monolayer QDIPs," in *Structural, Optical and Spectral Behaviour of InAs-based Quantum Dot Heterostructures*, Springer, 2018, pp. 43–58.

[21] K. C. G. Kumari, H. Ghadi, D. R. M. Samudraiah, and S. Chakrabarti, "Indigenous development of 320× 256 focal-plane array using InAs/InGaAs/GaAs quantum dots-in-a-well infrared detectors for thermal imaging," *Curr. Sci.*, pp. 1568–1573, 2017.

[22] S. Chakrabarti, A. D. Stiff-Roberts, P. Bhattacharya, and S. W. Kennerly, "High responsivity AlAs/InAs/GaAs superlattice quantum dot infrared photodetector," *Electron. Lett.*, vol. 40, no.

3, p. 1, 2004.

[23] A. V Barve *et al.*, "Confinement enhancing barriers for high performance quantum dots-in-a-well infrared detectors," *Appl. Phys. Lett.*, vol. 99, no. 19, p. 191110, 2011.

[24] X. Tang and Q. Hao, "Towards dual-band shot-wave and mid-wave infrared focal plane array by using colloidal quantum dots," in *Seventh Symposium on Novel Photoelectronic Detection Technology and Applications*, 2021, vol. 11763, pp. 168–174.

[25] X. Tang, M. M. Ackerman, and P. Guyot-Sionnest, "Flexible infrared electronic eyes for multispectral imaging with colloidal quantum dots," in *2019 International Conference on Optical Instruments and Technology: IRMMW-THz Technologies and Applications*, 2020, vol. 11441, pp. 90–96.

[26] S. Zhang, M. Chen, G. Mu, J. Li, Q. Hao, and X. Tang, "Spray-Stencil Lithography Enabled Large-Scale Fabrication of Multispectral Colloidal Quantum-Dot Infrared Detectors," *Adv. Mater. Technol.*, vol. 7, no. 6, p. 2101132, 2022.

[27] A. Rogalski, "Progress in quantum dot infrared photodetectors," *Quantum Dot Photodetectors*, pp. 1–74, 2021.

[28] H. G. Yousefabad, S. Matloub, and A. Rostami, "Ultra-broadband optical gain engineering in solution-processed QD-SOA based on superimposed quantum structure," *Sci. Rep.*, vol. 9, no. 1, pp. 1–11, 2019.

[29] S. Matloub, P. Amini, and A. Rostami, "Switchable Multi-color Solution-processed QD-laser," *Sci. Rep.*, vol. 10, no. 1, pp. 1–14, 2020.

[30] A. Trellakis, A. T. Galick, A. Pacelli, and U. Ravaioli, "Iteration scheme for the solution of the two-dimensional Schrödinger-Poisson equations in quantum structures," *J. Appl. Phys.*, vol. 81, no. 12, pp. 7880–7884, 1997.

[31] W. M. M. Lin *et al.*, "Recombination Dynamics in PbS Nanocrystal Quantum Dot Solar Cells Studied through Drift–Diffusion Simulations," *ACS Appl. Electron. Mater.*, vol. 3, no. 11, pp. 4977–4989, 2021.

[32] R. Dalven, "Electronic Structure of PbS, PbSe, and PbTe," in *Solid State Physics*, vol. 28, Elsevier, 1974, pp. 179–224.

[33] F. Toyama, "Interband and Intraband Optical Studies of CdSe Colloidal Nanocrystal Films," *Physics (College. Park. Md).*, 2004.

[34] F. Demangeot, D. Simeonov, A. Dussaigne, R. Butté, and N. Grandjean, "Homogeneous and inhomogeneous linewidth broadening of single polar GaN/AlN quantum dots," *Phys. Status Solidi Curr. Top. Solid State Phys.*, vol. 6, no. SUPPL. 2, pp. S598–S601, Jun. 2009, doi: 10.1002/pssc.200880971.

[35] H. Dortaj, M. Faraji, and S. Matloub, "High-speed and high-contrast two-channel all-optical modulator based on solution-processed CdSe/ZnS quantum dots," *Sci. Rep.*, vol. 12, no. 1, p. 12778, 2022, doi: 10.1038/s41598-022-17084-4.

[36] H. Dortaj *et al.*, "High-speed and high-precision PbSe/PbI2 solution process mid-infrared camera," *Sci. Rep.*, vol. 11, no. 1, pp. 1–11, Jan. 2021, doi: 10.1038/s41598-020-80847-4.

[37] J. N. Schulman, H. J. De Los Santos, and D. H. Chow, "Physics-based RTD current-voltage equation," *IEEE Electron Device Lett.*, vol. 17, no. 5, pp. 220–222, 1996.

[38] H. Dortaj and S. Matloub, "Design and simulation of two-color mid-infrared photoconductors based on intersubband transitions in quantum structures," *Phys. E Low-dimensional Syst. Nanostructures*, p. 115660, 2023, doi: https://doi.org/10.1016/j.physe.2023.115660.

[39] M. Peng *et al.*, "High-performance flexible and broadband photodetectors based on PbS quantum dots/ZnO nanoparticles heterostructure," *Sci. China Mater.*, vol. 62, no. 2, pp. 225–235, 2019.

[40] X. Tang, M. M. Ackerman, and P. Guyot-Sionnest, "Colloidal quantum dots based infrared electronic eyes for multispectral imaging," in *Optical Sensing, Imaging, and Photon Counting: From X-Rays to THz 2019*, 2019, vol. 11088, pp. 8–14.

[41] Z. Xie, Z. Deng, X. Zou, and B. Chen, "InP-based near infrared/extended-short wave infrared dual-band photodetector," *IEEE Photonics Technol. Lett.*, vol. 32, no. 16, pp. 1003–1006, 2020.

Role of Doping Concentration of n- and p-Strip Regions on Optoelectronical Characterization in IBC-SHJ Solar Cell

line 1: Pegah Paknazar
line 2: *dept. electrical and computer engineering*
line 3: *Jundi-Shapur University of Technology,*
line 4: Dezful, Iran
line 5: P.paknazar@jsu.ac.ir

line 1: Maryam Shakiba
line 2: *dept. electrical and computer engineering*
line 3: *Jundi-Shapur University of Technology,*
line 4: Dezful, Iran
line 5: shakiba@jsu.ac.ir

Abstract— **In this research, the photocarrier transmission mechanism and the effect of the doping concentration of n- and p-strip regions in interdigitated back contact silicon heterojunction (IBC-SHJ) cell efficiency have been studied. In this regards, short-circuit current density, open-circuit voltage, fill factor and cell efficiency values have been evaluated using J–V curves for different conditions. The doping concentration of the n- and p-strip regions have been changed using trial-and-error method to achieve improved efficiency in IBC-SHJ solar cell. The improved IBC-SHJ have no extra ARCs and more structural periodicity. Thus, a simple structure with improved conversion efficiency is proposed. The results have been shown that the n- and p-strip doping concentration were the most effective parameters on efficiency improvement. According to the results, the best doping concentration of Emitter and BSF regions to achieve improved efficiency is equal to 2×10^{19} cm^{-3} and 4.3×10^{18} cm^{-3} respectively.**

Keywords—Doping Concentration, n- and p-strip regions, IBC-SHJ solar cell, Optoelectronical characterization

I. Introduction

Widespread deployment of solar photovoltaics (PVs) is crucial to meeting the world's growing energy demand and preservation of the environment in the future [1]. So far, extensive research has been done to increase the efficiency of silicon solar cells with common finger contact structure [2-3]. In this way, design of tandem cells and new technologies of materials and alloys used in different layers have increased efficiency [4-5]. On the other hand, IBC solar cells is one of the configurations of rear contact solar cells. This feature reduced shading on the front of the cell and thus it is specifically useful in applications such as high current cells like concentrator PVs [6-7]. Also IBCs are so useful in large area installations. Because these cells can be interconnected easier and there is no need for any space between the cells in the module [8]. In this configuration of solar cell the electron hole pairs generated at the front surface of the cell can be collected at the rear of the cell due to the spatial electric field distribution [9].

During the last 10 years, research have been carried out to increase the efficiency of IBC-SHJ cells. In 2013, Silvaco presented the IBC-SHJ cell with an efficiency of 20.43% [9]. In 2016, M. Belarbi and colleagues achieved an efficiency of 23.20% by improving cell deposition parameters [10]. By adding more layers and anti-reflection coatings in the front side and repeating the periodicity of the structure on the back side of the cell, an efficiency of 26.30% for the IBC-SHJ structure was presented by K. Yoshikawa et al in 2017 [11]. By considering the periodicity of the IBC-SHJ structure and at the same time adding more layers and anti-reflection coatings in the front side and also using TCO layers in the back part of the cell, the efficiency of 27.41% was obtained by J. Bao and colleagues in 2020 [12]. But It should be noted that increase in efficiency in researches [11] and [12] will require an increase in the complexity of the structure and, as a result, an increase in the challenges of the fabrication process. In this research, the purpose is to design and simulate an improved silicon heterojunction solar cell with interdigitated back contacts, in a way that has the simplest structure among different structures and obtain the least challenge in the fabrication process. This research aims to improve the efficiency of IBC-SHJ solar cells by improving the doping concentration of these regions. The study on changing doping concentrations to enhance cell performance is important for advancing solar cell technology. The manuscript provides a comprehensive analysis of the impact of doping concentration on various cell parameters including short-circuit current density (Jsc), open-circuit voltage (Voc), fill factor (FF), and cell efficiency (η). In the second part of this research, the investigation and analysis of photo carrier transfer mechanisms in the IBC-SHJ cell is presented. In the third part, the mathematical relations governing the particle transfer and modeling of light radiation to the cell have been investigated. In the fourth part, the effect of different doping concentration of the n- and p-strip regions on the solar cell outputs including Jsc, Voc, FF and η have been analyzed and investigated. In the fifth section, the most important results are expressed.

II. ANALYSIS OF CARRIER TRANSPORT

In order to create anode and cathode electrodes, a structure of n- and p-type layers of hydrogenated a-Si are deposited on the back side of the IBC-SHJ cell so that the layers are one in between. Depending on the doping concentration of c-Si substrate, these n- and p-type layers, play the role of back surface field (BSF) and emitter respectively. As shown in Fig. 1, the width of the elementary structure (pitch) of the cell is equal to the distance between the opposite electrodes, considering the gap between them. It is possible to create periodicity in more complex structures, which will obviously face more challenges in fabrication process. In this research, the dimensions and types of materials used in different layers have been selected based on reproducibility, so that the simulation results are effective in improving the fabrication process [13].

The substrate used to optimize deposition parameters is an n-type c-Si wafer with a thickness of 300 μm and a width of 1750 μm, which is equivalent to one cell pitch. The resistivity of the substrate is equal to 2 Ω.cm. Also, using n-type c-Si substrate will lead to an increase in Voc as a result of increasing efficiency [12] [14]. On the front surface of the cell, a SiN$_x$ layer with a thickness of 75 nm has been used as an anti-reflection coating, for this purpose, the physical model of concentration mobility (CONMOB) has been used in the simulation. The maximum photo generation rate occurs in the upper part of the cell and is defined according to (1) [9]:

$$G = \alpha N_0 e^{-\alpha x} \qquad (1)$$

In equation (1), G is the photo generation rate, N$_0$ is the photon flux at the surface and x is the depth of the corresponding photo generation value. α represents the absorption coefficient and is expressed according to the following relation [9]:

$$\alpha = 4\pi k / \lambda \qquad (2)$$

In relation (2), k and λ express the extinction coefficient and the wavelength of the radiation light, respectively.

Therefore, due to not choosing the same type of impurity for ARC layers and c-Si substrate, the existence of an electric field on the top of the cell that leads to more drift of the photocarriers and prevents their recombination will be inevitable. According to Fig. 1, the presence of anti-reflection coating with the opposite doping concentration of the substrate leads to the creation of a space charge region in the front part of the cell. The internal electric field created in this space charge region will prevent the recombination of the generated photocarriers and move them towards the back contacts of the cell. Also, near the back contacts, due to the presence of the back p-n junction, another space charge region is created, which facilitates the drift of the carriers towards the anode and cathode electrodes, and by preventing the recombination of photocarriers, it will increase the photocurrent of the cell. In fact, the design of two space charge areas in the upper and lower parts of the cell, in addition to preventing the recombination of photocarriers, leads to drift them towards the interdigitated back contacts.

As shown in Fig. 2, the photocarriers generated in front part of the cell, drift towards the contacts with the similar doping concentration, under influence of electric field of space charge region. Photocarriers move vertically, towards the back contacts via the body part, and then laterally, nearby the contacts. Meanwhile, the larger the cell width, the more lateral distance the carrier travel to accumulate at the contact of the same name; therefore, as the cell width increases, the recombination probability increases, in addition the cost of the fabrication increases [15].

It is worth noting that in Fig. 1 and 2, the thickness of SiN$_x$, n-type c-Si, defective c-Si, i-a-Si, Emitter/a-Si, BSF/a-Si and Electrodes/Al layers are 75 nm, 150 μm, 1 nm, 6 nm, 20 nm, 20 nm, 0.2 μm respectively.

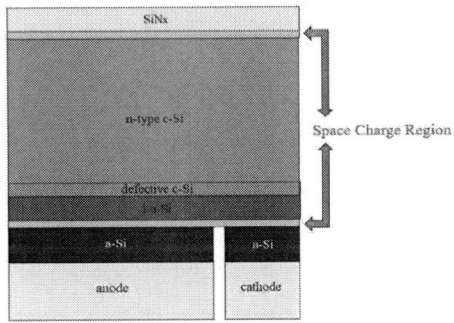

Fig. 1. Schematic of the simulated IBC-SHJ structure along with the approximated regions of space charge in the upper and lower areas of the cell

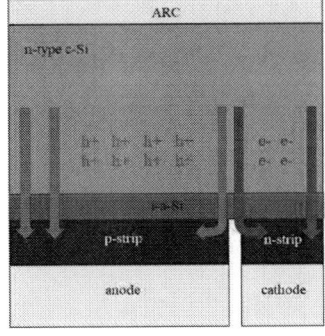

Fig. 2. Vertical and lateral travelling of minority and majority carriers in IBC-SHJ cell

III. PHYSICAL MODELS AND EQUATIONS

A. Physical Models

In order to match the numerical modeling and experimental results as much as possible, in this research, we have tried to use accurate physical models to describe the behavior of IBC-SHJ cells. In this regard, the band gap narrowing (BGN) model has been used to reduce the energy gap [9]. In fact, the narrowing of the band gap shows how applying a high doping concentration (greater than 10^{18} cm^{-3}) changes the band gap by reducing the energy of the conduction band and increasing the energy of the valence band [9]. On the other hand, in order to model different recombination mechanisms, Fermi, Shockley-Read-Hall (SRH), Auger and surface recombination models are also included in the simulation. Meanwhile, carrier recombination is also modeled as a function of doping concentration via SRH and Auger mechanisms. Disordered materials such as amorphous silicon contain a large amount of defect states in their energy band gap. Therefore, the density of state (DOS) model in the energy bandgap is used to model amorphous silicon devices. The density model of the defect states is described as a combination of the tail of the band using the exponential decay function and the mid of the band using the Gaussian distribution function. The energy distribution of exponential band tail and Gaussian distribution of trap states in the middle of the energy gap are key parameters for high accuracy simulation.

For the c-Si/a-Si interface on the back surface, the thermionic emission model has been used, in which the distribution function of the defect states at the interface of two layers, one for holes and the other for electrons, is modeled. In order to model the interface between the defects-free crystalline silicon layer and the amorphous silicon layer as much as possible, a defective c-Si thin layer has been used between the two layers [16]. The Sopra database is also used for the refractive index of a-Si layers [9]. AM1.5G solar spectrum has been used to simulate sunlight in standard conditions with a light intensity of 1000 w/m² and a temperature of 26°C.

B. Equations

The specialized software used in this research to design and simulation the IBC-SHJ solar cell is Silvaco software. The Athena tool in this software models and simulates the fabrication process in the Monte Carlo method and in conditions very similar to the new technologies presented in the field of semiconductor devices fabrication. In addition, the Atlas tool simulates the electrical behavior of cell designed in Athena. It uses the drift-diffusion model based on the discretization of differential equations and solving equations using numerical solutions such as Gumel, Newton-Raphson and Block methods. Some of the output parameters of Atlas tool are Fill Factor (FF) and Efficiency (η) obtained from relations (3) and (4) respectively [9]:

$$FF = \frac{V_{mpp}\, I_{mpp}}{V_{oc}\, I_{sc}} \qquad (3)$$

$$Cell\ Efficiency\ (\eta) = \frac{P_{out}}{P_{in}} = \frac{V_{oc}\, I_{sc}\, FF}{P_{light}} \qquad (4)$$

In relation (3), V_{mpp} and I_{mpp} represent the voltage and current at the maximum power point, respectively.

IV. RESULTS AND DISCUSSION

In this research, in order to improve the characteristics of the IBC-SHJ cell, the properties of the different layers, including the doping concentration of emitter region and the doping concentration of BSF region are optimized.

In the following, the influence of the mentioned parameters on the output characteristics of Jsc, Voc, FF and η is investigated to improve the efficiency of the IBC-SHJ solar cell. It should be noted that to investigate the effect of each parameter, other parameters of the cell are considered constant.

A. Effect of Emitter (p-strip) Doping Concentration

The current-voltage curve and Jsc, Voc, FF and η graphs considering the change of doping concentration of the emitter region are presented in Fig. 3.

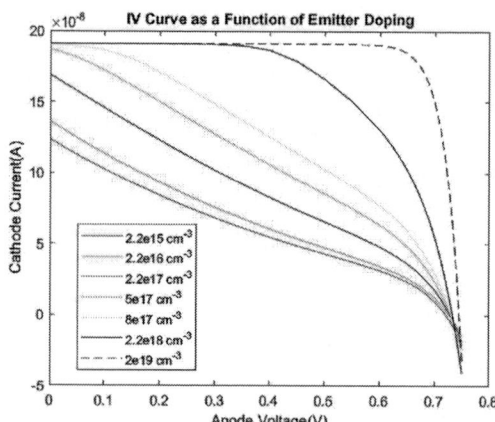

Fig. 3(a). Effect of emitter doping concentration on current-voltage curve

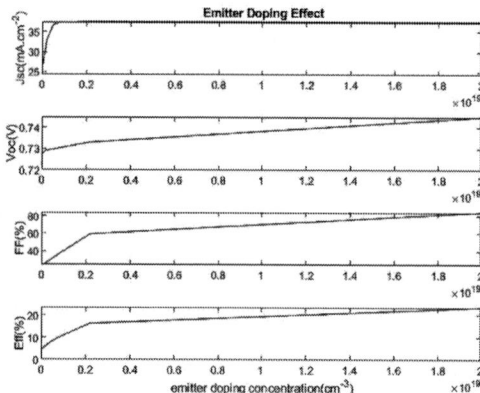

Fig. 3(b). Effect of emitter doping concentration on Jsc, Voc, FF and η

According to Fig. 3(a), increasing the doping concentration of emitter region leads to increase in cell current, but has little effect on the voltage. As shown in Fig. 3(b), increasing the doping concentration of emitter region, increases Jsc. In this situation, the number of free carriers and photocurrent increases. As concluded from the current-voltage curves in Fig. 3(a), increasing the doping concentration of emitter region increases Voc with a very small slope. According to the FF graph, FF increases. Because with the increase of emitter region doping concentration, V_{mpp} and I_{mpp} increases and the curve is completely S-shaped at the concentration of 2×10^{19} cm^{-3}. Therefore, according to relation (3), FF increases. On the other hand, the cell efficiency increases due to increase in Jsc, Voc and FF. As a result, the most improved value for the doping concentration of emitter region is equal to 2×10^{19} cm^{-3}.

B. Effect of BSF (n-strip) Doping Concentration

The current-voltage curve and Jsc, Voc, FF and η graphs considered with the change of BSF region doping concentration are presented in Fig. 4.

979-8-3503-6020-2/23 $31.00 © 2023 IEEE   38

The 5th Iranian International Conference on Microelectronics (IICM2023)

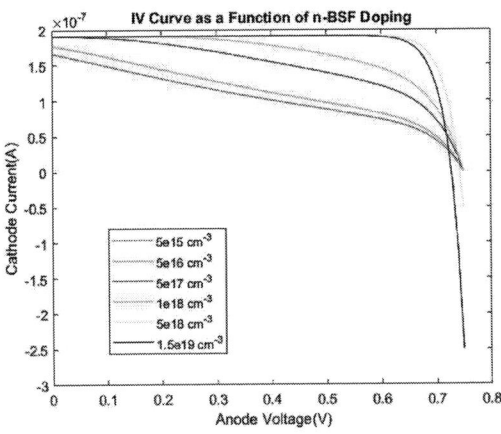

Fig. 4(a). Effect of BSF doping concentration on current-voltage curve

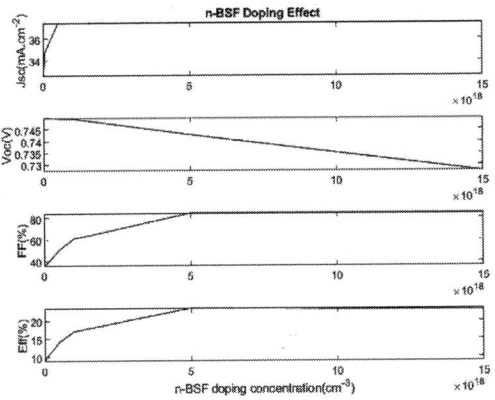

Fig. 4(b). Effect of BSF doping concentration on Jsc, Voc, FF and η

According to Fig. 4(a), it can be seen that increasing the doping concentration of BSF region increases the cell current. Also, as the concentration of this region increases, the voltage decreases. In this situation due to increase of free carriers and photocurrent of the cell Jsc increases and Voc decreases and the series resistance decreases. On the other hand, with the increase of doping concentration in the BSF region, the FF increases. Because according to Fig. 4(a), V_{mpp} and I_{mpp} increase and according to the relation (3) FF increases. Also, the cell efficiency increases up to 4.3×10^{18} cm^{-3} and decreases after that. In fact, to reduce the series resistance behind the cell, the doping concentration of the BSF region should be increased as much as possible. But on the other hand, the excessive increase of doping concentration in the BSF region leads to the increase of scattering mechanisms for the carriers, which reduces the cell efficiency. Therefore, in the proposed IBC-SHJ solar cell, the most improved value for the doping concentration of BSF region is equal to 4.3×10^{18} cm^{-3}. proposed values of IBC-SHJ cell efficiency at Different emitter and BSF doping concentrations are given in Tables I(A) and I(B), respectively. Also in Table II, the doping concentrations of emitter and BSF regions and Efficiency of the proposed IBC-SHJ solar cell compared to previous research.

TABLE I(A). IBC-SHJ CELL EFFICIENCY AT DIFFERENT EMITTER DOPING CONCENTRATION

Row	Parameters	
	Emitter doping (cm^{-3})	*Efficiency (%)*
1	2.2e15	4.31
2	2.2e16	4.75
3	2.2e17	6.54
4	5e17	8.44
5	8e17	10.08
6	2.2e18	16.29
7	2e19	23.52

TABLE I(B). IBC-SHJ CELL EFFICIENCY AT DIFFERENT BSF DOPING CONCENTRATION

Row	Parameters	
	BSF doping (cm^{-3})	*Efficiency (%)*
1	5e15	8.74
2	5e16	9.68
3	5e17	14.15
4	1e18	17.06
5	4.3e18	23.52
6	5e18	23.49
7	1.5e19	22.99

TABLE II. THE DOPING CONCENTRATIONS OF EMITTER AND BSF REGIONS AND EFFICIENCY OF THE PROPOSED IBC-SHJ SOLAR CELL COMPARED TO PREVIOUS STUCTURES

Ref.	Parameters		
	Emitter doping (cm^{-3})	*BSF doping (cm^{-3})*	*Efficiency (%)*
[9]	2e19	1.5e19	20.43
[6]	1e18	1e18	23.31
[Proposed structure]	2e19	4.3e18	23.52

V. CONCLUSION

In this research, the effect of doping concentration of Emitter (p-strip) and BSF (n-strip) in the IBC-SHJ cell were investigated. Our improved IBC-SHJ solar cell yields a short-circuit current density of 37.42 mA/cm² and open-circuit voltage of 745 mV with cell efficiency of 23.52% having a fill factor of 84.30 under AM1.5G without any extra ARCs and more structural periodicity. Thus, a simple structure with improved conversion efficiency is proposed. Simulation shows that cell efficiency varies due to change in doping concentration of these layers and so these parameters are optimized accordingly. According to the results, the optimum doping concentration of Emitter and BSF regions is equal to 2×10^{19} cm^{-3} and 4.3×10^{18} cm^{-3} respectively. In fact, the design of two space charge areas in the upper and lower parts of the cell, in addition to prevent the recombination of

979-8-3503-6020-2/23 $31.00 © 2023 IEEE

photocarriers, leads to drift them towards the interdigitated back contacts.

REFERENCES

[1] C. N Kruse, S. Schafer, F. Haase, V. Mertens, H. Schulte-Huxel, B. Lim, B. Min, T. Dullweber, R. Peibst, R. Brendel, " Simulation-based roadmap for the intergration of poly-silicon on oxide contacts into screen-printed crystalline silicon solar cells", Nature research, vol. 11, pp. 996, 2021.

[2] M. Shakiba, A. Kosarian, E. Farshidi, "Effects of processing parameters on crystalline structure and optoelectronic behavior of DC sputtered ITO thin film", J Mater Sci: Mater Electron, vol. 28, pp. 787-797, 2016.

[3] L. A Zafoschnig, S. Nold, J. C Goldschmidt, "The race for lowest costs of electricity production: techno-economic analysis of silicon, perovskite and tandem solar cells", IEEE Journal of Photovoltaics, vol. 10, pp. 6, 2020.

[4] D. Yan, J. Cuevas, A. Stuckelberger, E. Wang, S. Pheng Phang, T. Choon Kho, J. I Michel, D. Macdonald, J. Bullock, "Silicon solar cells with passivating contacts: classification and performance", Progress in Photovoltaics Wiley, vol. 31, pp. 310-326, 2022.

[5] M. Shakiba, M. Shakiba, "Role of Critical Processing Parameters on Fundamental Phenomena and Characterizations of DC Argon Glow Discharge", Optoelectronical Nanostructures, vol. 7, pp. 67-91, 2022.

[6] A. R. M Rais, S. Sepeai, M. K. M Desa, M. A Ibrahim, P.J Ker, S. H Zaidi, K. Sopian, "Photo-generation profiles in deeply-etched, two-dimensional patterns in interdigitated back contact solar cells", Journal of Ovonic Research, vol. 17, pp. 283-289, 2021.

[7] V. Giglia, R. Varache, J. Veirman, E. Fourmond, "Influence of cell edges on the performance of silicon heterojunction solar cells", Solar Energy Materials and Solar Cells, vol. 238, pp. 111605, 2022.

[8] P. Procel, G. Yang, O. Isabella, M. Zeman, "Theoretical evaluation of contact stack for high efficiency IBC-SHJ solar cells", Solar Energy Materials and Solar Cells, vol. 186, pp. 66-77, 2018.

[9] https://silvaco.com, "2D IBC-SHJ solar cell simulation and optimization", 2013.

[10] M. Belarbi, M. Beghdad, A. Mekemeche, "Simulation and optimization of n-type interdigitated back contact silicon heterojunction (IBC-SiHJ) solar cell structure using Silvaco Tcad Atlas", Solar Energy, vol. 127, pp. 206-215, 2016.

[11] K. Yoshikawa, H. Kawasaki, W. Yoshida, T. Irie, K. Konishi, K. Nakano, T. Uto, D. Adachi, M. Kanematsu, H. Uzu, K. Yamamoto, "Silicon heterojunction solar cell with interdigitated back contacts for a photoconversion efiiciency", Nature Energy, vol. 2, pp. 17032, 2017.

[12] J. Bao, A. Liu, Y. Lin, Y. Zhou, "An insight into effect of front surface field on the performance of interdigitated back contact silicon heterojunction solar cells", Materials Chemistry and Physics, vol. 255, pp. 123625, 2020.

[13] M. Lu, U. Das, S. Bowden, S. Hegedus, R. Birkmire, "Optimization of interdigitated back contact silicon heterojunction solar cells by two-dimensional numerical simulation", Institute of Energy Conversion, Univesity of Delaware, Newark, DE 19716 U.S.A, IEEE. 2009.

[14] N. Jensen, R. M Hausner, R. B Bergmann, J. H Werner, U. Rau, "Optimization and characterization of amorphous/crystalline silicon heterojunction solar cells", Prog. Photovolt. Res. Appl, vol. 10, pp. 1–13, 2002.

[15] F. Granek, M. Hermle, C. Reichel, A. Grohe, O. Schultz-Wittmann, S. Glunz, "Positive effects of front surface field in high-efficiency back-contact back-junction n-type silicon solar cells", PVSC '08. 33rd IEEE, pp. 1-5.

[16] D. Diouf, J. P Kleider, C. Longeaud, "Two-dimensional simulations of interdigitated ack contact silicon heterojunctions solar cells", Chapter 15 of the book physics and technology of amorphous-crystalline silicon heterostructure solar cells, Springer. pp. 483-519, 2011.

The 5th Iranian International Conference on Microelectronics (IICM2023)

25 – 26 October 2023

Broadened graded asymmetric waveguide structure for low divergence 915nm diode laser

Seyed Peyman Abbasi
Semiconductor Group
Iranian National Center for Laser Science & Technology
Tehran, Iran
pabbasi2001@gmail.com

Arash Hodaei
Semiconductor Group
Iranian National Center for Laser Science & Technology
Tehran, Iran
ahodaei@gmail.com

Abstract—**Multi-chip beam combining is applied for high-power laser diodes. In this method, fast collimation of the laser diode is an important process. Diode laser beam divergence is the primary parameter to couple the fiber optic and beam shaping. Expanding the waveguide layer thickness in epitaxial design is the usual approach to reduce the beam divergence. The 915nm broadened graded asymmetric structure is presented in this study, to reduce the divergence without raising the threshold current (I_{th}). The asymmetric waveguide shifted the vertical optical field to the n-section, which has smaller free carrier loss than the p-section. Furthermore, the linearly graded refractive index was used to control the modes. The main goal of this research is to decrease the divergence. In the proposed structure, fast divergence (FWHM) of 24 degrees and I_{th} density of less than 650A/cm² was achieved. Decreasing the beam divergence can increase the coupling efficiency to the optical fiber without increasing the I_{th}.**

Keywords— *Graded refractive index, Asymmetric structure, Beam divergence, Laser diode.*

I. INTRODUCTION

The 915 nm high-power laser diodes (HPLD) have been broadly utilized in the optical pumping of solid-state lasers like fiber optic lasers [1]. Nowadays, the 915nm fiber-coupled HPLDs are expanding and becoming popular among users [2]. In most of these lasers, the multi-chip is coupled by polarized beam combination or spatial beam combination methods. In this method, the beam of each laser diode chip is collimated and shaped individually, and the combined output of a single emitter beam is shaped into a single beam. Then an aspheric lens is utilized to couple the combined beam into an optical fiber [3]. The improvement of characterization of the diode laser and the optical component arrangement are the main issues to optimize this method.

Relying on the laser diode application, the design of the waveguide (WG) (as well as material and layer thickness) can be improved for laser parameters [4], [5]. For instance, a large optical cavity (LOC) decreased the divergence in some researches [4]. Applying this method in most cases reduces the confinement factor and thus increases the I_{th}. The internal loss is reduced by using asymmetric structures to compensate I_{th} increase [5]. The asymmetric structures were previously used to design WG and cladding layer [1], [6]. On the other hand, beam quality improvement and divergence decrease are vital issues to couple the laser diode beam to the optical fiber [7]. WG design is predicated on selecting correct parameters i.e.,

WG thicknesses and refractive indexes difference in the WG and cladding layer [8], [9].

The step-index profile difference is typically applied in waveguide design. Nevertheless, a graded refractive index profile has also been used. the graded refractive index separate confinement hetero-structure (GRINSCH) in non-broadened [10], [11] and broadened waveguides for HPLD [12] have also applied by some researchers. Some calculations reveal that assuming a constant active layer thickness, the I_{th} in the GRINSCH structure reduces comparing with the popular lasers [10].

Beam divergence depends on the near-field mode size. The near field determined by the normalized waveguide thickness, which depends on the waveguide thickness, waveguide-layer and cladding-layer indices, and the vacuum wavelength [13].

Increasing the waveguide layer thickness in step epitaxial design is the conventional method to decrease fast axis beam divergence. This method decreases the confinement factor, and increases the I_{th}, which are undesirable. In this research, the asymmetric structure was proposed that two linear graded refractive indices with different thicknesses and slopes are used for two n-waveguide (n-WG) and p-waveguide (p-WG) sections. Furthermore, the refractive index of the n-clad layer was selected higher than the p-clad layer by changing the aluminum mole fraction in two cladding layers. The effect of thickness and slope of linearly graded refractive index is investigated for laser diode parameters consisting of I_{th} and divergence. The double quantum well (DQW) was used in the active region to compensate the reduction in the confinement factor.

II. GRADED REFRACTIVE INDEX WAVEGUIDE

There are two conventional epitaxial structure designs for laser diode waveguides, step refractive index waveguide, and graded refractive index waveguide. Theory indicates that the I_{th} of a graded refractive index structure may be expected to be smaller than that in a conventional structure [10]:

- First, the intensity of the electric field induced in a varying bandgap semiconductor region is in proportion to the gradient of the bandgap; this dramatically raises the efficiency with which the thin well captures the carriers.

- Secondly, a waveguide with a graded refractive index is more effective in guiding electromagnetic waves.

979-8-3503-6020-2/23 $31.00 © 2023 IEEE

The efficiency with which the potential well captures the carriers is an important factor limiting the possibility of reducing the I_{th} in a single potential well laser, and if it is too low, the lasing action may not even occur at all [10].

The theory of refractive index dielectric waveguides is reviewed in [14]. The index in the graded-index region changes continuously. The ray follows a curved trajectory and turns continuously. In some models, the graded index region is a stack of thin layers. Each layer is so thin that the index variation in the layer can be neglected. In other words, each layer is used like a constant index region.

The structure of the broad-area diode laser is comprised of epitaxial-grown layers of semiconductor materials. Fig. 1 displays the diode laser structure with an asymmetric linearly graded refractive index waveguide. The waveguide layers are sandwiched between two cladding layers with a lower refractive index. Broad area laser diode stripe supports lateral multimode output beam. Fig. 1 depicts the vertical profile of a laser diode refractive index structure.

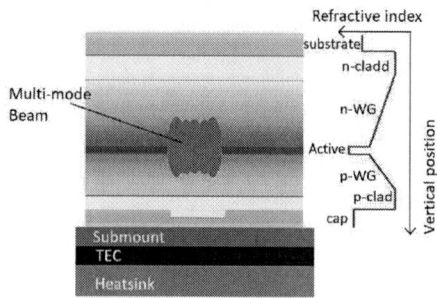

Fig. 1. The layer structure grown in the diode laser and the vertical refractive index.

III. LASER STRUCTURE

The epitaxial layer structure is displayed in Table I. The stripe width is 0.1 mm, and the length of the cavity is 0.1 cm. The front and rear mirror reflectivities are 5% and 95%, respectively.

TABLE I.   THE LAYER STRUCTURE OF THE LASER DIODE

Layer	Material	X (Al molar)	Thickness (nm)		Doping (cm⁻³)
Capping	GaAs	---	200	p	5×10^{19}
p-Clad	$Al_xGa_{1-x}As$	0.45	900	p	3×10^{18}
p-WG	$Al_xGa_{1-x}As$	0.15-0.45	850		---
barrier	$Al_xGa_{1-x}As$	0.05	10		---
QW	$In_xGa_{1-x}As$	0.08	7.5		---
barrier	$Al_xGa_{1-x}As$	0.05	10		---
QW	$In_xGa_{1-x}As$	0.08	7.5		---
barrier	$Al_xGa_{1-x}As$	0.05	10		---
n-WG	$Al_xGa_{1-x}As$	0.2-0.15	2.65		---
n-Clad	$Al_xGa_{1-x}As$	0.2	1	n	5×10^{17}
Substrate	GaAs	---	---	n	3×10^{18}

To increase the carrier capture, two $In_{0.08}Ga_{0.92}As$ (DQW) are used in the structure. The barrier heights are 91meV and 73meV for the electron and hole quantum well, respectively. The $Al_xGa_{1-x}As$ material is used for waveguide and cladding layers. The band gap energy and refractive index at temperature=300K are explained by [15]:

$$n_r = (13.1 - 3.1x)^{0.5} \qquad (1)$$

$$E_g = 1.424 + 1.247x \ (eV) \qquad (2)$$

where x is the aluminum mole fraction. the refractive index profile is shown in Fig. 2. Two parameters, refractive index differences between the n-clad and n-WG layers, Δn, and thickness of the n-waveguide, $d_{n\text{-WG}}$, are investigated in this research.

The refractive index and thickness of n-WG layers determine the vertical near-field (NF) beam size at the mirror facet. The fast beam divergence can be controlled by NF beam size. It should be commented that to minimize the beam divergence, the near-field pattern size must be expanded. Increasing the near-field size requires increasing the waveguide thickness. Therefore, the following three strategies are applied to the structure:

- Using a thicker n-WG layer than the p-WG layer.

- The asymmetric coefficient increases by raising the difference between the refractive indices of n- and p-clad. layers.

- The slope of the linearly graded refractive index in the n-WG layer is less than the p-WG layer.

Increasing the WG thickness decreases the confinement factor which increases the I_{th} because it is in proportion to the inverse of the confinement factor [7]. In the designed structure, the graded refractive index is used for self-focusing of the ray and increasing the confinement factor to avoid I_{th} increase. Two linearly graded refractive index profiles for n-WG and p-WG layers are utilized with different slopes.

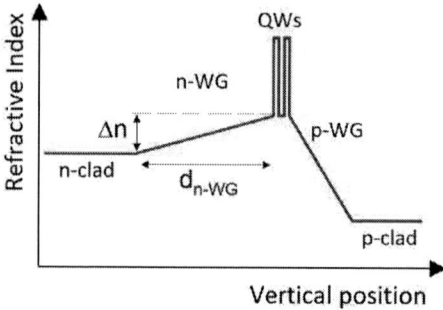

Fig. 2. Refractive index profile of linear profile of graded refractive index parameters.

The dependence of the beam divergence and the confinement factor on two parameters of Δn and $d_{n\text{-WG}}$ are shown in Fig. 3 and Fig. 4, respectively. As displayed in Fig. 3, increasing the WG thicknesses in a constant Δn decreases the slope of the linear n-WG graded index profile and the spot size can be enlarged so that the beam divergence is decreased. The value of the divergence and the confinement factor reduction rate regarding the n-WG thickness is 2 °/nm and 0.003 per nm, respectively. Similarly, the changes of these two parameters at constant n-WG thickness are shown in Fig. 4.

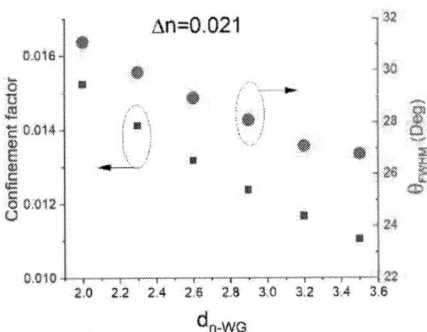

Fig. 3. Confinement factor and fast divergence vs. the n-waveguide thickness in a constant refractive index difference (Δn=0.021).

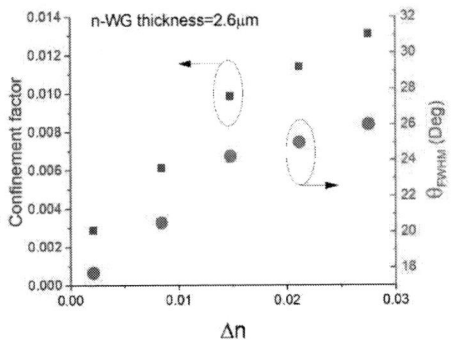

Fig. 4. Confinement factor and fast divergence vs. the different refractive index in a constant n-WG thickness (d=2.6 microns).

Decreasing the confinement factor has limitations, as it raises the I_{th}. Due to the exponential dependence of gain, it is good to use more than one QW to enhance the confinement factor. This dependence results from the gain saturation as the carrier density is increased to nearly fill the lowest set of states [9]. By distributing the carriers over the number of wells (N_{QW}), the gain per well is lowered by less than N_{QW} times, but the modal gain is still multiplied by nearly N_{QW} times this value. The I_{th} density, J_{thMQW}, in this multiple-QW laser diode is [9]:

$$J_{thMQW} \cong \frac{qN_{QW}d_{QW}BN_{tr}^2}{\eta_i} e^{\frac{2(\alpha_i+\alpha_m)}{N_{QW}\Gamma g_0}} \quad (3)$$

Where q and B are electron charge and Einstein coefficient, respectively. The d_{QW} and η_i are the summation of QWs thicknesses and internal quantum efficiency of the diode laser, respectively. The g_0 and N_{tr} are gain parameters of QWs and transparency carrier concentration in QW, respectively. The two parameters, α_i, and α_m, are internal loss and mirror loss of laser diode, respectively. It is noted that the optimum number of wells in constant confinement factor is the number that minimizes Eq. (3). For the designed DQW structure, the optimum confinement factor is near 0.011. Corresponding to Fig. 4, the optimum n-WG thickness is 2.6 microns.

IV. SIMULATION RESULTS

In this work, the laser structure was simulated by PICS3D software [16] which is validated in our previous work [17]. The simulation used the 8*8 k.p approach to calculate the band structure of the active area. In the simulation, nine lateral optical modes were calculated. According to Fig. 3 and 4, the optimized structure (for the beam divergence less than 30 degrees) is considered by choosing 2.6 microns and 0.21 for the n-WG thickness and the refractive index difference, respectively. The optimization is based on the confinement factor value, which should be more than 0.0125 to avoid increasing the I_{th}.

Total optical loss in a laser diode contains two principal terms consisting of the external optical loss that is related to the mirror (useful) and the internal (not useful) losses. The main contribution to the amount of internal loss is related to free carrier loss. The free-carrier absorption causes electron and hole transitions within a single band which expands as the free-carrier density and the absorbed light wavelength increases [18]. As the initial step, the internal loss is determined for the simulation. Internal loss is the main parameter for the I_{th} and the optical power. The free carrier loss is calculated by the sum of layer free carrier losses [18]:

$$\alpha_{internal} = \sum_i \Gamma_i(\sigma_{ni}n_i + \sigma_{pi}p_i) \quad (4)$$

where n_i and p_i are the densities of electrons and holes in the absorption cross-section of the i-th layer, respectively. σ_{ni} and σ_{pi} are the free electron and hole absorption cross-sections, respectively and Γ_i is the optical confinement factor for the i-th layer.

Fig. 5 and 6 show the optical field overlap with layers and the carrier density, respectively. The confinement factors of each layer can be determined in Fig. 5. The confinement factor of QWs (active layer) was near 1%, so this structure has the advantage of the low modal gain (LMG) structure. In the LMG structure, the filamentation is reduced, so the beam quality is improved [7]. The cross-section of hole absorption is more than two times higher than the cross-section of the electron absorption. The values $\sigma_{nWG} = 3 \times 10^{-18}$ cm^2 and $\sigma_{nWG} = 7 \times 10^{-18}$ cm^2 were used. According to Fig. 5, the optical field moves to the n-doped layer where the absorption cross-section is low. Fig. 6 shows that the carriers (electrons and holes) concentration at n-clad and p-clad layers is higher than WG layers. The overlap of the vertical optical field in these two layers is significantly lower than in other layers, so the internal loss is decreased in this structure. The free carrier loss in each layer is shown in Table 2 separately. The value of the internal loss is about 1.645 cm^{-1}.

TABLE II.   LOSSES OF LASER DIODE LAYERS

Layer	Γ_i	Loss (cm^{-1})
p-Clad	0.00003	0.00011
p-WG	0.13	0.226
barriers	0.02	0.033
n-WG	0.84	1.386
n-Clad	0.0004	0.00055
SUM	0.99	1.645

The 5th Iranian International Conference on Microelectronics (IICM2023)

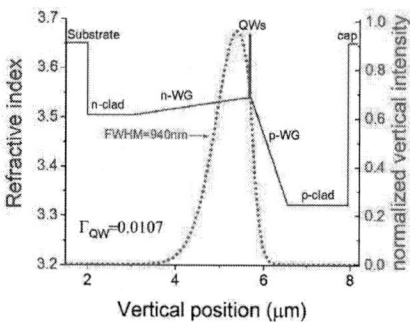

Fig. 5. Vertical refractive index profile and vertical optical field profile (confinement factor for fundamental mode =0.0107)

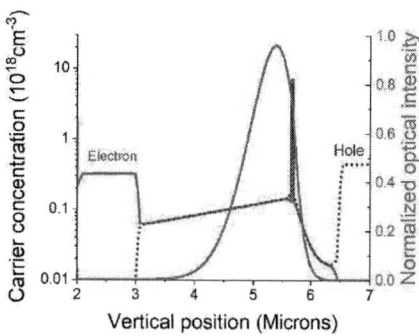

Fig. 6. Vertical profile of the carrier density and the optical field distribution perpendicular to the growth of layers.

The diode laser chip is mounted on a heatsink (thermal conductance 20 W/mK and a fixed temperature 300 K). The simulated cooling temperature conditions are based on the heatsink constant temperature. The characterization results of the laser structure are displayed in Fig. 7.

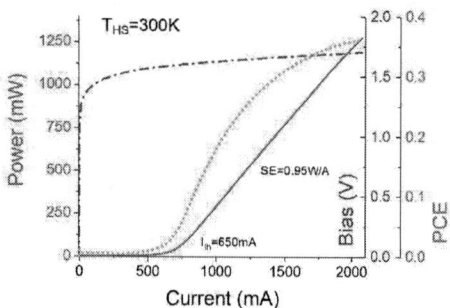

Fig. 7. Electro-optical characteristics of the laser. Optical power (solid red), bias (dash-dotted blue), and power conversion efficiency (dotted green)

The 2D near-field profile and the 1D lateral profile in 2A operating current are shown in Fig. 9. This current corresponds to the wavelength emission of 915nm. Reasons for the low quality of broad-area diode lasers are the number of lateral modes, mode mixing, filamentation, Thermal lens effect, and other non-linear effects [19]. According to Fig. 9, the intensity profile peak is near the average intensity (about 8% higher than the average intensity).

The slope efficiency and the I_{th} are 0.95 W/A, and 650 mA, respectively. The location of the p-n junction, the series resistance, especially in the p-type layers, and the losses due

to the free-carrier absorption depend on the doping profile and the doping level. Thus, reasonable control of these features is required. The series resistance in this structure is 78 mΩ. It shows that interfaces between AlGaAs layers (WG and clad layers) are suitable.

A graded profile between QW and barriers changes the position of the subbands in the QWs [10]. For the determination of QWs lasing wavelength, the cooling system and stripe width are mentioned. Alterations of the active area lasing wavelength and temperature are depicted in Fig. 8. Corresponding to this figure, the operation current is chosen by the wavelength. The lasing wavelength changes with the temperature equals to 0.28nm/K.

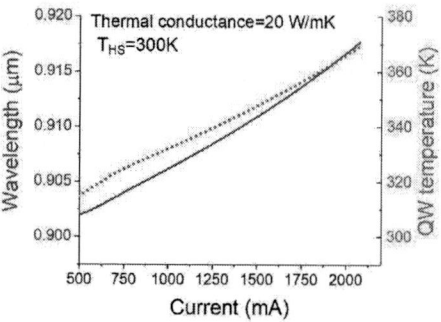

Fig. 8. Lasing wavelength and active area temperature alterations vs. the injected current. (Solid blue line for the wavelength and dotted red line for QW temperature)

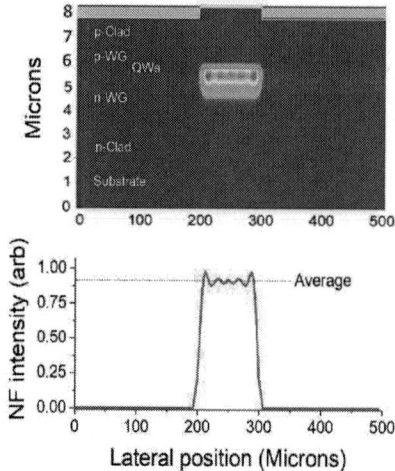

Fig. 9. The 2D pattern (Up) and The 1D lateral profile (down) of NF on the facet of the laser diode with a normalized arbitrary unit.

The near-field profile in Fig. 9 shows some lateral modes contributing to the final laser output optical power. The optical power of each mode is different in each current. Fig. 10 shows the optical power of each mode separately.

979-8-3503-6020-2/23 $31.00 © 2023 IEEE

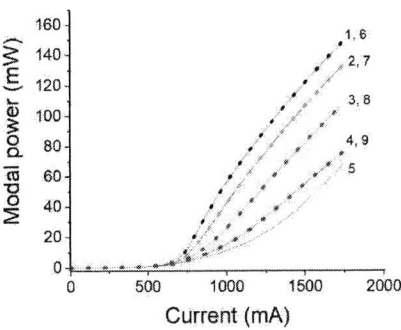

Fig. 10. All modes total power and the fundamental mode power vs. the injected current in the laser diode.

Finally, the intensity profile of the transversal (fast axis) far field is shown in Fig. 11. The divergence in FWHM and 95% of total power are 24 and 38, respectively.

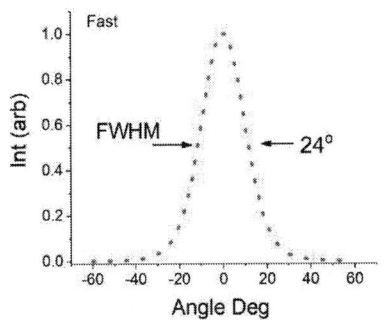

Fig. 11. Vertical profile of the optical far field for determination of the fast axis divergence.

V. CONCLUSION

In this research, an asymmetric graded refractive index is recommended to achieve lower divergence without increasing the I_{th} density. In this structure, two linear profiles with different slopes were used for the n- and p-WG layers and the refractive index of the n-clad layer is selected as higher than the p-clad layer. In this structure, the thickness and the slope of the linearly graded refractive index of the n-WG have been investigated. The slope of the graded region can be used to control vertical beam sizes in the near-field, and the carrier captures in the QWs. The first parameter can control the vertical divergence of the laser beam and the mode number. The second parameter can control the I_{th} density and slope efficiency. The results show that in the desired divergence, the WG layer can be adjusted with the thickness and the linear slope of the graded refractive index of the WG. The confinement factor must be considered to achieve the low modal gain without increasing the I_{th} density. To improve the gain power, DQW was used in the active region. Finally, the 2.65 microns n-WG with a slope of 0.019 per micron for the linear graded refractive index was selected. The results also reveal that the divergence in this structure is 24 degrees, which is better than the famous commercial products [20].

In addition, the value of the slope efficiency and the I_{th} density is 0.96 W/A, and 650 A/cm², respectively, which shows that the reduction of the divergence did not significantly increase the I_{th}. Decreasing the beam divergence would reduce the beam spot size after beam shaping. This issue is used to couple the laser diode beam to a small-diameter optical fiber (100microns) or confine the beam in the central section of the optical fiber with a higher core diameter (200 microns), which is important in fiber laser optical pumping.

REFERENCES

[1] A. Malag, E. Dabrowska, M. Teodorczyk, G. Sobczak, A. Kozłowska, J. Kalbarczyk, "Asymmetric heterostructure with reduced distance from active region to heatsink for 810-nm range high-power laser diodes", IEEE J. Quantum Electron., vol. 48, pp. 465–471, 2012.

[2] F. Bachmann, P. Loosen, R. Poprawe (Eds.), "High Power Diode Lasers Technology and Applications", first ed., Springer-Verlag Press, 2007.

[3] Y. Qi1, P. Zhao2, Y. Wu1, Y. Chen1, and Y. Zou1, "Design of 20 W fiber-coupled green laser diode by Zemax", Journal of Semiconductors, Vol. 38, No. 9, 2017.

[4] B.S. Ryvkin, E.A. Avrutin, "Asymmetric non-broadened large optical cavity waveguide structures for high-power long-wavelength semiconductor lasers", J. Appl. Phys., vol. 97 (6), pp. 123103–123106, 2005.

[5] D.A. Vinokurov, S.A. Zorina, V.A. Kapitonov, A.V. Murashova, D.N. Nikolaev, A.L. Stankevich, M.A. Khomylev, V.V. Shamakhov, A.Yu. Leshko, A.V. Lyutetski, T.A. Nalyot, N.A. Pikhtin, S.O. Slipchenko, Z.N. Sokolova, N.V. Fetisova, I.S. Tarasov, "High-power laser diodes based on asymmetric separate-confinement heterostructures", Semiconductors, vol. 39, pp. 370-373, Mar 2005.

[6] B.S. Ryvkin, E.A. Avrutin, "Improvement of differential quantum efficiency and power output by waveguide asymmetry in separate confinement structure diode lasers", IEE Proc. – Optoelectron., vol. 151, pp. 232-236, 2004.

[7] R. Deihl (Ed.), "High Power Diode Lasers, Fundamentals, Technology, Applications", first ed., Springer-Verlag, 2001.

[8] B.S. Ryvkin, E.A. Avrutin, "Effect of carrier loss through waveguide layer recombination on the internal quantum efficiency in large-optical-cavity laser diodes", J. Appl. Phys., vol. 97, pp. 113106-113106-5, 2005.

[9] L.A. Coldren, S.W. Corzine, M.I. Masanovic, "Diode Lasers and Photonic Integrated Circuits", John Wiley & Sons, 2012.

[10] B. Mroziewicz, M. ej Bugajski, W. Nakwaski, "Physics of Semiconductor Lasers", North-Holland, 1991.

[11] C. T. Hung, T. C. Lu, "830nm AlGaAs-InGaAs graded index double barrier separate confinement heterostructures laser diode with improved temperature and divergence characteristics", IEEE J. Quantum Electron., vol. 49 (1), pp. 127-132, 2013.

[12] S. P. Abbasi, M. H. Mahdieh, "Improvement of AlGaInAs/AlGaAs laser diode electro-optics characteristics by graded refractive index profile broadened waveguide", Opt. and Laser Technol., vol. 116, pp. 155–161, 2019.

[13] D. Botez, "Design considerations and analytical approximations for high continuous-wave power, broad-waveguide diode lasers", Appl. Phys. Lett., vol. 74, pp. 3102-3104, 1999.

[14] C. L. Chen, "Foundations for Guided-Wave Optics", A JOHN WILEY & SONS, INC., New Jersey, 2007, pp. 25-50.

[15] S. Adachi, "Properties of Aluminium Gallium Arsenide", London, INSPEC, the Institution of Electrical Engineers, 1993.

[16] Crosslight Software Inc. (2012). LASTIP, Vancouver, BC, Canada.

[17] A. Hodaei and S. P. Abbasi, "Effect of the QW number to produce circular single-mode beam in 1060 nm laser diodes," 2021 Iranian International Conference on Microelectronics (IICM), Tehran, Iran, Islamic Republic of, 2021, pp. 1-5.

[18] N.A. Pikhtin, S.O. Slipchenko, Z.N. Sokolova, I.S. Tarasov, "Internal optical loss in semiconductor lasers", Semiconductors, vol. 38 (3), pp. 360–367, 2004.

[19] M. Winterfeldt, P. Crump, H. Wenzel, G.Erbert, and G. Tränkle, "Experimental investigation of factors limiting slow axis beam quality in 9xx nm high power broad area diode lasers", J. of Appl. Phys., vol. 116, pp. 063103-063103-14, 2014.

[20] (2023). [Online]. Available: https://ii-vi.com

Base Transit Time Investigation of InP/InGaAs HBT Optoelectronic Mixer Using Different Base Doping Profiles

Hassan Kaatuzian
Electrical Engineering Department,
Photonics Research Lab.(PRL)
AmirKabir University of Technology
(AUT)
hsnkato@aut.ac.ir

Mehrdad Ghasemi
Electrical Engineering Department,
Photonics Research Lab.(PRL)
AmirKabir University of Technology
(AUT)
mehrdad.ghasemi@aut.ac.ir

Mahdi NoroozOliaei
Electrical Engineering Department
K. N. Toosi University of Technology
(KNTU)
mahdi.norooz@email.kntu.ac.ir

Abstract—**Heterojunction Bipolar Transistors (HBTs) are a type of bipolar transistors that have fabulous advantages for employing them in electronic circuits such as their fast speed. A key parameter of HBTs is transit time which determines the capabilities of these devices. Employing of lower transit time (better frequency response) is feasible by considering the base doping different profiles. In this paper, the behavior of the transit time parameter has been investigated for the mentioned near Gaussian and continuously decreasing doping profiles. It can be used to obtain even higher frequency responses in single or cascade configurations for 1.55-micron wavelength optoelectronic mixers. The idea in this study will be applied to HBTs in modulating the frequency range between gigahertz to terahertz.**

Keywords— Base Doping Gaussian Profile, Base Width, Current Gain, Heterojunction Bipolar Transistors (HBTs), Transit time.

I. Introduction

Heterojunction bipolar transistors (HBTs) are one type of bipolar transistors in which the emitter transition assumes a heterojunction structure. In other words, there is a broadband gap material in the emitter region, and a narrow band gap material is used for the base region. Several semiconductor materials are employed for both the emission and collector basis. These types of transistors have attracted a lot of attention due to their ability to use very fast devices and they are one of the most promising technologies for using in electronic devices. The faster switching speeds of HBTs are one of their advantages concerning the silicon bipolar transistors. It happens due to the reduced base resistance and collector-to-substrate capacitance [1]-[7].

HBTs have the capability of better performance concerning the Bipolar Junction Transistors (BJTs) in terms of emitter injection efficiency, base resistance, base-emitter capacitance, and cutoff frequency. HBTs also offer good linearity, low phase noise, and high power-added efficiency. The most important advantages of HBTs and their consequences are as follows:

1- Lower base resistance or equivalently higher base doping concentration causes lower forward transit time. So, the cutoff frequency (f_r) will be increased considerably.

2- The higher β (current gain) results in better intrinsic device linearity.

3- Low collector-substrate capacitors (C_{cs}) in HBTs due to the use of semi-insulating substrate types.

4- High efficiency due to the ability to turn off devices completely with a small base voltage change and the extremely small turn-on voltage variation between devices.

5- Good wide-band impedance matching due to the resistive nature of the input and output impedances.

6- Low cost and potential for high throughput.

Several works have been done including HBTs for different frequency applications especially high frequencies towards terahertz.

The connection type for HBTs for in-phase and quadrature-phase is applied in [8]. It caused distortion cancellation, especially in high frequencies. In this research, frequency analysis of feedforward cells is being investigated as well. In [9], a series connection triplet is being introduced for the connection type of the proposed mixer to get a significant improvement in dynamic range. The transformer technique or the antiphase magnetic coupling (Inductance) or coupling with 90 degrees phase difference between emitter and base junctions is being proposed in [10]; consequently, the designed mixers demonstrated the highest intermodulation intercept point (IIP3) with positive conversion gain in their frequency range. Double heterojunction bipolar transistor (DHBT) Subharmonic Mixer (SHM) topology with a matching circuit is being proposed by [11]. The mixer illustrated high bandwidth and appropriate conversion gain. A low-frequency noise model for HBTs has been investigated in [12] both experimentally and numerically. Quarter wave within differential configuration for better suppression of the unwanted signals and proper transient signals are being proposed in [13]-[15]. The base resistance reduction or equivalently noise figure reduction is being investigated in [16] for a type of HBTs.

In this paper, an analysis for the InP/InGaAs HBT based on the different near Gaussian profile base doping has been investigated for employing in the nonlinear structures such as mixers especially in the high frequencies as it has the proper transit time and current gain. The sinusoidal form of the doping profile is an approximation of the Gaussian distribution. It is being employed in this form for its simplicity.

II. HBT Model and the Configuration of its Mixer

The early HBT configuration and its schematic have been illustrated in Fig. 1. This figure shows the epitaxial layers of the InP/InGaAs HBT transistor used in the mixer structure.

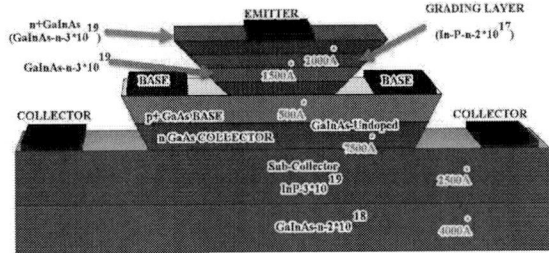

Fig. 1. Schematic of epitaxial layers of HBT Structure.

The single configuration of the opto-electronic mixer has been depicted in Fig. 2.

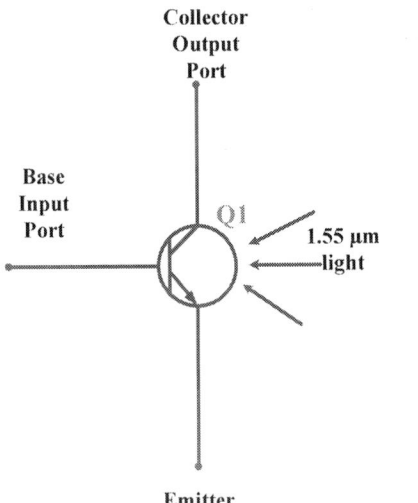

Fig. 2. Schematic of single configuration for Opto-Electronic mixer

The schematic diagram of the experimental arrangement for the cascade configuration of the Opto-Electronic mixer is depicted in Fig. 3.

A feedback emitting at 1.55 μm was modulated by a Mach Zehnder modulator [17]. The modulated light was amplified by erbium-doped or Raman fiber amplifiers and then the light was focused on the optical window that is located on the base region of HBT.

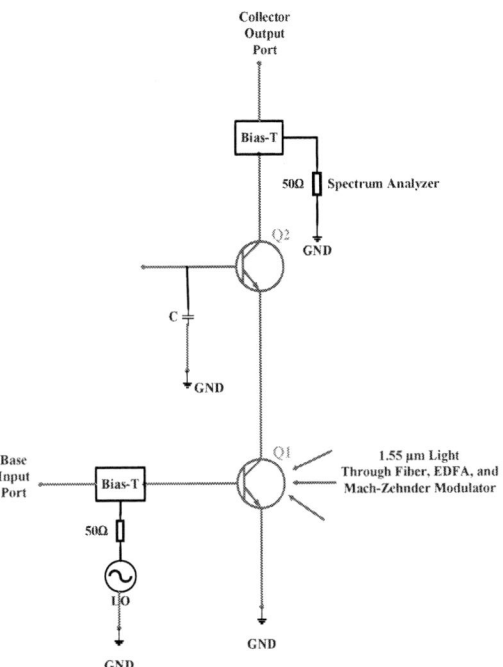

Fig. 3. Schematic of cascade configuration for Opto-Electronic mixer

III. Theory and Formulation

The roles of base resistance and junction capacitance of the base-emitter and base-collector are very important for determining the frequency response (related to the transit time) and current gain as discussed briefly in the previous section.

As mentioned earlier, base doping increment causes base resistance reduction and its decrement consequents in junction capacitances and tunneling leakage current reduction.

The meeting criteria for transit time is using different near Gaussian profile doping for base concentration.

Kroemer's double integration is giving the base transit time of InP/InGaAs HBTs by [18]:

$$\tau_b = \int_0^{W_b} \frac{n_i^2(x)}{N_a(x)} \left[\int_x^{W_b} \frac{N_a(y)dy}{D_n n_i^2(y)} + \frac{N_a(W_b)}{V_{sat} n_i^2(W_b)} \right] dx \quad (1)$$

Which W_b is the neutral base width, N_a is the p type of base doping concentration, D_n is the electron diffusion coefficient in the base, and V_{sat} is the saturation velocity of the minority carrier in the base.

The Einstein's equation says:

$$D_n(InGa\,As) = \frac{kT}{q} \mu_n(InGa\,As) \quad (2)$$

Which the $\mu_n(InGaAs)$ is electron mobility in $InGaAs$ base, K is Boltzmann constant ($1.380649 \times 10^{-23}\ [J/K]$), T is the absolute temperature and q is electronic charge ($1.60217662 \times 10^{-19}$ [Coulombs]).

The mobility of InGaAs can be expressed by the mole fraction of Indium [18]:

$$\mu_n(\text{InGa As}) = y_t \mu_n(InAs) + (1 - y_t)\mu_n(GaAs) \quad (3)$$

which y_t is the total indium content in the base region.

There is a relation between mobility, temperature, and doping concentration and it can be presented by the equation [19]:

$$\mu_T = \mu_{min} + \frac{\mu_{max(300K)}(300/T)^{\theta_1} - \mu_{min}}{1 + \left(\frac{N_{ref}}{1.3 \times 10^7 \times (T/300K)^{\theta_2}}\right)^{\lambda}} \quad (4)$$

Which the parameters of the above equation can be written for InGaAs. μ_{min} and $\mu_{max(300K)}$ are $300 \; cm^2/v - s$ and $14000 \; cm^2/v - s$ correspondingly. N_{ref} at $300K$ is $1.3 \times 10^{17} cm^{-3}$ and $\theta_1 = 1.59, \theta_2 = 3.68, \lambda = 0.48$ are being considered [20].

The current gain of InP/InGaAs HBT is brought into [21]:

$$\beta = \frac{2v_B}{W_b f_R \left[1 + \left(\frac{v_B}{v_{sat}}\right)\right]} = \frac{1}{f_R\left[\left(\frac{W_b^2}{2D_n}\right) + \left(\frac{W_b}{v_{sat}}\right)\right]} \quad (5)$$

which, v_B is effective electron velocity through the neutral base region, f_R is the recombination factor. The recombination factor can be attained by [21]:

$$f_R = C_{AP}N_a^2 + (C_{BB} + C_{SRH})N_a \quad (6)$$

Which C_{AP} is the Auger coefficient for holes, C_{BB} is the band-to-band radiative coefficient and C_{SRH} is the Shockley-Read-Hall coefficient. Here f_R is calculated as $1.6 \times 10^9 s^{-1}$ [21] considering N_a as $10^{19} \; cm^{-3}$, base width (W_b) is taken as $50nm$ and v_{sat} is taken as $8 \times 10^6 \; cm/s$ [22].

The base doping profile is considered as follows:

$$N_a(x) = N_p sin\left(\frac{(3.12 - \alpha)x}{W_b} + \alpha\right) \quad (7)$$

Which $0 \le x \le W_b$ and α varies from 0.05 to 1.57 and

$$\alpha = sin^{-1}\left(\frac{N_0}{N_p}\right) \quad (8)$$

N_0 is the doping concentration near the emitter, and N_p is the peak concentration in the base or the base width. At the base-emitter junction (i.e. at $x = 0$), $N_0 = N_p sin(\alpha)$, and at the base-collector junction (i.e. at at $x = W_b$) the doping concentration is $N_0 = N_p sin(3.12) = 0.22N_p$.

Replacing this profile in the equation (1) results into:

$$\tau_b = \frac{-1.998W_b^2 log\left(sin\frac{\alpha}{2}\right)}{D_n(3.12 - \alpha)^2} + \frac{0.043W_b}{v_{sat}(3.12 - \alpha)} + \frac{-0.022W_b\left(log(tan(\alpha/2))\right)}{v_{sat}(3.12 - \alpha)} \quad (9)$$

The first term inside the bracket of (5) represents the base transit time for the uniform profile. The current gain of InP/InGaAs HBT for a non-uniform base doping profile of (8) can be calculated from equation (5) by replacing the first term inside the bracket with the expression for transit time (τ_b) obtained in (9).

The parameters of the reference transistor have been summarized in Table I.

TABLE I. PARAMETERS OF REFERENCE TRANSISTOR

Parameters	Values
W_b	$50 \; [nm]$
f_R	$1.6 \times 10^9 s^{-1}$
$N_{ref}(300K)$	$1.3 \times 10^{17} cm^{-3}$
N_a	$10^{19} \; cm^{-3}$
v_{sat}	$8 \times 10^4 \; m/s$

The equation for uniform base doping by (1) will be presented here:

$$\tau_b = \frac{W_b^2}{2D_n(\text{InGaAs})} \quad (10)$$

The base transit time of the reference transistor is obtained by using the information in Table I and equations (9) and (10).

IV. RESULTS AND DISCUSSIONS

A. Base doping profile

The profile of base doping has been presented in Fig. 4 concerning the equation (7).

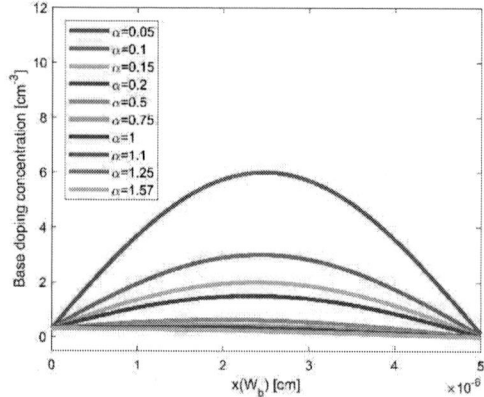

Fig. 4. Base doping profile of Heterojunction Bipolar Transistor (HBT) for miscellaneous α index, α=0.05 and 0.1 (near Gaussian) to α=1.57.

The profile is in the best form of near Gaussian profile in the cases of $\alpha = 0.05$ and $\alpha = 0.1$.

The Gaussian shape base doping profile will be decreased as the index (α) will be increased gradually. So, it is in the best performance in lower α values for example $\alpha = 0.05$ and $\alpha = 0.1$ as depicted in Fig. 4.

B. Base Transit time

The transit time diagram of the reference HBT has been illustrated in Figure 5; it is being calculated based on equations (9) and (10).

The transit time of the near uniform profile is better than the transit time of the near Gaussian one as shown in Fig. 5.

The transit time (which is related directly to frequency response) depends on many parameters such as α, v_{sat}, and W_b. The effect of changing in α has been shown in Fig. 5. It also indicates as the W_b will be increased for a determined α, the transit time will be increased as well. So, the operating point for W_b should be its lower values.

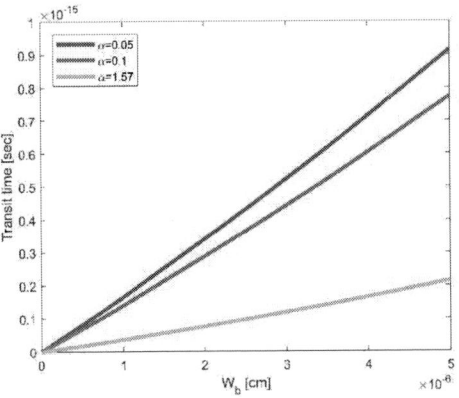

Fig. 5. Transit time of Heterojunction Bipolar Transistor (HBT) for different profiles, α=0.05 and 0.1 (near Gaussian) and α=1.57.

Fig. 6 shows the diagram of transit time concerning the changing of saturation velocity V_{sat}. It is indicated as it will be increased, the transit time will be decreased slightly.

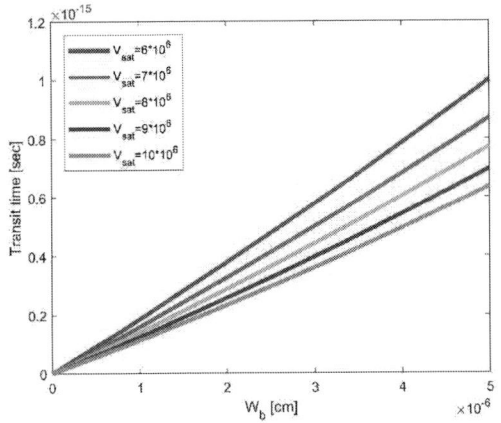

Fig. 6. Transit time (τ_b) of Heterojunction Bipolar Transistor (HBT) for different V_{sat} [cm/s].

V. USING IMPROVED HBT IN THE STRUCTURE OF OPTO-ELECTRONIC MIXER

Two configurations of large signal nonlinear models for simulating both single and cascade ones are being presented in this section similar to [1]-[5],[17]. Figures 7 and 9 show the equivalent circuit structure of the two models. The utilized parameters and their values have been illustrated in Tables II and III, correspondingly.

The frequency response from 1 GHz to 10 GHz has been shown in Figures 8 and 10.

A. Designed HBT Opto-Electronic Mixer in Single Configuration

The frequency response of a single configuration indicates higher down conversion gain in low frequencies concerning the upper ones at the determined frequency band (1GHz to 10GHz). In Fig. 8, redundant software results slightly differ, because of some parasitic ignored capacitances.

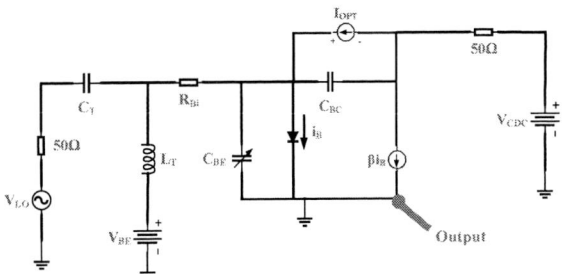

Fig. 7. Single Configuration of HBT Mixer.

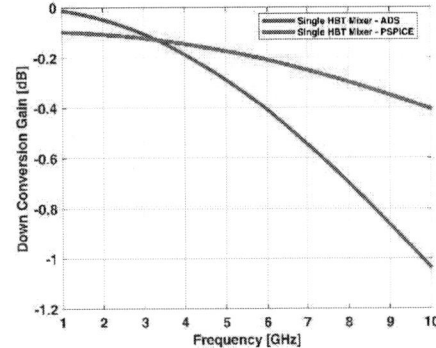

Fig. 8. Frequency Response of Single Configuration for HBT Mixer Down Conversion Gain (dB).

TABLE II. PARAMETERS OF SINGLE CONFIGURATION

Parameters	C_T (mF)	L_T (mH)	R_{B_i} (Ω)	C_{BE} (fF)	C_{BC} (fF)
Values	1	1	100.5	100	43
Parameters	β	I_{OPT} (fA)	V_{LO} (V)	V_{BE} (V)	V_{CDC} (V)
Values	130	43.2	1	0.8	2

It is noteworthy that the β parameter can be tuned even for higher values to get more appropriate results.

B. Designed HBT Opto-Electronic Mixer in Cascode Configuration

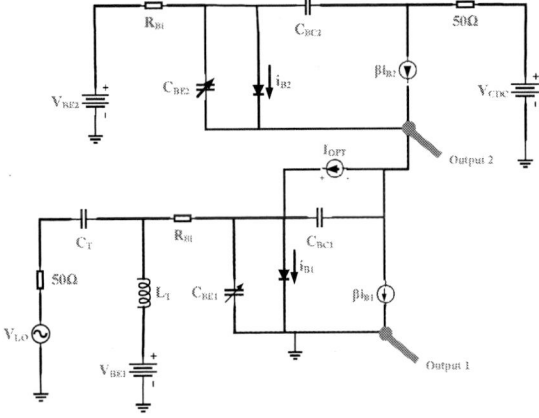

Fig. 9. Cascade Configuration of HBT Mixer.

979-8-3503-6020-2/23 $31.00 © 2023 IEEE

The 5ᵗʰ Iranian International Conference on Microelectronics (IICM2023)

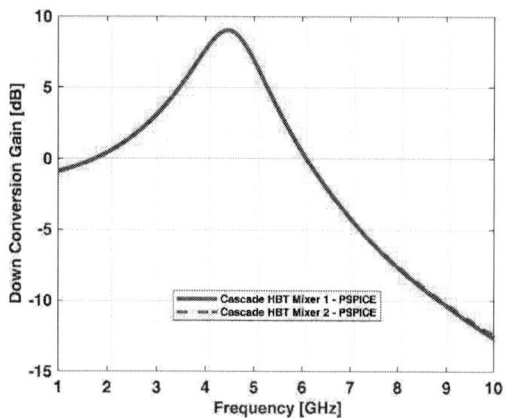

Fig. 10. Frequency Response of Cascade Configuration for HBT Mixer Down Conversion Gain (dB).

The 3dB bandwidth of Cascade HBT Mixer in Output 2 is around 1.457GHz by the PSPICE software results in Figures 9 and 10.

TABLE III. PARAMETERS OF CASCADE CONFIGURATION

Parameters	C_T (mF)	L_T (mH)	R_{B_i} (Ω)	C_{BE_1} (fF)	C_{BE_2} (fF)
Values	1	1	100.5	100	100
Parameters	C_{BC_1} (fF)	C_{BC_2} (fF)	V_{LO} (V)	V_{BE_1} (V)	V_{BE_2} (V)
Values	43	43	1	0.8	0.8
Parameters	β	I_{OPT} (fA)	V_{CDC} (V)	$I_{SPSPICE}$ (nA)	
Values	130	43.2	2	2.682	

VI. CONCLUSIONS

In this paper, a study was done for changing the base doping profile of Heterojunction Bipolar Transistors (HBTs) to find out the effects of these variations on the key parameters, i.e. transit time (τ_b). The transit time was most dependable on the α index of base different doping Gaussian profiles. Two single and cascade configurations were investigated to show improved performance for the cascade one. The cascade configuration presented a 3dB bandwidth of 1.457GHz for down conversion gain in its PSPICE model.

REFERENCES

[1] H. Kaatuzian, K. Farhang Razi, "Conversion Gain Improvement of InP/InGaAs HBT Opto-Electronic Mixer Using Nearly Gaussian Doping Profile and Linear Grading of Composition in Base Region", Electrical and Electronic Engineering, Vol. 6, No. 2, pp. 30-38 2016.

[2] H. Kaatuzian., H. Dehghan Nayeri, M. Ataei, and A. Zandi, "Structural parameters improvement of an integrated HBT in a cascode configuration opto-electronic mixer", Journal of Semiconductors, Vol. 34, No. 9, September 2013.

[3] H. Kaatuzian, "Theory and technology of manufacturing semiconductor devices", 2nd edition, AmirKabir Press, 2022.

[4] H. Livingston "A Survey of Heterojunction Bipolar Transistor (HBT) Device Reliability", IEEE Transactions on Components And Packaging Technologies, Vol. 27, No. 1, March 2004.

[5] William Liu, "Fundamentals of III-V Devices, HBTs, MESFETs, and HFETs/HEMTs", Wiley-Interscience, 1999.

[6] W. Liu, D. Costa, J.S. Harris Jr, "Derivation of the emitter-collector transit time of heterojunction bipolar transistors", Solid-State Electronics, Vol. 35, No. 4, pp. 541-545, April 1992.

[7] Y. Chou, R. Ferro, "Heterojunction Bipolar Transistors", Physics, 1997

[8] Yatao Peng, Lijun Zhang, Jun Fu, and Yudong Wang, "Analysis and Design of a Broadband SiGe HBT Image-Reject Mixer Integrating Quadrature Signal Generator", IEEE Transactions on Microwave Theory and Techniques, Vol. 64, No. 3, March 2016.

[9] Jonathan P. Comeau, and John D. Cressler, "A 28-GHz SiGe Up-Conversion Mixer Using a Series-Connected Triplet for Higher Dynamic Range and Improved IF Port Return Loss", IEEE Journal of Solid-State Circuits, Vol. 41, No. 3, March 2006.

[10] Jian Zhang, Mingquan Bao, Dan Kuylenstierna, Member, Szhau Lai, and Herbert Zirath, "Transformer-Based Broadband High-Linearity HBT Gm-Boosted Transconductance Mixers", IEEE Transactions on Microwave Theory and Techniques, Vol. 62, No. 1, January 2014.

[11] Tom K. Johansen, Jens Vidkjær, Viktor Krozer, Agnieszka Konczykowska, Muriel Riet, Filipe Jorge, and Torsten Djurhuus, "A High Conversion-Gain Q-Band InP DHBT Subharmonic Mixer Using LO Frequency Doubler", IEEE Transactions on Microwave Theory and Techniques, Vol. 56, No. 3, March 2008.

[12] Mattia Borgarino, Corrado Florian, Pier Andrea Traverso, and Fabio Filicori, "Microwave Large-Signal Effects on the Low-Frequency Noise Characteristics of GaInPGaAs HBTs", IEEE Transactions on Electron Devices, Vol. 53, No. 10, October 2006.

[13] Saeed Zeinolabedinzadeh, Ickhyun Song, Uppili S. Raghunathan, Nelson E. Lourenco, Zachary E. Fleetwood, Michael A. Oakley, Adilson S. Cardoso, Nicolas J.-H. Roche, Ani Khachatrian, Dale McMorrow, Stephen P. Buchner, Jeffrey H. Warner, Pauline Paki-Amouzou, and John D. Cressler, "Single-Event Effects in a W-Band (75-110 GHz) Radar Down-Conversion Mixer Implemented in 90 nm, 300 GHz SiGe HBT Technology", IEEE Transactions on Nuclear Science, Vol. 62, No. 6, December 2015.

[14] Saeed Zeinolabedinzadeh, Ahmet C. Ulusoy, Farzad Inanlou, Hanbin Ying, Yunyi Gong, Zachary E. Fleetwood, Nicolas J.-H. Roche, Ani Khachatrian, Dale McMorrow, Stephen P. Buchner, Jeffrey H. Warner, Pauline Paki, and John D. Cressler "Single-Event Effects in a Millimeter-Wave Receiver Front-End Implemented in 90 nm, 300 GHz SiGe HBT Technology", IEEE Transactions on Nuclear Science, 2016.

[15] Ebrahim M. Al Seragi, Subhra Dash, K. Muthuseenu, John D. Cressler, Hugh J. Barnaby, Ani Khachatrian, Stephen P. Buchner, Dale McMorrow and Saeed Zeinolabedinzadeh, "Radiation Hardened Millimeter-Wave Receiver Implemented in 90nm, SiGe HBT Technology", IEEE Transactions On Nuclear Science, 2021.

[16] Jae-Sung Rieh, Basanth Jagannathan, David R. Greenberg, Mounir Meghelli, Alexander Rylyakov, Fernando Guarin, Zhijian Yang, David C. Ahlgren, Greg Freeman, Peter Cottrell, and David Harame, "SiGe heterojunction bipolar transistors and circuits toward terahertz communication applications", IEEE Transactions on Microwave Theory and Techniques, Vol. 52, No. 10, October 2004.

[17] Y. Betser, J. Lasri, V. Sidorov, Sh. Cohen, D. Ritter, M. Orenstein, G. Eisenstein, A. J. Seeds, and A. Madjar, "An integrated hetero junction bipolar transistor cascode opto-electronic mixer", IEEE Transactions on Microwave Theory and Teqniques, Vol. 47, No. 7, July 1999.

[18] P. Rinaldi, H. SchÄattler "Minimization of the base transit time in semiconductor devices using optimal control", in Proc 4th Int. conf. Dynamical systems and differential equations, Wilmington, NC, USA 742-751, 2002

[19] Pransejit Saha ; Sukla Basu, "A study of base transit time and gain of InP/InGaAs HBTs for uniform and nearly Gaussian base doping profile", Journal of electron devices", Vol.15, pp.1254-1259, 2012

[20] M. Sotoodeh, A. H. Khalid, and A. A. Rezazadeh, "Empirical low-fieldmobility model for III-V compounds applicable in device simulation codes", J. Appl. Phys. 87, 2890-2900 (2000).

[21] Juan M. LoÂ pez-GonzaÂ lez, Pau Garcias-SalvaÂ, LluõÂs Prat "Bulk recombination in the neutral base region of abrupt InP/InGaAs HBTs", Solid-State Electronics Letter 43, 1307-1311 (1999).

[22] S. M. Frimel and K. P. Roenkera, "Gummel–Poon model for Npn heterojunction bipolar photo-transistors", J. Appl. Phys. 82, 3581-3592 (1997).

Current-Mode Wideband Frontends With Linearity Enhancement for 5G Receivers

Adibeh Rahmani
Department of Electrical and Computer Engineering
Urmia University
Urmia, Iran
st_ad.rahmani@urmia.ac.ir

Mortaza Mojarad
Department of Electrical and Computer Engineering
Urmia University
Urmia, Iran
m.mojarad@urmia.ac.ir

Seyed Sadra Kashef
Department of Electrical and Computer Engineering
Urmia University
Urmia, Iran
s.kashef@urmia.ac.ir

Abstract— In this article, new receiver frontends have been proposed for direct conversion receivers for 5G communication standard. The receiver consists of a low-noise transconductance amplifier (LNTA), a current-mode passive mixer, and a transimpedance amplifier (TIA). Two different designs have been presented for implementing the LNTA with a particular focus on employing linearization methods. The receiver frontends have been simulated in a 0.18 μm CMOS process using ADS. The frontends utilize common-gate LNTAs with the Multiple-Gate Transistors linearization. The first presented frontend achieves 30.1 dB conversion gain, 7.9 dB noise figure (NF), +2 dBm IIP3, and consumes 16.2 mW power. The other receiver frontend makes use of a combined NMOS and PMOS linearization and achieves 23 dB conversion gain, 21.5 dB NF, and +6 dBm IIP3 with the power consumption of 21.42 mW. Both of the frontend implementations use a single 1.8 V supply voltage and the maximum S_{11} is -9 dB.

Keywords—Receiver, Frontend, Low-noise transconductance amplifier, LNTA, Transimpedance Amplifier

I. INTRODUCTION

Modern technology has brought about a wide range of applications including Internet of Things, smart vehicles, virtual gaming, augmented and virtual reality, and healthcare. The second and third generations of wireless standards are inadequate to satisfy the need for the growing demand for wireless systems, leading to the development of subsequent standards such as 4G, 4.5G, and most recently, 5G. The high-data-rate wireless links provided by 5G standard can cover a wide range of applications.

For instance, for applications such as healthcare and intelligent transportation, the demand for reduced latency, higher bit-rate, and enhanced reliability is imperative [1]. Moreover, cost must be regarded as a crucial parameter in the design of frontend circuits. As a result, efforts have been made to decrease the cost of transceivers. Traditional transceivers employ a SAW filter in order to eliminate unwanted blockers which accompany the desired signal. The major disadvantages of using these filters are higher cost and larger area, as well as deteriorated sensitivity of the receiver due to the insertion loss of the filter. As a solution to this issue, SAW-Less receivers have been presented [2]. Various techniques, such as active RF filtering [2], [3] and mixer-first

direct conversion structures are widely discussed in the literature [4], [5]. The main disadvantages of these methods are elevated power consumption and higher noise figure. The current-mode configuration has been employed in some previous works which includes low-noise transconductance amplifiers (LNTAs), current-mode passive mixers, and transimpedance amplifiers (TIAs). The TIA acts also as a first-order low-pass filter which can effectively eliminate undesired out-of-band interferences and significantly enhance the linearity of the circuit [6]. Other design strategies incorporate advanced linearization techniques. The low-noise amplifier (LNA) is required to be of a wide bandwidth in order to cover the designated bandwidth range of 5G, i.e. from 0.7 GHz to 7.2 GHz [7]. There are different methods for implementation of broadband LNAs, such as shunt-feedback method [8] and exploiting distributed amplifiers [9]. A wideband LNA can be implemented using a Common-Source (CS) amplifier with multiple band-pass filters [10]. Alternatively, Common-Gate (CG) topology can be utilized because of superior input matching and higher linearity. Nonetheless, CS amplifiers achieve much better noise performance [10]. Conventional noise cancellation techniques to improve noise performance increase complexity and cost and reduce voltage swings which may degrade linearity [11]. Another important LNA design metric is linearity, which trades off with other design parameters. Designing wideband LNAs with high linearity is quite challenging. There are different methods to increase linearity. The derivative superposition (DS) method uses an auxiliary transistor to curtail the non-linearity of the main transistor, in which one transistor is often in the strong inversion region and the other is in the weak inversion. The problem of this technique is difficulty to accurately bias each transistor and also it is prone to device mismatches [12]. Reference [13] uses a bipolar transistor to carry out MOSFET linearization, nonetheless, it causes the bandwidth to shrink.

In this paper, current-mode highly-linear wideband frontends have been proposed which include LNTAs, current-mode passive mixers, and transimpedance amplifiers. The paper is organized as follows. In Section II, the proposed frontend is introduced and the circuit implementations are presented. In Section III, the simulation results are given, and Section IV concludes the paper.

979-8-3503-6020-2/23 $31.00 © 2023 IEEE

II. THE RECEIVER ARCHITECTURE AND IMPLEMENTATIONS

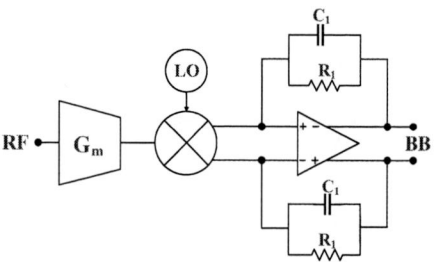

Fig. 1. Receiver frontend architecture in [6].

The structure of the current-mode direct conversion receiver is shown in Fig. 1. The higher the transconductance of the LNTA, the lower NF the receiver can achieve. On the other hand, there are always tradeoffs between NF, bandwidth, IIP3 and input matching. Herein, the G_m stage or the LNTA converts the input voltage signal into a current which flows through the passive mixer. The current signal is then down-converted by the mixer. The small input impedance of the TIA reduces the voltage swing at the output node of the mixer which increases linearity. The TIA finally transforms the current into a voltage signal at the output of the frontend and also performs a first-order filtering.

A. Low Noise Transconductance Amplifiers

In this paper, two LNTAs have been proposed, as presented in Fig. 2. and Fig. 3. Since the output of the LNTA is current, in the proposed designs, the current is conveyed by current mirrors. Furthermore, to acquire a wideband input matching, the CG topology is used to realize the amplifier.

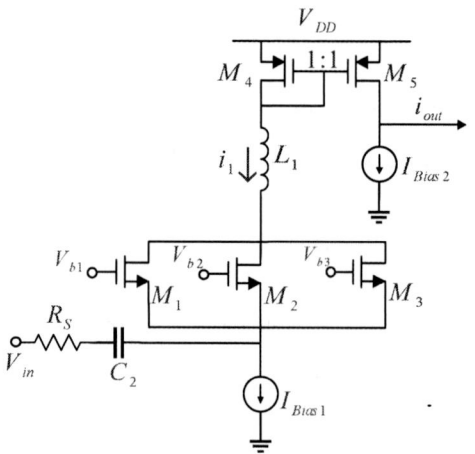

Fig. 2. The proposed CG LNTA.

In Fig. 2, the schematic of the first proposed LNTA has been depicted. The transistors M_1-M_3 form the CG input stage which provides a broadband input matching. The input impedance at resonance can be approximately calculated by:

$$Z_{in} \approx \frac{1}{g_{m1}} \left\| \frac{1}{g_{m2}} \right\| \frac{1}{g_{m3}}. \tag{1}$$

where g_m represents the transconductance of the transistor. Transistors M_1-M_3 have different aspect ratios and gate biasing voltages. The advantage of this work is the high linearity of the circuit because the third harmonic distortion of the drain currents of transistors M_1-M_3 which are connected in parallel, cancel each other and the small signal current flowing through L_1 and the current mirror does not have a third harmonic distortion. This results in an improved IIP3 for the LNTA. As a consequence of the limited impedances of the parasitic capacitances at higher frequencies, the gain of the circuit will decrease as frequency is raised. This implies a decrease in the overall bandwidth of the LNTA and the frontend. The inductor L_1 has been employed in order to compensate the bandwidth shrinkage and to obtain a flat gain over the desired frequency range [14]. The small-signal current i_1 is amplified and directed towards the output node via the current mirror which is comprised of transistors M_4 and M_5. The transconductance of the circuit is equal to:

$$G_{m,1} \approx \frac{g_{m5}}{2g_{m4}}(g_{m1} + g_{m2} + g_{m3}) \tag{2}$$

The NF of this LNTA can be given by

$$NF_{LNTA,1} = 1 + \left(\frac{\gamma \left((\frac{g_{m1} + g_{m2} + g_{m3}}{4} + g_{m4})(\frac{g_{m5}}{g_{m4}})^2 + g_{m5} \right)}{R_S \frac{(g_{m1} + g_{m2} + g_{m3})^2}{4}(\frac{g_{m5}}{g_{m4}})^2} \right) \tag{3}$$

From (3) it can be deduced that for smaller g_m's and aspect ratios for M_4 and M_5, the NF of this amplifier can be reduced to less than 3 dB.

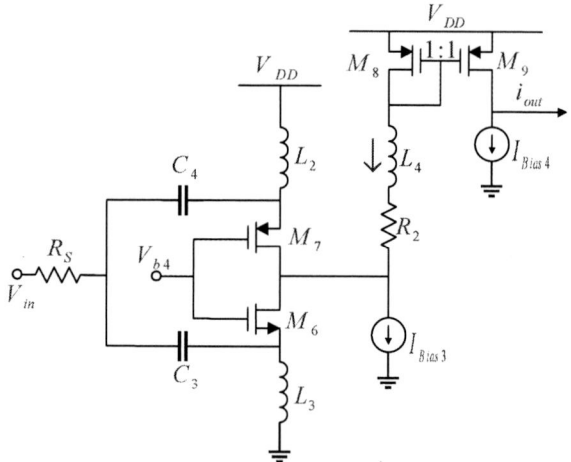

Fig. 3. Proposed CG LNTA with combined NMOS and PMOS for nonlinearity cancellation.

In this paper, another LNTA based on the input CG stage and current conveyors has been proposed which is depicted in Fig. 3. Herein, capacitors C_3 and C_4, inductors L_2 and L_3, and the transconductances of transistors M_6 and M_7 guarantee the input impedance matching.

The implementation of both NMOS and PMOS transistors M_6 and M_7 as in Fig. 3. reduces the input impedance at the resonant frequency giving rise to better matching. Moreover, it enhances the gain and also performs third-order distortion cancellation. Therefore, the attainable third-order intercept point (IIP3) can be increased [15]. The inductor L_4 has been used to achieve flat gain as was practiced for the LNTA of Fig. 2. The input impedance of this circuit is approximately equal to

$$Z_{in} \approx \frac{1}{g_{m6} + g_{m7}} . \tag{4}$$

The transconductance of the circuit is given by

$$G_{m,2} \approx \frac{g_{m9}}{2g_{m8}}(g_{m6} + g_{m7}) . \tag{5}$$

The circuit exhibits an enhancement in transconductance when compared to the conventional single-transistor CG. The NF of this circuit can be derived as follows:

$$NF_{LNTA,1} = 1 + \left(\frac{\gamma\left(\left(\frac{g_{m6} + g_{m7}}{4} + g_{m8}\right)\left(\frac{g_{m9}}{g_{m8}}\right)^2 + g_{m9}\right)}{R_S \frac{(g_{m6} + g_{m7})^2}{4}\left(\frac{g_{m9}}{g_{m8}}\right)^2} \right) \tag{6}$$

Similar to the previous LNTA, the transconductances of M_8 and M_9 should be chosen small. In order to expand the bandwidth, the parasitic capacitance at drain node of M_6 and M_7 or C_{par}, has been nullified by the inductor L_4. This leads to a design constraint given by

$$g_{m8} \approx \sqrt{\frac{C_{par}}{L_4}} . \tag{7}$$

The linearity analysis for similar configurations for the input transistors are given in [15].

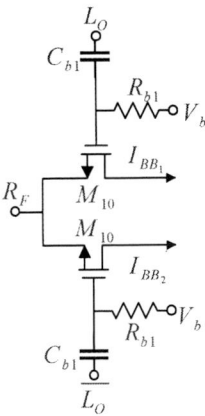

Fig. 4. Current-mode passive mixer.

The passive current-mode mixer has been shown in Fig. 4. The input signal is the output current from the LNTA and the down-converted output is fed to the TIA to be converted to a voltage signal for further processing. The biasing voltage V_b should be chosen carefully to prevent the transistors M_{10} from operating in the saturation region in order to improve low-frequency noise performance.

The TIA in Fig. 5. is composed of transistors M_{11}-M_{18} and the feedback resistors and capacitors R_1 and C_1. The important point which should be taken into account is that the opamp used in the TIA should have both high DC gain and -3-dB bandwidth to reduce the input impedance of the TIA over the required frequency range. This makes the output node of the mixer almost a virtual ground and the distortion of the frontend due to the passive mixer is reduced significantly. Therefore, a two-stage amplifier has been used to provide sufficient gain.

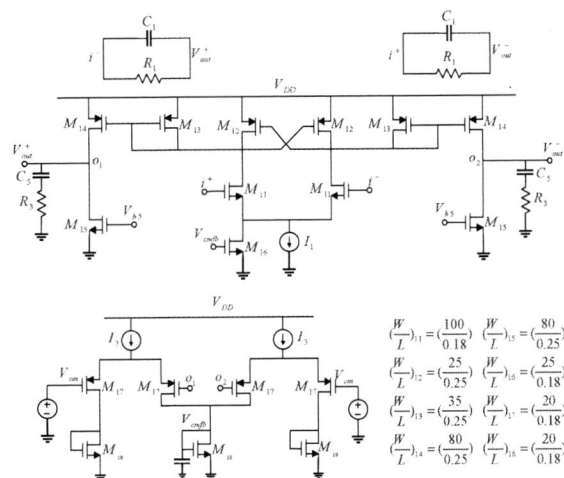

Fig. 5. TIA circuit with CMFB.

The first stage is a low-gain amplifier that includes the transistors M_{11}-M_{13}, the second stage includes M_{14} and M_{15}. The common-mode feedback circuit formed by transistors M_{17} and M_{18} has been also shown in Fig. 5. The gain of the first stage is approximately given by

$$A_V \approx \frac{g_{m11}}{g_{m13} - g_{m12}} \tag{8}$$

From (8) it can be deduced that the first stage exploits a gain boosting technique and as g_{m12} is raised, the gain of the first stage increases. However, this increase also leads to an increase in the output impedance of the first stage, which in turn pushes the poles associated with the output node towards the origin, degrading the overall stability of the amplifier. The gain of the second stage which is a common-source amplifier is equal to $g_{m14}(r_{o14}\|r_{o15})$. By conducting a small-signal analysis, the frequencies of the dominant and non-dominant poles have been obtained as follows:

$$\omega_{p,dom} \approx \frac{1}{(r_{o14}\|r_{o15}) \times C_5}$$

$$\omega_{p,nondom} \approx \frac{1}{\frac{1}{g_{m13} - g_{m12}} \times C_5} \tag{9}$$

where r_{o14} and r_{o15} represent the small-signal output (drain-source) resistances of transistors M_{14} and M_{15}, respectively, and C_{par} is the lumped parasitic capacitance at the output node of the first stage.

Herein the stability is ensured by a new cost-efficient compensation technique. It is worth mentioning that the pole associated with the output node of the first stage (non-dominant pole) is far from origin because this node does not have high impedance but the output node of the second stage is of a large resistance. The negative phase shift due to the non-dominant pole has been compensated by the positive phase shift of a left-half plane (LHP) zero. The LHP zero is generated by the series connection of R_3 and C_5 connected to the output node of the amplifier. The frequency of the zero is given by

$$\omega_Z \approx \frac{1}{R_3 C_5} \tag{10}$$

It is interesting to note that since this pole-zero cancellation takes place at the frequency beyond the unity-gain frequency of the amplifier, it is resilient to PVT variations and mismatches. This compensation leads to a sufficient phase margin and does not degrade the bandwidth. Therefore gain boosting is carried out without degrading the frequency response.

III. SIMULATION RESULTS

The proposed receiver frontends have been designed and simulated in a standard 0.18 μm CMOS technology. Simulations have been carried out for the two proposed frontends based on the architecture in Fig. 1, and utilizing each of the newly presented LNTAs, the passive mixer in Fig. 4, and the TIA in Fig. 5.

The simulation results for S_{11} which is a measure of the input impedance matching are shown in Fig. 6. It is apparent that the S_{11} for both designs is less than -9 dB. Fig. 7. shows the double side-band (DSB) NF of the frontends and the two-tone test results and the values of IIP3 are reported in Fig. 8 and Fig. 10. The frontend which uses the LNTA with all-NMOS input stage of Fig. 2. achieves the DSB NF of 7.9 dB and its IIP3 is equal to +2 dBm. For the frontend which uses the LNTA with NMOS-PMOS input stage shown in Fig. 3. the NF and IIP3 are equal to 20.5 dB and 6 dBm, respectively. Fig. 9. shows the simulation results for the conversion gain. For the frontend with all-NMOS input stage the conversion gain is 30 dB while consuming 16.2 mW power. Also, for the frontend using the LNTA with NMOS-PMOS input stage, the conversion gain is 23 dB with the power consumption of 21.4 mW. The gain of the both implementations are flat and can cover a broad spectrum of frequencies. The open-loop frequency response of the proposed amplifier is shown in Fig. 11. The DC gain of the amplifier is 48 dB, the unity-gain frequency is almost equal to 1.1 GHz with the phase margin of 74 degrees.

The comparison of this work against previously reported works has been provided in Table. 1. It can be concluded that the proposed circuits outperform the previous works in terms of bandwidth and linearity.

IV. CONCLUSION

In this article, broadband receiver frontends for 5G standard are proposed. The current-mode direct conversion receiver consists of a low noise transconductance amplifier (LNTA), a current-mode passive mixer, and a transimpedance amplifier (TIA). Two different circuits have been proposed for realizing the LNTA adopting novel bandwidth extension and linearization methods. Moreover, a new high-gain and wide-bandwidth amplifier has been proposed to develop the TIA. This amplifier reduces the input impedance of the TIA making the frontend more linear. The extensive simulation results prove the efficacy of the proposed techniques to improve the overall performance of the wideband 5G receivers.

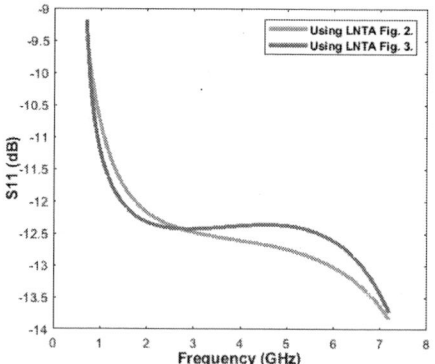

Fig. 6. Simulated S11 of proposed receiver.

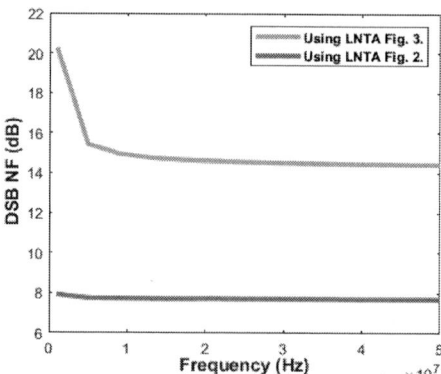

Fig. 7. Simulated DSB NF of proposed receiver.

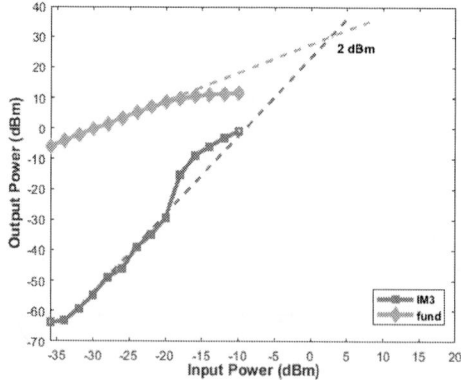

Fig. 8. Simulated IIP3 of proposed receiver using LNTA Fig. 2.

The 5ᵗʰ Iranian International Conference on Microelectronics (IICM2023)

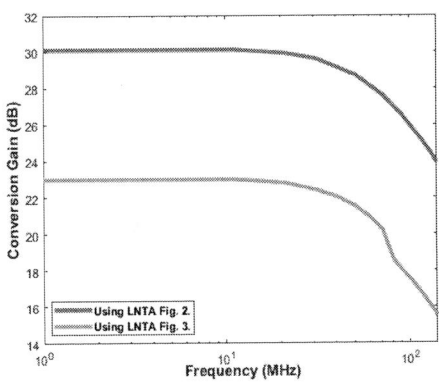

Fig. 9. Simulated Conversion Gain of proposed receiver.

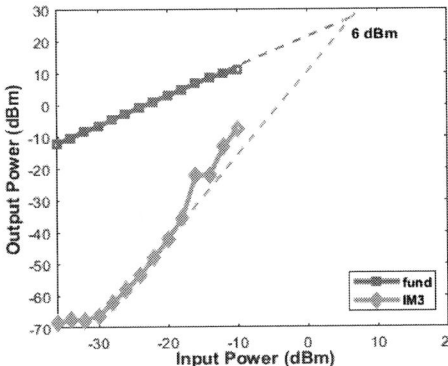

Fig. 10. Simulated IIP3 of proposed receiver using LNTA Fig. 3.

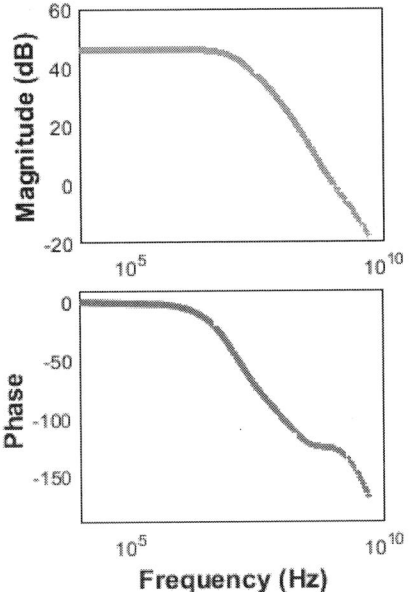

Fig. 11. The simulated open-loop frequency response for the amplifier used in TIA.

Table. 1. Comparison against previously reported works

	[5]	[6]	[8]	[14]	Receiver Using LNTA of Fig. 2.	Receiver Using LNTA of Fig. 3.
Tech.	0.13μm	0.13μm	0.13μm	0.18μm	0.18μm	0.18μm
Frequency [GHz]	5.15–5.825	1-5.2	0.1-0.93	2.4-9.5	0.7-7.2	0.7-7.2
Gain [dB]	26	22.4-24.3	13	10.4	30.1	23
S11 [dB]	<-5	<-5	<-10	<-9.4	<-9	<-9
NF [dB]	3.5	6.5-8.3	4	-8.8	7.9	20.5
IIP3 [dBm]	-2	≥-1.5	-10.2	4-11	2	6
P_DC [mW]	72	13 35	0.6	9	16.2	21.42
FoM [dB]	2.8	12.5	2.4	12.1	21.3	3.7

REFERENCES

[1] A. O. Watanabe, M. Ali, S. Y. B. Sayeed, R. R. Tummala, and M. R. Pulugurtha, "A Review of 5G Front-End Systems Package Integration," IEEE Trans. Components, Packag. Manuf. Technol., vol. 11, no. 1, pp. 118–133, 2021.

[2] H. Darabi, "A blocker filtering technique for wireless receivers," Dig. Tech. Pap. - IEEE Int. Solid-State Circuits Conf., vol. 42, no. 12, pp. 84–86, 2007.

[3] A. Ghaffari, E. A. M. Klumperink, M. C. M. Soer, and B. Nauta, "Tunable high-q N-Path Band-Pass filters: Modeling and verification," IEEE J. Solid-State Circuits, vol. 46, no. 5, pp. 998–1010, 2011.

[4] G. Pini, D. Manstretta, and R. Castello, "Analysis and Design of a 260-MHz RF Bandwidth +22-dBm OOB-IIP3 Mixer-First Receiver with Third-Order Current-Mode Filtering TIA," IEEE J. Solid-State Circuits, vol. 55, no. 7, pp. 1819–1829, 2020.

[5] M. Valla, G. Montagna, R. Castello, R. Tonietto, and I. Bietti, "A 72-mW CMOS 802.11a Direct Conversion Front-End With 3.5-dB NF and 200-kHz 1/f Noise Corner," vol. 40, no. 4, pp. 970–977, 2005.

[6] J. Kim and J. Silva-Martinez, "Low-power, low-cost CMOS direct-conversion receiver front-end for multistandard applications," IEEE J. Solid-State Circuits, vol. 48, no. 9, pp. 2090–2103, 2013.

[7] R. Fujimoto, K. Kojima, and S. Otaka, "A 7-GHz 1.8dB NF CMOS low noise amplifier," Eur. Solid-State Circuits Conf., vol. 37, no. 7, pp. 49–52, 2001.

[8] S. B. T. Wang, A. M. Niknejad, and R. W. Brodersen, "Design of a sub-mW 960-MHz UWB CMOS LNA," IEEE J. Solid-State Circuits, vol. 41, no. 11, pp. 2449–2456, 2006.

[9] F. Zhang and P. R. Kinget, "Low-power programmable gain CMOS distributed LNA," IEEE J. Solid-State Circuits, vol. 41, no. 6, pp. 1333–1343, 2006.

[10] A. Bozorg and R. B. Staszewski, "A 0.02-4.5-GHz LN(T)A in 28-nm CMOS for 5G Exploiting Noise Reduction and Current Reuse," IEEE J. Solid-State Circuits, vol. 56, no. 2, pp. 404–415, 2021.

[11] A. Bozorg and R. B. Staszewski, "A 20 MHz-2 GHz Inductorless Two-Fold Noise-Canceling Low-Noise Amplifier in 28-nm Two-Fold Noise-Canceling Low-Noise Amplifier in 28-nm CMOS," IEEE Trans. Circuits Syst. I Regul. Pap., vol. 69, no. 1, pp. 42–50, 2022.

[12] S. Ganesan, E. Sánchez-Sinencio, and J. Silva-Martinez, "A highly linear low-noise amplifier," IEEE Trans. Microw. Theory Tech., vol. 54, no. 12, pp. 4079–4085, 2006.

[13] C. Xin and E. Sánchez-Sinencio, "A linearization technique for RF low noise amplifier," Proc. - IEEE Int. Symp. Circuits Syst., vol. 4, pp. 3–6, 2004.

[14] A. Bevilacqua and A. M. Niknejad, "An ultrawideband CMOS low-noise amplifier for 3.1-10.6-GHz wireless receivers," IEEE J. Solid-State Circuits, vol. 39, no. 12, pp. 2259–2268, 2004.

[15] B. K. Kim, D. Im, J. Choi, and K. Lee, "A highly linear 1 GHz 1.3 dB NF CMOS low-noise amplifier with complementary transconductance linearization," IEEE J. Solid-State Circuits, vol. 49, no. 6, pp. 1286–1302, 2014.

979-8-3503-6020-2/23 $31.00 © 2023 IEEE

The 5th Iranian International Conference on Microelectronics (IICM2023)

25 – 26 October 2023

Design of a Calibration Circuit for Adaptive Phase-Locked Loop in the 5GHz Range Using CMOS 180nm Technology

Reza MirAlvandi
Faculty of Electrical and Computer Engineering
Khajeh Nasir Toosi University of Technology
Tehran, Iran
miralvandi@email.kntu.ac.ir

line 1: 2nd Mahdi Ehsanian
Faculty of Electrical and Computer Engineering
Khajeh Nasir Toosi University of Technology
Tehran, Iran
ehsanian@kntu.ac.ir

Abstract—In this article, we present the design of a phase-locked loop (PLL) incorporating a digitally controlled calibration oscillator circuit for frequency adaptation. The adaptation process is carried out digitally, guided by the performance of the oscillator. When the oscillator experiences a deviation from its specified frequency range, the active calibration circuitry detects the lock state and applies an appropriate digital code to the capacitor bank. This eliminates the need for user intervention to reprogram the PLL for frequency adjustment. The proposed circuit comprises a voltage-controlled oscillator (VCO) equipped with a capacitor bank and a digital calibration circuit, allowing for a broader range of output frequencies to be covered. The designed PLL circuit operates effectively within the frequency range of 3 to 6.7GHz, utilizing CMOS 180nm technology with 1.8V supply voltage and 15.5 mW power consumption, and has undergone rigorous simulation. The design successfully achieves a 200 MHz frequency range variation for the reference frequency, with a phase noise value of -122.71dBc/Hz at 10MHz. The innovation in this paper lies in the capability of the proposed design for tracking the reference frequency. In the event of a change in the reference frequency, there is no need for user intervention to reprogram the phase-locked loop (PLL). The system automatically performs the relocking operation. This feature can be utilized in unmanned aerial vehicles (UAVs or drones) to implement frequency hopping capabilities.

Keywords— Calibration, frequency synthesizer, adaptive phase-locked loop (PLL)

I. Introduction

Phase-locked loops (PLLs) are considered fundamental components in wireless communications, finding extensive applications in generating various frequencies for radio channels [1]. With the rapid expansion of telecommunications and electronics, the necessity of determining diverse frequency bands for different applications has become paramount. PLLs play a significant role in telecommunication circuits for the selection of desired frequency channels. Furthermore, phase-locked loops can serve as frequency synthesizers, making them pivotal components within various segments of communication system transmitters and receivers.

In today's electronic and communication systems, having a reference signal is crucial for tasks such as encoding, decoding, data recovery, and clock signal generation, all of which fall under the purview of frequency synthesizers. In the

realm of phase-locked loops, two critical parameters stand out: lock time and lock range.

The lock time of a phase-locked loop (PLL) assumes critical significance, especially in applications like frequency hopping. In systems designed to counter jamming, rapid frequency hopping at intervals as short as milliseconds, or even microseconds, becomes a necessity. Consequently, if the lock time of the PLL is excessively high, the transmitted data may be entirely lost from the receiver's perspective. Thus, in such applications, the lock time holds paramount importance.

Another common phenomenon encountered in the realm of telecommunications and electronic systems is frequency deviation. To effectively address this issue, the phase-locked loop should possess a broader lock range, ensuring it maintains its lock state even when frequency deviation occurs. This article introduces a phase-locked loop equipped with the capability to autonomously adjust the output frequency in response to changes in the reference frequency. This innovation aims to tackle challenges associated with the integration of phase-locked loops into circuits.

In this design, following a change in the input frequency, the loop's tuning model undergoes modification. Once the loop loses its lock state, the circuit initiates the generation of a control voltage within the calibration voltage circuit, proportionate to the new frequency. Subsequently, the loop is re-locked. In essence, this design not only capitalizes on the voltage-controlled oscillator's ability to operate across a wide frequency spectrum but also empowers it to execute the locking operation at any frequency within this range without requiring user intervention, thus optimizing the output frequency.

II. Review of Phase-Locked Loop Structure

In Fig. 1, a schematic of a basic circuit for a phase-locked loop consisting of a charge pump circuit can be seen.

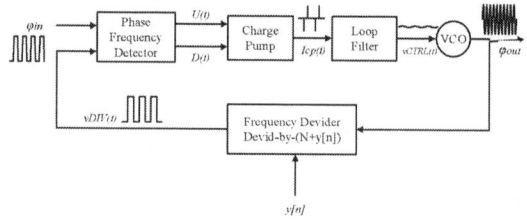

Fig. 1. Basic Block Diagram of a PLL [1]

979-8-3503-6020-2/23 $31.00 © 2023 IEEE

As we know, the transfer function of the system can be written in the form of (1).

$$\frac{\varphi_{out}(s)}{\varphi_c(s)} = \frac{sK_{PD}F(s)}{s + \frac{K_{PD}K_{vco}F(s)}{N}} \qquad (1)$$

The operation of a phase-locked loop is facilitated by a voltage-controlled oscillator (VCO) responsible for generating a variable-frequency signal, which is controlled by phase detector signals. Achieving calibration in a phase-locked loop typically involves three categories:

1. **VCO Calibration**: This entails calibrating the VCO to extend its lock range.

2. **Bandwidth Calibration**: Adjusting the bandwidth parameters.

3. **Lock State Calibration**: Ensuring the lock state is accurately established.

What enables the operation of a phase-locked loop is a voltage-controlled oscillator (VCO) that generates a variable frequency signal and is controlled by phase detector signals. To have a calibrated phase-locked loop, calibration can be pursued in three categories: 1) VCO calibration, 2) bandwidth calibration, and 3) lock state calibration. In this design, the objective is to calibrate the VCO to achieve a wider lock range.

In the context of this design, the focus is on VCO calibration to broaden the lock range.

When we refer to oscillator calibration, it implies that the frequency range of the oscillator is fine-tuned using a suitable capacitor bank, thereby modifying its characteristic curve. Fig. 2 illustrates the impact of utilizing a capacitor bank on the oscillator.

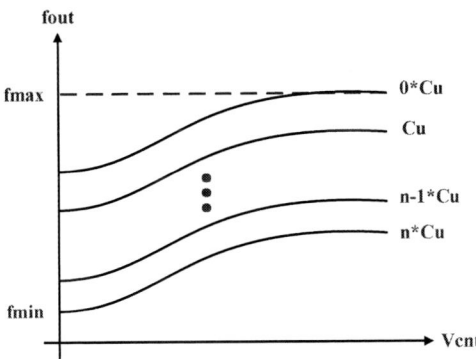

Fig. 2. The Effect of Using a Capacitor Bank on the Oscillator

In the proposed approach, the phase-locked loop's lock status is assessed through the utilization of an auxiliary circuit that operates in conjunction with the primary loop. If the lock conditions are found to be compromised, the phase-locked loop is disengaged, and the oscillator undergoes adjustments facilitated by the auxiliary circuit. Once the error at the output of the phase-frequency detector converges to zero, the primary loop is re-engaged, and the circuit seamlessly resumes its regular operation.

The comprehensive structure of the proposed design is depicted in Fig. 3.

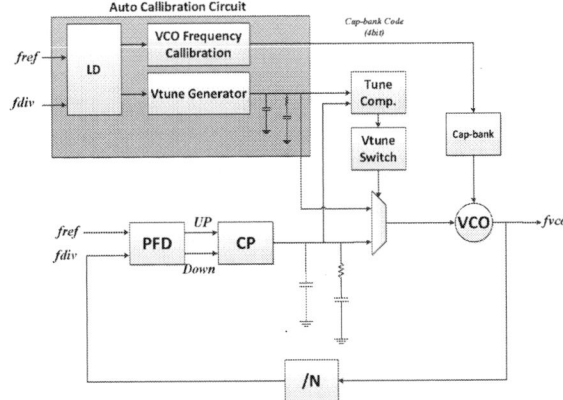

Fig. 3. Proposed design

The critical processes for swiftly reestablishing the loop's lock state are executed within the Tune block. The initial phase involves the detection of any lock deviation within the loop. Subsequently, the multiplexer situated at the oscillator's input disengages the chosen tuning voltage from the primary loop and redirects it to the Tune circuit. Within the Tune circuit, the oscillator's output undergoes adjustments via an internal capacitor bank, all while continuously monitoring the lock status.

Upon the loop successfully regaining its lock state, the multiplexer reconnects the primary phase-locked loop, transmitting the finely-tuned voltage generated by the loop filter back to the oscillator's input.

III. Designing Different Components of a Phase-Locked Loop

A. Designing the Voltage-Controlled Oscillator (VCO)

When designing a voltage-controlled oscillator (VCO), a crucial factor to consider is its capability to choose a specific frequency band. Therefore, the VCO architecture should include a capacitor bank that can be manipulated through a 4-bit binary code. The inclusion of a capacitor bank in the VCO design is essential to ensure that variations in the control voltage alone do not result in a broad spectrum of frequencies at the oscillator's output. In simpler terms, we employ the capacitor bank to select the desired frequency channel, while relying on voltage variations for precise frequency tuning. In this design, we have implemented the LC structure for the oscillator. Fig. 4 provides an overview of the VCO's overall structure, including the output buffer circuit and the employed capacitor bank.

The 5th Iranian International Conference on Microelectronics (IICM2023)

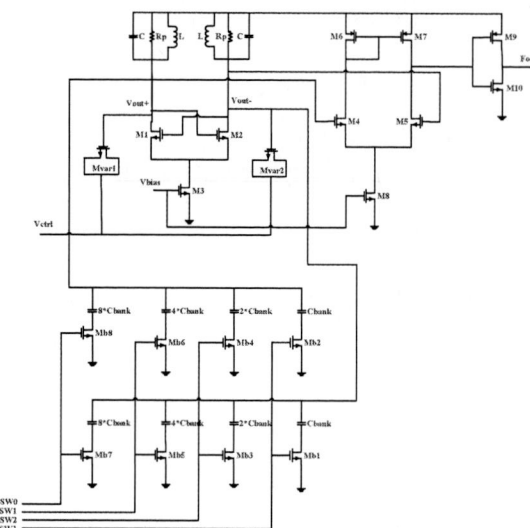

Fig. 4. Voltage-Controlled Oscillator (VCO) Circuit

Binary codes SW0 to SW3 are employed to govern the switching transistors. To generate these codes, a binary search circuit is utilized, leveraging the reference frequency as its basis.

B. Frequency Divider

The frequency divider must deliver high-frequency performance, necessitating the utilization of TSPC (True Single Phase Clock) structures in its implementation. Fig. 5 illustrates the configuration of the employed counter.

This structure should be positioned at the input of the frequency divider, enabling the division of the input signal by 2 within high-frequency ranges. This, in turn, facilitates the use of low-frequency models in the subsequent stages of the frequency divider [2].

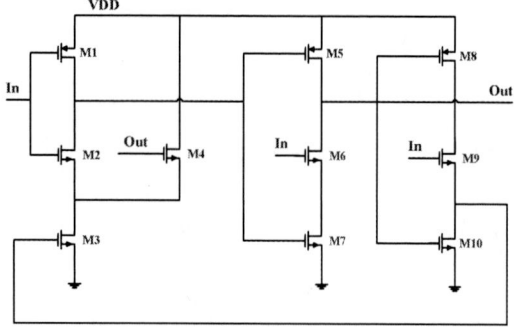

Fig. 5. The TSPC counter architecture

Basic structures are also applicable for detector and charge-pump circuits. However, in this article, our primary emphasis is directed towards the control and calibration of voltage-controlled oscillators.

C. Loop Lock Detector and Counter Circuit

To detect when a loop departs from the lock state, we require an LD (Lock Detect) circuit. The suggested configuration for the LD circuit is depicted in Fig. 6.

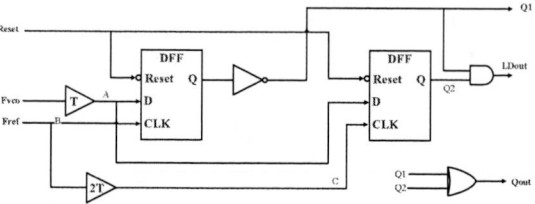

Fig. 6. Circuit Diagram of the Lock Detector for the Loop [8]

The LD block controls when the voltage control switch should change its state and the phase detector path should become active. Once the input from the isolator is disconnected from the main loop, two mechanisms need to be employed. First, an appropriate capacitor bank must be selected for the isolator, and then an optimal control voltage should be generated for it. In the following section, the proposed circuit for regulating the output frequency of the isolator is presented. In the output waveform, you can observe the lock state detection of the loop.

To choose binary codes corresponding to the reference input frequency, a digital counter with a 4-bit configuration based on D flip-flops is essential. Fig. 7 and Fig. 8 provide a visual representation of the 4-bit counter structure and the clock generation scheme employed for the counter circuit.

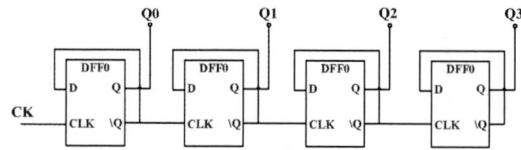

Fig. 7. Binary code finder circuit

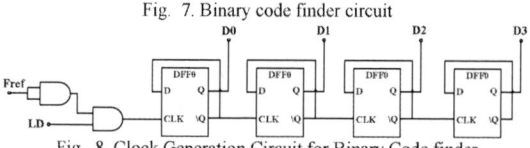

Fig. 8. Clock Generation Circuit for Binary Code finder

The system functions as follows: When the loop is not in the locked state, the 4-bit counter initiates counting, and it continues counting until the loop reverts to the locked state. Essentially, in the absence of lock, the counter is driven by clock pulses, causing the binary code sent to the capacitor bank to change, thus leading to variations in the values of the vector capacitors. Consequently, these adjustments modify the output frequency of the oscillator. Fig. 9 provides a visual representation of the signals associated with the counter and the lock detection circuit.

The 5th Iranian International Conference on Microelectronics (IICM2023)

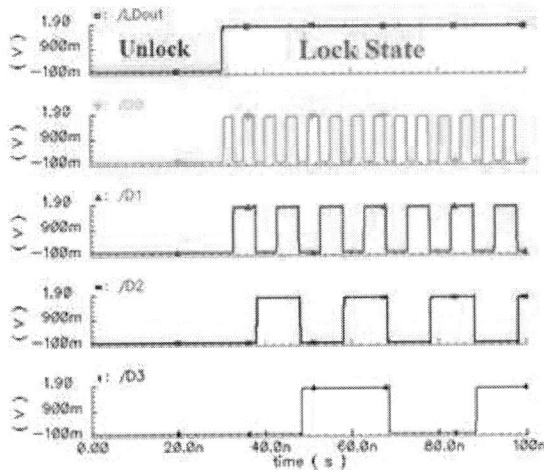

Fig. 9. The output of the binary code finder

IV. SIMULATION RESAULT

Before conducting simulations for the phase-locked loop (PLL) in conjunction with the calibration circuit, we initially evaluate the performance of the loop in the absence of the calibration circuit. Fig. 10 illustrates the output signal of the oscillator while in the locked state.

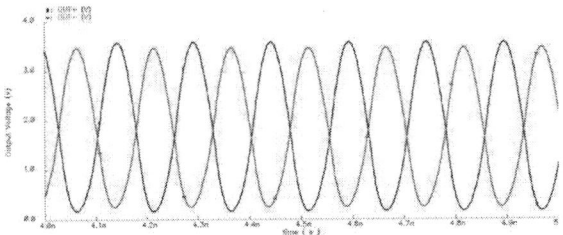

Fig. 10. The normal operation of the phase-locked loop (PLL) without the presence of the calibration circuit

The oscillator's output frequency demonstrates fluctuations within an approximate range of 200 MHz, which constrains the extent to which the reference frequency can be altered, consequently reducing the overall range of variations. In this specific circuit, the phase noise is measured -94.95dBc@1MHz, and you can observe its graphical representation in Fig. 11.

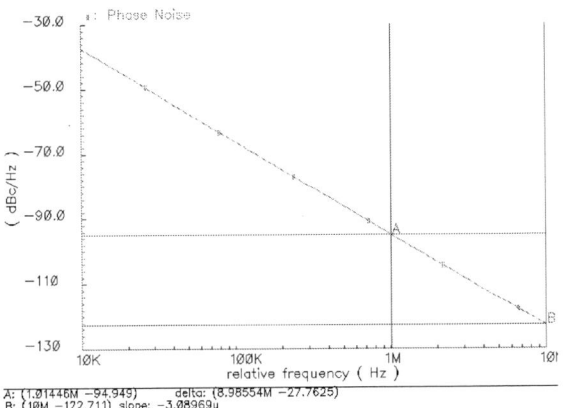

A: (1.01446M −94.949)   delta: (8.98554M −27.7625)
B: (10M −122.711) slope: −3.08969u

Fig. 11. The phase noise plot of an oscillator

To simulate the proposed design effectively, we needed to manipulate the reference frequency to observe the behavior of the oscillator's tuning voltage. Consequently, we conducted a parametric simulation, varying the input reference frequency within the range of 400-500MHz. For clarity, this frequency range was divided into four segments, and the system's performance was analyzed at frequencies of 400MHz, 428MHz, 461MHz, and 500MHz. Fig. 12 provides a visual representation of the tuning voltage of the oscillator across this frequency range.

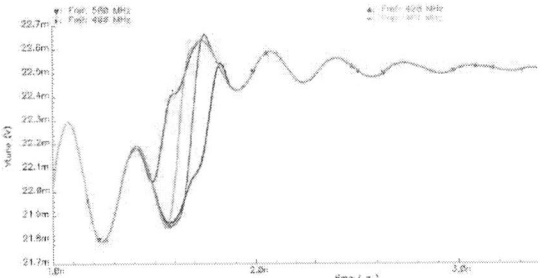

Fig. 12. The behavior of the tuning voltage in the presence of the calibration circuit

In Fig. 13 we observe a Monte Carlo simulation for the output frequency of the VCO.

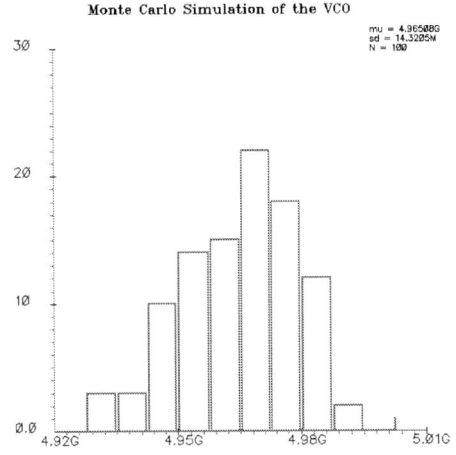

Fig. 13. Monte Carlo simulation of the VCO

Simulation results pertaining to Process, Voltage, and Temperature (PVT) conditions have been illustrated in Fig. 14 through Fig. 16.

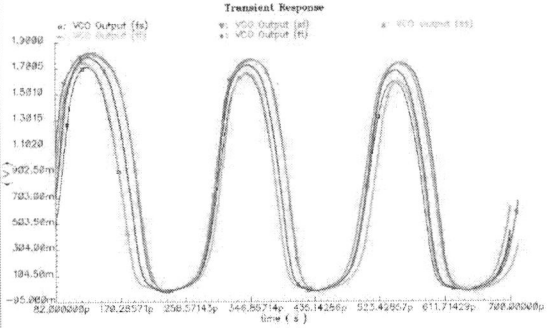

Fig. 14. Process Variation

The 5ᵗʰ Iranian International Conference on Microelectronics (IICM2023)

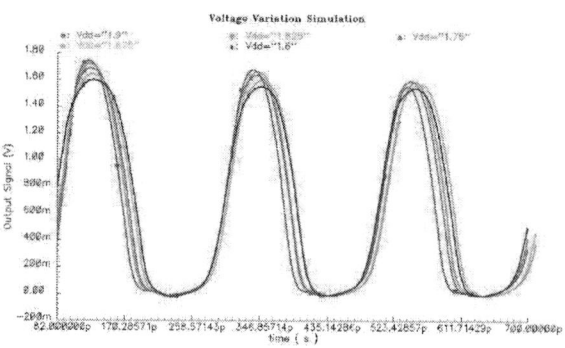

Fig. 15. Voltage Variation Simulation

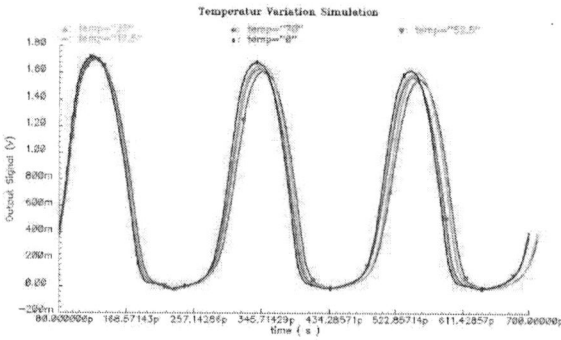

Fig. 16. Temperature Variation Simulation

Table 1 compares the specifications and simulation results of the proposed design with several other works in the same field.

Table 1. Performance Comparison

	This work	[1]	[2]	[3]
Process (nm)	180	65	65	65
Freq. (GHz)	3~6.7	0.1~5	0.96~2.06	1.9~2.6
Ref. Freq. (MHz)	400~500	15~50	40	32
Power (mW)	15.5	14~20.5	NA	0.53
Supply (V)	1.8	1.2	1.2	1.2
Phase noise (dBc/Hz@1MHz)	-94.95	-126.6	NA	-140.5 (@10MHz)
Calibration Time (µs)	1.2	1.25~1.86	4.03	NA

Fig. 17 depicts the oscillator alongside the calibration circuit. As you may be aware, in the 180nm technology, the inductors occupy a considerable amount of chip space.

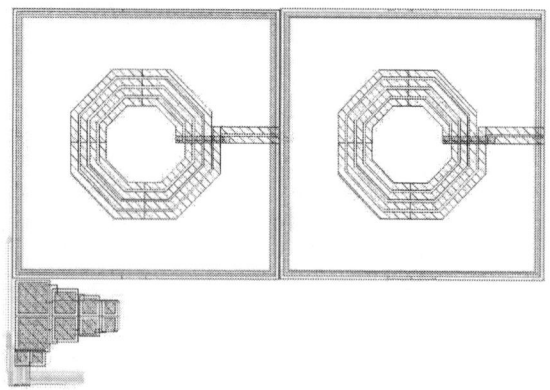

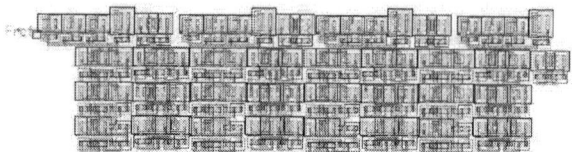

Fig. 17. The capacitance bank, along with the CMOS switches Layout

V. CONCLUSION

A calibration circuit was meticulously designed and subjected to simulation for the adaptive phase-locked loop. This study predominantly revolved around enhancing frequency adaptability from the standpoint of the oscillator's frequency. It's important to note that when the reference frequency undergoes changes, the phase-locked loop can potentially lose its lock status. Such occurrences can transpire for two primary reasons: firstly, the oscillator may venture out of its designated frequency range, and secondly, fluctuations in phase noise can compromise the loop's stability, leading to it losing lock.

In this article, our primary aim was to address the initial issue, specifically, preventing the oscillator from deviating outside its specified frequency range. We approached this challenge with a straightforward solution.

A pivotal constraint in conventional phase-locked loops is the limited range within which the oscillator's output frequency can be varied. To expand the output frequency range, adjustments to the capacitors within the oscillator's capacitor bank are imperative. When the loop loses lock, the lock detection circuit springs into action, activating the clock for the digital counter within the calibration circuit. Subsequently, an appropriate digital code is computed to be applied to the capacitor bank, and the corresponding tuning voltage is furnished to the oscillator. This process continues until the frequency disparity between the input and output reaches its minimum value, at which point the calibration circuit is disengaged from the loop.

In essence, traditional phase-locked loops face a constraint that restricts changes in the reference frequency to a mere 100MHz range. To overcome this limitation, it becomes imperative to devise an auxiliary circuit for oscillator calibration capable of adjusting the oscillator's output when confronted with substantial shifts in the reference frequency. Such a solution allows the loop to remain locked, even in the face of substantial changes. This article presents the design and simulation of the phase-locked loop, in tandem with the calibration circuit, within the confines of the CMOS 180nm technology.

REFERENCES

[1] Zhang, Zhao, et al. "A fast auto-frequency calibration technique for wideband PLL with wide reference frequency range." 2018 IEEE Asian Solid-State Circuits Conference (A-SSCC). IEEE, 2018.

[2] Ryu, Hyuk, et al. "Fast automatic frequency calibrator using an adaptive frequency search algorithm." IEEE Transactions on Very Large Scale Integration (VLSI) Systems 25.4 (2016): 1490-1496.

[3] Chen, Peng, et al. "A 529-µW fractional-N all-digital PLL using TDC gain auto-calibration and an inverse-class-F DCO in 65-nm CMOS." IEEE Transactions on Circuits and Systems I: Regular Papers 69.1 (2021): 51-63.

[4] Su, Pin-En, and Sudhakar Pamarti. "Fractional-$ N $ Phase-Locked-Loop-Based Frequency Synthesis: A Tutorial." IEEE Transactions on Circuits and Systems II: Express Briefs 56.12 (2009): 881-885.

[5] B. Razavi, Design of CMOS phase-locked loops: from circuit level to architecture level. Cambridge, United Kingdom: *Cambridge University Press*, 2020.

[6] Stephens, Donald R. Phase-locked loops for wireless communications: digital, analog and optical implementations. Springer Science & Business Media, 2007.

[7] Da Dalt, N. and Sheikholesami, A., n.d. *Understanding Jitter And Phase Noise.*

[8] Pawar, Shobha N., and Pradeep B. Mane. "Wide band PLL frequency synthesizer: A survey." 2017 International Conference on Advances in Computing, Communication and Control (ICAC3). IEEE, 2017.

[9] Jamali, Babak, and Aydin Babakhani. "A 0.2-2.6 GHz instantaneous frequency-to-voltage converter in 90nm CMOS." 2016 IEEE Radio and Wireless Symposium (RWS). IEEE, 2016.2

[10] Shizhen, Huang, Lin Wei, and Fenglin Gao. "A wide band and low PN PLL design for digital tuner." APCCAS 2008-2008 IEEE Asia Pacific Conference on Circuits and Systems. IEEE, 2006.

[11] Kuo, Ko-Chi, and Chi-Wei Wu. "An fast lock technique for wide band PLL frequency synthesizer design." 2014 International Conference on Information Science, Electronics and Electrical Engineering. Vol. 2. IEEE, 2014.

[12] Kim, Nakyoon, and Yong Moon. "A study on wide-band frequency synthesizer for advanced wireless communication." 2011 International SoC Design Conference. IEEE, 2011.

[13] Lin, Han-Bo, Tzu-Chao Yan, and Chien-Nan Kuo. "A 1–5 GHz frequency-to-voltage converter using limiting amplifier." 2016 IEEE International Symposium on Radio-Frequency Integration Technology (RFIT). IEEE, 2016.

[14] Ting, Guo, et al. "A 20.5 GHz wide-band programmable divide-by-N frequency divider." 2014 International Symposium on Integrated Circuits (ISIC). IEEE, 2014.

The 5th Iranian International Conference on Microelectronics (IICM2023)

25 – 26 October 2023

A 0.9-8 GHz Highly Linear SAW-Less Direct-Conversion Receiver Front-End for 5G Communication Standard

Erfan Salighe
Deptartment of Electrical and Computer Engineering
Urmia University
Urmia, Iran
st_e.salighe@urmia.ac.ir

Mortaza Mojarad
Deptartment of Electrical and Computer Engineering
Urmia University
Urmia, Iran
m.mojarad@urmia.ac.ir

Abstract—**A broadband SAW-less direct conversion receiver front-end has been designed and implemented in a 0.18 μm CMOS process for fifth generation new radio (5G NR) communication standard. The proposed receiver consists of a single-ended low-noise transconductance amplifier (LNTA) driving a current-mode passive mixer terminated by a transimpedance amplifier (TIA). The common source (CS) topology has been employed for LNTA while utilizing a new structure to increase the input matching bandwidth without affecting the noise figure (NF) and linearity of the LNTA. A new technique has been proposed for bandwidth extension of LNTA's transconductance (G_m). The linearity of the front-end has been improved by utilizing a linearity enhancement technique and using the current-mode topology for the proposed receiver. A high-gain and high-bandwidth operational transconductance amplifier has been proposed to reduce the input impedance of the TIA for improving the linearity. The proposed front-end consumes only 13.5 mW. Over 0.9 GHz – 8 GHz frequency range the receiver achieves a conversion gain more than 27.75 dB, maximum NF of 7.85 dB, maximum S_{11} of -10 dB, and 10.4 dBm of IIP3.**

Keywords—*direct-conversion receiver, 5G communication, highly linear receiver, wideband RF front-ends, Current mode receiver, transconductance*

I. INTRODUCTION

Today, there is an increasing demand for high-data-rate and dependable wireless communication circuits and systems to satisfy the stringent requirements of fifth-generation (5G) mobile communication standards. To achieve the frequency range of 5G standard, these systems should include wideband radio frequency (RF) receivers. In addition, the RF receiver to satisfy the 5G performance requirements should have low power consumption, low cost, low noise, and high linearity [1]. Therefore, the surface acoustic wave (SAW)-less receiver architectures are often employed to improve the noise performance, reduce cost, and obtain more flexibility for receivers. Furthermore, direct conversion receivers are widely used in multiband RF receivers for their high level of integration [2], [3]. The current mode SAW-less direct-conversion receiver shown in Fig. 1 which utilizes a low-noise transconductance amplifier (LNTA) driving passive current-mode mixers followed by a transimpedance amplifier (TIA) is becoming a popular topology to implement wideband front-ends [4]-[6].

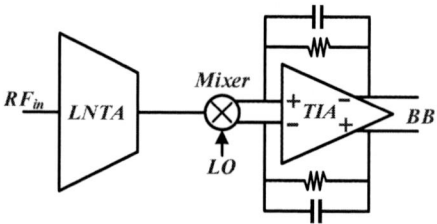

Fig. 1. Current mode SAW-less receiver

Several RF receivers with various architectures have been proposed in [5]-[11]. In [7] and [8] off-chip SAW filters are replaced with on-chip N-path bandpass filters to attenuate the blockers. These structures degrade noise figure (NF) and increase power consumption. The mixer-first receivers improve the linearity by eliminating the low-noise amplifier (LNA) [9], [10]. However, the absence of LNAs cause higher LO leakage, higher NF, and increased sensitivity to PVT (Process-Voltage-Temperature) variations. In [5], the Frequency-Translational Noise-Cancelling receivers (FTNC-RX) are illustrated. Although, they can achieve ultra-low NF by neutralizing thermal noise by utilizing multiple passive mixer-based down-conversion paths, they suffer from LO leakage, high power consumption, and cost. Active feedback has been employed in [6] and [11] to improve receiver sensitivity by suppressing blockers. Since these topologies are of high sensitivity to device mismatches, they conduct imperfect blocker rejection.

A low-noise transconductance amplifier (LNTA) is the most critical circuit block in the current-mode receiver topology and the frequency range of the receiver is determined by the bandwidth of its LNTA. Therefore, the LNTA must achieve wideband input impedance matching and a wideband transconductance. Meanwhile, the LNTA design must consider tradeoffs between input matching, noise, bandwidth (BW), linearity, and transconductance (G_m). Moreover, a sufficiently large G_m is required to minimize the system noise figure (NF) [12]-[14]. The LN(T)As can be realized by common-source (CS) [15], common-gate (CG) [16], and CG-CS topologies[17], [18]. Although CS structures have typically lower NF, the attainable bandwidth is small. Therefore, methods such as resistive or active feedback techniques [13]. [14] and

979-8-3503-6020-2/23 $31.00 © 2023 IEEE

distributed amplifiers [19] have been employed to extend the bandwidth. Common-gate or cascode topologies can achieve broadband input matching and high linearity. Nevertheless, these structures endure high NF compared to CS. Hence, CG amplifiers combined with noise canceling or G_m enhancement techniques have been developed [16].

In this paper, a low-cost, highly linear, and wideband receiver front-end for 5G NR applications has been proposed. In the proposed direct-conversion receiver, a new method has been presented for the LNTA to extend its bandwidth and improve the linearity. In addition, a compensation method has been proposed to increase gain and bandwidth of the operational transconductance amplifier (OTA) used in TIA. This paper is organized as follows. Section II describes the structure of the proposed broadband front-end. Section III provides simulation results, and section IV concludes the paper.

II. Proposed Structure for Broadband Front-end

A. Structure of The Proposed Broadband LNTA

The schematic of the proposed broadband LNTA is depicted in Fig. 2. The proposed LNTA has been designed based on the CS topology and include complementary linearity enhancement by using nonlinear active feedbacks [20]. Conventional common-gate and resistive shunt feedback CS amplifiers have been used as most common topologies to design broadband LNTAs. The input impedance of shunt feedback CS and CG amplifiers is almost equal to $1/g_m$, where g_m is the transconductance of a transistor. The conventional circuit realizations are unable to satisfy stringent requirement of input matching and bandwidth for NR band of 5G standard. The transistors M_1, M_2 and the inductor L_D in proposed LNTA are used to establish a broadband input matching. Moreover, the equivalent circuit of the proposed matching network has

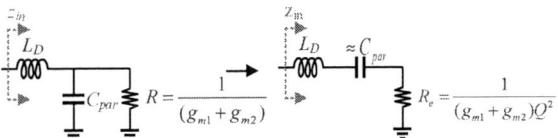

Fig. 3. Equivalent circuit of the proposed matching network

been shown in Fig. 3 that the input impedance in the resonant frequency of L_D and C_{par} is given by

$$Z_{in} = \frac{1}{(g_{m1} + g_{m2})Q^2} , \quad Q = \omega C_{par} R \qquad (1)$$

where g_{m1} and g_{m2} are the transconductances of M_1 and M_2, respectively, and C_{par} is the parasitic capacitance at the drain node of M_1 and M_2. As mentioned before, a broadband LNTA must achieve wideband transconductance to have a flat gain. The impedance of the parasitic capacitances at the output node of the LNTA decreases when frequency is raised. Therefore, at high frequencies, the small-signal output current of the LNTA flows into the ground, leading to gain reduction at high frequencies. This implies the shrinkage of the bandwidth of the LNTA. As shown in Fig. 2, the Inductor-Less G_m peaking blocks have been utilized to increase gain at higher frequencies. This gain peaking neutralizes the gain decrement caused by the parasitic capacitances which results in the bandwidth extension of the proposed LNTA. Here, in order to investigate the proposed Inductor-Less G_m Peaking technique in improving the bandwidth of the proposed LNTA, the transconductance has been obtained for both low and high frequencies in (2) and (3), respectively, based on the equivalent circuit for G_m calculation shown in Fig. 4.

$$G_{m,LF} = \frac{i_{sc}}{V_1} = \frac{g_{m4}}{1 + \frac{g_{m4}}{g_{m6}}} + \frac{g_{m3}}{1 + \frac{g_{m3}}{g_{m5}}} \qquad (2)$$

$$G_{m,HF} = \frac{i_{sc}}{V_1} = \left(\frac{g_{m4}}{1 + \frac{g_{m4}}{g_{m6}}} + \frac{g_{mb4}\frac{g_{m10}}{g_{m8}}}{1 + \frac{g_{m4}}{g_{m6}}} \right) + \left(\frac{g_{m3}}{1 + \frac{g_{m3}}{g_{m5}}} + \frac{g_{mb3}\frac{g_{m9}}{g_{m7}}}{1 + \frac{g_{m3}}{g_{m5}}} \right)$$

$$(3)$$

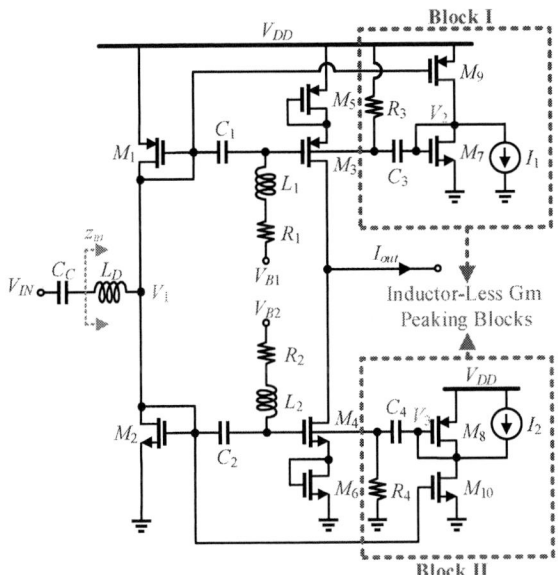

Fig. 2. The proposed broadband LNTA

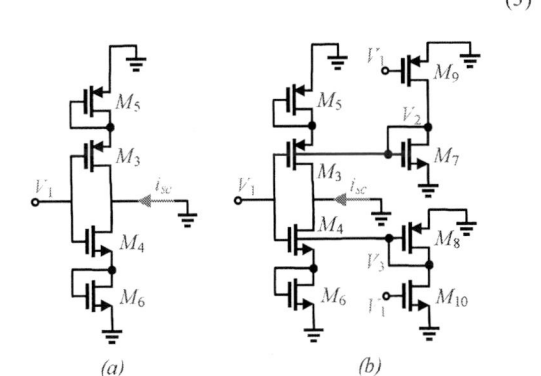

Fig. 4. Equivalent circuit for G_m calculation a) at low frequencies b) at high frequencies

979-8-3503-6020-2/23 $31.00 © 2023 IEEE

The 5th Iranian International Conference on Microelectronics (IICM2023)

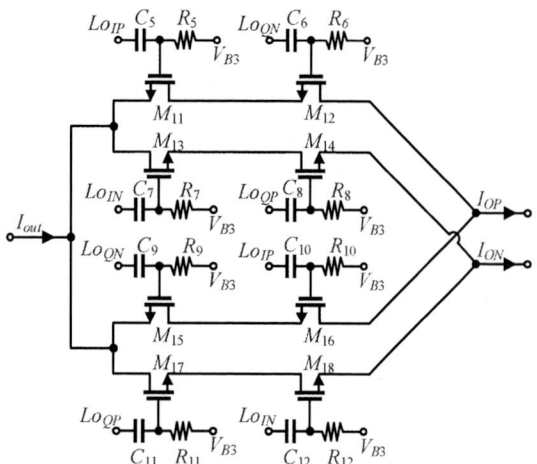

Fig. 5. Double balanced current mode passive mixer circuit

B. Current Driven Passive Mixer

The passive mixer circuit has been depicted in Fig. 5 in which it is driven by a 50% duty cycle local oscillator (LO). The two switches connected in series produce an overall 25% duty cycle for the LO. The double balanced topology improves LO feedthrough rejection and noise figure. Furthermore, it helps mitigate self-mixing effect. In addition, DC gate bias of the mixer's switches ensures that the transistors in the passive mixer do not operate in the saturation to further improve the low-frequency noise. Due to the DC gate biasing voltages, the transistors has been biased at either weak inversion or triode. Moreover, in the active mixers the linearity is improved by increasing the DC bias current while in the passive mixers there is no static power consumption. The switches resistance is nonlinear but their linearity is much better than active mixer. The linearity of the current-mode passive mixers is more than voltage mode passive mixers because the voltage mode passive mixers have significant voltage swing across the switches while the current mode passive mixers have insignificant signal voltage swing across the switches. Hence, the linearity of current mode passive mixers is substantially better.

C. Proposed Structure for TIA

The proposed 4-stage operational transconductance amplifier (OTA) with common mode feedback circuit used to implement the TIA has been shown in Fig. 5. The feed forward compensation has been employed which add zeros for increasing bandwidth and gain of OTA simultaneously. The Miller compensation has high amount of power consumption compared with feedforward compensation for obtaining high gain and large bandwidth. High gain-bandwidth product (GBW) provides noticeable small input impedance for a large frequency range that causes to attenuate voltage swing at the input of TIA which can enhance the linearity of front-end. The frequency response of the proposed OTA is shown in Fig. 6. The gain of the proposed amplifier is equal to 85.7 dB. It achieves 2.52 GHz unity-gain frequency and the phase at this frequency is equal to -127.74 degrees which results in 52.26 degrees phase margin.

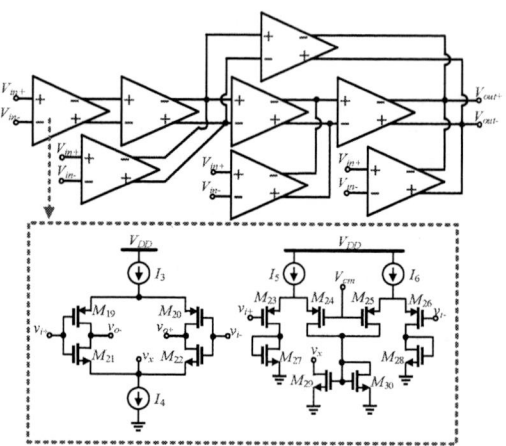

Fig. 5. Operational transconductance amplifier (OTA) used in TIA

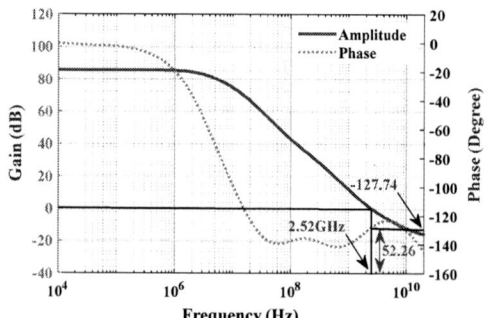

Fig. 6. Simulation results for the frequency response of the OTA

III. SIMULATION RESULTS

The wideband front-end receiver using the proposed techniques for increasing LNTA bandwidth with linearity enhancement has been designed in a standard 0.18 μm CMOS process using ADS. The simulated transconductance of the proposed LNTA has been shown in Fig. 7. The variation of LNTA G_m is less than 3 mS in 0.9 GHz to 9 GHz frequency range. Moreover, the simulated S_{11} for the front-end also been simulated for different process corners over a temperature variation spanning from - 40 ∘C to 85 ∘C has been shown in Fig. 8. The presented receiver front-end achieves maximum S_{11} of -10 dB from 0.9 GHz to 8 GHz. Fig. 9 (a) shows the conversion gain at 50 MHz baseband frequency versus LO frequency. The minimum gain conversion occurs at 2 GHz LO frequency and is equal to 27.78 dB and maximum gain conversion is 35 dB at 7 GHz LO frequency. The conversion gain versus baseband frequency has been shown in Fig. 9(b). Fig. 10 shows the simulated NF_{DSB} versus baseband frequency for different process corners over a temperature variation. The minimum NF is equal to 7.45 dB for TT corner process. The result from a two-tone test simulation for the proposed font-end is given in Fig. 11 in which the power of the fundamental signals at the output as well as the corresponding IM products are extrapolated at the Third Input Intercept Point (IIP3). Therefore, Fig. 11 demonstrates IIP3 equal to 10.4 dBm. In addition, IIP3 Monte Carlo simulations have been shown in Fig. 11. Table I presents comparison between the proposed receiver front-end and state-of-the-art receivers.

979-8-3503-6020-2/23 $31.00 © 2023 IEEE

The 5th Iranian International Conference on Microelectronics (IICM2023)

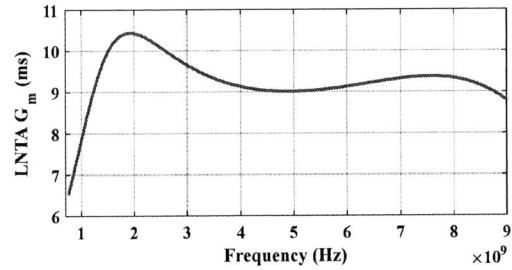

Fig. 7. The transconductance of proposed LNTA driven with passive mixer followed by TIA

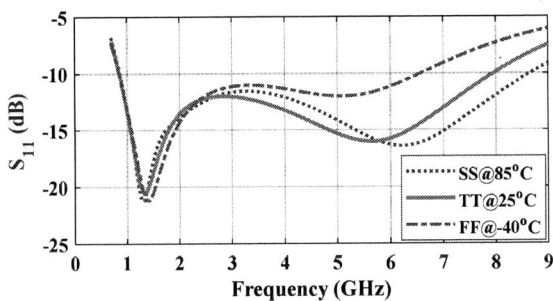

Fig. 8. Simulated S_{11} for the proposed receiver

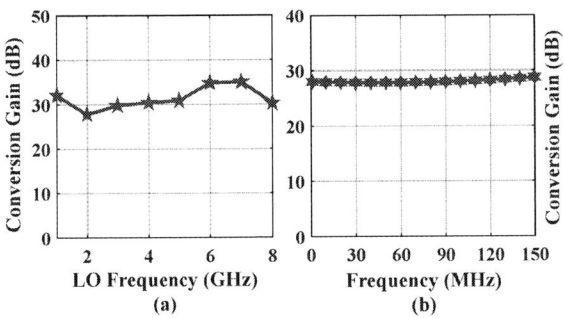

Fig. 9. (a). Simulated conversion gain at 50 MHz baseband frequency of the proposed receiver and (b) simulated conversion gain at 2 GHz LO frequency of the proposed receiver versus down-converted frequency.

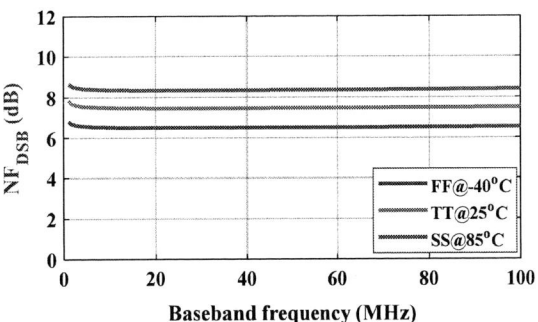

Fig. 10. Simulated DSB NF of the proposed receiver versus baseband frequency.

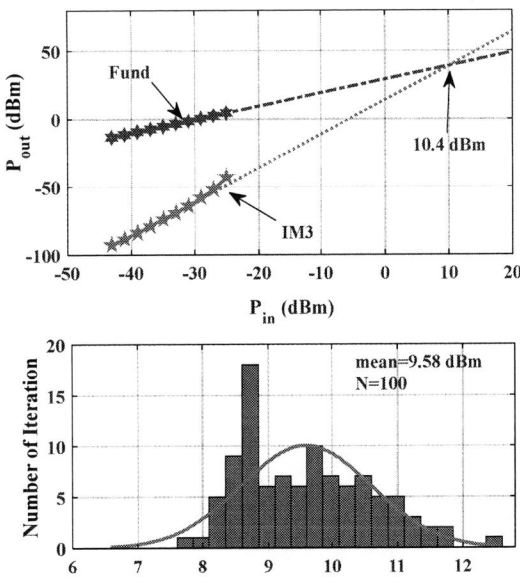

Fig. 11. Simulated IIP3 for the proposed receiver and IIP3 Monte Carlo simulation result

TABLE I.   COMPARISON TO PREVIOUSLY REPORTED WORKS

	Technology	Frequency [GHz]	Gain [dB]	IIP3 [dBm]	NF [dB]	Power [mW]	Supply [V]
[21]	90 nm	2.0-5.8	18-23	≥-1.5	12.2-13.8	85	2.7
[22]	130 nm	1-5.2	22.4-24.3	≥-1.5	6.5-8.3	48	1.5/1.2
[23]	180 nm	3.1-8	21	(-5.6)-(-2.6)	5-6.6	44.85	2.3
[24]	130 nm	0.9-2.3	34.5-35.5	4-11	9.5-11.5	30-36	1.5
[25]	180 nm	1.55-2.3	22.5-25	7	7.7-9.5	10	2
[26]	90 nm	0.1-3.85	20	-3.23	8.4-11.5	9.8	1.2
[27]	90 nm	0.8-6	3-36	-3.5	5-5.5	29	2.5
[28]	65 nm	2.4-2.5	49.5	-25.75	8.2	2.16	0.8
[29]	130 nm	5.9-7.1	35	-31	5.5	7.5	1.2
[30]	65 nm	7-9	21-24	-7.5	6.2-8	316	N/A
This work	180 nm	0.9-8	27.75-35	10.4	7.45	13.5	1.8

979-8-3503-6020-2/23 $31.00 © 2023 IEEE

As shown in Table I, the proposed receiver demonstrates outstanding performance compared with recently published receivers and proves the effectiveness of utilized techniques for linearity enhancement and frequency band expansion.

IV. CONCLUSION

A broadband direct-conversion receiver front-end has been presented in this paper. The proposed front-end has been designed in a standard 0.18 μm CMOS technology. The proposed receiver consists of a single-ended LNTA driving a current mode passive mixer followed by a low-input impedance TIA. It can provide high linearity while consuming low power. A new structure has been proposed to increase input matching bandwidth. Moreover, a new technique is proposed to provide broadband transconductance for the LNTA in addition to employing linearity enhancement technique. The current mode passive mixer has been utilized to attenuate noise and self-mixing effects and also boosting linearity. In addition, in order to lower the input impedance of TIA, a new OTA with feed forward compensation has been proposed. The new OTA achieves high bandwidth, high gain and low power consumption to reduce the input impedance of TIA. The proposed receiver dissipates about 13.5 mW from a 1.8-V supply. The simulation results show that the proposed receiver exhibits 0.9 GHz – 8 GHz frequency range and high baseband bandwidth consuming small amount of power. These features make the receiver well-suited for utilizing for 5G NR standards.

REFERENCES

[1] S. Onoe, "Evolution of 5G mobile technology toward 1 2020 and beyond," in IEEE Int. Solid-State Circuits Conf. (ISSCC), 2016, pp. 23–28.

[2] J. Ryynanen, K. Kivekas, J. Jussila, L. Sumanen, A. Parssinen, and K. A. I. Halonen, "A single-chip multimode receiver for GSM900, DCS1800, PCS1900, and WCDMA," IEEE J. Solid-State Circuits, vol. 38, no. 4, pp. 594–601, Apr. 2003.

[3] J. Laskar, B. Matinpour, and S. Chakraborty, Modern Receiver FrontEnds: Systems, Circuits, and Integration. Hoboken, NJ: Wiley, 2004.

[4] Z. Ru, N. Moseley, E. Klumperink, and B. Nauta, "Digitally Enhanced Software-Defined Radio Receiver Robust to Out-of-Band Interference," IEEE J. Solid-State Circuits, vol. 44, no. 12, pp. 3359-3375, Dec. 2009.

[5] D. Murphy, H. Darabi, A. Abidi, A. Hafez, A. Mirzaei, M. Mikhemar, and M.-C. Chang,"A blocker-tolerant, noise-cancelling receiver suitable for wideband wireless applications," IEEE J. Solid-State Circuits, vol. 47, no. 12, pp. 2943-2963, Dec. 2012.

[6] S. Youssef, R. van der Zee, and B. Nauta, "Active feedback technique for rf channel selection in front-end receivers," IEEE J. Solid-State Circuits, vol. 47, no. 12, pp. 3130-3144, Dec. 2012.

[7] Y. Xu and P. R. Kinget, "A switched-capacitor RF front end with embedded programmable high-order filtering," IEEE J. Solid-State Circuits, vol. 51, no. 5, pp. 1154–1167, May 2016.

[8] M. Darvishi, R. van der Zee, and B. Nauta, "Design of active N-path filters," IEEE J. Solid-State Circuits, vol. 48, no. 12, pp. 2962–2976, Dec. 2013.

[9] Y.-C. Lien et al., "Enhanced-selectivity high-linearity low-noise mixerfirst receiver with complex pole pair due to capacitive positive feedback," IEEE J. Solid-State Circuits, vol. 53, no. 5, pp. 1348–1360, May 2018.

[10] C. Andrews and A. C. Molnar, "A passive mixer-first receiver with digitally controlled and widely tunable RF interface," IEEE J. SolidState Circuits, vol. 45, no. 12, pp. 2696-2708, Dec. 2010.

[11] J. Zhu, H. Krishnaswamy, and P. R. Kinget, "Field-programmable LNAs with interferer-reflecting loop for input linearity enhancement," IEEE J. Solid-State Circuits, vol. 50, no. 2, pp. 556–572, Feb. 2015.

[12] L. Zhang, Y. Xu, K. Tripurari, P. R. Kinget and H. Krishnaswamy, "Analysis and Design of a 0.6- to 10.5-GHz LNTA for Wideband Receivers," in IEEE Transactions on Circuits and Systems II: Express Briefs, vol. 62, no. 5, pp. 431-435, May 2015.

[13] B. G. Perumana, J.-H. Zhan, S. S. Taylor, B. R. Carlton, and J. Laskar, "Resistive-feedback CMOS low-noise amplifiers for multiband applications," IEEE Trans. Microw. Theory Techn., vol. 56, no. 5, pp. 1218–1225, May 2008.

[14] M. T. Reiha and J. R. Long, "A 1.2 V reactive-feedback 3.1–10.6 GHz low-noise amplifier in 0.13 m CMOS," IEEE J. Solid-State Circuits, vol. 42, no. 5, pp. 1023–1033., May 2007.

[15] H. Yu, Y. Chen, C. C. Boon, C. Li, P.-I. Mak, and R. P. Martins, "A 0.044-mm2 0.5-to-7-GHz resistor-plus-source-follower-feedback noise-cancelling LNA achieving a flat NF of 3.3±0.45 dB," IEEE Trans. Circuits Syst. II, Exp. Briefs, vol. 66, no. 1, pp. 71–75, Jan. 2019

[16] F. Belmas, F. Hameau, and J. Fournier, "A low power inductorless LNA with double Gm enhancement in 130 nm CMOS," IEEE J. Solid-State Circuits, vol. 47, no. 5, pp. 1094–1103, May 2012.

[17] A. Bozorg and R. B. Staszewski, "A 0.02–4.5-GHz LN(T)A in 28 nm CMOS for 5G exploiting noise reduction and current reuse," IEEE J. Solid-State Circuits, vol. 56, no. 2, pp. 404–415, Feb. 2021.

[18] B. Shirmohammadi and M. Yavari, "A linear wideband CMOS balunLNA with balanced loads," IEEE Trans. Circuits Syst. II, Exp. Briefs, vol. 69, no. 3, pp. 754–758, Mar. 2022

[19] F. Zhang and P. R. Kinget, "Low-power programmable gain CMOS distributed LNA," IEEE J. Solid-State Circuits, vol. 41, no. 6, pp. 1333–1343, Jun. 2006.

[20] E. Salighe, and M. Mojarad. "A linearity enhancement technique for low-noise transconductance amplifiers in SAW-less receivers." AEU-International Journal of Electronics and Communications 160 (2023): 154499.

[21] J.-H. C. Zhan, B. R. Carlton, and S. S. Taylor, "A broadband low-cost direct-conversion receiver front-end in 90 nm CMOS," IEEE J. SolidState Circuits, vol. 43, no. 5, pp. 1132–1137, May 2008

[22] J. Kim and J. Silva-Martinez, "Low-power, lost-cast CMOS directconversion receiver front-end for multi-standard applications," IEEE J. Solid-State Circuits, vol. 48, no. 9, pp. 2090–2103, Sep. 2013.

[23] M. Ranjan and L. E. Larson, "A Low-Cost and Low-Power CMOS Receiver Front-End for MB-OFDM Ultra-Wideband Systems, " in IEEE J. Solid-State Circuits, vol. 42, no. 3, pp. 592-601, March 2007.

[24] N. Poobuapheun, W.-H. Chen, Z. Boos, and A. M. Kiknezad, "A 1.5-V 0.7–2.5-GHz CMOS quadrature demodulator for multiband direct-conversion receivers," IEEE J. Solid-State Circuits, vol. 42, no. 8, pp. 1669–1677, Aug. 2007.

[25] N. Kim, V. Aparin, and L. E. Larson, "A resistively degenerated wideband passive mixer with low noise figure and high ," IEEE Trans. Microw. Theory Tech., vol. 58, no. 4, pp. 820–830, Apr. 2010.

[26] A. Amer, E. Hegazi, and H. F. Ragaie, "A 90-nm wideband merged CMOS LNA and mixer exploiting noise cancellation," IEEE J. SolidState Circuits, vol. 42, no. 2, pp. 323–328, Feb. 2007.

[27] R. Bagheri, A. Mizraei, S. Chehrazi, M. E. Heidari, M. Lee, M. Mikhemar, W. Tang, and A. Abidi, "An 800-MHz–6 GHz software-defined wireless receiver in 90-nm CMOS," IEEE J. Solid-State Circuits, vol. 41, no. 12, pp. 2860–2876, Dec. 2006.

[28] B. Park and K. Kwon, "2.4-GHz Bluetooth low energy receiver employing new quadrature low-noise amplifier for low-power low-voltage IoT applications," IEEE Trans. Microw. Theory Techn., early access, Dec. 10, 2020.

[29] N. Shams, A. Abbasi and F. Nabki, "A 6 GHz 130 nm CMOS Harmonic Recombination RF Receiver Front-End Using N-Path Filtering," 2020 18th IEEE International New Circuits and Systems Conference (NEWCAS), Montreal, QC, Canada, 2020.

[30] N. Li et al., "A High-Linearity 7–9-GHz Receiver Front End With Strength-Oriented Nonlinearity Cancellation," in IEEE Microwave and Wireless Technology Letters, vol. 33, no. 8, pp. 1231-1234, Aug. 2023.

A New Design for 1.75 to 2.55 GHz GaN Power Amplifier with More Than 40 dBm Output Power and 12 dB Maximum Gain

Marzieh Chegini
Electrical and Computer Engineering Department.
University of Tehran (UT)
Tehran, Iran
marziyeh.chegini@ut.ac.ir

Mahmoud Kamarei
Electrical and Computer Engineering Department.
University of Tehran (UT)
Tehran, Iran
kamarei@ut.ac.ir.com

HojjatAllah Nemati
Sobhe Shaghayegh Engineering Co.
Iran University of science and technology (IUST)
Tehran, Iran
sobhe.shaghayegh94@gmail

Abstract—**This paper presents the design, implementation, and measurement results for a 1.75~2.55 GHz Power Amplifier. The PA was Designed and fabricated with a CGH40010-F GaN device from Wolfspeed. The manufactured PA shows 42.2-dBm peak output power, with more than 47% average drain efficiency over the 1.75~2.55 GHz frequency range.**

Keywords—*class AB power amplifier, GaN HEMT, 2.45 GHz, CGH40010F, 10 watt PA.*

I. INTRODUCTION

In recent years, wireless communication standards have covered several frequency bands in one system. So, since the use of a separate amplifier for each frequency band increases the complexity and cost of the system, the demand for wideband and multi-band power amplifiers has increased as an integral part of a telecommunication system.

Although many works have been done on wideband harmonic-tuned power amplifiers [1-4], these amplifiers have very complex matching networks to achieve high efficiency due to the adjustment of harmonic impedances. In cases where a medium efficiency is required, the design of a linear power amplifier that follows the simplicity of the design due to the avoidance of harmonic impedances adjustment is of interest [5-6].

2.45 GHz frequency band covers a wide area. this band is used for mobile phones, wifi, IoT smart devices, CCTV cameras, and medical equipment. Small devices usually do not need a lot of Wi-Fi bandwidth, but they need a wider and better range and coverage area. Since the 2.45 GHz frequency easily passes through walls and hard objects and covers a large area, it can solve this need. It should also be noted that the frequency of 2.45 GHz is the frequency at which water boils and is therefore used in microwave ovens.

In this paper, we present a wideband power amplifier that covers the 2.45 GHz frequency band and a little bit around it. We present a 10 watt class AB GaN power amplifier, that works in a 1.75~2.55 GHz frequency band with 0.8 GHz bandwidth. The PA was designed using a CGH40010-F

transistor from Wolfspeed on 30 mil FR4 PCB. This paper is organized as follows. Section II describes the design approach in two steps. First, load/source pull strategy, and next, design matching networks. Section III includes the measurement results of the fabricated PA and finally, section IV provides a Conclusion.

II. PROPOSED DESIGN APPROACH

The first step in designing a power amplifier is always to set the DC operating point. After selecting the bias point of the device, the most important step is designing the input and output matching networks. To design matching networks, it is necessary to know what impedances should be realized from the drain and source of the device. Depending on the purpose of the design, these impedances can be impedances that show the highest possible power or the highest possible efficiency from the device, or they can be impedances that lead to a trade-off between power and efficiency.

A. Load/Source Pull Strategy

In this paper, the design process is based on the load/source pull simulation technique. In this method, it is enough to find the optimal load and source fundamental impedances for several frequency points within the target bandwidth, after choosing the appropriate bias point. For this purpose, we choose three frequency points, ie. 1.85, 2.15, and 2.55 GHz, and 200 mA drain DC current for bias of the device.

Since in real devices, the center of the power contours and the center of the efficiency contours in the load/source pull simulation never exactly coincide, it is never possible to obtain the maximum power and the maximum possible efficiency of the device simultaneously.

Here, our goal is to access as much power as possible from the device. Therefore, in the simulation of the load/source pull, we try to find the regions corresponding to our goal. If we consider the minimum acceptable power of 10 watts, the contours of power and efficiency in three frequencies within the bandwidth will be as shown in Fig. 1.

979-8-3503-6020-2/23 $31.00 © 2023 IEEE

The 5th Iranian International Conference on Microelectronics (IICM2023)

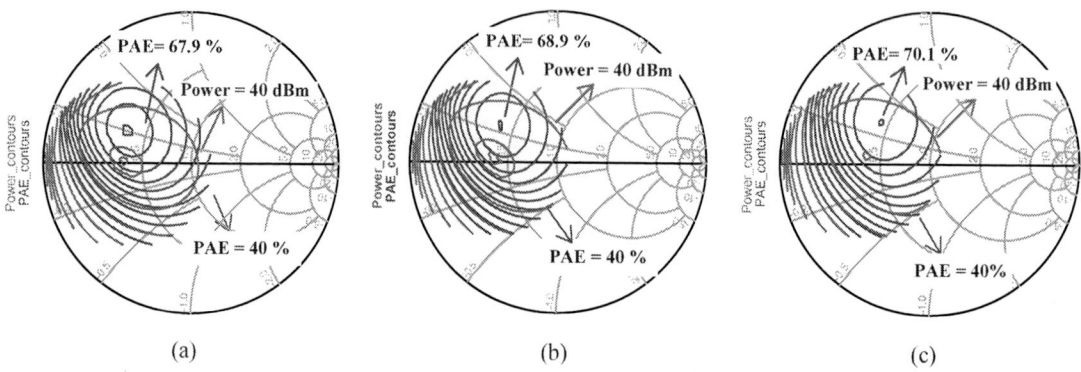

(a)   (b)   (c)

Fig. 1.   Contours of PAE and output power at some frequency inside the bandwidth. a) 2.45 GHz, b) 2.15 GHz and c) 1.85 GHz.

In this figure, the contour equivalent to 10 watts (40 dBm) power is specified. The regions inside this contour are the geometric location of the load impedances, which if realized in the design of the output matching network (OMN) of this device, we can achieve more than 10 watts output power. The minimum and maximum possible of PAE, inside this contour is also determined. For example, according to Fig. 1 (a), if we choose the load impedance inside the contour equivalent to the power of 10 watts, we will achieve a minimum efficiency of 40% and a maximum efficiency of 67.9%. This analysis for two frequencies 2.15 and 1.85 is also shown in Figs. 1 (b) and (c), respectively. An important point is that these numbers shown by the contours in Fig. 1 for power and efficiency are ideal. It means that assuming that it is possible to design a matching network that can realize these impedances and does not have any losses, these numbers will be realized in a practical design. However, due to the losses caused by the substrate and the use of real elements that are never 100% ideal, it is predictable that in the final design, the performance of our PA will be slightly less than these numbers.

Finally, the optimal region for the load impedances is considered inside the contours equivalent to 10 watts power in Fig. 1, and this region is painted in a checkered pattern in Fig. 2.

As we get closer to the central contour of the region with the checkered pattern in Fig. 2, we will have load impedances that are equivalent to more power. Therefore, we consider the impedances equivalent to the most central contour as the primary goal of the design, and in the process of designing the matching networks, we try not to exceed these checkered regions. So, the impedances that are Equivalent with the most

central power contour and their equivalent source impedances are listed in Table I.

TABLE I.   OPTIMUM LOAD AND SOURCE IMPEDANCES

Freq. (GHz)	load impedance	Source impedance
1.85	19.8 + j*2.5	1.2 - j*1.7
2.15	18.5 + j*1.2	1.6 - j*1.2
2.45	20 - j*0.4	1.4 - j*4.1

It should be noted that in the design of a linear power amplifier, harmonic impedances are not adjusted. So, in load/source pull simulation, the second and third harmonic of load, assumed equal to 50 ohms, for complete matching, and also, harmonic impedances of source, assumed to short circuit, To prevent non-linearity transfer from input to output [7].

B. Matching Networks Design

To design the matching networks, assuming a suitable basic topology according to Fig. 2, a software optimization was done in ADS software. in this optimization process, we tried to realize the optimum load and source impedances, listed in table I.

As seen in this Fig., stubs with lengths of λ/4 in gate and drain bias path, were used to avoid leakage of RF signal to the bias circuit. Also, in Fig. 3 (b), resistors R-stab1 and R-stab2 and capacitor C-stab were used to establish the unconditional stability of the PA.

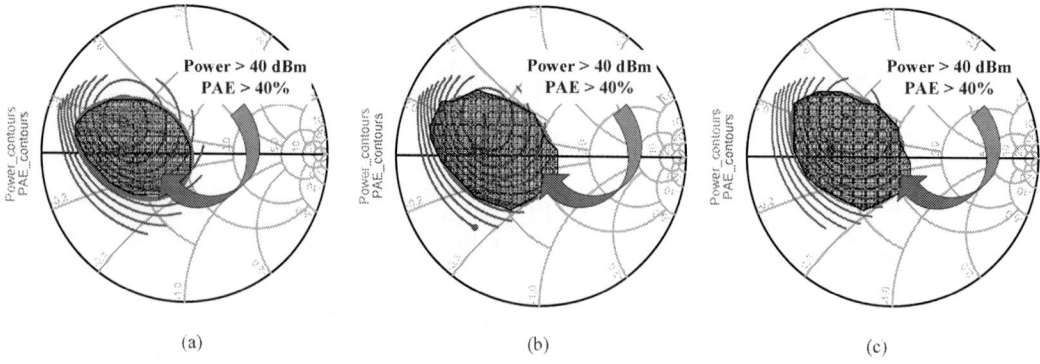

(a)   (b)   (c)

Fig. 2.   The region of load impedances, that is equivalent to an output power more than 10 watts at some frequency inside the bandwidth. a) 2.45 GHz, b) 2.15 GHz and c) 1.85 GHz

979-8-3503-6020-2/23 $31.00 © 2023 IEEE

It is worth noting that the dimensions of the vertical open-ended stubs, which were considered against each other, were chosen equal. Therefore final circuit has a symmetrical structure and, as a result, undesirable electromagnetic effects are minimized. Also, a complete schematic of the designed PA is shown if Fig. 4.

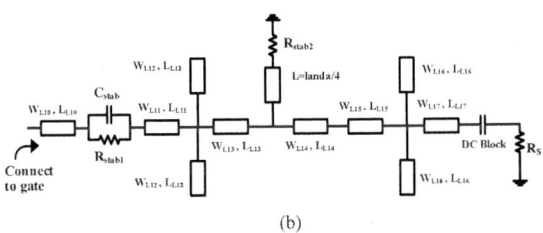

(a)

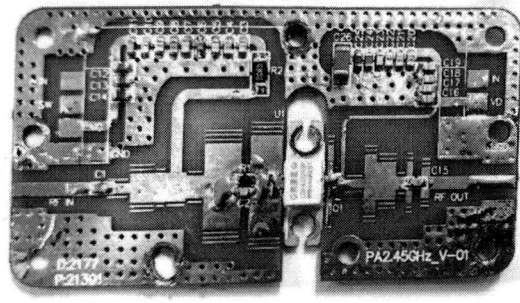

(b)

Fig. 3. Basic structure of matching networks a) output matching b) input matching

I. IMPLEMENTATION, MEASUREMENT AND RESULT

The designed power amplifier was implemented using a CGH40010-F transistor on FR4 PCB with 0.8 mm thickness is shown in Fig. 5.

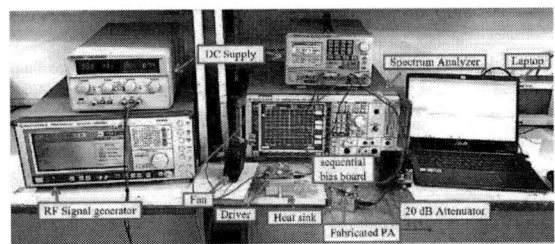

Fig. 4. Fabricated PA

Also, the measurement set-up is shown in Fig. 6 and large signal measurement results under continuous wave excitation are depicted in Figs. 7 to Fig. 9.

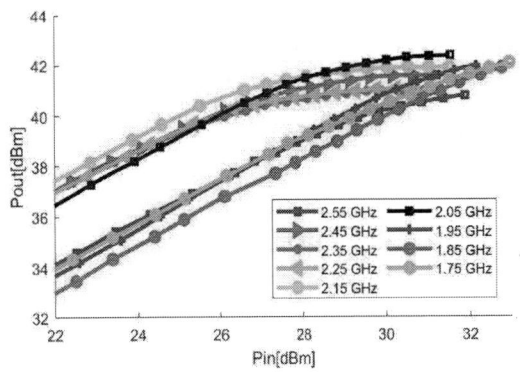

Fig. 5. Designed set-up for the PA measurement.

Output power versus input power at several frequency points within the target bandwidth is shown in Fig. 7. According to this figure, this PA can display the output power of 40 dBm and more in the entire bandwidth of 1.75 to 2.55 GHz.

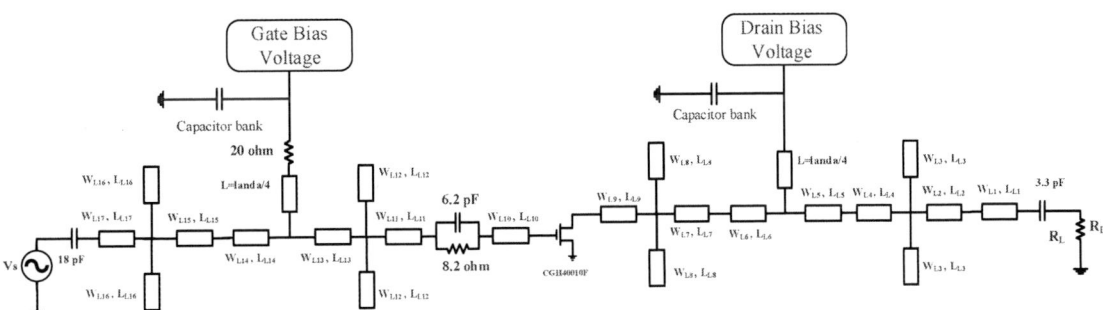

Fig. 6. Large signal measurement result of designed PA. output power vs. input Power.

The large signal frequency Performance of PA is shown in Fig. 8, at 30 dBm input power. As shown here, the fabricated PA has maximum drain efficiency of 64.1% in 2.45 GHz frequency, maximum power gain 12.2 dB, and maximum output power 42.2 dBm in 2.05 GHz frequency. Also, this PA has more than 40 dBm output power and more than 40% drain efficiency in the 1.75–2.55 GHz frequency band.

Fig. 7. Complete schematic of designed PA

979-8-3503-6020-2/23 $31.00 © 2023 IEEE

The 5th Iranian International Conference on Microelectronics (IICM2023)

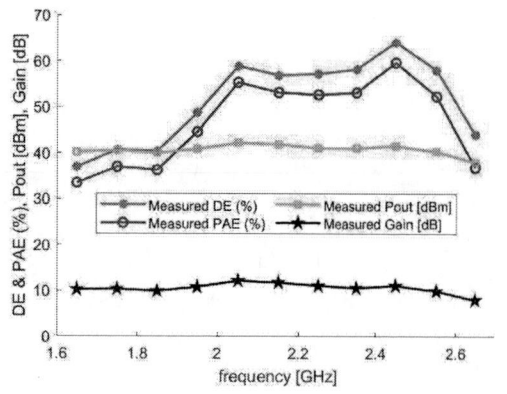

Fig. 8. measurement results of designed PA. Drain efficiency, PAE, output power, and power gain of the PA vs. the frequency at 30 dBm input power.

Fig. 9, shows the large signal gain and efficiency of designed PA vs. output power at several frequencies inside the bandwidth.

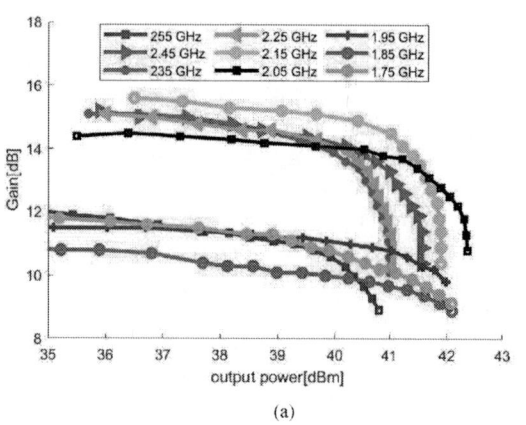

(a)

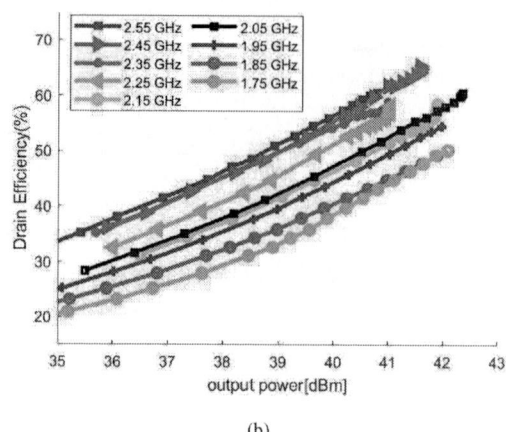

(b)

Fig. 9. Large signal measurement of designed PA vs. output Power. a) Gain and b) Efficiency.

The performance of the Fabricated PA is compared with that of the state-of-the-art PAs in Table. II, twice. One in 2.45 GHz frequency and the other in 1.75 ~ 2.55 GHz frequency band.

II. CONCLUSION

This paper described the design and fabrication process of a class AB power amplifier in frequency band 1.75 to 2.55 GHz, which covers several frequency bands in the 4th generation LTE standard. The measurement results show that the designed PA has more than 40 dBm output power and more than 40% drain efficiency in the 1.75–2.55 GHz frequency band.

TABLE II. COMPARISON WITH STATE-OF-THE-ART PAS.

Ref.	Freq. (GHz)	Gain (dB)	Pout (dBm)	PAE (%)	DE (%)	Num. of stages	Technology	Fab./Sim.
[8]	2.45	17	24.5	16	NA	2	SiGe Bipolar	Fab.
[9]	2.4	16.7	NA	49.5	NA	1	Pseudomorphic HEMT	Fab.
		30.5		53.1		2		
[10]	2.4	NA	11.2	NA	35	1	28 nm CMOS	Fab.
[11]	2.4	34.1	22.5	42.2	NA	2	0.5um GaAs PHEMT	Sim.
[12]	2.4	11.5	19.2	26	NA	1	CMOS	Fab.
[13]	2.4	14.2	NA	53.1	NA	1	NA	Fab.
[14]	1.5 ~ 2.5	24	36	40	NA	2	GAN	Fab.
[15]	2.5	27	32.7*	57	NA	2	SOI-CMOS	Fab.
[16]	2.4	NA	12.3	61.8	63.4	1	65nm CMOS	Sim.
[17]	2.4	32	31.5	25	NA	3	65nm CMOS	Fab.
[18]	2.4	13.6	45.3	78.8	82	1	GaN	Fab.
[19]	2.2	13.9	41.7	74.4	NA	1	GaN	Fab.
[20]	2.7	9.5	40.8	NA	46	1	GaN	Fab.
This work	2.45	11.1	41.6	60	64.1	1	GaN	Fab.
	1.75 ~ 2.55	10 ~ 12.2	40~ 42.2	37 ~ 60	41 ~ 64.1	1	GaN	Fab.

* Graphically estimated

ACKNOWLEDGMENT

The authors would like to thank the "Sobhe Shaghayegh Engineering Co." for supporting this research financially and Spiritually.

REFERENCES

[1] M. Moshfegh, and H. Miar-Naimi, "High-efficiency broadband class-e pas for more than one octave: analysis and design," IEEE Trans. Microw. Theory Tech., vol. 70, no. 7, pp. 3534–3547, July. 2022.

[2] Y.M.A. Latha, K. Rawat, "Design of ultra-wideband power amplifier based on extended resistive continuous class b/j mode," IEEE Trans Circuits Syst II ., vol. 69, no 2, pp. 419–423, July 2021.

[3] M. Chegini, J. Yavandhasani and M. Kamarei, "A new design for mode transfer-based harmonic tuned power amplifier (MHPA)," AEU - Int J Electron Commun 2022., vol. 155, 154335.

[4] S. Aghajani, M. Kamarei and M. Chegini, "High efficiency continuous class J/B power amplifier design with 130% Fractional Bandwidth," 2023 31th International. Conf. on Electr. Eng. ICEE 2023, to be published.

[5] M. Chegini, H. Nemati and M. Kamarei, "A new 10 watt power amplifier for gsm 900 mhz base stations with 44% bandwidth," 2023 31th International. Conf. on Electr. Eng. ICEE 2023, to be published.

[6] M. Chegini, H. Nemati and M. Kamarei, "A new 10 watt 1.6 ghz linear power amplifier with more than 11 db gain," 2023 31th International. Conf. on Electr. Eng. ICEE 2023, to be published.

[7] S. K. Dhar et al., "Investigation of input–output waveform engineered continuous inverse class f power amplifiers". IEEE Trans. Microw. Theory Techn., vol. 67, no.9, pp. 3547–3561, Jul. 2019.

[8] J.H. Kim et al., "A 2.4 GHz SiGe bipolar power amplifier with integrated diode linearizer for WLAN IEEE 802.11b/g applications," in Proc. IEEE Radio and Wireless Symposium, San Diego, CA, USA, 2006.

[9] A. Ali, S. W. Haider Shah and K. Iqbal, " Design of an Efficient Single-Stage and 2-Stages Class-E Power Amplifier (2.4GHz) for Internet-of-Things," in Proc. International Conference on Frontiers of Information Technology (FIT), Islamabad, Pakistan, 2018.

[10] A. Seidel, M. Kreißig and F. Ellinger, "An Ultra-Compact 0.17mm2 2.4GHz Low-Voltage Class-E Power Amplifier in 28nm CMOS," in Proc. IEEE-APS Topical Conference on Antennas and Propagation in Wireless Communications (APWC), Cartagena, Colombia, 2018.

[11] V. S. Girnale and H. B. Patil, "Design and performance analysis of 2.4 GHz power amplifier for wireless sensor network," in Proc. International Conference on Inventive Systems and Control (ICISC), Coimbatore, India, 2018.

[12] M. Gilasgar, A. Barlabe and L. Pradell, "A 2.4 GHz CMOS Class-F Power Amplifier With Reconfigurable Load-Impedance Matching," IEEE Trans Circuits Syst I ., vol. 66, no 1, pp. 31–42, July 2018.

[13] A. Sengupta, M. Kanti Mandal and R. Pal, "Design of a Class-B Power Amplifier at 2.4 GHz with Improved Harmonic Suppression," in Proc. IEEE MTT-S International Microwave and RF Conference (IMARC), Kolkata, India, 2018.

[14] Ashish Jindal et al., "1.5-2.5 GHz Measurement Based Power Amplifier Using SSPL GaN HEMT Device," in Proc. IEEE MTT-S International Microwave and RF Conference (IMARC), Mumbai, India, 2019.

[15] A. Giry et al., "A 2.5GHz LTE Doherty Power Amplifier in SOI-CMOS Technology," in Proc. 17th IEEE International New Circuits and Systems Conference (NEWCAS), Munich, Germany, 2019.

[16] M. N. Sasikanth and T. K. Bhattacharyya., "A High efficiency body injected differential power amplifier at 2.4GHz for low power applications," in Proc. 31st International Conf. on VLSI Design and 2018 17th International Conf. on Embedded Systems (VLSID), Pune, India, 2018.

[17] A. Afsahi, A. Behzad and L. E. Larson., "A 65nm CMOS 2.4GHz 31.5dBm Power Amplifier with a Distributed LC Power-Combining Network and Improved Linearization for WLAN Applications," in Proc. IEEE International Solid-State Circuits Conference - (ISSCC), San Francisco, CA, USA, 2010.

[18] A. Suzuki and S. Hara., "2.4GHz high efficiency GaN power amplifier using matching circuit less design," in Proc. 4th Australian Microwave Symposium (AMS), Sydney, NSW, Australia, 2020.

[19] P. Zurek, T. Cappello and Z. Popovic., "A Concurrent 2.2/3.9-GHz Dual-Band GaN Power Amplifier," in Proc. IEEE Topical Conference on RF/Microwave Power Amplifiers for Radio and Wireless Applications (PAWR), Orlando, FL, USA, 2019.

[20] B. M. Abdelrahman, H. N. Ahmed., "Triband 1.2/1.8/2.7 GHz Doherty Power Amplifier Using Novel Output Combining Network," in Proc. 47th European Microwave Conference (EuMC), Nuremberg, 2017.

Design and Performance Analysis of a Diplexer for Simultaneous GSM and Bluetooth Communication

1st Hamid Rahimpour
Physics & Accelerators Research School, Nuclear Science and Technology Research Institute
Tehran, Iran
hrahimpour@aeoi.org.ir

2nd Sajjad Mohammadian
research and development center, Sepahan Hamrah Commercial services development
Tehran, Iran
sajjad.mohammadian.86@gmail.com

3rd Reza Nemati
research and development center, Sepahan Hamrah Commercial services development
Tehran, Iran
reza_nemati@yahoo.com
rd.head@sepahanhamrah.com

Abstract— In this paper, we propose a novel diplexer structure connected to an antenna, enabling simultaneous operation of GSM/GPRS and Bluetooth bands. The compact design employs microstrip open-ended resonators to reject the Bluetooth band in the GSM/GPRS path and vice versa in the alternate path. The proposed structure is designed and simulated on a 10-mil FR-4 substrate. Measurement results confirm the functionality of the proposed diplexer, demonstrating insertion loss of less than 1 dB across all GSM/GPRS bands and providing at least 25 dB isolation from the Bluetooth band. By combining this structure with a dual-band antenna covering the GSM/GPRS/Bluetooth bands, we achieve improved range, directivity, and signal integrity. Additionally, this approach eliminates antenna length mismatch problems that are typically encountered with conventional monopole antennas in the Bluetooth bands.

Keywords—Diplexer, GSM/GPRS, Bluetooth, open-ended resonator, microstrip filter, band-reject filter, antenna.

I. INTRODUCTION

Feature-phones, commonly known as basic mobile phones, are affordable and simple devices that offer essential mobile functionalities. These phones are usually targeted toward specific markets seeking cost-effective communication solutions. Feature-phones comprise various components, each serving specific functions such as integrated on-chip processors, RF front-end, user interface peripherals, batteries, and antennas [1], [2]. Bluetooth antennas in feature-phones are usually connected separately from the GSM antenna through a long-wired monopole antenna [3]. Using monopole antennas in older mobile phones has been associated with limitations and challenges, including bandwidth constraints and limited radiation patterns [3].

In recent years, integrating separate antennas for different functionalities, such as GSM, Bluetooth, and FM radio, has posed challenges and drawbacks [4]. The need to accommodate multiple antennas within small devices like feature-phones can lead to increased complexity, larger form factors, and potential interference issues [5]. One approach to achieving this integration involves the utilization of multi-band antennas, designed to operate efficiently across a multitude of frequency bands. In this context, a novel design of an electrically small multi-band antenna, complete with matching circuits tailored for mobile terminals, is proposed [6], [7]. This design aims to cover a wide range of frequency bands, including GSM, LTE, DCS, and more.

Employing a diplexer is essential for sharing the same antenna among multiple signal paths, enabling the simultaneous use of multiple communication systems without significant interference [8]. Combining different antenna functionalities into a single multi-band solution and utilizing diplexers to share a common antenna path are approaches that have been explored to address the challenges associated with separate antennas for GSM, Bluetooth, and other applications. Recently, various types of diplexers have been introduced to facilitate sharing the same antenna for multi-standard applications[9]–[11].

In this paper, a single antenna is employed for multiple standards such as FM, GSM, GPRS, and Bluetooth. A simulated multi-band inverted-F antenna is designed to cover all the mentioned standards. We propose a novel, compact-sized, and cost-effective diplexer to integrate GSM/GPRS, FM, and Bluetooth functionalities using simple microstrip filters. The detailed design will be discussed in the upcoming sections.

II. CONVENTIONAL STRUCTURE OF GSM/GPRS CELLULAR PHONE

The conventional structure of the RF front-end in a cellular phone is illustrated in Fig. 1 [12]. Based on this figure, it is evident that a basic feature-phone employs a unified antenna for both FM and GSM/GPRS bands, while a distinct monopole antenna serves the purpose of Bluetooth connectivity [4]. This unified antenna approach allows users to enjoy FM radio without needing to connect a headphone cable. However, in the Bluetooth path, the utilization of a wire as an antenna brings challenges such as length mismatches and assembly issues. Conversely, it is worth noting that the

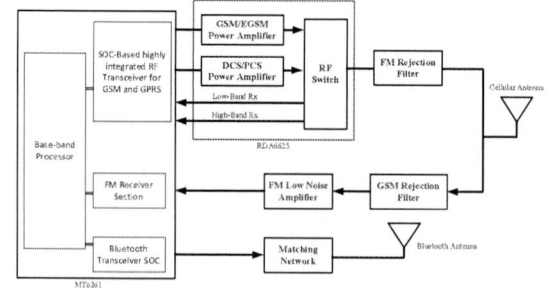

Fig. 1. Conventional structure of a cellular feature-phone.

The 5th Iranian International Conference on Microelectronics (IICM2023)

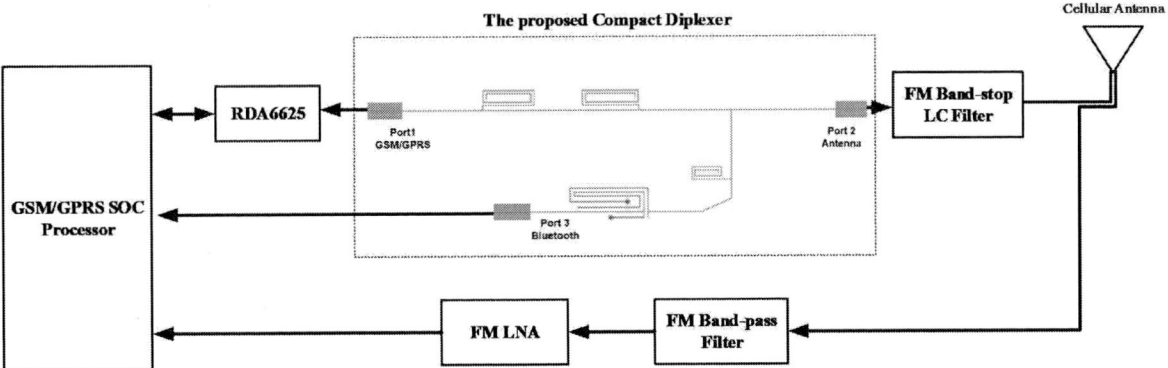

Fig. 2. The proposed block diagram of mobile cell-phone based on the novel diplexer.

antenna gain, pattern, and directivity of this arrangement are not particularly significant when compared to the conventional GSM/GPRS antenna [3].

To enhance the isolation between the FM receiver and the GSM/GPRS transceiver, a series LC band-stop filter with a center frequency of 97MHz is employed, effectively rejecting the FM band within the GSM/GPRS path. Similarly, a parallel LC tank is employed within the FM path to reject unwanted frequency bands. Despite these measures, separate antennas are still utilized in these-feature phones. A System on Chip (SoC) based processor is utilized to handle the whole function of all peripherals. This processor has different parts consisting of a GSM/GPRS transceiver, FM Tuner, Bluetooth transceiver, charge circuit, and audio parts.

While integrating separate antenna functionalities offers advantages such as reduced space requirements and improved cost-effectiveness, significant technical challenges must be addressed. Ensuring adequate isolation between combined bands, minimizing interference, and maintaining efficient performance across all bands are key design considerations. These issues are comprehensively addressed in the proposed design, which leverages several microstrip resonators and filters.

III. THE PROPOSED STRUCTURE OF DIPLEXER-BASED MOBILE RF FRONT-END

The block diagram of the feature-phone with a single antenna for multiple standards is depicted in Fig. 2. Notably, a novel diplexer is employed to connect both Bluetooth and GSM/GPRS transceivers to the same antenna. It is important to highlight that the feature-phone's antenna has been subjected to a detailed study, resulting in the selection of a suitable structure capable of covering the desired bandwidths.

According to Fig. 2, the FM receiver is connected to one port of the antenna, while the other port is connected to the sum port of the diplexer. To enhance isolation between these two antenna ports, a band-pass filter for the FM band is integrated into the FM path, with a corresponding band-reject filter placed in the other port. The critical isolation between the Bluetooth and GSM/GPRS paths is achieved through the use of the proposed diplexer.

As explained before, the primary purpose of utilizing a diplexer in Fig. 2 is to remove an extra antenna for Bluetooth. The proposed diplexer consists of a band reject filter in one path and an inter-digital bandpass filter in the other path. For

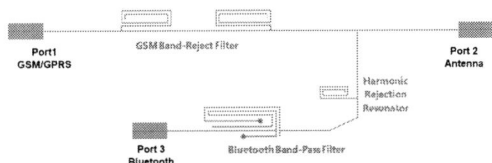

Fig. 3. The proposed PCB layout of the compact diplexer.

designing a band-reject filter adding several open-ended resonators to the microstrip path is one way to shape the frequency bandwidth [13]. This method is used in this paper to reject the Bluetooth band in the GSM/GPRS path (Port 1 and Port 2 of Fig. 1). Also, a semi-interdigital bandpass filter is designed for the other path to only pass the Bluetooth bandwidth (Port 2 and Port 3) [14]. As an interdigital bandpass filter suffers from a second harmonic rejection issue, an open-ended resonator at a center frequency of 5GHz is added to improve the out-of-band rejection.

The PCB layout of this innovative diplexer is shown in Fig. 3. As per this figure, two open-loop resonators are implemented to reject Bluetooth frequency bands from 2.2 to 2.6 GHz. It is essential to maintain the GSM/GPRS insertion loss below 0.5 and 3.5 dB at lower and upper bands respectively to achieve the desired maximum transmit power. According to Fig. 3, the shared antenna should be connected to Port 2 and the GSM and Bluetooth path should be connected to Port 1 and Port 3, respectively.

IV. SIMULATION AND MEASUREMENT RESULTS

The proposed structure is simulated by the Advanced Design Systems simulator on a 10-mil FR-4 substrate. Electromagnetic simulations were conducted to account for all parasitic effects of the substrate, ensuring the design's validity prior to fabrication.

A triple-band inverted-F antenna (IFA) was chosen for its compact size, making it a prevalent choice in feature-phone applications. Due to its reduced size and compact nature, the IFA was selected as the internal handset antenna for sharing multiple standards. The layout of the employed antenna, along with the simulated radiation pattern, is depicted in Fig. 4 and Fig. 5, respectively. This proposed antenna effectively covers

979-8-3503-6020-2/23 $31.00 © 2023 IEEE

four frequency bands, encompassing FM, GSM, GPRS, and Bluetooth.

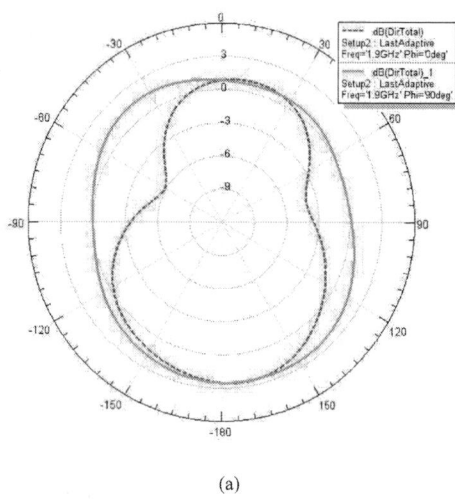

Fig. 4. The model structure of triple-band feature-phone antenna.

E-plane and H-plane cuts of the antenna radiation pattern at the GSM band is shown in Fig. 5.

The simulated frequency response of the GSM/GPRS band reject filter is illustrated in Fig. 6. As previously explained, this filter is designed to reject the Bluetooth bandwidth. According to the figure, the proposed filter effectively rejects the frequency range of 2.2 to 2.6 GHz by more than 20 dB. Additionally, the insertion loss for the GSM/GPRS bandwidth is shown to be less than 1 dB for both the mobile's lower and upper bands, complying with GSM performance standard requirements.

The frequency response of the Bluetooth band-pass filter is depicted in Fig. 7. In this case, the insertion loss for the Bluetooth frequency range of 2.4 to 2.5 GHz is less than 3 dB. Furthermore, the proposed filter exhibits strong rejection of GSM and GPRS bands, surpassing 25 dB. As shown in Fig. 3, a complementary resonator is employed to reject the second harmonic of the interdigital band-pass filter. With a center frequency around 5 GHz, this resonator effectively eliminates any harmonics of the filter up to 6 GHz. Additionally, the proposed filter demonstrates rejection levels exceeding 20 dB for the GSM and GPRS bands.

The fabricated PCB of the proposed diplexer and the measurement results are displayed in Fig. 9, Fig. 10, and Fig. 11, respectively. According to these figures, the difference between design and simulation results is less than 80MHz and all design requirements are met for the desired application. Notably, the overall size of the diplexer is compact enough to be seamlessly integrated with a mobile phone motherboard, without incurring any excessive costs.

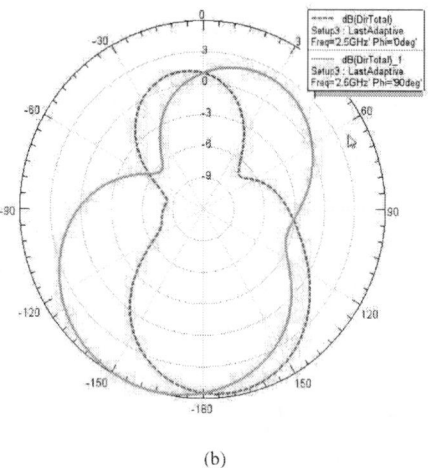

(a)

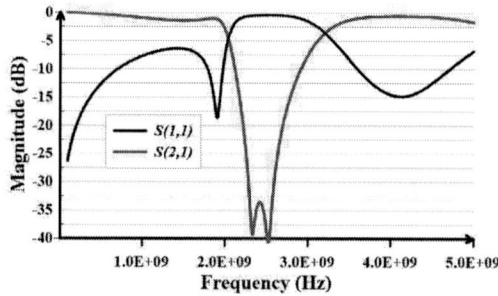

Fig. 6. Simulated frequency response of the GSM/GPRS path filter.

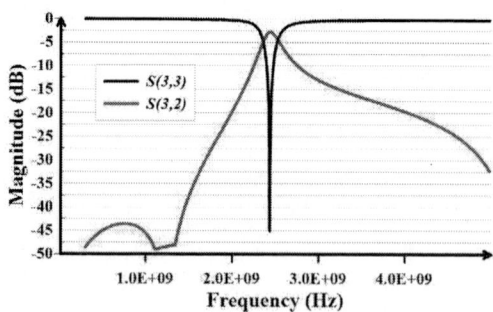

Fig. 7. Simulated frequency response of the Bluetooth band-pass filter.

(b)

Fig. 5. Radiation pattern of the IFA antenna in GSM band.

The 5th Iranian International Conference on Microelectronics (IICM2023)

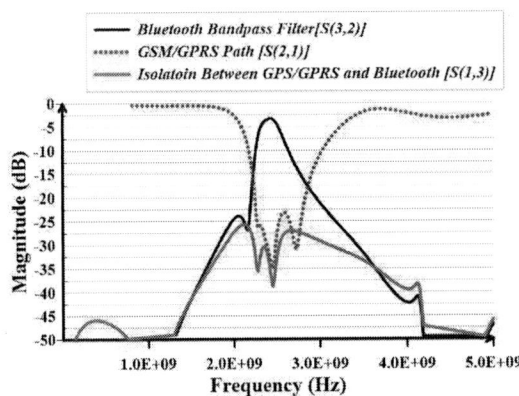

Fig. 8. Simulated frequency response of the proposed diplexer.

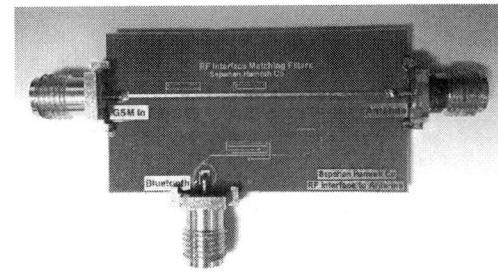

Fig. 9. Fabricated PCB layout of the proposed diplexer on 10mil FR-4 Substrate.

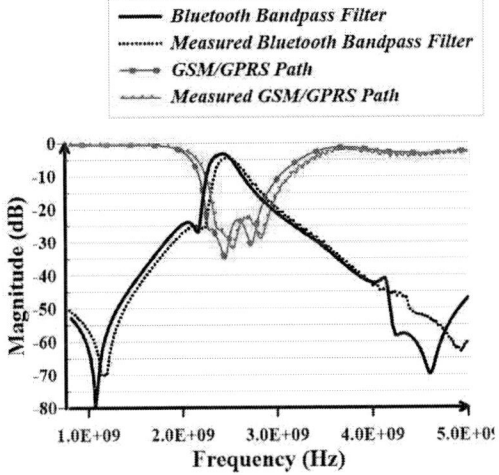

Fig. 10. Measured frequency response of the Bluetooth band-pass filter and GSM/GPRS path.

Bluetooth standards. The proposed diplexer comprises a band-reject filter and a band-pass filter. Based on the fabrication results, the proposed structure exhibits an insertion loss of less than 1 dB and an isolation of over 25 dB between the GSM and Bluetooth ports.

Table I shows a comparison result of the proposed diplexer with other similar works in almost the same frequency bandwidth.

TABLE I. Comparison with Similar Works

	Dimentions (mm²)	Maximum Insertion loss (dB)	Out of Band Rejection (dB)	Min Isolation (dB)	Year
[15]	70*80	<2	>25	-	2018
[9]	~55*42	<3	>15	20	2018
[11]	45.2*25	<3.5	>30	28	2007
[8]	~ 45*30	<4	>30	30	2019
This Work	26*15	<2	>30	25	2023

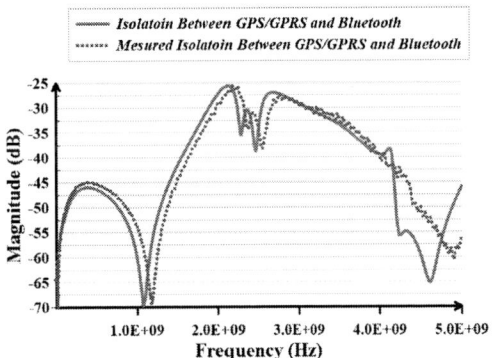

Fig. 11. Measured and simulated frequency response of the isolation between Bluetooth and GSM/GPRS.

VI. ACKNOWLEDGMENT

The authors intend to sincerely thank the management of Sepahan Hamrah Commercial Services Development, for his unparalleled support and assistance to their research endeavors. They are thankful for his facilitation of laboratory testing facilities and equipment throughout their research journey. His collaborations have significantly contributed to the successful execution of this research.

V. CONCLUSION

is to integrate multiple standard bands for use in feature-phones, In this paper, we have designed, simulated, and fabricated a novel, compact-sized diplexer. The primary purpose of the proposed diplexer streamlining them through a single antenna. Additionally, a multi-band inverted-F antenna was simulated, effectively covering FM, GSM, GPRS, and

REFERENCES

[1] S. Dessai and K. Ramakrishna, "Implementation of Video Capture and Playback in Mobile Systems," *International Journal of Reconfigurable and Embedded Systems (IJRES)*, vol. 2, no. 3, Nov. 2013, doi: 10.11591/ijres.v2i3.4028.

[2] M. Tiits and T. Kalvet, "Nordic small countries in the global high-tech value chains: the case of telecommunications systems production in Estonia," *The Other Canon Foundation and Tallinn University of Technology Working Papers in Technology Governance and Economic Dynamics*, 2012, Accessed: Aug. 07, 2023. [Online]. Available: https://ideas.repec.org/p/tth/wpaper/38.html

[3] Z. Zhang, "Antenna Design for Mobile Devices, 2nd Edition | Wiley." https://www.wiley.com/en-in/Antenna+Design+for+Mobile+Devices%2C+2nd+Edition-p-9781119132325 (accessed Aug. 06, 2023).

[4] A. Baschirotto *et al.*, "Baseband analog front-end and digital back-end for reconfigurable multi-standard terminals," *IEEE Circuits and Systems Magazine*, vol. 6, no. 1, pp. 8–28, 2006, doi: 10.1109/MCAS.2006.1607635.

[5] M. Brandolini, P. Rossi, D. Manstretta, and F. Svelto, "Toward multistandard mobile terminals - Fully integrated receivers requirements and architectures," *IEEE Trans Microw Theory Tech*, vol. 53, no. 3 II, pp. 1026–1038, Mar. 2005, doi: 10.1109/TMTT.2005.843505.

[6] A. Cihangir, F. Ferrero, C. Luxey, and G. Jacquemod, "A novel multi-band antenna design with matching network for use in mobile terminals," in *2012 6th European Conference on Antennas and Propagation (EUCAP)*, IEEE, Mar. 2012, pp. 1667–1671. doi: 10.1109/EuCAP.2012.6206603.

[7] H. Rhyu, Byungwoon Jung, F. J. Harackiewicz, and Byungje Lee, "Design of a Multi-Band Internal Antenna Using an Open Stub," in *2005 Asia-Pacific Microwave Conference Proceedings*, IEEE, 2005, pp. 1–4. doi: 10.1109/APMC.2005.1606771.

[8] X. Guan, P. Gui, W. Huang, S. Xie, B. Ren, and X. Zhang, "Novel Reconfigurable Diplexers Based on Short-Circuited Stub-Loaded Resonator," *2019 International Conference on Microwave and Millimeter Wave Technology, ICMMT 2019 - Proceedings*, May 2019, doi: 10.1109/ICMMT45702.2019.8992231.

[9] N. Pandit, R. K. Jaiswal, and N. Prasad Pathak, "A compact dual mode diplexer for wireless applications," *IEEE MTT-S International Microwave and RF Conference, IMaRC 2018*, Nov. 2018, doi: 10.1109/IMARC.2018.8877242.

[10] K. H. Li, C. W. Wang, and C. F. Yang, "A miniaturized diplexer using planar artificial transmission lines for GSM/DCS applications," *Asia-Pacific Microwave Conference Proceedings, APMC*, 2007, doi: 10.1109/APMC.2007.4555114.

[11] S. T. G. Bezerra and M. T. De Melo, "Microstrip diplexer for GSM and UMTS integration using ended stub resonators," *SBMO/IEEE MTT-S International Microwave and Optoelectronics Conference Proceedings*, pp. 954–958, 2007, doi: 10.1109/IMOC.2007.4404413.

[12] N. Ghittoril, A. Vignal, P. Malcovati2, S. D'amico, and A. Baschirotto, "Analog Baseband Channel for GSM/UMTS/,WLAN/Bluetooth Reconfigurable Multistandard Terminals."

[13] H. Rahimpour and N. Masoumi, "High-Resolution Frequency Discriminator for Instantaneous Frequency Measurement Subsystem," *IEEE Trans Instrum Meas*, vol. 67, no. 10, pp. 2373–2381, Oct. 2018, doi: 10.1109/TIM.2018.2816804.

[14] R. Divya Bharathi, J. Evangeline Yamini, A. Evangeline, and D. Badri Narayanan, "Design and analysis of interdigital microstrip bandpass filter for centre frequency 2.4 GHz," *ICONSTEM 2017 - Proceedings: 3rd IEEE International Conference on Science Technology, Engineering and Management*, vol. 2018-January, pp. 930–933, Jun. 2017, doi: 10.1109/ICONSTEM.2017.8261339.

[15] D. J. Simpson, R. Gomez-Garcia, and D. Psychogiou, "Planar RF Duplexer with Multiple Levels of Transfer-Function Reconfigurability," in *2018 48th European Microwave Conference (EuMC)*, IEEE, Sep. 2018, pp. 535–538. doi: 10.23919/EuMC.2018.8541603.

979-8-3503-6020-2/23 $31.00 © 2023 IEEE

The 5th Iranian International Conference on Microelectronics (IICM2023)

25 – 26 October 2023

High-level synthesis-based approach for CNN acceleration on FPGA

Adib Hosseiny
*Department of Electronics and
Communication Engineering
University of Kurdistan*
Sanandaj, Iran
0009-0002-6142-0431
adib.hosseiny@uok.ac.ir

Hadi Jahanirad
*Department of Electronics and
Communication Engineering
University of Kurdistan*
Sanandaj, Iran
0000-0001-8586-6281
h.jahanirad@uok.ac.ir

Abstract— **This paper presents a comprehensive approach to implementing Convolutional Neural Networks (CNNs) on Field-Programmable Gate Arrays (FPGAs). CNNs have become a cornerstone in numerous fields, enabling breakthroughs in areas such as computer vision, natural language processing, and speech recognition. CNNs comprise multiple layers designed to perform various computations. In this research, we propose a general methodology using High-level synthesis(HLS) tools for implementing CNNs on FPGAs and provide several use cases demonstrating competitive FPGA resource utilization in comparison to state-of-the-art works. Our experimental results demonstrate a significant reduction in resource utilization for DPS units, amounting to approximately 80% when compared to other neural network accelerators deployed on FPGAs. Furthermore, we have accomplished a noteworthy 50% reduction in Look-Up Table (LUT) usage compared to alternative accelerators, alongside an overall superior performance in comparison to CPU or GPU implementations.**

Keywords—CNN, FPGA, HLS, Deep learning

I. INTRODUCTION

In the realm of object detection, a plethora of algorithms have been employed, including R-CNN (region-based convolutional neural networks), You Only Look Once (YOLO)[15,16,17], Spatial Pyramid Pooling Network (SPP) [20], as well as other R-CNN-based approaches like Fast R-CNN [22], and adapted variants such as Faster R-CNN [23] and other network structures like DenseNet and cross-stage partial networks in [18,19]. Distinguishing itself from the rest, YOLO leverages a solitary CNN to anticipate both the bounding boxes encircling objects and their associated probabilities of being an object and belonging to a specific class. Numerous research endeavors have been undertaken to integrate diverse object detection and deep learning algorithms into embedded platforms, with a particular focus on FPGAs [1-7,24].

In the current investigation, we followed a specific form of CNN to elucidate the manifold attributes of a CNN and

thoroughly examine each constituent during the implementation phase. We opted for object detection neural networks as the foundational CNN topology due to its inclusion of diverse mathematical operations and a multitude of network layers, thereby providing a rigorous test for the computational capabilities of FPGAs.

As per [12], when considering various network topologies for object detection, it is evident that the YOLO series exhibits superior suitability for real-time object detection algorithms compared to other alternatives. The implementation of YOLO in FPGA systems proves to be significantly more feasible in contrast to other algorithms due to the presence of numerous recursive function calls within the network structure. These recursive calls contribute to computationally expensive operations, particularly in terms of memory management and addressing the substantial memory demands inherent to such architectures.

In the neural network inference stage, numerous read/write operations and memory accesses occur. As stated in a previous study [8], an optimal read/write operation entails the synthesized Register Transfer Level (RTL) commencing the writing process immediately after the reading operations, without requiring any interruptions for write permissions. CNNs possess a sequential attribute, wherein each input feature map slice undergoes computations within a layer, and the resulting output becomes the input for the subsequent layer. This inherent sequential behavior is a fundamental aspect of CNN design.

Nevertheless, [9] utilizes a streaming data methodology to ascertain the movement of the sliding cube across network layers and the acceptance of new inputs by each layer. When designing such hardware, it is essential to account for the latencies associated with various layers within the network, ensuring that the requirements of each layer (such as latency, resource utilization, etc.) are adequately fulfilled. In a related approach, the technique proposed in [10] adopts a fixed-point implementation to accelerate the execution of the compact YOLOv2 algorithm.

979-8-3503-6020-2/23 $31.00 © 2023 IEEE

The 5ᵗʰ Iranian International Conference on Microelectronics (IICM2023)

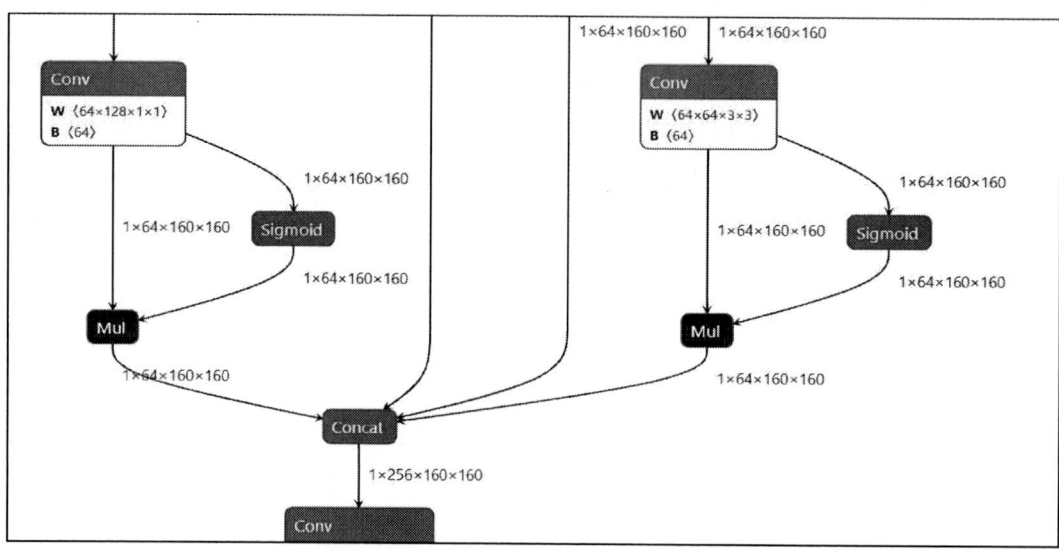

Fig. 1.A section of a neural network for object detection shown in Netron

The adoption of this methodology possesses the capability to reduce resource consumption. Nonetheless, it is crucial to acknowledge that this advantage is accompanied by a trade-off in algorithmic precision. Alternative methodologies, such as the approach described in a previous study [11], employ quantization techniques to optimize activation function outputs and convolutional layer calculations.

In the realm of accelerating CNNs on FPGAs, both HLS and RTL approaches exhibit distinct merits and compromises. HLS offers expedited development time and user-friendliness, rendering it suitable for swift prototyping endeavors and the exploration of diverse algorithmic designs. Moreover, HLS facilitates the FPGA acceleration of CNNs even for applications with high computational complexity and huge memory requirements. Conversely, RTL design confers finer-grained control over hardware implementation, enabling tailored optimizations geared towards the specific CNN model or target FPGA device. Consequently, HLS design is frequently employed for applications with high memory requirements and for the deployment of neural networks or situations where maximum performance and low development time are prerequisites. Therefore, for the deep structure of neural networks, HLS design can be a more practical way of implementation.

The architecture proposed in reference [10] employs a scaled version of the YOLO family, which introduces a General Matrix Multiplication (GeMM) core designed within an OpenCL environment. Utilizing quantization significantly enhances the network's speed and resource efficiency compared to networks employing 16-bit or 32-bit floating point data types. However, adopting quantized neural networks necessitates modifications during the training phase, including training with quantized values for activation functions and other layers. This adjustment is crucial since the network's structure remains fixed during the inference stage, and inadequate design choices can result in accuracy degradation.

II. PROPOSED METHOD

Initially, the network undergoes training utilizing the Python programming language for coding and training the network. Subsequently, the resulting model file is saved with a ".pt" extension, denoting a Pytorch model file that encompasses the entirety of the trained model's attributes and data. Following this, the Pytorch file is converted into an open neural network exchange (ONNX) file format to facilitate a comprehensive examination of all model properties. To accomplish this task, we employ Netron, a dedicated tool designed for visualizing neural networks, deep learning, and machine learning models. For reference, a depiction showcasing the utilization of Netron software to visualize the model file is presented in Fig.1.

The ONNX-exported network model comprises several convolutional layers, consisting of different kernel-size convolutional layers. Additionally, there are multiple max-pooling layers with varying strides and paddings. Each convolutional layer is followed by an activation function. The network includes two up-scaling layers for up-sampling, as well as some concatenation layers that combine the preceding layer's feature maps. Reshape layers are employed to transform the output matrix dimensions of the previous layer into 2 dimensions, yielding class outputs and probabilities. Finally, three transpose layers are used to rearrange the matrix to achieve the desired format.

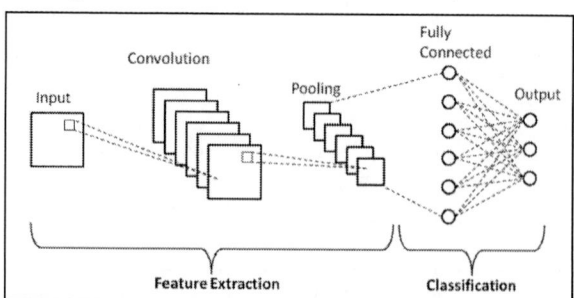

Fig. 2. A sample deep convolutional neural network structure

Transpose layers perform matrix reallocation on the received feature map. However, these layers are taken into account for total latency calculation in the HLS code, but their total latency is negligible according to those of convolutional layers. The exported Network ONNX model for our experiment on a YOLO network contains various types of layers. Upon exporting ONNX, the initial step in the design process involves simulating the network in HLS-C++ and verifying its functionality. Once the simulation is successful in Vitis-HLS, the subsequent step is to synthesize the HLS code, generating the desired hardware. This stage is the most intricate due to the scale of the network and the sequential processing of numerous layers to produce an output matrix. The output matrix corresponds to detecting 80 objects and determining their locations within the input image as the network is trained on a dataset for detecting those number of classes in the output. A sample convolutional neural network architecture is shown in Fig.2.

The dimensions of convolutional layers vary concerning their inputs, weights, and biases. Additionally, convolutional layers possess distinct properties such as padding, dilation, stride, and kernel size. These attributes are configured within the network structure during the training stage, with their respective values stored in the corresponding model file. Certain pooling layers in the Network can be executed concurrently due to the utilization of the same input feature map. Our implementation techniques for pooling layers in HLS involve shifting and obtaining the absolute value of the results, as well as defining helper functions like the maximum function and utilizing functions from the "hls_math" library.

III. DESIGN OPTIMIZATIONS

The network comprises distinct regions where a single tensor can be reused to store multiple layer outputs. This approach may lead to conflicts such as write-after-read (WAR) or read-after-write (RAW) when implementing certain loops within each layer using HLS. Nevertheless, reusing these tensors significantly reduces the memory requirements of the final design. Additionally, our findings indicate that it can slightly improve latency. The network's learned parameters are constant values that act as multiplication or addition factors in the design.

The HLS compiler treats them as read-only memory (ROM). However, storing learned parameters with 32-bit IEEE floating-point precision is not feasible due to the limited resources of the Programmable Logic (PL) component in the system-on-chip (SoC). To address this limitation, Vitis-HLS provides the "ap_fixed.h" library, which allows the placement of floating-point values in the FPGA using user-defined data type precisions. Leveraging this library facilitates the use of arbitrary precision datatypes within the HLS design and optimizes resource utilization, such as BRAM usage.

One fundamental strategy for the academic design of pooling and convolutional layers in HLS involves setting the iteration length of the loops to the total length, while simultaneously validating the indices to prevent any invalid values. An alternative solution is to reverse the kernel loops and utilize the absolute value of each index. This latter approach has demonstrated a substantial reduction in the overall latencies of the layers.

In general, when implementing neural network inference, we can employ arbitrary precision (AP) datatypes provided by Xilinx Vitis HLS to minimize the memory usage of the accelerator. The network parameters must be stored either in on-chip memories, such as registers or BRAMs or in off-chip memories, like DDR. By selecting an appropriate bit width for the network parameters, the overall memory requirements can be reduced in comparison to a 32-bit implementation on CPU or GPU platforms. However, reducing the bit-width of network parameters introduces the possibility of accuracy loss, as it affects all weight multiplications and bias additions. To determine the suitable bit-width, we first evaluate the design for the convolutional layers, which contain the weight and bias components. By assessing the accuracy loss in comparison to the original model, we can decide the trade-off between the accuracy and memory requirements of the design.

In the context of the HLS project, various levels of optimization can be implemented. The coding process for neural networks within the HLS environment primarily revolves around iterations, commonly known as loops. Different types of optimizations can be applied when iterating over these loops. Firstly, it is essential to determine whether the loop under consideration is perfect, meaning it is a nested loop where operations are confined to the innermost loop. For instance, in the case of convolutional layers, two operations are executed within a single iteration of the outer loop. One operation involves adding biases from the preceding layer, while the other operation entails multiplying the corresponding weights. The optimization of convolutional layers primarily focuses on loops involving operations.

When it comes to pipelining these loops, it is advisable to prioritize pipelining bias loops if an initiation interval (II) of 1 can be achieved. This preference stems from the fact that bias loops within convolutional layers solely consist of assignment and addition operations, which demand minimal resources like DSPs (Digital Signal Processors) compared to alternative methods involving loop unrolling. Conversely, for computational loops such as the multiplication of input feature maps with layer weights, it is recommended to unroll the loop using small factors. This strategy aims to minimize latency in the computation of the loop. By employing these optimization techniques, the HLS project can enhance the efficiency and performance of the neural network coding process within the HLS environment.

Various optimization techniques are specifically designed to improve the performance of computation-specific regions within a network. These regions can consist of different layers in the network. In the case of convolutional layers, we propose a set of design techniques in HLS to enhance the overall performance of these layers in the final design. In a CNN structure, there is an immediate activation layer following the convolutional layers. Thus, considering these two layers together for optimization techniques can be advantageous as it consumes fewer tensors for storing the immediate output feature maps.

A convolutional layer involves two computation loops; one for adding bias to the output feature map and another for multiplying the input feature map with the layer's weights. While there are other loops in the structure of a convolutional layer, such as iterating over all elements in the

Platform	Network	LUT	FF	BRAM	DSP	Power	Latency	
Zynq-7035	FPGA YOLO	47k	40k	787	409	7.518	-	[13]
ZCU102	Lightweight YOLOv2	135k	370k	1706	377	4.5	-	[14]
VC707	Sim-YOLOv2	155k	115k	1144	272	18.29	339	[10]
Zedboard	YOLOv3-tiny	25.9k	46.7k	185	160	3.36	532	[12]
XC7Z7045	Angel-Eye	182.6k	127.6k	486	780	9.63	-	[15]
XC7Z7100	YOLOv7-tiny	59k	9k	1092	16	10.79	15	This work

Table.1. Results for hardware acceleration of CNN on FPGAs

input feature map (a 3-dimensional matrix), the most critical loops for performance are the bias and weight loops. Combining the activation function with the bias loop increases the number of operations in the bias layer, while also potentially reducing the overall resource usage and latency of the layer. Since almost all convolutional layers in CNN structures have a non-zero padding value, it is essential to consider out-of-bounds memory accesses in HLS design. An effective implementation to address this issue is by utilizing the abs() function from the hls_math library, which can invert the kernel arguments and eliminate the need for out-of-bound checks in the kernel loops.

Another optimization technique for convolutional layers is the utilization of HLS pragmas such as pipeline and unroll. By considering the computation amount in a layer and the layer's characteristics (e.g., kernel size, padding, and stride), we can choose which pragmas to use for that specific layer. Pooling layers are another commonly used layer in state-of-the-art CNN structures. Generally, pooling layers have a zero padding value. However, in certain architectures, including the one we are studying for object detection, there are pooling loops with non-zero padding values. This leads to increased computational requirements for iterating over all input feature map values.

In CNNs, there are two major types of operations: layers with assignment operations and computation layers. While CNN operations can be classified based on various criteria, the most useful classification for HLS design is the one mentioned earlier. All loops within a pooling layer in a CNN can be classified as either assignment loops or simple iteration loops. Through our experiments with pooling layers in HLS design, we have found that the most efficient computation optimization pragma is pipelining with an initiation interval of 1. If a pooling layer does not have enough room for an initiation interval of 1, alternative design techniques should be considered. Other layers in the CNN structure, such as up-sampling and concatenation layers, are less computationally expensive compared to the aforementioned layers. Therefore, for these layers, we consider using HLS pragmas for memory management, such as alignment and flattening of the memory structure. Another crucial aspect to consider in HLS design is the presence of parallel-friendly regions within the network structure. As mentioned earlier, the overall structure of CNNs is sequential. However, these networks often include residual

or identity connections that contribute to the overall network performance.

In these regions, it is important to carefully allocate memory to the layers. For example, if there are two layers that both accept the same input feature map but are fed into different layers, running them in parallel can be beneficial. While HLS design does not directly support running two functions in parallel, we can achieve improved performance by allocating memory in a way that allows HLS to infer parallel execution. Additionally, reallocating tensors across different layers can be applied to reduce the overall memory requirement of the accelerator. Optimizing pooling layers involves a trade-off between resource usage and latency. When considering the unrolling of loops, the outermost loop requires more resources and power compared to the inner loops. If resource usage is not a concern, the outermost loop, iterating over each feature map, can also be unrolled with a factor of a quarter of the loop size. This approach is viable because the calculations for each slice of the feature map are independent of one another, striking a suitable balance between resource usage and latency. Consequently, unrolling the innermost loop proves to be an efficient optimization. However, it is important to note that unrolling loops in all layers with different factors is not the optimal choice for optimization.

In the Vitis HLS project, a pragma exists that enables the HLS compiler to pipeline loops with a trip count greater than the specified threshold [21]. This facilitates meeting the timing requirements of the design by incorporating user-defined optimizations. When the pipeline style is set to Free-Running Flushable Pipelining (FRP), it improves delay requirements and reduces pipeline control fanout. However, FRP leads to increased power consumption since pipeline registers are clocked even when no data is present [21].

Another approach mentioned in [21] is using the config_array_partition command, which instructs the HLS compiler to partition arrays in the design along any dimension to reduce resource usage and accelerate the design. The Vitis HLS compiler recommends avoiding recursive function calls and refraining from nesting function calls within other functions. This avoids significant design complexity and minimizes the time complexity associated with implementing each layer of the neural network. Annotating the output image with bounding boxes and class probabilities may not always be necessary. Although there is a slight latency involved in annotating the output image, if the application only requires a Boolean value indicating the presence of a specific object (e.g., a cat) in the image, annotation can be skipped, and a simple Boolean value can be passed to the next stage. Table 1 presents the implementation results.

IV. RESULTS

These results demonstrate favorable performance considering the network's size and memory requirements as shown in Table 1. The achieved latency of approximately 15ms in synthesis highlights the design's capability for real-time implementation. The overall performance of our design using XC7Z7100 is satisfactory for real-time object detection applications. Furthermore, the design's latency supports a 30-FPS input rate and even higher input rates. Additionally, [14] proposes a lightweight implementation of the YOLOv2 network on an FPGA using the more expensive ZCU102 board, with a less complex network. However, both accelerators in [10, 14] require more resources compared to our accelerator. Considering the reduced number of neural network parameters in scaled versions of YOLOv2 and YOLOv3 in [10, 13, 14], along with their corresponding resource usage, we can conclude that our proposed approach for Network on the XC7Z7100, which includes a more intricate network structure and a substantial number of parameters, can be successfully implemented for real-time applications on mid-budget SoCs such as XC7Z7045.

V. CONCLUSION

This paper presented a general methodology for implementing convolutional neural networks on FPGAs. By leveraging the parallel processing capabilities of FPGAs, the proposed approach enables efficient computation of different components of the network. The experimental results indicate competitive FPGA resource utilization and improved performance compared to existing works. The findings of this research contribute to the advancement of FPGA-based CNN implementations and provide valuable insights for researchers and practitioners in the field.

REFERENCES

[1] Shawahna, Ahmad, Sadiq M. Sait, and Aiman El-Maleh. "FPGA-based accelerators of deep learning networks for learning and classification: A review." IEEE Access 7 (2018): 7823-7859.

[2] Dias, Mauricio A., and Daniel AP Ferreira. "Deep learning in reconfigurable hardware: A survey." In 2019 IEEE International Parallel and Distributed Processing Symposium Workshops (IPDPSW), pp. 95-98. IEEE, 2019.

[3] El-Shafie, Al-Hussein A., and Serag ED Habib. "Survey on hardware implementations of visual object trackers." IET Image Processing 13, no. 6 (2019): 863-876.

[4] Wang, Jichen, Jun Lin, and Zhongfeng Wang. "Efficient hardware architectures for the deep convolutional neural network." IEEE Transactions on Circuits and Systems I: Regular Papers 65, no. 6 (2017): 1941-1953.

[5] Babu, Praveenkumar, and Eswaran Parthasarathy. "Hardware Acceleration of Image and Video Processing on Xilinx Zynq Platform." Intelligent Automation & Soft Computing 30, no. 3 (2021).

[6] Pestana, Daniel, Pedro R. Miranda, João D. Lopes, Rui P. Duarte, Mário P. Véstias, Horácio C. Neto, and José T. De Sousa. "A full-featured configurable accelerator for object detection with YOLO." IEEE Access 9 (2021): 75864-75877.

[7] Babu, Praveenkumar, and Eswaran Parthasarathy. "Optimized object detection method for FPGA implementation." In 2021 Sixth International Conference on Wireless Communications, Signal Processing and Networking (WiSPNET), pp. 72-74. IEEE, 2021.

[8] Zeng, Kai, Qian Ma, Jia Wen Wu, Zhe Chen, Tao Shen, and Chenggang Yan. "FPGA-based accelerator for object detection: a comprehensive survey." The Journal of Supercomputing 78, no. 12 (2022): 14096-14136.

[9] Nguyen, Duy Thanh, Tuan Nghia Nguyen, Hyun Kim, and Hyuk-Jae Lee. "A high-throughput and power-efficient FPGA implementation of YOLO CNN for object detection." IEEE Transactions on Very Large Scale Integration (VLSI) Systems 27, no. 8 (2019): 1861-1873.

[10] Yap, June Wai, Zulkalnain bin Mohd Yussof, Sani Irwan bin Salim, and Kim Chuan Lim. "Fixed point implementation of tiny-YOLO-v2 using OpenCL on FPGA." International Journal of Advanced Computer Science and Applications 9, no. 10 (2018).

[11] Günay, Bestami, Sefa Burak Okcu, and Hasan Şakir Bilge. "LPYOLO: Low Precision YOLO for Face Detection on FPGA." arXiv preprint arXiv:2207.10482 (2022).

[12] Nguyen, Duy Thanh, Tuan Nghia Nguyen, Hyun Kim, and Hyuk-Jae Lee. "A high-throughput and power-efficient FPGA implementation of YOLO CNN for object detection." IEEE Transactions on Very Large Scale Integration (VLSI) Systems 27, no. 8 (2019): 1861-1873.

[13] Wei, Guangju, Yanzhao Hou, Qimei Cui, Gang Deng, Xiaofeng Tao, and Yuan Yao. "YOLO acceleration using FPGA architecture." In 2018 IEEE/CIC International Conference on Communications in China (ICCC), pp. 734-735. IEEE, 2018.

[14] Nakahara, Hiroki, Haruyoshi Yonekawa, Tomoya Fujii, and Shimpei Sato. "A lightweight YOLOv2: A binarized CNN with a parallel support vector regression for an FPGA." In Proceedings of the 2018 ACM/SIGDA International Symposium on field-programmable gate arrays, pp. 31-40. 2018.

[15] Redmon, Joseph, and Ali Farhadi. "Yolov3: An incremental improvement." arXiv preprint arXiv:1804.02767 (2018).

[16] Bochkovskiy, Alexey, Chien-Yao Wang, and Hong-Yuan Mark Liao. "Yolov4: Optimal speed and accuracy of object detection." arXiv preprint arXiv:2004.10934 (2020).

[17] Wang, Chien-Yao, Alexey Bochkovskiy, and Hong-Yuan Mark Liao. "YOLOv7: Trainable bag-of-freebies sets new state-of-the-art for real-time object detectors." arXiv preprint arXiv:2207.02696 (2022).

[18] Wang, Chien-Yao, Hong-Yuan Mark Liao, Yueh-Hua Wu, Ping-Yang Chen, Jun-Wei Hsieh, and I-Hau Yeh. "CSPNet: A new backbone that can enhance the learning capability of CNN." In Proceedings of the IEEE/CVF conference on computer vision and pattern recognition workshops, pp. 390-391. 2020.

[19] Huang, Gao, Zhuang Liu, Laurens Van Der Maaten, and Kilian Q. Weinberger. "Densely connected convolutional networks." In Proceedings of the IEEE conference on computer vision and pattern recognition, pp. 4700-4708. 2017.

[20] He, Kaiming, Xiangyu Zhang, Shaoqing Ren, and Jian Sun. "Spatial pyramid pooling in deep convolutional networks for visual recognition." IEEE transactions on pattern analysis and machine intelligence 37, no. 9 (2015): 1904-1916.

[21] Xilinx. "Vitis High-Level Synthesis User Guide." (2022).

[22] Girshick, Ross. "Fast r-cnn." In Proceedings of the IEEE international conference on computer vision, pp. 1440-1448. 2015.

[23] Ren, Shaoqing, Kaiming He, Ross Girshick, and Jian Sun. "Faster r-cnn: Towards real-time object detection with region proposal networks." Advances in neural information processing systems 28 (2015).

[24] Hosseiny, A., Jahanirad, H. Hardware acceleration of YOLOv7-tiny using high-level synthesis tools. J Real-Time Image Proc 20, 75 (2023). https://doi.org/10.1007/s11554-023-01324-5.

The 5th Iranian International Conference on Microelectronics (IICM2023)

Neural networks & logistic regression for FPGA hardware Trojan detection

Milad Pazira
Faculty of Electrical Engineering
Noshirvani University of Technology
Babol,Iran
Miladpazira1994@gmail.com

Yasser Baleghi
Faculty of Electrical Engineering
Noshirvani University of Technology
Babol,Iran
y.baleghi@nit.ac.ir

Mohammad-Ali Mahmoodpour
Faculty of Electrical Engineering
Shahid Beheshti University
Tehran, Iran
m.mahmoodpour@alumni.sbu.ac.ir

Hossein Jafari
Faculty of Electrical Engineering
Shahid Beheshti University
Tehran, Iran
hos.jafari@mail.sbu.ac.ir

Abstract—**A hardware Trojan is a malware of increasing importance due to the increase in the growing number of digital circuits. A hardware Trojan can enter the circuit at any stage of chip manufacturing. The first step to deal with this malware is to detect the presence of this withering factor inside an integrated circuit. Since the appearance of this malware, various methods including thermal image processing have been proposed to detect hardware Trojans. The first challenge for research in this field is the lack of a database available for thermal images of Trojan chips. Accordingly, different images were taken from a Trojan-affected FPGA using a T4 thermal camera in this study. Our database includes 12 series of thermal images which are captured for each chip 55 seconds after programming. Then, two different methods have been proposed to detect hardware Trojans in the created database. In this paper, we propose a thermal image processing-based Hardware Trojan detection method on FPGA chips using neural networks, assuming the availability of a golden chip. Results demonstrate that if our method is combined with the previous method, the detection rate can be increased significantly.**

Keywords— Hardware Trojan, fault detection, thermal image processing, FPGA, Neural Networks

I. INTRODUCTION

Hardware Trojan (HT) is one of the most important malwares that exists in digital circuits which degrades or alters the performance of integrated circuits due to human intervention or error. The HT can be introduced during different stages like designing, fabrication process, etc. The only way to prevent the entry of this malware into a circuit is to utilize the designing ability of trusted designers in reliable factories. Hardware Trojans cause modification in the functionality and specification of digital circuits, denial of service, and leakage of information [1-3]. To deal with this malware, a method should be proposed to prevent the entry of such error or an extra circuit in the chip to prevent undesired factors [4]. Since the appearance of this issue, many detection methods have been proposed which are generally categorized into two classes: functional testing and side-channel effects. In the functional testing manner, input terminals of a chip are triggered and the output patterns will be monitored until

fabrication errors are detected. If logical output values do not match with the original pattern, a defect in the chip function or the occurrence of a hardware Trojan can be detected. Side-channel analysis methods can use side-channel signals, including latency and power [5,6]. One of the ways to detect Trojans by side-channel analysis is to use heat emitted from the chip. In the proposed method, we utilized a classification network that is trained by thermal images obtained from digital circuits in working mode. The following sections are as follows: section II contains related works in hardware Trojan detection, and section III, defines the formation of our dataset which contains thermal images. In section IV our proposed method will be introduced in detail. The experimental results are demonstrated in terms of accuracy metric in section V.

II. RELATED WORKS

Secure and reliable hardware has been a very important research topic in recent years. In 2007, researchers published a paper on the detection of Trojans using fingerprints in integrated circuits [7]. In 2008, Wang et al. provided a comprehensive overview on the classification of hardware Trojans, based on physical characteristics, excitation circuit type, and malware circuit performance [8]. In method [5], hardware Trojan detection is performed using the delay parameter. They used the DES Core circuit for their test, which had 64 outputs. The researchers applied 163 test patterns to the inputs and measured 64 x 163 delays. Then, the 10432 obtained latencies were reduced to three values using PCA. Finally, they claimed that these three values could separate healthy and Trojan circuits. To substantiate their claim, they placed Trojan chips in different locations, and the results showed success in detecting Trojans. In 2014, Nowroz, et al [6], presented a new method of detecting hardware Trojans based on thermal images [9] from electronic chips. Their paper proposed two supervised and unsupervised learning methods for detecting Trojans. In their supervised manner, the Euclidean distance of ten large values of eigenvalues between the two golden and the test chips can indicate the presence or absence of hardware Trojans on the ICs. In their unsupervised method, it is first necessary to map

979-8-3503-6020-2/23 $31.00 © 2023 IEEE

the thermal images into a power consumption map [10] by a transformation matrix. Then, the difference between the power consumption map of the two golden and the test chips is known as the residual power consumption map. Using PCA, two features of this mapping have been obtained and with the help of the DBSCAN clustering method, hardware Trojans will be detectable.

In 2016, an article [11] explored hardware Trojan detection using power analysis, which eventually led to Trojan detection. A practical method for detecting hardware Trojans using electromagnetic radiation is presented in [12]. Also approaches to increase the detection sensitivity of hardware Trojans using the side-channel are reviewed in [12]. It provides a general method for modeling noise to detect Trojans. In 2018, Zhong et al. [13] proposed a method for analyzing Trojan occurrence behavior on a chip. They utilized a NAND gate on a 50 MHz clock frequency to model the Trojan. After using a Kalman filter to decrease the noise effects, they used two values of the thermal difference between the two golden and test chips at different time intervals, and their results showed that the proposed method was successful for detecting Trojans.

III. DATASET FORMATION

In the present approach, we attempted to find hardware Trojans by using thermal images. Due to the lack of a standard database with thermal images of circuits with and without Trojans, we tried to create this database using T4 FLIR thermal cameras. This camera has a thermal resolution of 160 x 120. We used 4 hardware Trojan benchmarking models available on the valid trust-hub website to which Basic-RSA circuits are prone so that we can use them to build our database and model hardware Trojan. As we know, RSA algorithm is one of the first public key encryption methods that is widely used to secure data transmission, These 4 types of Trojans can create destructive effects on the performance of this algorithm. We will have 5 circuits consisting of 4 circuits with Trojans and one healthy circuit. Each circuit is run for 55 seconds, and thermal images are taken at 5-second intervals until we have 12 images. These experiments were repeated 8 times for each circuit to increase reliability which eventually reached 40 repetitions.

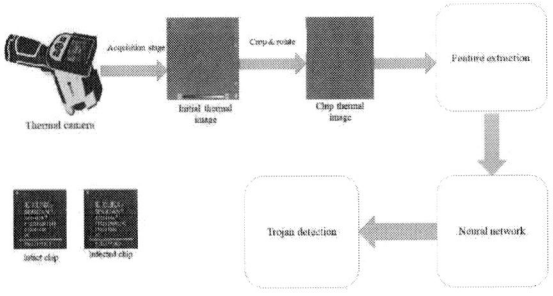

Fig. 1. Flowchart of proposed method

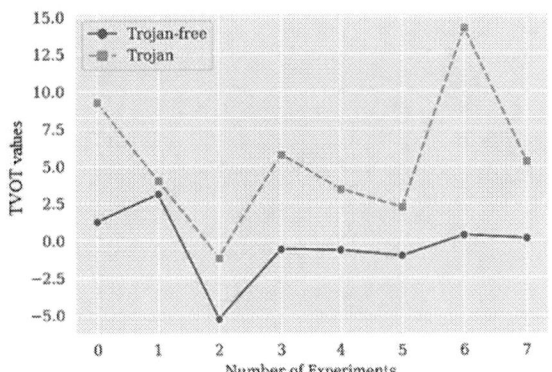

Fig 2. TVOT values in free and t100 basic-rsa circuit

IV. PROPOSED METHOD

In this experiment, we sought to find features to distinguish between healthy and infected chips. As mentioned in the previous section, in each of the 40 tests, 12 thermal photographs were taken with the same time interval, from healthy and unhealthy circuits, in a period of 55 seconds.

A. Proposed Features

To detect hardware Trojan, we have provided two sets of features, referred as temperature variation over time steps (TVOT) and overheated pixel occurrences (OPO). By taking advantage of them and feeding them to a shallow neural network, the existence of hardware Trojan on an FPGA will be classified. The operating temperature values have been calculated at different time intervals ($t = 5(n - 1)|\ n\epsilon\mathcal{N}, 1 \leq n \leq 12$) for test chip which is demonstrated as T_{test}. Temperature difference for test chip at each calculated time is considered as follows: $\Delta T_{test(n)} = T_{test(n+1)} - T_{test(n)}$ where $T_{test(n)}$ is the average temperature calculated at time n. Then, the difference of the 11 values obtained with the corresponding values in the golden chip (ΔT_{gold}) gives ΔT_{diff}. Due to the fact that hardware Trojan chips emit more temperature than their corresponding golden chips to lessen the impact of outside noise, we first sort ΔT_{diff}, then sum three middle values and select them as a feature. Results of TVOT feature is shown in Fig.2 which is tested on Basic_rsa_t100 circuit 16 times. As it shown, TVOT can somewhat discriminate between Trojan and Trojan-free circuits as Trojan circuits have a higher temperature than Trojan-free circuits.

Two characteristics are suggested in our technique recommendation to identify Hardware Trojans on an operating FPGA. The average difference between each thermal picture and the one before it was used to calculate TVOT. Due to the employment of additional components, this characteristic should logically be greater in Trojan-infected circuits than Trojan-free ones. For this purpose, we determined the average difference between each taken image and its prior one in terms of the thermal value of the pixels, and then we subtracted the above from the corresponding

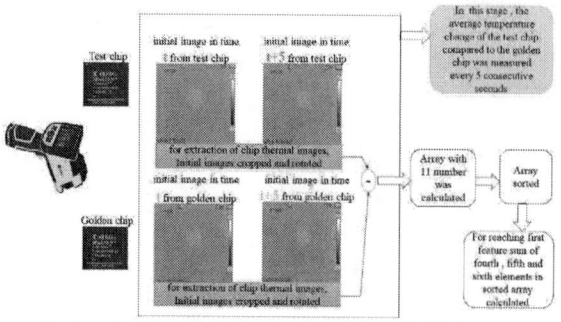

Fig 3. Proposed method for capturing TVOT values

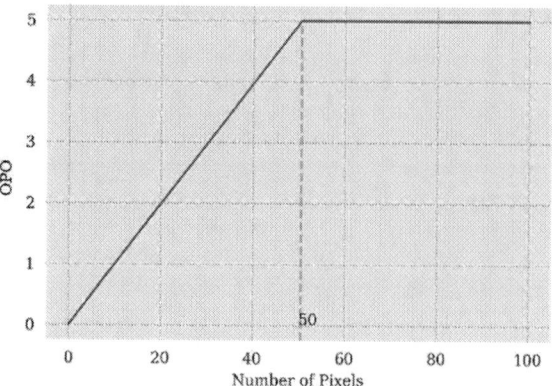

Fig. 4. Function used to OPO feature

values extracted from the golden chip. For each experiment, we obtained 12 thermal pictures, from which we computed 11 differential values. We sorted the collected values to minimize the damaging impact of any noise in the experiment setting, and the fifth, sixth, and seventh sorted numbers were then added to provide the first experiment feature. Fig.3 shows our methodology for extracting TVOT values from 12 thermal images in our experiment.

For the OPO, we've counted the number of pixels whose temperature 0.2 centigrade higher than the golden chip during execution time. First, temperature changes over the entire test time are calculated on the test chip. Given the corresponding values in the golden chip, if the pixel changes in the thermal images of the test chip are more than 0.2 ° C relative to the changes in the golden chip, that pixel is prone to hardware Trojan. Then, the number of pixels that were prone to hardware Trojan was included as input values for extracting of OPO. The threshold value has been empirically determined through multiple experiments conducted at room temperature. According to the observations, the temperature variations resulting from process variation are significantly lower than this threshold value, and it can be safely disregarded. Finally, from this value, the function of Figure 4 is used to obtain the final value of the OPO.

B. Nowroz Method

In [6] authors suggested a technique for Hardware Trojan detection with thermal imaging. To get better results, we used their procedure on our dataset in this work. The authors selected eigenvectors corresponding to the first ten largest eigenvalues of the thermal image matrices as the optimal projection axes. Then the average thermal map of 1000 valid chips were used to extract the golden feature matrix. For each chip under test, the distance of its feature matrix and the golden feature matrix was calculated. They showed how, as the amount of process variations increases, it becomes harder to discern between true chips and chips that have been Trojan-infected. The distance between the testing feature matrix B_i and the authentic feature matrix B was calculated by equation 1:

$$Distance = \|B_i - B\| \qquad (1)$$

C. Combining methods to improve detection performance

In this section, three features, including two features introduced and the normalized features of Nowruz method feature, are used to detect hardware Trojans. It is expected that with more features, higher accuracy can be achieved in

detecting Trojans. The network used in the proposed method consists of three fully connected layers, each containing 10 neurons. For training the network, the input features used are OPO and TVOT. The training was conducted over 20 epochs with a learning rate of 0.01 using the gradient descent optimizer and the Cross-Entropy loss function.

V. EXPERIMENTAL RESULTS

Our suggested technique ,was tested on IR thermal images captured during 40 distinct tests using 4 different type of HT in Basic_rsa circuits from trust-hub.org which each test contains 12 thermal image. Then we used the combination of mentioned features to train a neural network for Trojan detection. We used two different types of neural networks in our experiments. Logistic regression has been used to obtain a threshold for the feature introduced by Nowruz and a neural network with three fully connected layers, each containing 10 neurons, has been used for the proposed and combined feature. Training, validation and test data include 70%, 5% and 25% of the dataset, respectively. The training was performed 10 times in such a way that the data were randomly separated. In table 1 the result of our training is shown.

TABLE I. AVEARAGE ACCURACY OF TEST DATA FOR CLASSIFICATION OF HEALYHY AND INFECTED CHIPS

Method	Neural Network Type	Accuracy
Nowruz	Logistic Regression	80%
Proposed	3 FC layer – 10 Neuron each	82%
Nowruz + Proposed	3 FC layer – 10 Neuron each	**86%**

Table 1 illustrates the achieved accuracy using various features. According to the table, utilizing both OPO and TVOT features has resulted in higher accuracy compared to the Nowruz method. Additionally, it can be observed that by combining the proposed features with the Nowruz approach, it becomes feasible to attain higher accuracy.

VI. CONCLUSION

In this work, we propose a new way for detecting Hardware Trojan. In our test, thermal cameras with a resolution of

120x160 were used to find the overhead of the chip with the hardware Trojan compared to the original chip. Two new differential features have been introduced to detect hardware Trojans. To model the circuits of the hardware Trojan, help has been taken from Trust_hub [14,15]. From the circuits in this benchmark, the basic_rsa circuit with 4 different attack types was used in this experiment. To classify healthy chips and chips with Trojans, we used a fully connected neural network with 3 hidden layers and 10 neurons in each layer. The results showed that the proposed method for detection performed well and if the features of past works are also used, the detection performance will be improved.

REFERENCES

[1] S. Deb Paul, "Hands-on Learning of Hardware and Systems Security," Advances in Enginering Education, vol. 9, no. 2, 2021, doi: 10.18260/3-1-1153-07602.

[2] R. Naveenkumar, N. M. Sivamangai, A. Napolean, and S. Sridevi Sathayapriya, "Review on Hardware Trojan Detection Techniques," National Academy Science Letters, May 2023, doi: 10.1007/s40009-023-01247-6

[3] J. Cruz, P. Slpsk, P. Gaikwad, and S. Bhunia, "TVF: A Metric for Quantifying Vulnerability Against Hardware Trojan Attacks," IEEE Transactions on Very Large Scale Integration (VLSI) Systems, vol. 31, no. 7, pp. 969–979, Jul. 2023, doi: 10.1109/tvlsi.2023.3270866.

[4] E. Love, Y. Jin and Y. Makris, "Enhancing security via provably trustworthy hardware intellectual property," 2011 IEEE International Symposium on Hardware-Oriented Security and Trust, 2011, pp. 12-17, doi: 10.1109/HST.2011.5954988.

[5] Yier Jin and Y. Makris, "Hardware Trojan detection using path delay fingerprint," 2008 IEEE International Workshop on Hardware-Oriented Security and Trust, 2008, pp. 51-57, doi: 10.1109/HST.2008.4559049

[6] A. N. Nowroz, K. Hu, F. Koushanfar and S. Reda, "Novel Techniques for High-Sensitivity Hardware Trojan Detection Using Thermal and Power Maps," in IEEE Transactions on Computer-Aided Design of Integrated Circuits and Systems, vol. 33, no. 12, pp. 1792-1805, Dec. 2014, doi: 10.1109/TCAD.2014.2354293

[7] D. Agrawal, S. Baktir, D. Karakoyunlu, P. Rohatgi and B. Sunar, "Trojan Detection using IC Fingerprinting," 2007 IEEE Symposium on Security and Privacy (SP '07), Berkeley, CA, USA, 2007, pp. 296-310, doi: 10.1109/SP.2007.36.

[8] Xiaoxiao Wang, M. Tehranipoor and J. Plusquellic, "Detecting malicious inclusions in secure hardware: Challenges and solutions," 2008 IEEE International Workshop on Hardware-Oriented Security and Trust, Anaheim, CA, USA, 2008, pp. 15-19, doi: 10.1109/HST.2008.4559039.

[9] M. Pazira, Y. Baleghi and A. Akbari, "Hardware Trojan Detection Using Thermal Imaging in FPGAs with Combined Features," 2021 7th International Conference on Signal Processing and Intelligent Systems (ICSPIS), Tehran, Iran, Islamic Republic of, 2021, pp. 1-5, doi: 10.1109/ICSPIS54653.2021.9729357.

[10] S. Reda, A. N. Nowroz, R. Cochran, and S. Angelevski, "Post-silicon power mapping techniques for integrated circuits," Integr. VLSI J., vol. 46, no. 1, pp. 69–79, 2013

[11] Qiang Sui, Zhikai Wu, Jun Li and Shaoqing Li, "A detection method of Hardware Trojan based on two-dimension calibration," 2016 2nd IEEE International Conference on Computer and Communications (ICCC),2016, pp. 2795-2799, doi: 10.1109/CompComm.2016.7925207

[12] Su, Ting, et al. "Part I: Evaluation for hardware Trojan detection based on electromagnetic radiation." Journal of Electronic Testing 36.5 (2020): 591-606

[13] Zhong, Jingxin, and Jianye Wang. "Thermal images based Hardware Trojan detection through differential temperature matrix." Optik 158 (2018): 855-860

[14] H. Salmani, M. Tehranipoor, and R. Karri, "On Design vulnerability analysis and trust benchmark development", IEEE Int. Conference on Computer Design (ICCD), 2013

[15] B. Shakya, T. He, H. Salmani, D. Forte, S. Bhunia, M. Tehranipoor, "Benchmarking of Hardware Trojans and Maliciously Affected Circuits", Journal of Hardware and Systems Security (HaSS), April 2017

The 5th Iranian International Conference on Microelectronics (IICM2023)
25 – 26 October 2023

Ultra-Low Power SRAM-PUF for IoT Devices Based on CNTFETs

Alireza Shafiei
Dep. of Electrical and Computer Engineering
Graduate University of Advanced Technology
Kerman , Iran
alibitw73@gmail.com

Mehrnaz Monajati
Dep. of Electrical and Computer Engineering
Graduate University of Advanced Technology
Kerman , Iran
m.monajati@gmail.com

Abstract—With the rapid advancement of artificial intelligence and machine learning in Industry 4.0 and cyber-physical systems, security poses a significant challenge for humans. To address this, Physical Unclonable Functions (PUFs) have emerged as a promising and lightweight solution for securing IoT devices. The demand for low-power and secure crypto-devices has become critical in the context of the Internet of Things (IoT) and its emerging technologies. Although IoT has enabled battery-powered devices to transmit sensitive data, it has also introduced issues of high power consumption and security vulnerabilities. In this paper, we propose an investigation into the use of adiabatic logic with carbon nanotube field-effect transistors (CNTFETs) for designing lightweight IoT devices that tackle these challenges. This computing platform offers potential benefits in terms of security and energy efficiency for IoT applications.

Keywords— Physical unclonable function (PUF), Adiabatic, Carbon nanotube field-effect transistor (CNTFET), Lightweight, SRAM-PUF

I. INTRODUCTION

The Internet of Things (IoT) represents a revolutionary technological concept, aiming to create a global network connecting various devices and objects. Being recognized as a pivotal field of future technology, the IoT has captured the attention of numerous industries [1]. However, to be successful, IoT devices must address several challenging aspects, including low energy consumption, lightweight design, and robust security measures to counter potential threats.

One promising approach for enhancing security in IoT devices is the use of Physical Unclonable Functions (PUFs), which can be likened to digital fingerprints for both silicon and non-silicon chips. PUFs offer an economical means of generating secret bits for secure systems, especially in the context of IoT devices [2]. Nevertheless, designing a reliable and energy-efficient PUF presents a significant hurdle [2].

Silicon-based devices possess inherent physical variations, such as internal resistance, capacitors, leakage, and oxide thickness, which are challenging to control during the manufacturing process. Similarly, carbon nanotube transistors, like MOSFETs, feature parameters that can be altered during manufacturing, such as nanotube diameter, pitch, and tox, leading to substantial impacts on the threshold voltage. Various PUF topologies, including

Ring-Oscillator (RO), Arbiter PUF, Butterfly PUF, Glitch PUF, and SRAM PUF, each have their advantages and disadvantages, such as high power consumption or limited challenge-response pair (CRP) sets [3]. In this article, we focus on the SRAM-PUF, which relies on random startup values.

The following sections detail the background of adiabatic logic and carbon nanotube transistors in Section II. Section III introduces the proposed CNTFET SRAM-PUF cell and the architecture built using this cell. Subsequently, Section IV presents security metrics and the power consumed by the SRAM-PUF. Finally, Section V concludes and explores future research directions related to the proposed PUF.

II. BACKGROUND

A. Adiabatic Logic

Adiabatic logic is a clocking technique that enables the design of ultra-low power circuits. It achieves this by efficiently recycling the charge stored in the load capacitor, leading to a reduction in overall power consumption. The fundamental concept behind adiabatic logic is depicted in Fig. 1. However, one of the main limitations of adiabatic logic is that these circuits can only operate at frequencies lower than 1 GHz. Additionally, utilizing multi-phase clocking introduces an overhead for adiabatic logic-based circuits [2].

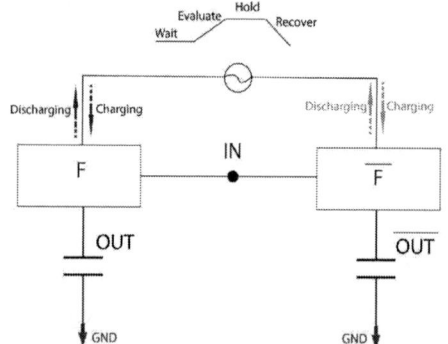

Fig. 1. Adiabatic charging/discharging technique

The dissipated energy of the adiabatic depends on the constant time (τ), while τ is the evaluate/recover phase of the capacitor. By lengthening the constant time ($\tau >> RC$), adiabatic logic dissipates energy substantially less than

979-8-3503-6020-2/23 $31.00 © 2023 IEEE

conventional CMOS logic. The dissipated energy of the adiabatic can be expressed as follows [2]:

$$E_{adiabatic} = \frac{RC}{\tau} C V_{dd}^2 \qquad (1)$$

B. Carbon Nanotube Transistors

Carbon nanotube field-effect transistors (CNTFETs) present a viable alternative to conventional CMOS technology [4]. Fig. 2 illustrates the structure of a CNTFET, where carbon nanotubes serve as the channel located beneath the gate. These carbon nanotubes are essentially rolled graphene layers with specific chiral vectors determining their electrical properties, such as conductive or semi-conductive characteristics. CNTFETs offer faster operation compared to MOSFETs and consume less power.

The threshold voltage of a CNTFET can be easily changed by changing the diameter of the nanotube. The following formulas are used to compute a CNTFET's threshold voltage [4]:

$$D_{cnt(nm)} = \frac{\sqrt{3}a_0}{\pi}\sqrt{n_1^2 + n_1 n_2 + n_2^2} = 0.0783\sqrt{n_1^2 + n_1 n_2 + n_2^2} \qquad (2)$$

$$V_{th(v)} \approx \frac{E_{bg}}{2e} = \frac{\sqrt{3}}{3}\frac{aV_\pi}{eD_{cnt}} \approx \frac{0.43}{D_{cnt(nm)}} \qquad (3)$$

where D_{cnt} is the CNT diameter, n_1 and n_2 are the chiral vector integers, e is the unit electron charge, Ebg is the CNT bandgap, a_0 ($\approx 0.142nm$) is the interatomic distance between each carbon atom and its neighbor, V_π ($\approx 3.033eV$) is the carbon π-π bond energy in the tight bonding model, and a ($\approx 2.49\text{Å}$) is the carbon to carbon atom distance [4].

The use of CNTFETs in PUF circuits introduces several advantages. Leveraging the unique properties of carbon nanotube transistors, these PUFs are designed to be more secure and resilient against environmental variations [4]. This makes them an attractive choice for enhancing the security of IoT devices and other applications requiring robust authentication mechanisms.

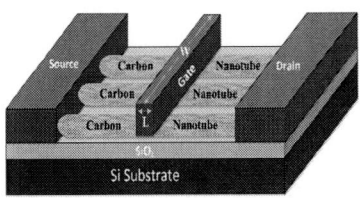

Fig. 2. Schematic of a carbon nanotube transistor (CNTFET) [5]

III. PROPOSED PUF DESIGN

The proposed ultra-low power adiabatic logic-based carbon nanotube transistor SRAM-PUF circuit topology is depicted in Fig. 3. In this topology, Transistor M0 functions as the enable/disable PUF cell, controlling the operation of the circuit. Transistors M1, M2, M3, and M4 together form the bistable structure, which plays a crucial role in generating random bits.

The generation of random bits is achieved through process variation arising from differences in threshold

voltage within the bistable structure. These inherent variations lead to the generation of random challenge bits (Vcb). Additionally, the circuit provides two complementary outputs (R and R-) as part of its functionality. These complementary outputs are instrumental in generating the response bits of the SRAM-PUF.

This circuit design utilizes the advantages of adiabatic logic and carbon nanotube transistors to achieve ultra-low power consumption while ensuring the generation of secure and random responses for PUF-based applications.

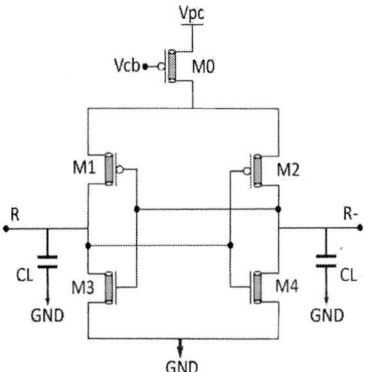

Fig. 3. Proposed SRAM-PUF cell

A. Operation of the proposed design

The operation of the PUF cell in adiabatic logic with four phases, namely, wait, evaluate, hold, and recover, is as follows:

- Wait Phase: When the challenge bit (Vcb) is low (Vcb=0), the PUF cell becomes active, and the transistor M0 is turned on. In this phase, the PUF cell is ready to respond to incoming challenges.

- Evaluate Phase: When the challenge bit is high (Vcb=1), the cell operates in the evaluate phase of the clock, and it becomes inactive. During this phase, both PCNFETs (M1, M2) start conducting. Due to the variation in threshold voltage between these transistors, one of them conducts current more quickly than the other, causing the related load capacitor to rapidly charge. This differential charging results in complementary outputs, with one output leading to logic "1" and the other leading to logic "0".

- Hold Phase: In this phase, the PUF cell holds a stable response. The outputs generated during the evaluate phase are maintained, and the PUF response remains constant and secure.

- Recovery Phase: As the clock enters the recovery phase, the voltage begins to decrease from Vdd to ground, and the load capacitor discharges back to the power clock source. This prepares the PUF cell for the next challenge.

The operation of the circuit in adiabatic logic is depicted in Fig. 4, which illustrates the different phases and the behavior of the PUF cell during each phase. This design aims to achieve ultra-low power consumption and secure PUF responses through the clever use of adiabatic logic and carbon nanotube transistors.

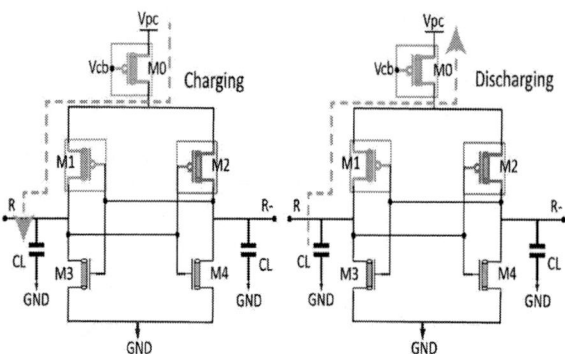

Fig. 4. Operation of the circuit during the evaluate and recovery phases with transistor M1 exhibiting a lower threshold voltage

B. Design of 4-bits SRAM-PUF

Fig. 5 showcases the architecture of a 4-bit cascaded SRAM-PUF, an advanced configuration that integrates multiple PUF cells in a cascading arrangement. Each individual PUF cell operates with its own dedicated power clock, metieulously set with a 90-degree phase difference relative to the adjacent cell. This phase offset ensures precise coordination and orchestration of the PUF cells throughout their operational cycles.

For instance, while the first PUF cell enters the wait phase, the subsequent cell in the same local PUF progresses to the evaluate phase. Similarly, the other two PUFs concurrently undertake their respective phases, with one in the hold phase and the other in the recovery phase [3].

This innovative cascaded design allows for efficient utilization of resources and maximizes the parallel processing capability of the PUF cells. By harmonizing the individual PUFs with carefully timed phase shifts, this architecture aims to optimize performance, achieve robustness, and enhance the overall security and reliability of the SRAM-PUF system.

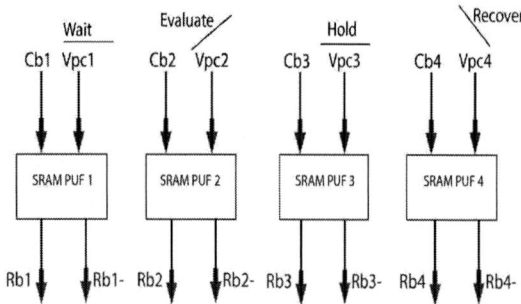

Fig. 5. 4-bit SRAM-PUF architecture

IV. SIMULATION RESULTS

In this study, we utilized the Stanford library model [6] for the baseline CNTFET with 32nm technology to analyze the 4-bit SRAM-PUF The parameters of the CNTFET model and the values used in the SRAM-PUF design are detailed in Table I for reference. Our simulations were executed within HSPICE environment. Following the initial simulations, we undertook additional post-processing steps using MATLAB software. The analog output values generated by the circuit were first extracted using HSPICE and subsequently transformed into digital values through MATLAB for further analysis.

To comprehensively investigate the behavior and performance of the chips under varying conditions, we employed Monte Carlo simulations. This advanced simulation technique allowed us to emulate characteristics such as threshold voltage (with a variation of up to 150%) and temperature, providing a comprehensive understanding of their performance across different scenarios. For the evaluation of PUF metrics, we took a meticulous approach. We manipulated the parameters that exert the most significant influence on the threshold voltage, as defined in (3). Specifically, we varied the values of n1 and n2, which represent the chiral vector integers in the CNTFET model. Monte Carlo simulations were performed using HSPICE software to generate output data for these PUF parameters. Subsequently, we meticulously evaluated these simulation results using MATLAB software. This included calculations to assess PUF metrics, such as uniqueness and reliability. This meticulous approach enabled us to thoroughly analyze and assess the performance of our SRAM-PUF design. By taking into account the effects of various parameters and environmental conditions, we achieved a comprehensive understanding of its behavior and capabilities.

TABLE I.   PARAMETERS OF THE CNTFET MODEL AND VALUES EMPLOYED IN SRAM-PUF DESIGN

Parameters	Description	Value
Lch	*Length of Gate/Drain/Source*	*32nm*
Lgeff	*Length of mean free path length of intrinsic CNT channel*	*100nm*
Tox	*Oxide thickness*	*4nm*
K	*Dielectric Constant*	*16*
Pitch	*distance between the centers of two adjacent CNTs*	*20nm*
m,n(M0)	*Chiral vector of M0*	*(19,0)*
m,n(bistable)	*Chiral vector of M1,M2,M3,M4*	*(13,0)*
Efi	*Fermi level energy of S/D Tube*	*0.6eV*
Tubes	*The number of tubes in the device*	*3*
Csub	*Coupling Capacitance*	*40pF/m*

The power supplies employed in our simulations have a swing range of 0 to 0.9 V. The frequency of the challenge bit and the power clock were set at 10 MHz and 100 MHz, respectively. We obtained our results using a reference temperature of 27°C and capacitances of 10 fF. In our simulations, we employed 10fF capacitors as a common reference point for the capacitance values within the SRAM-PUF cells. While CNTFETs introduce unique characteristics that may influence the overall capacitance, we chose this standard value to maintain compatibility with industry-standard CMOS technology for SRAM cells. This choice enables meaningful comparisons with existing empirical data and industry benchmarks, facilitating a foundational assessment of the SRAM-PUF's performance.

Fig. 6 illustrates the waveforms of the challenge bit, power supply, and response bit during our experiments. These waveforms offer valuable insights into the behavior of the SRAM-PUF under specific conditions.

To assess the performance of the SRAM-PUF, we examined important metrics such as energy dissipation, uniqueness, and reliability. These evaluation metrics provide crucial information about the efficiency, security, and robustness of the proposed 4-bit SRAM-PUF architecture. By analyzing these metrics and conducting comprehensive simulations, we aim to gain deeper insights into the behavior and effectiveness of the SRAM-PUF under various scenarios, thereby contributing to the advancement and optimization of PUF-based security systems for IoT and other applications.

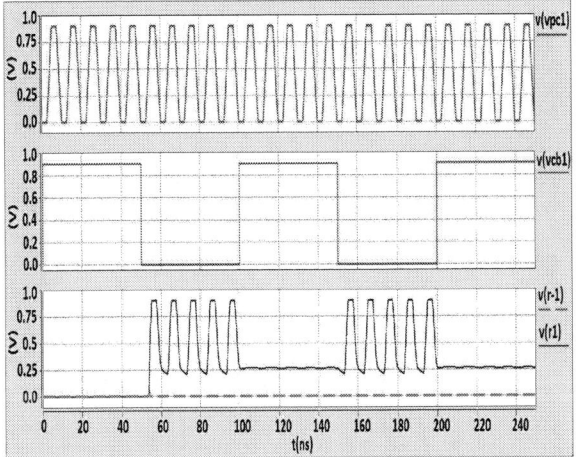

Fig. 6. Input and output signals of the SRAM-PUF cell

Our validation approach leverages Monte Carlo simulations, which excel in capturing the statistical behavior of complex systems, addressing inherent randomness and variations within our SRAM-PUF design. By systematically varying key parameters, including nanotube characteristics, temperature fluctuations, and stochastic factors, we generate a diverse range of possible outcomes, mimicking real-world operational scenarios. Statistical analyses of the simulation data enable a robust evaluation of performance metrics, including power consumption, uniqueness, and reliability, under varying conditions. The alignment between Monte Carlo simulations and deterministic simulations reinforces the credibility of our results. Sensitivity and robustness analyses offer insights into the design's reliability, further strengthening the validity of our findings. Despite relying on simulations, our comprehensive approach provides a reliable basis for the presented outcomes.

A. Power dissipation

The primary rationale behind the adoption of both adiabatic logic and CNTFET technology in the SRAM-PUF design is the pursuit of substantial power consumption reduction. This synergistic integration targets the mitigation of energy usage, a pivotal aspect in modern electronic systems, particularly for low-power applications such as IoT.

In Fig. 7, we present a comprehensive visualization of power consumption over time. This graphical representation offers a clear depiction of how the combined benefits of adiabatic logic and CNTFET technology contribute to the overarching goal of minimizing power utilization within the SRAM-PUF. By showcasing the dynamic fluctuations in power consumption throughout different operational phases, this figure substantiates the efficacy of our chosen approach in achieving enhanced energy efficiency and sustainability.

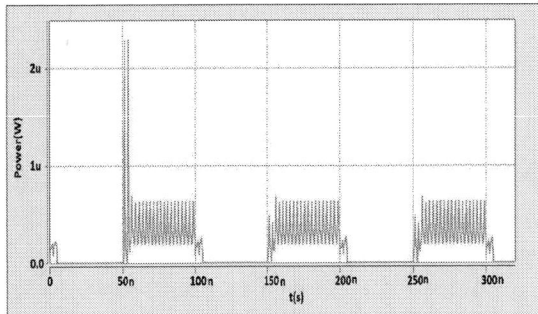

Fig. 7. Power consumption

B. Uniqueness

PUF's ability to distinguish a particular Integrated Circuit (IC) from others with the same structure using the same challenge C is measured through its uniqueness. When two chips, i and j (where $i \neq j$), receive the same challenge C, and their responses are denoted as Ri and Rj, the average inter-device uniqueness can be expressed as follows [7]:

$$Uniqueness = \frac{2}{d(d+1)} \sum_{i=1}^{d-1} \sum_{j=i+1}^{d} \frac{HD(RiRj)}{n} \times 100\% \quad (4)$$

Where d represents the number of devices (ICs) being compared. n denotes the bit length of the PUF responses. $HD\ (Ri, Rj)$ signifies the Hamming Distance between the responses of the two distinct PUFs (Ri and Rj) [7].

The ideal value for uniqueness is 50%, indicating that each PUF response is entirely different from the others, resulting in perfect discrimination capability among individual ICs. A higher uniqueness percentage reflects a stronger ability of the PUF to distinguish between different devices, adding to its effectiveness and security in applications such as authentication and anti-counterfeiting measures.

C. Reliability

The PUF design is expected to exhibit the capability to consistently reproduce the same response bit R when presented with the same challenge bit C, even in the presence of varying environmental conditions like supply voltage and temperature. The reliability of the PUF can be quantified by calculating the average intra-device Hamming Distance (HD) using the following equation [7]:

$$HD_{intra} = \frac{1}{d} \sum_{t=1}^{d} \frac{HD(R_i R'_{i,t})}{n} \times 100\% \quad (5)$$

Where R_i represents the response of the chip i measured under nominal operating conditions. $R'_{i,t}$ denotes the t-th sample of the response R_i, extracted under different supply voltage and temperature conditions [7]. n represents the bit size of the PUF response (in this study, $n = 4$). d is the number of devices (chips) used in the analysis ($d = 30$ in this study). Additionally, the temperature range considered in our analysis spans from -40°C to 100°C.

$$Reliability = 100\% - HD_{intra} \quad (6)$$

The calculated uniqueness and reliability results, based on Equations (4) and (6), respectively, are presented graphically in Fig. 8 and Fig. 9. These figures illustrate the variations in uniqueness and reliability under threshold voltage and temperature variations, respectively. The study utilizes a bit size of $n = 4$ and a total of 30 devices ($d = 30$) for the analysis.

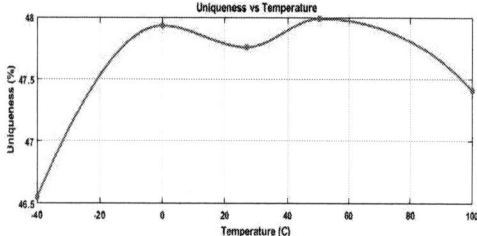

Fig. 8. Uniqueness of the SRAM-PUF under the threshold voltage variation

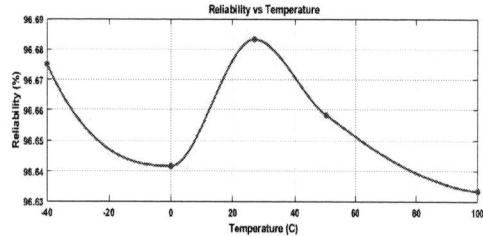

Fig. 9. Reliability of the SRAM-PUF under the threshold voltage variation

By analyzing uniqueness and reliability metrics, this study aims to assess and validate the robustness and security of the proposed PUF design in real-world scenarios, taking into account the impact of environmental variations on its performance. In Table II, we present a comparative analysis of various PUFs reported in the literature alongside our work. In the table, "NA" denotes data that is not available. We assess the key characteristics and performance metrics of each PUF design to highlight the strengths and advantages of our proposed approach. This comparison provides valuable insights into the uniqueness, reliability, and energy efficiency of different PUF designs, showcasing the superiority and effectiveness of our ultra-low power SRAM-PUF based on CNTFETs for IoT devices.

TABLE II. COMPARISON OF PUF PAREMETR

Parameter	PUF Comparison				
	[3]	*[2]*	*[9]*	*[8]*	*This work*
Tech-Process (nm)	CMOS-180	CMOS-45	Tristate-180	FinFET-45	CNTFET-32
Bit-length (bit)	*4*	*128*	*128*	*4*	*4*
Start-up power (nw)	*3080*	*NA*	*157.5*	*65.69*	*29.62*
Uniqueness (%)	*40.50*	*49.41*	*50.27*	*49.46*	*47.41*
Reliability (%)	*96.20*	*99.60*	*99.82*	*99.47*	*96.63*

V. CONCLUSION AND FUTURE WORK

In this paper, we proposed an innovative adiabatic logic-based approach to design an efficient SRAM-PUF using only five carbon nanotube transistors. Our simulation results demonstrated successful implementation and promising performance. Comparing with other reported PUFs, our design achieved a remarkable 99.04% reduction in start-up power consumption compared to PUF [3] and a 55.03% reduction compared to PUF [8]. These significant power improvements make our ultra-low power SRAM-PUF based on CNTFETs a superior choice for energy-efficient IoT devices. We highlight the potential of carbon nanotube transistors for future PUF advancements and recommend exploring Ferroelectric CNTFET technology for even greater power optimization and performance gains [10]. Overall, our work contributes valuable insights and sets the stage for more efficient and secure VLSI circuits and IoT applications.

REFERENCES

[1] I. Lee and K. Lee, "The internet of things (IOT): Applications, investments, and challenges for enterprises," *Business Horizons*, vol. 58, no. 4, pp. 431–440, 2015. doi:10.1016/j.bushor.2015.03.008.

[2] S. D. Kumar and H. Thapliyal, "Design of adiabatic logic-based energy-efficient and reliable PUF for IOT devices," *ACM Journal on Emerging Technologies in Computing Systems*, vol. 16, no. 3, pp. 1–18, 2020. doi:10.1145/3390771.

[3] S. D. Kumar and H. Thapliyal, "QUALPUF: A Novel Quasi-Adiabatic Logic based Physical Unclonable Function," *Proceedings of the 11th Annual Cyber and Information Security Research Conference*, 2016. doi:10.1145/2897795.2897798.

[4] H. Momeni, A. Ghazizadeh, and F. Sharifi, "Multi-valued logic arbiter PUF designs based on cntfets," *Computers and Electrical Engineering*, vol. 102, p. 108295, 2022. doi:10.1016/j.compeleceng.2022.108295.

[5] F. Zahoor *et al.*, "Carbon Nanotube Field Effect Transistor (CNTFET) and resistive Random access memory (RRAM) based ternary combinational logic circuits," *Electronics*, vol. 10, no. 1, p. 79, 2021. doi:10.3390/electronics10010079.

[6] Stanford University CNFETModelWeb site. (2008). [Online]. Available:http://nano.stanford.edu/model.php?id=23.

[7] A. Al-Meer and S. Al-Kuwari, "Physical unclonable functions (PUF) for IOT devices," *ACM Computing Surveys*, vol. 55, no. 14s, pp. 1–31, 2023. doi:10.1145/3591464.

[8] C. Monteiro and Y. Takahashi, "Ultra-low-power FinFETs-based TPCA-PUF circuit for secure IOT devices," *Sensors*, vol. 21, no. 24, p. 8302, 2021. doi:10.3390/s21248302.

[9] S. Hemavathy and V. S. Kanchana Bhaaskaran, "Design and Analysis of Secure Quasi-Adiabatic Tristate Physical Unclonable Function," *2020 IEEE International Symposium on Smart Electronic Systems (iSES) (Formerly iNiS)*, Chennai, India, 2020, pp. 109-114, doi: 10.1109/iSES50453.2020.00034.

[10] M. K. Q. Jooq, M. H. Moaiyeri and K. Tamersit, "A New Design Paradigm for Auto-Nonvolatile Ternary SRAMs Using Ferroelectric CNTFETs: From Device to Array Architecture," *IEEE Transactions on Electron Devices*, vol. 69, no. 11, pp. 6113-6120, Nov. 2022, doi: 10.1109/TED.2022.3207703.

A Novel Approach for Offline and Online Application-Dependent testing of FPGA interconnects

Ahmad Menbari
Dept. Electronics and Comm. Eng.
University of *Kurdistan*
Sanandaj, Iran
a.menbari@uok.ac.ir

Hemin Rahimi
Dept. Electronics and Comm. Eng.
University of *Kurdistan*
Sanandaj, Iran
hemn.rahimi@uok.ac.ir

Hadi Jahanirad
Dept. Electronics and Comm. Eng.
University of *Kurdistan*
Sanandaj, Iran
h.jahanirad@uok.ac.ir

Abstract— FPGA interconnection network is tested primarily using application-dependent testing, which tests only the resources used in a particular application instead of testing all FPGA resources. The majority of state-of-the-art application-dependent methods, involve reprogramming user-specified LUTs into single-term functions for testing stuck-at and bridging faults while interconnect configurations remain unchanged. In this paper, a novel approach for offline and online application-dependent testing of FPGA interconnects is presented. In this approach, test vectors are applied directly to user-specified designs to test FPGA interconnects without reconfiguring LUTs into other functions. As a result, it is possible to test an FPGA simultaneously in its normal mode of operation. However, 100% fault coverage cannot be achieved in online mode of test due to the presence of LUTs with uncontrollable inputs. According to simulation results, 100% fault coverage is achieved for almost all benchmark circuits for the offline mode of test.

Keywords—Application-dependent testing; field programmable gate array (FPGA); interconnect testing; online testing

I. INTRODUCTION

FPGAs (Field-Programmable Gate Arrays) are electronic devices that can be programmed to implement any logic design specified by the user. FPGAs are popular target devices for various applications due to their configurable feature, which allows them to be developed, debugged, and implemented in a quicker and less expensive manner than application-specific integrated circuits (ASICs) [1], [11], [12]. A further advantage of FPGA-based products is that they can be quickly upgraded in the field if any design errors are discovered after the product is released [4], [13], [14].

Generally, FPGA architectures include configurable logic blocks (CLBs) arranged in a 2D array, programmable interconnects (networks, switches) and input/output blocks (IOBs). A CLB is comprised of Look Up Tables (LUTs), Multiplexers and Flip Flops, and the input-output blocks are arranged at the edge of the grid and connected to the logic blocks through an interconnection network [2], [3].

An FPGA programmable interconnect resources account for almost 80% of its die area, making them highly prone to faults, such as open, stuck-at, and bridging. Therefore, it is essential to test interconnect resources in order to ensure the reliability of the system developed on FPGA [2], [3], [4]. Test time and fault coverage are critical parameters for an FPGA interconnect testing approach.

Methodologies for FPGA testing can be broadly categorized into application-dependent and application-independent testing. Application-independent testing, or manufacturing testing, involves testing all FPGA resources. In this approach, the entire FPGA interconnects are tested for all possible configurations. Application-dependent testing, on the other hand, does not test the entire interconnection network of an FPGA, rather only the resources used in a particular application are tested [5].

Application-dependent approach to FPGA interconnection testing is dominant approach in the field for two reasons. First, a plenty of time is required to test all possible interconnect configurations. Second, an FPGA may have a faulty part, but the rest that user-specified design is implemented on might be fault-free. Therefore, the majority of state-of-the-art methods present various approaches to application-dependent testing of interconnects. In these approaches, typically the configurations of logic blocks are modified by programming the FPGA while interconnect configurations remain unchanged. Then, during several configurations, the required test vectors are applied to test the interconnection network. Therefore, test time is dominated by the number of configurations in these approaches [3], [4], [5].

A majority of current methods of application-dependent testing involve reprogramming user-specified LUTs into single-term functions for testing stuck-at and bridging faults. These approaches have two main objectives. The first objective is to increase the fault coverage or introduce new faults in the interconnection network. As a second objective, they want to reduce the number of test configurations in order to decrease test time.

These test approaches can only be performed when the design under test is sufficiently idle. Therefore, online testing, which is satisfactory for highly fault-sensitive and mission-critical applications, cannot be enabled for these designs since the LUTs must be reconfigured for testing.

In this paper, a novel approach for application-dependent testing of FPGA interconnects is proposed. Using this

979-8-3503-6020-2/23 $31.00 © 2023 IEEE

approach, stuck-at and bridging faults of interconnects can be detected without modifying the functionality of the LUTs. Therefore, the design programmed on FPGA can be tested online when it performs in its normal mode of operation. However, some LUT inputs cannot be controlled in the online mode, so it is not possible to achieve 100% fault coverage. Testability of LUTs including uncontrollable inputs can be increased by using unused IOs [6]. In the case that the required number of these IOs is not available, uncontrolled LUT inputs are controlled only by one unused IO pin in offline mode. To test FPGA interconnection network, several rules are developed to activate stuck-at and bridging faults of interconnects for an LUT4 (4-input LUT) with a random functionality in order to analyze testability of a random LUT first. Then, an algorithm based on these rules is used to find efficient test vectors for the user-specified design in order to achieve the desired fault coverage. The algorithm includes a number of simple steps of sensitization and justification. The main contributions of this paper are highlighted as follows:

1- A novel approach for application-dependent testing of FPGA interconnects is proposed, which supports both offline and online tests.

2- Test vectors are directly applied to user-specified design in order to test FPGA interconnects without reconfiguring LUTs into other functions. Therefore, it is feasible to test FPGA simultaneously in its normal mode of operation.

3- In the offline mode of test, only one of the unused IO pins can be used to increase the testability of the LUTs. Therefore, almost all benchmark circuits are covered with 100% fault coverage in this test mode.

II. ULITATURE REVIEW

FPGA interconnect testing approaches are classified as application-independent and application-dependent. This paper focus on application-dependent testing; therefore, these testing approaches are reviewed in this section. Fault coverage and the number of configurations required to complete test are important parameters to evaluate an application-dependent testing approach.

Das and Touba [1] proposed an application-dependent testing scheme for detecting pairwise bridging faults in an FPGA interconnection network. The number of test configurations is increased because only the fanout branches of one net are tested during each configuration. Each configuration transforms logic blocks from the user-specified design into transparent logic in LUTs followed by flip-flops to construct scan chains.

Tahoori first introduced the concept of single-term functions for FPGA testing [2]. The general form of a single-term function is a logic AND or a logic OR function with possibly some inversions at the inputs and/or the output. As such, it is a logical function that includes only one maxterm or one minterm. The combination of inputs corresponding to a single-term function is defined as activating input, which results in the detection of any sensitized faults. Fig. 1 illustrates one example of a single-term function by implementing an AND function on a LUT with the first bit inverted. When the activating input (0111) is applied to the LUT, A stuck-at-1 (A/1), B/0, C/0, and D/0 faults can be

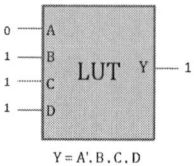

Fig. 1. Single-term function with activating input vector.

detected. Also, bridging fault between A, B, and A, C, and A, D can be detected.

The authors of [3] have proposed an application-dependent method which performs testing by modifying only the configuration of logic blocks that have been utilized in a specific design. The LUTs are reconfigured as AND function and OR function in order to detect stuck-at-1 and stuck-at-0 faults, respectively. Because there are only two test configurations in this approach, high-fault coverage cannot be achieved considering stuck-at faults are the only faults covered.

In [4], all LUTs that are reconfigured as minterm or maxterm functions are tested for stuck-at and bridging faults. In each configuration, the Walsh code generates several activating input vectors, which are applied to the net in order to detect all possible faults in an LUT. When interconnection network consists of N nets, its Walsh code is given by its binary representation as a number (its width is calculated by (Log_2^{N+2})). To test ISCAS'89 benchmark circuits using this method, approximately 10 test configurations are required on average.

Kumar and Lombardi [8] have proposed a method for generating test configurations to detect stuck-at and bridging faults using a heuristic-based algorithm. This approach is based on reprogramming the LUTs into AND or OR functions to sensitize faults by activating input vectors. In the configuration generation process, a polynomial and greedy heuristic algorithm is used for net selection in order to reduce the number of configurations required to achieve 100% fault coverage.

The method presented in [6] generates one test configuration based on walking-1 approach to test FPGA interconnection network. To perform this method, more test vectors are required, and unused IOs must be available. Authors of [7] add two columns to existing Walsh code tables to test bridging faults. To generate test configurations, an application-dependent test based on satisfiability and Walsh code is generated. Authors of [5] modify the functionality of LUTs for generating test configurations in order to detect stuck-at and different types of bridging faults. The proposed work uses Boolean satisfiability to minimize the number of test configurations.

III. PROPOSED METHOD

A. Test rules for LUTs

In this paper, we aim to test FPGA interconnects without modifying the functionality of the LUTs. Fig. 2 illustrates a design that is implemented using FPGA with eight LUTs of different functionalities that are connected in a network. In order to detect stuck-at and bridging faults of interconnects, the most efficient test vectors must be selected and applied. A straight forward approach is to find test vectors for each

The 5th Iranian International Conference on Microelectronics (IICM2023)

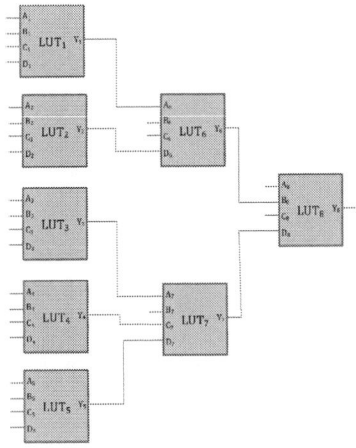

Fig. 2. An example for network of LUTs.

I	ABCD	Y
I_0	0000	Y_0
I_1	0001	Y_1
I_2	0010	Y_2
I_3	0011	Y_3
I_4	0100	Y_4
I_5	0101	Y_5
I_6	0110	Y_6
I_7	0111	Y_7
I_8	1000	Y_8
I_9	1001	Y_9
I_{10}	1010	Y_{10}
I_{11}	1011	Y_{11}
I_{12}	1100	Y_{12}
I_{13}	1101	Y_{13}
I_{14}	1110	Y_{14}
I_{15}	1111	Y_{15}

Fig. 3. Truth table of LUT4.

single fault individually. Suppose A6 has a stuck-at-1 fault. The first step to detect the fault is to find an input vector in LUT6 that can activate this fault. Therefore, it is essential to develop several rules to activate interconnects faults (stuck-at and bridging) for an LUT4 with random functionality.

The truth table of an LUT4 is illustrated in Fig. 3, where there are 216 possible functionalities for the LUT. Assume that we want to find an input vector to activate D/1 (D-stuck-at-1) for an LUT4 with a random functionality. Let the outputs of first two input vectors (0000 and 0001) be different ($Y0 \neq Y1$). By applying 0000 to the LUT, D/1 is detected since the output of LUT for fault-free input (0000) is unequal with the output of faulty input (0001) based on the assumption. Obviously, the second input vector (0001) detects D/0 for the same reason. This vector also detects A dom D, B dom D, and C dom D as D =1 is unequal with A=B=C=0. A dominant bridging fault (e.g., A dom b) is defined as a short-circuit defect where one signal line (A) consistently overrides and forces another signal line (B) to its logic state. Therefore, I_0 and I_1 can activate this set of faults {D/0, D/1, A dom D, B dom D, C dom D}. Similarly, by applying a pair of input vectors {(I_2, I_3) or (I_4, I_5) or (I_6, I_7) or (I_8, I_9) or (I_{10}, I_{11}) or (I_{12}, I_{13}) or (I_{14}, I_{15})} the mentioned set of faults are detected if and only if their pair of corresponding outputs are unequal {$Y_2 \neq Y_3$ or $Y_4 \neq Y_5$ or $Y_6 \neq Y_7$ or $Y_8 \neq Y_9$ or $Y_{10} \neq Y_{11}$ or $Y_{12} \neq Y_{13}$ or $Y_{14} \neq Y_{15}$}.

The following rules are applied for an LUT4 in order to detect every stuck-at and dominant bridging faults of interconnects:

1. If {$Y_0 \neq Y_1$ or $Y_2 \neq Y_3$ or $Y_4 \neq Y_5$ or $Y_6 \neq Y_7$ or $Y_8 \neq Y_9$ or $Y_{10} \neq Y_{11}$ or $Y_{12} \neq Y_{13}$ or $Y_{14} \neq Y_{15}$}, the set of faults {D/0, D/1, A dom D, B dom D, C dom D} are detected by the corresponding pair of test vectors {(I_0, I_1) or (I_2, I_3) or (I_4, I_5) or (I_6, I_7) or (I_8, I_9) or (I_{10}, I_{11}) or (I_{12}, I_{13}) or (I_{14}, I_{15})}.

2. If {$Y_0 \neq Y_2$ or $Y_1 \neq Y_3$ or $Y_4 \neq Y_6$ or $Y_5 \neq Y_7$ or $Y_8 \neq Y_{10}$ or $Y_9 \neq Y_{11}$ or $Y_{12} \neq Y_{14}$ or $Y_{13} \neq Y_{15}$}, the set of faults {C/0, C/1, A dom C, B dom C, D dom C} are detected by the corresponding pair of test vectors {(I_0, I_2) or (I_1,

I_3) or (I_4, I_6) or (I_5, I_7) or (I_8, I_{10}) or (I_9, I_{11}) or (I_{12}, I_{14}) or (I_{13}, I_{15})}.

3. If {$Y_0 \neq Y_4$ or $Y_1 \neq Y_5$ or $Y_2 \neq Y_6$ or $Y_3 \neq Y_7$ or $Y_8 \neq Y_{12}$ or $Y_9 \neq Y_{13}$ or $Y_{10} \neq Y_{14}$ or $Y_{11} \neq Y_{15}$}, the set of faults {B/0, B/1, A dom B, C dom B, D dom B} are detected by the corresponding pair of test vectors {(I_0, I_4) or (I_1, I_5) or (I_2, I_6) or (I_3, I_7) or (I_8, I_{12}) or (I_9, I_{13}) or (I_{10}, I_{14}) or (I_{11}, I_{15})}.

4. If {$Y_0 \neq Y_8$ or $Y_1 \neq Y_9$ or $Y_2 \neq Y_{10}$ or $Y_3 \neq Y_{11}$ or $Y_4 \neq Y_{12}$ or $Y_5 \neq Y_{13}$ or $Y_6 \neq Y_{14}$ or $Y_7 \neq Y_{15}$}, the set of faults {A/0, A/1, A dom C, B dom C, D dom C} are detected by the corresponding pair of test vectors {(I_0, I_8) or (I_1, I_9) or (I_2, I_{10}) or (I_3, I_{11}) or (I_4, I_{12}) or (I_5, I_{13}) or (I_6, I_{14}) or (I_7, I_{15})}.

In Table. 2, the highest possible fault coverage is reported for all LUT4s is according to the abovementioned concepts. According to this table, when applying above-mentioned rules to 2^{16}=65,536 possible functions in LUT4, 100% fault coverage can be achieved for 64,338 LUTs. Therefore, there is a high probability that LUT6 can detect A6/1 through its input vectors.

TABLE I. FAULTI COVERAGE FOR ALL 2^{16} = 65536 LUTs BASED ON THE PROPOSED RULES

Number of LUT functionalities	Fault coverage (%)
64338 (98.17%)	100
1128 (1.72%)	75
60 (0.09%)	50
10 (0.003%)	25

979-8-3503-6020-2/23 $31.00 © 2023 IEEE

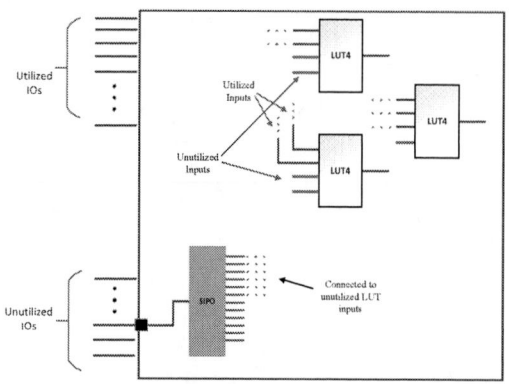

Fig. 4. Increase testability by controlling uncontrollable LUT inputs using an unused IO and an SIPO.

Fig. 5. process of programming a design on FPGA and testing its interconnection by the proposed method

B. FPGA interconnection testing

In the previous subsection, several rules were established to detect stuck-at and bridging faults of interconnects in a single LUT. In this subsection, the interconnections between various LUTs in a circuit programmed on FPGA are tested by following steps:

1- The interconnection faults in the circuit are collected in one set and one of them injected to the circuit. For example, $A_6/1$ is selected from the design of Fig. 2.

2- Activation: the established rules for LUTs are applied for the faulty LUT in order to activate the selected fault. For example, the rules are applied for LUT_6 in order to activate $A_6/1$. In the case that there is more than one input vector in the LUT to activate the selected fault, next steps are applied for all of them. There could be several LUT input vectors that activate the injected fault during this step. For each of them, the next steps have been taken.

3- Sensitization: It is necessary to assign the inputs to LUT_8 (A_8, C_8, and D_8) such that the output (Y_8) varies for the different values of B_8 ($=Y_6$). It is possible to satisfy the mentioned condition if LUT8 can cover $B_8/0$ or $B_8/1$. There is a high probability that all of the LUT4 faults can be activated, as reported in the previous section.

4- Justification: Since their outputs have been determined in the previous steps, the unassigned inputs of the LUTs (LUT1,2,3,4,5,7) are easily assigned in this step. As a result of this step, several test vectors can be achieved in order to cover the selected fault.

5- Fault simulation: finally, the selected test vectors achieved by the previous step are applied to the circuit to find out how many faults can be detected by each of them. The test vector with more fault coverage is selected as a test vector.

C. Uncontrollble LUT inputs determination

When a circuit is programmed on FPGA, some of the LUT inputs cannot be controllable using the circuit primary inputs as shown in Fig. 4. Therefore, the results of the Table I are valid for a fully controllable LUT4. Although, the proposed rules in subsection III.A can be applied for an LUT with two or three controllable inputs, 100% fault coverage cannot be achieved for these LUTs. Therefore, in the online mode of test, 100% fault coverage cannot be achieved.

TABLE II. SIMULATION RESULTS FOR THE PROPOSED OFFLINE AND ONLINE TESTS.

Benchmark circuits	Fault coverage (%)		Number of test vectors
	Offline	Online	
c432	100	82.11	12
c499	100	95.09	15
c880	100	89.76	56
c1355	98.23	94.76	61
c1908	99.45	86.66	67
c2670	100	87.65	98
c3540	99.34	90.09	131
c5315	100	91.23	145
c6288	100	94.34	58
c7552	100	87.45	211
s344	100	85.55	20
s349	100	87.12	15
s382	100	81.34	13
s400	98.77	88.33	20
s420	100	82.09	35
s444	99.56	88.33	21
s510	100	82.30	45
s526	100	80.92	17
s641	99.80	82.13	63
s713	98.67	84.23	69
s820	100	87.11	79
s953	100	88.78	87
s1196	100	88.12	75
s1238	99.60	85.44	77
s1423	99.50	85.89	54
s1488	100	86.66	89
s1494	100	87.45	92
s5378	99.08	84.45	167
s9234	99.34	81.34	172
s35932	100	84.10	399
s38417	99.23	80.23	622

979-8-3503-6020-2/23 $31.00 © 2023 IEEE

TALE III. SIMULATION RESULTS FOR COMPARATIVE ASSESMENT OF TEST CONFIGURATIONS.

Benchmark circuits	Number of test configurations			
	[5]	[8]	[6]	Proposed
s344	4	3	1	0
s349	4	3	1	0
s382	4	3	1	0
s400	4	3	1	0
s420	4	3	1	0
s444	4	3	1	0
s510	4	3	1	0
s526	4	3	1	0
s641	4	3	1	0
s713	3	3	1	0
s820	4	4	1	0
s953	4	4	1	0
s1196	4	4	1	0
s1238	4	4	1	0
s1423	4	4	1	0
s1488	4	4	1	0
s1494	4	4	1	0
s5378	4	4	1	0
s9234	4	4	1	0
s35932	-	3	1	0
s38417	-	4	1	0

TALE IV. SIMULATION RESULTS FOR COMPARATIVE ASSESMENT OF TEST VECTORS.

Benchmark circuits	Number of test vectors			
	[5]	[8]	[6]	Proposed
s344	4	3	14	20
s349	4	3	15	15
s382	4	3	9	13
s400	4	3	9	20
s420	4	3	23	35
s444	4	3	10	21
s510	4	3	30	45
s526	4	3	8	17
s641	4	3.	40	63
s713	3	3	39	69
s820	4	4	54	79
s953	4	4	68	87
s1196	4	4	52	75
s1238	4	4	-	77
s1423	4	4	36	54
s1488	4	4	69	89
s1494	4	4	67	92
s5378	4	4	116	167
s9234	4	4	118	172
s35932	-	3	290	399
s38417	-	4	475	622

In the offline mode, the inputs of all LUTs must become controllable in order to achieve 100% fault coverage. Therefore, unused IOs of FPGA must utilize in this case in order to increase testability of the LUTs. Therefore, by using one of the unused IO pins, whenever a test vector is applied to the circuit programmed on FPGA, its corresponding bitstream of the uncontrollable LUT inputs, memorized in a serial input/parallel output (SIPO), is applied to LUT inputs in order to test the FPGA interconnects.

IV. SIMULATION RESULTS AND COMPARISON

The whole process of programming a design on FPGA and testing its interconnection by the proposed method is illustrated in Fig. 5. After synthesizing the Verilog format of the circuit to netlist fie (in *blif* format), it must be optimized and mapped into the target architecture [13].

Then, the number of LUTs and uncontrollable LUT inputs are extracted by processing the netlist of the circuit. Finally, all of the test vectors and the bitstream that must be applied to uncontrollable LUT inputs are obtained in the test procedure. In fact, the test methodology proposed in the section III is applied to netlist in this step. Netlist processing and test procedure have been implemented using C++.

In the offline mode of test, these test vectors are applied to circuit in order to detect the possible interconnection faults. For online test, while some of the LUT inputs cannot be controlled and the test is performed when the FPGA operates in normal mode, the selected test vectors must be detected by a detector and apply to the circuit to determine fail or pass. Obviously, for this mode of test, 100% fault coverage cannot be achieved as all of the LUTs are not fully controllable. For example, about 56%, 22%, and 10% of the

LUTs in benchmark s5378 have four, three, and two controllable inputs, respectively.

In Table II, simulation results on ISCAS'85 and ISCAS'89 benchmarks are reported for the proposed online and offline test methods. Benchmark circuit are listed in the first column. Fault coverage and number of test vectors are reported in the following columns for offline and online test modes. In the offline test, 100% fault coverage is achieved for almost all benchmark circuits. However, the highest fault coverage is 90% for online test.

In Table. III and Table IV, the number of test configurations and the number of test vectors required for the proposed offline test method are compared against some of the previous methods. A test configuration is defined as reprogramming the user-specified LUTs in order to test the interconnection network. As discussed in the previous sections, FPGA interconnections are tested without modifying LUTs into other functions in the proposed method.

In [5] and [8], the number of test vectors and the number of test configurations are equal; because, only one test vector is applied to the circuit in each configuration. The proposed method and [6] apply more test vectors in order to test the circuit as the number of test configurations are less in these approaches. In the method proposed in [6], the user-specified LUTs are reprogrammed only one time to test the interconnects, while in the proposed method test vectors are applied to the circuit without modifying the functionality of LUTs. As shown in section III, the rules constructed for activating faults in LUT4s cannot achieve 100% fault coverage for all LUT4 functionalities. Therefore, for some of the benchmark circuits, 100% fault coverage cannot be achieved for the proposed offline method.

V. CONCLUSION

In this paper, a novel approach for offline and online application-dependent testing of FPGA interconnects has presented. The test vectors are directly applied to user-specified designs without reconfiguring LUTs in this approach, which eliminates the need for test configuration. This allows FPGA to be tested simultaneously in the normal mode of operation. However, 100% fault coverage cannot be achieved in online mode of test due to the presence of LUTs with uncontrollable inputs. According to simulation results, 100% fault coverage is achieved for almost all benchmark circuits for the offline mode of test.

REFERENCES

[1] D. Das and N. A. Touba, "A low cost approach for detecting, locating, and avoiding interconnect faults in FPGA based reconfigurable systems," in Proc. Int. Conf. VLSI Des., 1999, pp. 266–269.

[2] M. B. Tahoori, "Application-independent testing of FPGA interconnects," in Proc. 18th IEEE Symp. Defect Fault Tolerance VLSI Syst., Nov. 2003, pp. 409–416.

[3] M. B. Tahoori, E. J. McCluskey, M. Renovell, and P. Faure, "A multiconfiguration strategy for an application dependent testing of FPGAs," in Proc. 22nd IEEE VLSI Test Symp., Apr. 2004, pp. 154–159.

[4] M. Tahoori, "Application-dependent testing of FPGAs," IEEE Trans. Very Large Scale Integration (VLSI) Systems, vol. 14, no. 9, pp. 1024–1033, Sep. 2006.

[5] Banik, Shukla, Suchismita Roy, and Bibhash Sen. "Application-dependent testing of FPGA interconnect network." IEEE Transactions on Very Large Scale Integration (VLSI) Systems, vol. 27, no. 10, pp. 2296-2304, July 2019.

[6] T. N. Kumar, H. A. F. Almurib, and F. Lombardi, "Single-configuration fault detection in application-dependent testing of field programmable gate array interconnects," IET Comput. Digit. Techn., vol. 7, no. 3, pp. 132–141, May 2013.

[7] A. Cilardo, "New techniques and tools for application-dependent testing of FPGA-based components," IEEE Trans. Ind. Informat., vol. 11, no. 1, pp. 94–103, February 2015.

[8] Kumar, T. Nandha, and Fabrizio Lombardi. "A novel heuristic method for application-dependent testing of a SRAM-based FPGA interconnect." IEEE transactions on Computers, vol. 62, no. 1, pp. 163-172, December 2011.

[9] H. A. F. Almurib, T. N. Kumar, and F. Lombardi, "Scalable applicationdependent diagnosisof interconnects of SRAM-based FPGAs," IEEE Trans. Comput., vol. 63, no. 6, pp. 1540–1550, Junary 2014

[10] Nirmalraj, T., S. Radhakrishnan, and S. K. Pandiyan. "Automatic diagnosis of single fault in interconnect testing of SRAM - based FPGA." IET Computers & Digital Techniques 15, no. 5, pp. 362-371, september 2021.

[11] Boutros, Andrew, and Vaughn Betz. "FPGA architecture: Principles and progression." IEEE Circuits and Systems Magazine, vol. 21, no. 2, pp. 4-29, May 2021.

[12] Samantaray, Subhransu Ranjan, Sarita Nanda, and P. K. Dash. "A fast and adaptive dynamic phasor estimation algorithm implemented on field programmable gate array (FPGA)." IEEE Transactions on Industrial Electronics, vol. 69, no. 2, pp. 2088-2098, February 2021.

[13] Rahimi, H., and Hadi Jahanirad. "An evolutionary approach to implement logic circuits on three dimensional FPGAs." Expert Systems with Applications 174 (2021): 114780.

[14] Yazdanshenas, Sadegh, and Vaughn Betz. "Interconnect solutions for virtualized field-programmable gate arrays." IEEE Access 6 (2018): 10497-10507.

The 5ᵗʰ Iranian International Conference on Microelectronics (IICM2023)

25 – 26 October 2023

An Integrated Wearable Bio-Impedance Spectroscopy System for Remote Monitoring Heart Failure in 65nm CMOS Technology

Arman Ghouchani
Department of Electrical Engineering
Sharif University of Technology
Tehran, Iran
arman.ghoochani@sharif.edu

Mohammad Sharifkhani
Department of Electrical Engineering
Sharif University of Technology
Tehran, Iran
msharifk@sharif.edu

Abstract— **A wearable bio-impedance spectroscopy (BIS) system designed for remote monitoring of heart failure patients is presented. Leveraging bio-impedance as a non-invasive signal, the system enables continuous monitoring of vital parameters, optimizing treatment strategies and patient care. The system achieves a high-resolution of 100mΩ, accompanied by low noise and distortion levels, ensuring accurate and reliable measurements. The implementation is carried out in 65nm technology with a 1.2V supply voltage and power consumption of 720μW. Operating within the frequency range of 10KHz to 1MHz, the system captures 16 distinct samples, facilitating the generation of the Cole-Cole plot for comprehensive tissue analysis.**

Keywords— Remote Monitoring, Wearable Devices, Bio-Impedance Spectroscopy, Instrumentation Amplifier, Cole-Cole Plot, Heart Failure.

I. INTRODUCTION

Utilizing remote monitoring represents an effective approach for overseeing heart failure patients and mitigating the risk of critical incidents such as heart attacks. This novel technique holds promise for the future of healthcare by employing sensor-based monitoring to track patient well-being, enhance the capabilities of healthcare professionals, and optimize treatment strategies [1].

To facilitate early disease detection and monitor human health, non-invasive wearable devices are widely employed. Among these devices, bio-impedance serves as a non-invasive signal with diverse clinical applications, including the monitoring of congestive heart failure (CHF), diagnosis of heart and circulatory system conditions [1], assessment of hydration levels, and determination of body composition [2]. By analyzing resistance of bio-impedance, it becomes feasible to extract valuable biological tissue characteristics [3].

Fig. 1 depicts a bio-impedance spectroscopy (BIS) system comprising interconnected blocks operating from 10 KHz to 1 MHz [1]. In this setup, one pair of electrodes is responsible for delivering current to the sample, while another pair measures the induced voltage, effectively eliminating the influence of contact impedance terms during the measurement process.

In previous work, a portable system that measures heart rate using bio-impedance from the forearm was reported [4]; however, it was only measured at a single fixed frequency. Furthermore, in comparison to the AFE4300, an analog front-end chip manufactured by Texas Instruments, this research work showcases reduced power consumption. The proposed system introduces a continuous Bio-Impedance Spectroscopy (BIS) measurement approach specifically designed for chest monitoring, offering seamless integration into wearable devices.

The rest of the paper is organized as follows: a brief system-level design is presented in Section III, followed by the description of the analog front-end circuit-level design of the BIS system in Section IV. Then, Section V presents simulation results from the RC load which operates as a conceptual model of biological tissue.

II. BACKGROUND

The resistive properties of the intra-cellular fluid (ICF) and extra-cellular fluid (ECF) in the human body impede the flow of AC current, while the cell membrane exhibits capacitive behavior. At higher frequencies, the current passes through the ICF and ECF, as observed in Fig. 2 (left). To understand the relationship between bio-impedance and the frequency of an AC current, the Cole-Cole model is employed, as shown in Fig. 2 (right) [5].

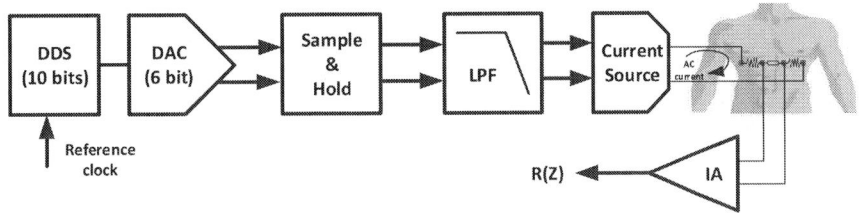

Fig. 1. BIS measurement system block diagram.

979-8-3503-6020-2/23 $31.00 © 2023 IEEE

97

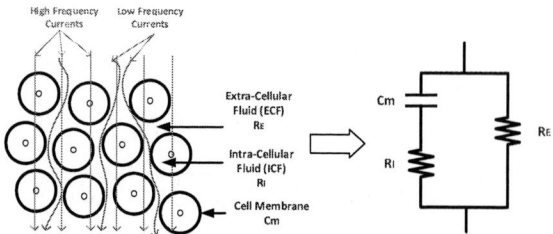

Fig. 2. The flow of current in body cells at a low and high frequency, and the equivalent circuit model.

This model includes a cell membrane capacitor (Cm) in series with ICF resistance (RI) and parallel to ECF resistance (RE). The Cole-Cole plot in Fig. 3 represents bioimpedance across frequencies by plotting reactance against resistance[6].

Bioimpedance, the measurement of electrical impedance in biological tissues, reveals the resistance encountered by electrical currents under applied voltage signals. The behavior of biological tissue in high and low-frequency electrical fields is explained by a conceptual model. At lower frequencies, the gradual charging and discharging of the cell membrane restrict current flow through the cells, while at higher frequencies, this effect becomes negligible, enabling freer current passage [1].

To measure bio-impedance, various techniques are employed, including single-frequency (SF), multi-frequency (MF), and impedance spectroscopy (IS). Single-frequency measurement offers a high signal-to-noise ratio (SNR). However, to accurately analyze the RC components of a complex bio-impedance network (Fig. 2 right), measurements at multiple frequencies are necessary to obtain comprehensive data.

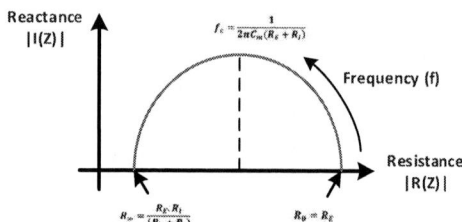

Fig. 3. Cole-Cole plot of bio-impedance.

III. SYSTEM LEVEL DESIGN

Fig. 1 illustrates the configuration of the Bio-Impedance Spectroscopy (BIS) system employed in this study. The system incorporates various components to facilitate the measurement process. A Direct Digital Synthesizer (DDS) is utilized to generate 16 discrete frequencies by manipulating a reference frequency. The digital signal produced by the DDS is subsequently converted into an analog signal using a Digital-to-Analog Converter (DAC). A Sample and Hold circuit, in conjunction with a Low Pass Filter (LPF), is employed to reconstruct a sinusoidal voltage waveform. Finally, a Current Source is employed to convert the sinusoidal voltage into a corresponding current, which is subsequently injected into the tissue of interest. To ensure accurate measurements, two separate pairs of electrodes are utilized: one pair for the current source and another pair for sensing the voltage via an Instrumentation Amplifier (IA). Implementing this tetra-polar electrode configuration mitigates the influence of unknown contact impedances, enhancing measurement accuracy and reliability.

The utilization of multiple frequencies provides the capability to construct a precise Cole-Cole plot (refer to Fig. 3), enabling accurate monitoring of patient status and facilitating informed treatment decisions. To achieve this, a Direct Digital Synthesizer (DDS) is employed to generate 16 discrete frequencies ranging from 10 kHz to 1 MHz, derived from a reference clock signal. The output of the DDS is subsequently converted into an analog signal by a Digital-to-Analog Converter (DAC). Notably, the DAC incorporates a C-2C structure, optimizing power consumption and catering to the power limitations of mobile devices, which is of utmost importance in this context.

In an ideal scenario, the Sample and Hold block should pass data from the DAC without introducing any distortion. However, due to the finite gain of the operational amplifier (OP-AMP), some limitations arise. To ensure acceptable precision, let's assume that only a 0.5-bit error is tolerable. In a 6-bit system, where 1/64 represents one least significant bit (LSB), the aim is to achieve a maximum error of 0.5 LSB. Therefore, the calculation of error is as follows:

$$\frac{1}{2} \times \frac{1}{64} \text{ LSB} = \frac{1}{128} = 0.78\% \text{ error} \tag{1}$$

Next, the task at hand involves finding the appropriate gain for the Sample and Hold block to keep the error below 0.78%. Assuming a 0.5% error is satisfactory, the equation can be expressed:

$$\left(1 + \frac{Cx}{Cs}\right) \frac{1}{A} = 0.5\% \text{ error} \tag{2}$$

Considering that Cx represents parasitic capacitors at node P (as shown in Fig. 5), Cs is the holding capacitor, and A denotes the OP-AMP gain, the solution for A can be derived:

$$\left(1 + \frac{0.1 \text{ PF}}{0.1 \text{ PF}}\right) \frac{1}{A} = \frac{5}{1000} \rightarrow A = 400 \tag{3}$$

Assuming the parasitic capacitor is equal to 0.1 pF, the desired OP-AMP gain for the Sample and Hold block is 400, which is equivalent to 52 dB.

To ensure the passage of the entire signal, the Low Pass Filter (LPF) should possess a bandwidth of 1 MHz. In a 6-bit system, the desired attenuation is approximately 38 dB. To determine the appropriate order of the LPF, the following calculation is performed:

$$\frac{Fs-2FB}{FB} = \frac{20\text{MHz}-(2 \times 1\text{MHz})}{1 \text{ MHz}} = 18 \tag{4}$$

Fs represents the clock frequency, and FB corresponds to the maximum signal frequency. Therefore, an antialiasing filter with an 18 MHz bandwidth is necessary. Considering the target attenuation of 38 dB, the LPF order can be determined as follows:

$$\text{LPF Order} = \frac{38 \text{ dB}}{20 \log(18)}$$
$$\rightarrow \frac{38}{20 \times 1.25} = 1.5^{\text{th}} \text{ order} \tag{5}$$

To achieve an improved output signal, it is advisable to employ a second-order LPF.

In order to convert voltage at different frequencies into current and inject it into the tissue, the Improved Howland current source proves to be an excellent choice. Among the crucial specifications for current sources, the output impedance holds utmost importance. The Improved Howland current source, by virtue of its structure, exhibits an infinite output impedance regardless of the output impedance of its operational amplifier (OP-AMP). The input and output impedances are defined as follows[7]:

$$R_i = \frac{R_1 R_{4b}(R_2 + R_{4a})}{[R_1(R_{4a} + R_{4b}) + R_{4a}R_3]} \qquad (6)$$

$$R_o = \frac{R_{4b}\left(1 + \dfrac{R_{4a}}{R_2}\right)}{\left[\dfrac{(R_{4a} + R_{4b})}{R_2} - \dfrac{R_3}{R_1}\right]} \qquad (7)$$

Theoretically, when both positive and negative feedbacks satisfy the equation (R4a + R4b)/R2 = R3/R1, an infinite output impedance can be achieved. However, due to tolerances present in resistors, the output impedance becomes finite. It is worth noting that the output impedance is also influenced by the drop voltage in the load, resulting in load dependency.

To attain a resolution of 100mΩ, the detection of voltage in the system relies on an instrumentation amplifier (IA). With a 100µArms injected current, the IA can detect the lowest voltage of approximately 2µV, accounting for a 1/5 ratio of the current reaching the desired tissue. This voltage detection ensures the achievement of the desired 100mΩ resolution. Additionally, in consideration of the signal processing algorithm, a minimum signal-to-noise ratio (SNR) of 10dB is set, corresponding to a maximum Readout Input-Referred Noise (IRN) of 200nV/√Hz.

IV. Circuit Level Design

A. DDS

Direct Digital Synthesis (DDS) is a digital technique that generates analog signals, including sine waves, using digitally stored values. By utilizing a "template" stored in memory, which contains amplitude values for each phase of the waveform, DDS directly synthesizes signals without the need for phase-locked loops. Through manipulation of phase values and signal operations like addition and scaling, DDS can produce various waveforms and precise frequencies. In this system, a 20 MHz clock reference generates 16 frequencies ranging from 1 kHz to 1 MHz, which are crucial for generating the Cole-Cole plot.

B. DAC

Capacitor arrays are widely utilized in the design of Digital-to-Analog Converters (DACs). Among various capacitive DAC architectures, the C2C (Capacitor-to-Capacitor) DAC is preferred due to its advantages such as small capacitance ratios, high conversion rate, and low power consumption. The schematic diagram of the C2C DAC is depicted in Fig. 4. The digital inputs to the DAC are controlled by a Direct Digital Synthesizer (DDS). DAC switches receive the digital bits as inputs and alternate the output voltage between the reference voltage (Vref) and ground (GND).

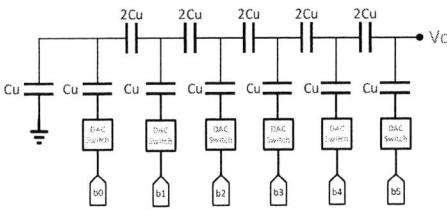

Fig. 4. 6-bit C2C DAC architecture.

The total capacitance spread of the C2C 6-bit DAC is 17Cu, which is significantly lower compared to a binary-weighted capacitor array DAC, which would have a total capacitance of 64Cu. The novelty of the designed C2C DAC lies in the selection of the unit capacitor value to achieve high performance while minimizing the effects of parasitic capacitance.

C. Sample & Hold

Fig. 5 illustrates the structure of the Sample and Hold circuit. During the acquisition mode (Φ1), switches S1-S4 are turned on, while switches S5 and S6 are turned off. This configuration resets the Operational Amplifier (OP-AMP), and the sampling capacitors Cs track the analog input voltage. When transitioning to the hold mode (Φ2), switches S3 and S4 are turned off first, followed by switches S1 and S2, and finally, switches S5 and S6 are turned on.

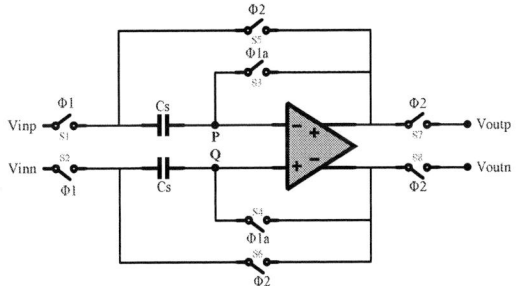

Fig. 5. CMOS sample-and-hold circuit architecture.

Although this switching sequence helps suppress input-dependent charge injection, the circuit faces a challenge in terms of long hold settling time because the differential output always starts from zero at the beginning of the hold mode. Additionally, the channel charge injection from switches reduces the precision. To address the first problem, two additional switches, S7 and S8, are introduced to separate the output of the operational amplifier from the hold capacitor. To tackle the second problem, a specific switching sequence is employed during the transition from the sampling to the hold mode. First, switches S3 and S4 are turned off (Φ1a), followed by a short delay of approximately 2 ns. Then, switches S1 and S2 are turned off as well (Φ1). Finally, in the last step, switches S5 and S6 are turned on, completing the transition. This approach ensures that the channel charge injection does not affect the accuracy of the circuit.

As mentioned in Section III, the operational amplifier (op-amp) requires a gain of 54 dB. Additionally, it is necessary for the op-amp to have Rail-to-Rail input and output capabilities to cover the full range of the signal and achieve maximum Signal-to-Noise Ratio (SNR). To enable Rail-to-Rail input, two pairs of NMOS and PMOS transistors are used at the input stage to accommodate the entire signal range. For the output stage, a class AB configuration is employed to achieve a full output swing. The schematic of the op-amp is depicted in Fig. 6. The op-amp design demonstrates excellent performance and stability, with a gain of 60 dB and a phase margin of 60 degrees.

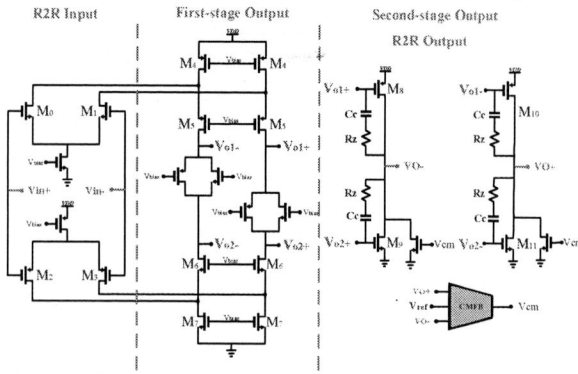

Fig. 6. Detailed schematic of the presented two-stage rail-to-rail amplifier.

D. Low Pass Filter

As mentioned in Section III, it is necessary to implement a second-order low-pass filter (LPF). To achieve this, a passive real pole can be introduced by including R_1 and C_2 in the input configuration, as illustrated in Fig. 7.

Additionally, capacitor C_2 can be differentially connected across the inputs, as shown by the solid lines in the diagram. Alternatively, for enhanced rejection of common-mode noise, two capacitors (each with twice the value) can be connected between each input or output and ground, as demonstrated by the dashed lines.

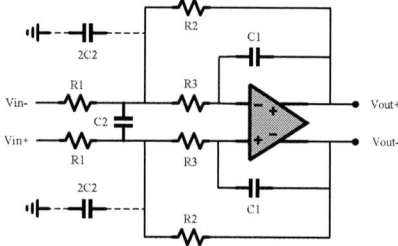

Fig. 7. Second-Order Low-Pass Filter.

It is necessary to limit the attenuation in the pass band that spans from 10kHz to 1MHz, allowing for consistency over the range of input signals. To calculate the maximum variation of gain/attenuation within this frequency range, the following equation can be used:

$$20 \log \left(1 - \frac{1}{64}\right) = -0.13 \text{ dB} \tag{8}$$

Here, 1/64 represents the maximum tolerable error associated with the 6-bit accuracy of the system, and the objective is to ensure that the error between the output of a 10kHz input and a 1MHz input is less than 1 LSB (Least Significant Bit). By performing the calculation, the maximum difference gain is equal to -0.13 dB.

E. Current Source

Fig. 8 illustrates the current source circuit. Based on the equation mentioned in section III, R_{4b} is selected as 1KΩ and R_{4a} as 99KΩ, while R_1 and R_2 are both set to 50KΩ. Consequently, R_3 is calculated as 100KΩ.

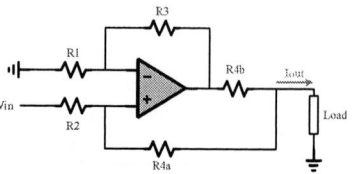

Fig. 8. Improved Howland current source circuit.

For safety considerations, this current source has a peak-to-peak current of 300µA. Additionally, the amplitude of the current remains constant across the frequency range of 10kHz to 1MHz and can accommodate load resistances ranging from 10 ohms to 120 ohms. The maintenance of a constant current is crucial for accurately determining the impedance of the tissue.

F. Instrumentation Amplifier

Fig. 9 depicts the circuit diagram of the Instrumentation Amplifier (IA). This block can be divided into an input stage and an output stage. The input stage consists of a differential pair composed of transistors M_1 and M_2, which is degenerated by resistor R_1. Each stage incorporates local feedback. In the input stage, this feedback is achieved through transistors M_3 to M_6, while in the output stage, it is implemented by transistors M_7 to M_{18} using a cascode topology to enhance the gain. A Common Mode Feedback (CMFB) structure is utilized to establish the DC voltage.

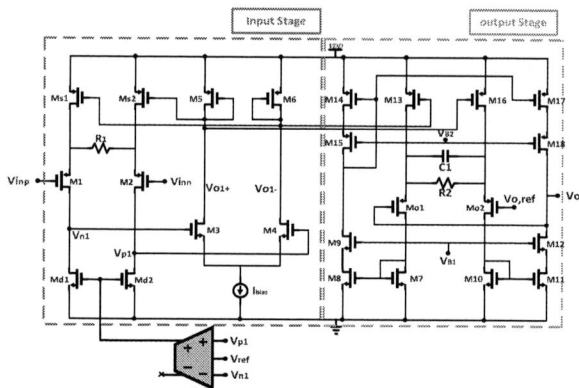

Fig. 9. IA transistor level schematic.

Due to the high gain of the feedback loop in the input stage, the voltages at the two nodes, Vp1 and Vn1, are approximately equal. Consequently, considering the current sources M_{d1} and M_{d2}, which possess the same gate-source voltage and drain voltage, the current in the two input differential branches becomes equal ($I_{DM1} = I_{DM2}$). As a result, any voltage generated at the input due to the identical current in transistors M_1 and M_2 is directly transferred to both ends of resistor R_1 with a gain close to one. This process is repeated in the differential pair M_{o1} and M_{o2} due to the high gain of the feedback loop in the second stage.

V. RESULTS

A system-level design and a circuit-level design have been presented for a remote monitoring system aimed at overseeing heart failure patients. In order to generate the Cole-Cole plot, 16 voltages at different frequencies are extracted from the output of the Instrumentation Amplifier (IA), along with 16 injected currents from the current source. These voltage and current data points are used to create the Cole-Cole plot using MATLAB code, enabling the validation of the circuit's performance.

Fig. 10, displays the Cole-Cole plot obtained for an RI and RE value of 40 ohms and 60 ohms, respectively along with a Cm value of 100nF (as depicted in Fig. 2). The obtained results exhibit a close alignment with the expected load variations, indicating the successful performance of the circuit. These plots serve as an important validation step, demonstrating that the implemented circuit accurately captures and reflects the impedance characteristics of the monitored tissue.

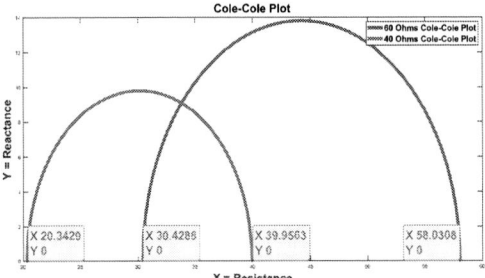

Fig. 10. Cole-Cole plot for RE and RI equal to 40 and 60 ohms.

The Instrumentation Amplifier (IA) in this design achieves an Input Referred Noise (IRN) of approximately 14.5nV/√Hz in the operation frequency range, which is well below the desired threshold of 200nV/√Hz. This indicates that the IA meets the noise requirement. Additionally, the Total Harmonic Distortion (THD) at the output signal should be -38 dB, to ensure the successful recovery of the 6-bit data from the Digital-to-Analog Converter (DAC).

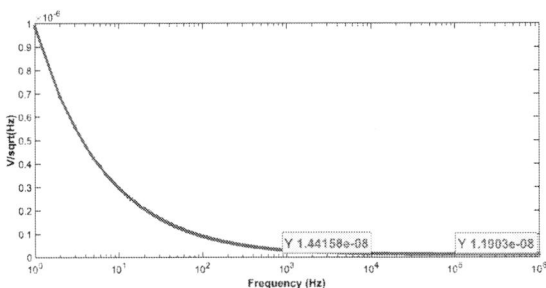

Fig. 11. Instrumentation Amplifier's Input Referred Noise.

In the following equation, the calculations related to the resolution and accuracy of the IA are presented. It is important to note that these calculations assume a detection bandwidth of 10 Hz and a required SNR value of 10 dB for effective signal processing.

$$\text{Resolution} = \frac{\sqrt{IRN^2 \times BW}}{\text{Injection Current}} = 0.16 \ m\Omega/\sqrt{Hz} \quad (9)$$

In Fig. 12, the Total Harmonic Distortion (THD) plot of the IA's output signal is depicted. In this test, the full chain of impedance measurement goes under test at full scale at 1MHz. The graph demonstrates that the maximum harmonic is the second harmonic as expected. This distortion is approximately -40dB below the signal tone. This signifies that the entire chain offers sufficiently accurate output for impedance spectroscopy application. Consequently, it can be inferred that

the measured data is fully recoverable without significant distortions, ensuring the accuracy and fidelity of the amplified signal.

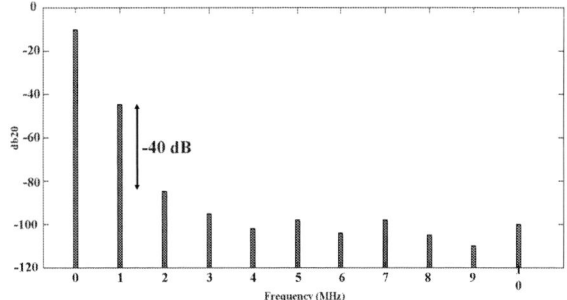

Fig. 12. Instrumentation Amplifier's THD.

Table 1 presents a comprehensive comparison between the AFE4300 from Texas Instrument, AD5933 from Analog Devices, and the proposed work across various aspects. The proposed implementation exhibits substantially lower power consumption, a crucial factor for battery-operated mobile devices. Leveraging the 65nm technology has further enabled us to achieve higher speeds and precision in measuring impedance, enhancing the overall performance of the system.

TABLE I.

COMPARISON BETWEEN AFE4300، AD5933 AND PROPOSED WORK

	Clock Frequency	Power	Supply Voltage	Technology	Error of Mag
AFE4300	1 MHz	3.2mW	3.3v	350 nm	2%
AD5933	16.7 MHz	33mW	3.3v	350 nm	3%
This Work	20 MHz	720µW	1.2v	65 nm	1.5%

VI. ACKNOWLEDGMENT

I would like to express our deepest gratitude to Mojtaba Rezapour for his invaluable guidance, and support.

REFERENCES

[1] G.L. Cuba Gyllensten, "Monitoring heart failure using noninvasive measurements of thoracic impedance," Ph.D. dissertation, Dept. Elect. Eng., Technische Universiteit Eindhoven, 2018.

[2] P.J. Yoo, D.H. Lee, T. Oh and E.J. Woo, "Wideband Bio-impedance Spectroscopy using Voltage Source and Tetra-polar Electrode Configuration," *Journal of Physics*, Conference Series 224, 2010.

[3] B. Ibrahim, D. A. Hall, and R. Jafari, "Bio-impedance spectroscopy (BIS) measurement system for wearable devices", *2017 IEEE Biomedical Circuits and Systems Conference (BioCAS)*, pp. 1-4, 2017.

[4] M. C. Cho, J. Y. Kim, and S. H. Cho, "A bio-impedance measurement system for portable monitoring of heart rate and pulse wave velocity using small body area", *2009 IEEE International Symposium on Circuits and Systems (ISCAS)*, pp. 3106-3109, 2009.

[5] U. G. Kyle, I. Bosaeus, A. D. D. Lorenzo, P. Deurenberg, M. Elia, J. M. Gómez, B. L. Heitmann, L. Kent-Smith, J.-C. Melchior, M. Pirlich, H. Scharfetter, A. M. W. J. Schols, and C. Pichard, "Bioelectrical impedance analysis—part I: a review of principles and methods" *Clinical Nutrition*, vol. 23, pp. 1226-1243, 2004.

[6] F. Seoane, J. Ferreira, J. J. Sanchéz and R. Bragós, "An analog front-end enables electrical impedance spectroscopy system-on-chip for biomedical applications" *Physiological Measurement*, vol. 29, p, 2008.

[7] S. A. Santos, T. Schlebusch, and S. Leonhardt, "Simulation of a current source with a cole-cole load for multi-frequency electrical impedance tomography" *35th Annual International Conference of the IEEE Engineering in Medicine and Biology Society (EMBC)*, pp. 6445-6448, 2013.

979-8-3503-6020-2/23 $31.00 © 2023 IEEE

The 5ᵗʰ Iranian International Conference on Microelectronics (IICM 2023)

October 25-26, 2023

2-D Axisymmetric Modeling of Circular PCB Coils and Solenoids in COMSOL Multiphysics

Farshad Gozalpour and Mohammad Yavari

Integrated Circuits Design Laboratory, Department of Electrical Engineering, Amirkabir University of Technology (Tehran Polytechnic), P.O. 15875-4413, Tehran 15914, Iran.

Emails: f.gozalpour@aut.ac.ir, myavari@aut.ac.ir

Abstract—In this paper, a 2-dimentional (2-D) axisymmetric modeling is presented for biomedical circular printed circuit board (PCB) coils and solenoids in COMSOL Multiphysics. With this model, the number of required meshes in finite element method (FEM) based simulations is reduced, which speeds up the iterative design procedure of inductive links in biomedical implants. While having a good accuracy, the presented 2-D axisymmetric modeling reduces the extraction time of electromagnetic parameters of PCB coils and solenoids by about 90%, compared to 3-D modeling. In order to investigate the accuracy of presented 2-D modeling, we have fabricated a PCB coil and a solenoid with suitable geometries for implantation in body. The measurement results of electromagnetic parameters have a good agreement with the 2-D modeling based simulation results, and they verify each other. Finally, using the presented 2-D modeling, a comprehensive parametric simulation has been performed to study the behavior of coupling coefficient (k) and the maximum achievable k of PCB coils as a function of geometric parameters.

Keywords—*2-dimentional (2-D) axisymmetric modeling, circular PCB coil, solenoid, biomedical inductive link, electromagnetic parameters, COMSOL Multiphysics.*

I. INTRODUCTION

Implantable medical devices (IMDs) with the main purpose of improving the body's organs with unusual performance have attracted special attention in recent years. They include cardiac pacemaker [1], brain-machine interface (BMI) [2], retinal prosthesis [3], cochlear implant [4], etc. The use of batteries to provide the power in IMDs requires periodic surgery due to the limited battery life-time. Meanwhile, wireless power transmission (WPT) provides a continuous and safe operation of IMDs without the need for wires passing through the skin that causes infection. As a result, the wireless operation of IMDs is essential to prevent infection and human comfortability. Inductive coupling is the oldest and predominant strategy of wireless transmission that can provide the power required by most biomedical systems (from a few milliwatts to several tens of milliwatts) with high reliability, safety, and efficiency. Circular printed circuit board (PCB) coils and solenoids play an important role in inductive links and their geometric parameters have a direct effect on the link characteristics.

In design of inductive links, in order to achieve the desired targets, usually an iterative procedure must be performed [5-12]. A general flow diagram of iterative design procedure in biomedical inductive links is illustrated in Fig. 1. Every

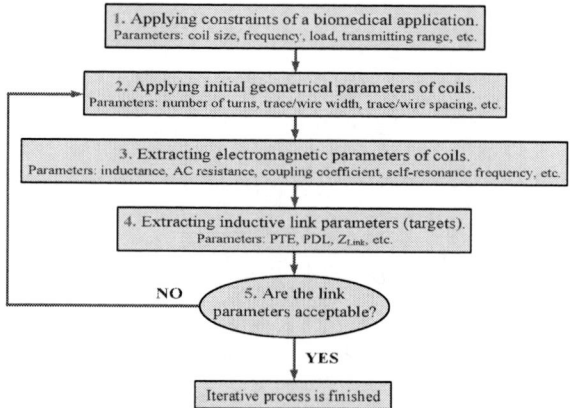

Fig. 1. A general flow diagram of iterative design procedure in biomedical inductive links.

biomedical application has its own design constraints and targets. The design constraints are mainly related to the working frequency, transmitting range, loading resistance, coil size, etc. First, using these constraints, initial values are applied for the geometric parameters of the coils such as number of turns, outer diameter, trace/wire width, trace/wire spacing, etc. Then, the electromagnetic parameters of the coils such as inductance, AC series resistance, parallel capacitance, coupling coefficient (k), etc., are derived and finally the parameters of the link performance such as power delivered to load (PDL) and power transmission efficiency (PTE) are obtained. If the achieved parameters are not acceptable in comparison with the design targets, this procedure should be repeated as far as the desired values are achieved. One of the factors that slows down this process is the finite element method (FEM) based simulations in extracting the electromagnetic parameters of the coils. Therefore, providing a two-dimensional (2-D) model for the coils can be an effective solution to this problem, considering that it converts 3-D volume of the geometry to 2-D domains and reduces the time required to solve the governing equations in defined geometry.

In this paper, using axisymmetric tools in COMSOL Multiphysics 5.5, a 2-D axisymmetric modeling for circular PCB coils and solenoids is proposed that can be utilized in most of previously published iterative design procedures of inductive links [5, 11, 12]. Since this model converts the 3-D volume of geometry to 2-D domains, FEM based simulations are performed more quickly, and as a result, the

979-8-3503-6020-2/23 $31.00 © 2023 IEEE

102

required time in iterative procedure of extracting the electromagnetic parameters and the optimization of the link performance is significantly reduced. Finally, the proposed 2-D model has been applied and a comprehensive step-by-step parametric simulation is performed in order to study the effect of geometric variables of the coils such as outer diameter, number of turns, trace width and trace spacing on the behavior of coupling coefficient and the maximum achievable coupling between the PCB coils.

The rest of the paper is organized as follows. Section II presents the proposed 2-D axisymmetric modeling of circular PCB coils and solenoids. In section III, the 2-D modeling based simulation results and also the measurement results of the coil parameters are presented. Finally, conclusion is given in section IV.

II. 2-D AXISYMMETRIC MODELING

Fig. 2(a) and Fig. 2(b) illustrate the top view of a circular PCB coil and a solenoid with their geometric parameters. In PCB coil, the parameters W, S, n, D_{out}, and D_{in} are trace width, trace spacing, number of turns, outer diameter, and inner diameter, respectively. In solenoid, the parameters n, p, l, D, and d are the number of turns, winding pitch, winding length, winding diameter, and wire diameter, respectively. The electrical model of PCB coil and solenoid is shown in Fig. 2(c), which includes the inductance L, AC series resistance R_S, AC parallel resistance R_P, and the parallel capacitance C_P. At low frequencies, the parallel resistance R_P can be ignored, and the self-resonance frequency and quality factor (Q) of PCB coil and solenoid are mainly determined by L, R_S, and C_P.

Fig. 3 shows the 2-D axisymmetric modeling of circular PCB coil and solenoid, where the rectangular and circular domains illustrate the cross-section area of conductor turns in PCB coil and solenoid, respectively. These domains are defined as a resistive-inductive-capacitive (RLC) coil group. They consider both in-plane and out-of-plane currents flowing in equilibrium condition that is required for 2-D modeling of the coils. The RLC coil group is a combination of magnetic model of a single-turn coil and a multi-terminal electrical model of in-plane current [13]. To define the arrangement for connection of domains, RLC group employs Bravais network that has two primary vectors **a** and **b** and provides two points $\mathbf{r}_1$ and $\mathbf{r}_2$: $\mathbf{r}_1-\mathbf{r}_2 = m_1\mathbf{a}+m_2\mathbf{b}$, where m_1 and m_2 are integer numbers. RLC coil group proposes $n+1$ electric potentials (V_0, V_1, ..., V_n), where V_0 is the reference voltage and n is the number of turns. The potential difference of the i-th turn is equal to:

$$V_i^d = V_i - V_{i-1} \tag{1}$$

where $i = 1, ..., n$. The potential difference for any turn gives the out-of-plane external current density ($\mathbf{J}_e$) as follows:

$$\mathbf{J}_e = \sigma \frac{V_i^d}{2\pi r}\mathbf{e}_\varphi \tag{2}$$

where $\mathbf{e}_\varphi$ is the out-of-plane unit vector, σ is the conductivity, and r is the vertical separation of the center of conductor cross-sections and the line $r = 0$. So, the out-of-plane current in i-th turn can be expressed by:

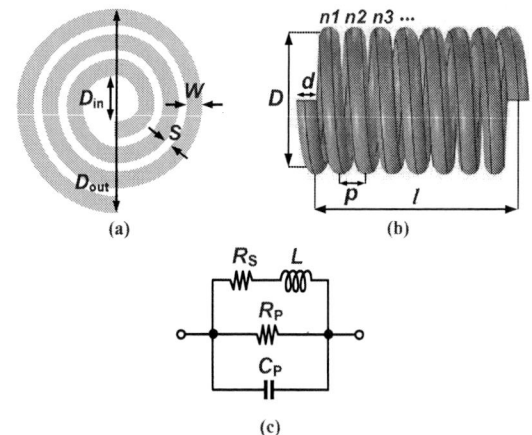

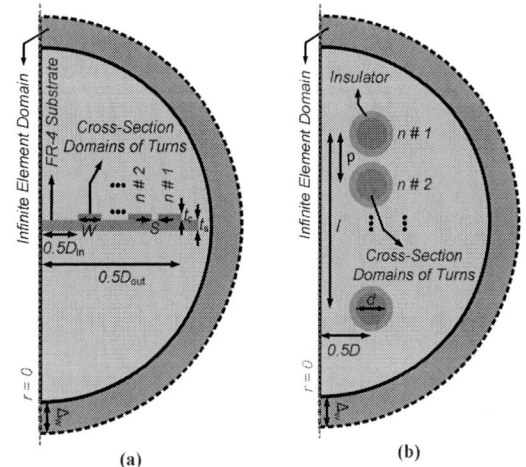

Fig. 2. (a) Circular PCB coil, (b) solenoid, (c) electrical model of PCB coil and solenoid.

Fig. 3. 2-D axisymmetric modeling of (a) circular PCB coil and (b) solenoid.

$$I_i = \int_{\Omega_i}\left(\mathbf{J}.\mathbf{e}_\varphi\right) \tag{3}$$

Equilibrium condition between in-plane current and difference of out-of-plane currents in two neighboring turns implies [13]:

$$\begin{cases} I_i - \int_{\partial\Omega_i}\left(\mathbf{J}.\mathbf{n}\right) - I_{i+1} = 0 & i = 1,...,n-1 \\ I_n - \int_{\partial\Omega_n}\left(\mathbf{J}.\mathbf{n}\right) - I_{Coil} = 0 & i = n \end{cases} \tag{4}$$

where **n** is in-plane unit vector and I_{Coil} is the coil current. The electric potential is considered to have a constant value in cross section domain of turns, and therefore, this variable must be discarded in the mentioned domains by employing Ampere's law. The Ampere's law considers a $\mathbf{J}_e$ orthogonally to the domains, and also establishes a voltage limitation on the boundary of every cross-section domains of the turns. The magnetization, conduction and dielectric models needed for Ampere's law is considered as follows:

$$\mathbf{B} = \mu_0\mu_r\mathbf{H}, \quad \mathbf{J} = \sigma\mathbf{E}, \quad \mathbf{D} = \varepsilon_0\varepsilon_r\mathbf{E} \tag{5}$$

where **B**, **H**, **D**, and **E** are magnetic flux density, magnetic field, electric field intensity, and electric flux density, respectively. Also, μ_r, μ_0, ε_r, and ε_0 are relative permeability,

The 5ᵗʰ Iranian International Conference on Microelectronics (IICM 2023)

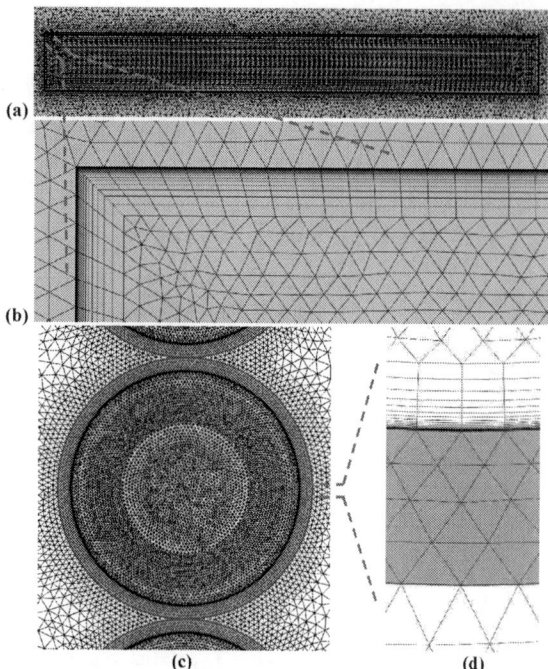

Fig. 4. Meshing of (a, b) rectangular cross-section domain and its boundary layer in PCB coil, (c, d) circular cross-section domain and its boundary layer in solenoid.

TABLE I   GEOMETRIES OF THE FABRICATED PCB COIL AND SOLENOID.

Parameter	D_{out} [mm]	D_{in} [mm]	W [mm]	S [mm]	n
Circular PCB Coil	25.5	11.1	0.3	0.3	12
Parameter	D [mm]	d [mm]	p [mm]	l [mm]	n
Solenoid	3.22	0.2	0.22	9.46	43

permeability of free space, relative permittivity, and permittivity of free space, respectively.

The characterization of infinite domains is a common challenge in FEM simulations. In these simulations, artificial boundaries are usually used to limit the model to an area of interest. These boundaries must not influence the simulations inside the area of interest. In this paper, we have used infinite element domain (IED). The IED employs coordinate stretching in a virtual layer around the area of interest. The stretching function is defined as [14]:

$$f(\xi) = \frac{\xi}{\gamma - \xi}\Delta_p, \quad \gamma = \frac{\Delta_s + \Delta_p}{\Delta_s} \qquad (6)$$

where ξ is the dimensionless coordinate ($0 < \xi < 1$), Δ_p is the pole distance, Δ_s is the scaled width of IED, and γ is a number larger than one. Δ_s and Δ_p are the primary parameters for stretching function. $f(\xi)$ returns a new stretched position, and therefore, the movement for stretching is defined as $\Delta\mathbf{x} = f_i(\xi) - \Delta_w \xi$, where Δ_w is the original width of infinite domain as shown in Fig. 3.

III. SIMULATION AND MEASUREMENT RESULTS

In order to investigate the accuracy of proposed 2-D axisymmetric modeling, we have fabricated PCB coil and solenoid with reported geometric parameters in Table I. The PCB coil is implemented on FR-4 PCB with substrate thickness (t_s) and copper thickness (t_c) of 1.6 mm and 0.035 mm, respectively. Also, the magnet copper wire is employed

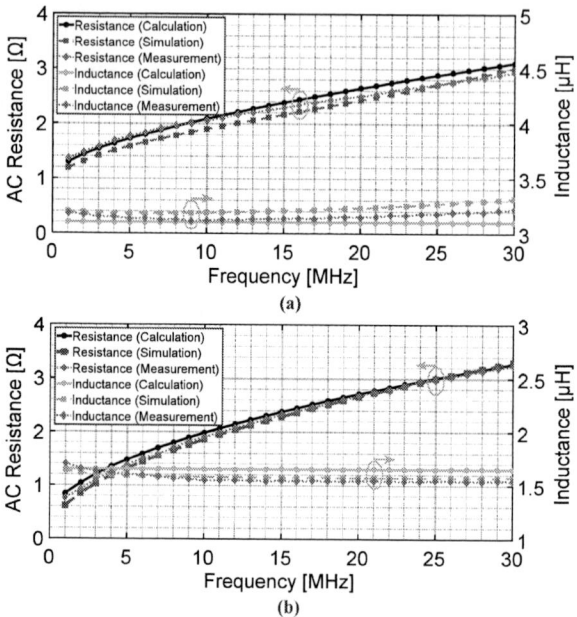

Fig. 5. Inductance and AC resistance of (a) circular PCB coil, (b) solenoid.

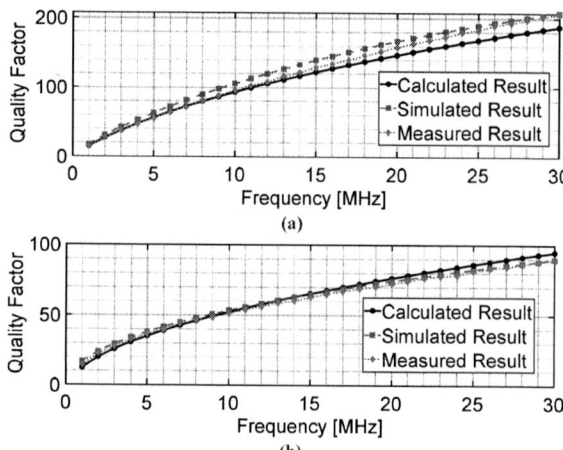

Fig. 6. Quality factor of (a) circular PCB coil, (b) solenoid.

for the solenoid. It should be noted that due to insulating layer of magnet wire in solenoid, p is slightly larger than d. In simulations, free triangular structure is used as the meshing of 2-D cross-section domains in geometry (Fig. 4). Also, boundary layer meshing is used for the boundaries of the cross-section domains of the turns. This is a meshing with compressed element distribution and is capable to estimate the skin effect on boundaries properly.

Fig. 5(a) and Fig. 5(b) depict the measured, simulated, and calculated inductance and AC series resistance in the range of 1 MHz to 30 MHz for circular PCB coil and solenoid, respectively. The measurement results have a good agreement with the 2-D modeling based simulation results, and they verify each other. It is worth mentioning that the measurement results of Fig. 5 have been obtained using the Wayne Kerr 6550P high frequency LCR meter with 1J1011 fixture connected to the front panel BNC sockets. Also, the calculations have been extracted by the conventional equations of PCB coils and solenoids [11, 15, 16].

The 5th Iranian International Conference on Microelectronics (IICM 2023)

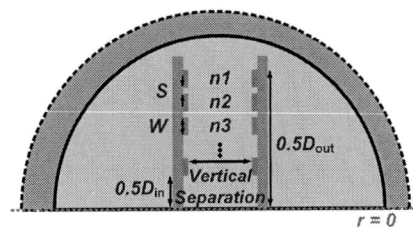

Fig. 7. 2-D simulation setup of coupling coefficient for PCB coils.

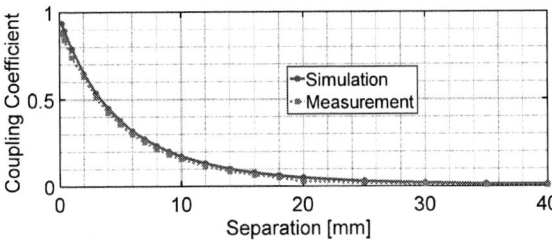

Fig. 8. Coupling coefficient of PCB coils with the geometries of Table I.

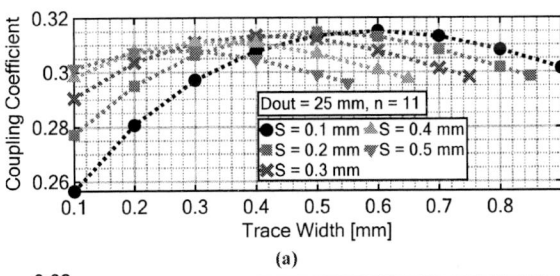

(a)

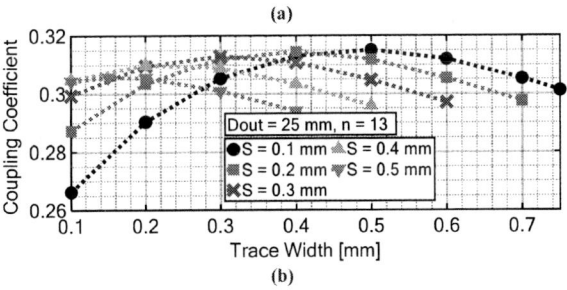

(b)

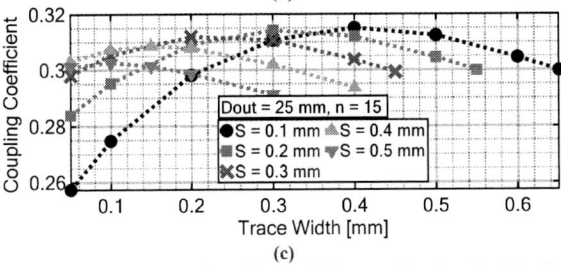

(c)

Fig. 9. Parametric sweep of coupling coefficient as a function of n, W and S: (a) $n = 11$, (b) $n = 13$, (c) $n = 15$.

Using the calculated, simulated, and measured inductance and AC series resistance, the calculated, simulated, and the measured quality factor ($Q = \omega L/R$) of PCB coil and solenoid in the range of 1 MHz to 30 MHz are extracted and illustrated in Fig. 6(a) and Fig. 6(b), respectively. The measurement results have a good agreement with the 2-D modeling based simulation results, and they verify each other.

The simulation results of coupling coefficient (k) are based on the 2-D modeling of Fig. 7. Also, the measurement

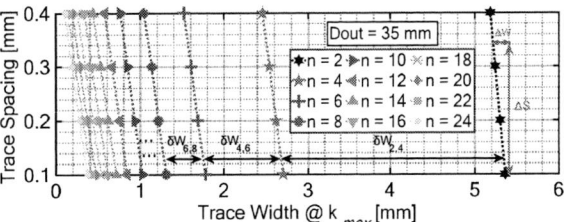

Fig. 10. Simulated S-$W_{k\text{-max}}$ lines for maximum coupling coefficient as a function of n ($D_{out} = 35$ mm).

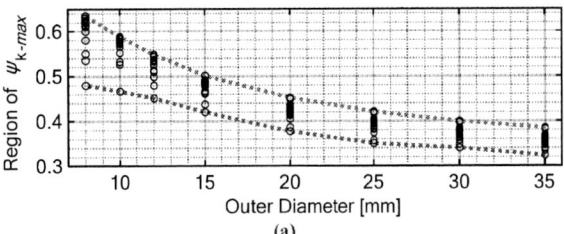

(a)

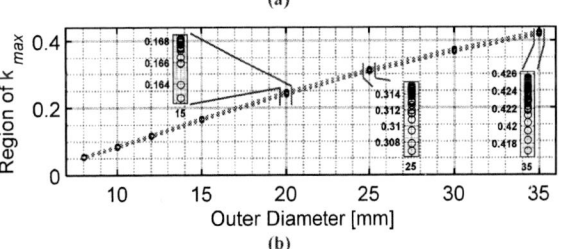

(b)

Fig. 11. (a) Region of $\Psi_{k\text{-max}}$ and (b) corresponding region of k_{max} for different D_{out}.

results are based on the measured voltage transfer function of inductive link with Rigol RSA3015N vector network analyzer. Fig. 8 illustrates the measured and simulated coupling coefficient between two circular PCB coils for the separation of 0.2 mm to 40 mm. The geometric parameters of both PCB coils are the same as Table I. According to Fig. 8, the measurement result has very good matching with simulation result and verifies the results of proposed 2-D model. It is worth mentioning that the simulation duration of coupling coefficient with 2-D model is only 25 sec, while it is equal to 350 sec for the conventional 3-D model. It can be seen that the simulation time has been drastically reduced by the proposed 2-D model.

In the next part of this section, a step-by-step parametric simulation is performed by 2-D axisymmetric modeling to study the behavior of maximum coupling coefficient of circular PCB coils as a function of geometric parameters such as D_{out}, n, W and S. In these simulations, the separation of PCB coils is kept as 6 mm. The simulation of Fig. 9 shows the k sweep for $D_{out} = 25$ mm and the different values of W and S. Also, the values of 11, 13, and 15 are selected for n in Fig. 9(a), Fig. 9(b), and Fig. 9(c), respectively. It can be observed that with increasing W, k is an increasing function of W to some extent, and then decreases. It is important to note that this increasing-decreasing behavior of k is independent of W, S and n. This behavior of k is also independent of D_{out}, which the corresponding results are reported in the next simulations.

In the next step, for the each S of 0.1 mm, 0.2 mm, 0.3 mm, and 0.4 mm, W is swept with much smaller steps to get the exact value of maximum k (k_{max}) and corresponding W

979-8-3503-6020-2/23 $31.00 © 2023 IEEE   105

$(W_{k\text{-}max})$. This simulation is performed for D_{out}=35 mm and different n, and the S-$W_{k\text{-}max}$ lines are shown in Fig. 10. Each S-$W_{k\text{-}max}$ lines corresponds to a specific n. $\delta W_{i,j}$ is the value of the shift between the lines $n = i$ and $n = j$. According to Fig. 10, $\delta W_{i,j}$ is approximately equal to:

$$\delta W_{i,j} = W_j = \frac{i}{j} W_i \tag{7}$$

where W_i and W_j are the $W_{k\text{-}max}$ for $n = i$ and $n = j$, respectively, and i and j can be any number of turns. This approximation is not limited to $D_{out} = 35$ mm and it holds for the all values of D_{out}. Also, the S-$W_{k\text{-}max}$ lines for a particular D_{out} have almost the same slopes ($\Delta S/\Delta W$).

In the final step, the parametric simulation of Fig. 10 is repeated for different values of D_{out} and the results are summarized in Fig. 11. $\Psi_{k\text{-}max}$ is the ratio of D_{in}/D_{out} where the k is maximized. This ratio is a function of coil geometric parameters such as W, S, n and D_{out}. Each of tiny circles in Fig. 11 corresponds to a specific D_{out}, W, S and n, where the k is maximized. The region of $\Psi_{k\text{-}max}$ as well as the corresponding region of k_{max} are shown in Fig. 11(a) and Fig. 11(b), respectively. It is observed that as D_{out} increases, the $\Psi_{k\text{-}max}$ decreases, meaning that k_{max} occurs at lower D_{in}/D_{out} ratios.

IV. CONCLUSION

In this paper, using axisymmetric tools in COMSOL Multiphysics, a 2-D axisymmetric modeling for biomedical circular PCB coils and solenoids is proposed that converts the 3-D volume of geometry to 2-D domains. Therefore, the FEM based simulations are performed more quickly, and as a result, the required time in iterative procedure of extracting the electromagnetic parameters and the optimization of the link performance is significantly reduced. As a comparison, the simulation of coupling coefficient with the proposed 2-D model and the conventional 3-D model is performed in 25 sec and 350 sec, respectively. The accuracy of 2-D based simulation results of electromagnetic parameters has been verified with the measurement results. Finally, the proposed 2-D model has been applied and a comprehensive step-by-step parametric simulation is performed in order to study the effect of geometric variables of the coils such as outer diameter, number of turns, trace width and trace spacing on the behavior of coupling coefficient and the maximum achievable coupling between the PCB coils.

REFERENCES

[1] U. Anwar, O. A. Ajijola, K. Shivkumar, and D. Marković, "Towards a leadless wirelessly controlled intravenous cardiac pacemaker," *IEEE Trans Biomedical Engineering*, vol. 69, no. 10, pp. 3074-3086, Oct. 2022.

[2] J. P. Uehlin *et al.*, "A Single-Chip Bidirectional Neural Interface with High-Voltage Stimulation and Adaptive Artifact Cancellation in Standard CMOS," *IEEE J. Solid-State Circuits*, vol. 55, no. 7, pp. 1749-1761, Jul. 2020.

[3] A. Akinin *et al.*, "An optically addressed nanowire-based retinal prosthesis with wireless stimulation waveform control and charge telemetering," *IEEE J. Solid-State Circuits*, vol. 56, no. 11, pp. 3263-3273, Nov. 2021.

[4] F. Gozalpour and M. Yavari, "An Improved FSK-Modulated Class-E Power and Data Transmitter for Biomedical Implants," *AEU-Int. J. Electronics and Communications*, p. 154786, 2023.

[5] A. Ibrahim, and M. Kiani, "A figure-of-merit for design and optimization of inductive power transmission links for millimeter-sized biomedical implants," *IEEE Trans Biomedical Circuits and Systems*, vol. 10, no. 6, pp. 1100-1111, Dec. 2016.

[6] Y. Jia *et al.*, "A mm-sized free-floating wirelessly powered implantable optical stimulation device," *IEEE Trans Biomedical Circuits and Systems*, vol. 13, no. 4, pp. 608-618, Aug. 2019.

[7] U.-M. Jow, and M. Ghovanloo, "Design and optimization of printed spiral coils for efficient transcutaneous inductive power transmission," *IEEE Trans Biomedical Circuits and Systems*, vol. 1, no. 3, pp. 193-202, Sept. 2007.

[8] A. Khalifa *et al.*, "The microbead: A 0.009 mm 3 implantable wireless neural stimulator," *IEEE Trans Biomedical Circuits and Systems*, vol. 13, no. 5, pp. 971-985, Oct. 2019.

[9] M. Kiani, U.-M. Jow, and M. Ghovanloo, "Design and optimization of a 3-coil inductive link for efficient wireless power transmission," *IEEE Trans Biomedical Circuits and Systems*, vol. 5, no. 6, pp. 579-591, Dec. 2011.

[10] A. K. RamRakhyani, S. Mirabbasi, and M. Chiao, "Design and optimization of resonance-based efficient wireless power delivery systems for biomedical implants," *IEEE Trans Biomedical Circuits and Systems*, vol. 5, no. 1, pp. 48-63, Feb. 2010.

[11] M. Schormans, V. Valente, and A. Demosthenous, "Practical inductive link design for biomedical wireless power transfer: A tutorial," *IEEE Trans Biomedical Circuits and Systems*, vol. 12, no. 5, pp. 1112-1130, Oct. 2018.

[12] M. Kiani and M. Ghovanloo, "A figure-of-merit for designing high-performance inductive power transmission links," *IEEE Trans Industrial Electronics*, vol. 60, no. 11, pp. 5292-5305, Nov. 2013.

[13] C. Multiphysics, "AC/DC Module User's Guide for COMSOL," *Stockholm, Sweden: COMSOL AB*, 2013.

[14] https://doc.comsol.com/5.5/doc/com.comsol.help.comsol/comsol_ref_definitions.12.115.html

[15] D. W. Knight, "An introduction to the art of solenoid inductance calculation with emphasis on radio-frequency applications," Feb. 2016, [Online]. Available: http://g3ynh.info/zdocs/magnetics/part_1.html

[16] D. W. Knight, "Selonoid Impedance and Q" Jan. 2016, [Online]. Available: http://g3ynh.info/zdocs/magnetics/solenz.html

The 5th Iranian International Conference on Microelectronics (IICM 2023)

October 25-26, 2023

An Asynchronous Strategy for Efficient Audio Processing for Better Perception in Cochlear Implants Based on Peak and Trough Detection

Amin Armin, Mohammad Yavari and Amir Kashi

Integrated Circuits Design Laboratory, Department of Electrical Engineering, Amirkabir University of Technology (Tehran Polytechnic), P.O. 15875-4413, Tehran 15914, Iran.

Emails: aminarmin77@aut.ac.ir, myavari@aut.ac.ir, kashi@aut.ac.ir

Abstract— In this paper, we propose a novel algorithm for processing acoustic signals in cochlear implants that is asynchronous and low power. Our algorithm stimulates the electrodes based on the frequency of the input sound signal and identifies its peaks and troughs. The algorithm behaves in such a way that until a peak or trough is identified, the subsequent blocks do not consume power. For example, the analog-to-digital converter is turned off until a peak or trough is detected. This ensures that the power consumption is proportional to the input sound signal and significantly lower compared to algorithms that continuously operate the entire system. Our approach has the potential to improve the perception of sound in cochlear implants and reduce power consumption, which are both important considerations in this field. We present simulation results to demonstrate the effectiveness of our algorithm in processing acoustic signals for cochlear implants.

Keywords—*Asynchronous Strategy, Cochlear Implant, Low Power, Temporal Fine Structure (TFS), Peak and Trough Detection, Sound signal processing, Amplitude and phase extraction, Sound perception.*

I. INTRODUCTION

According to the latest announcement by the World Health Organization (WHO), approximately 7% of the world's population (500 million people) suffer from hearing loss. Also, according to the WHO, World Hearing Report, published on March 2, 2021, it warns that nearly 2.5 billion people worldwide - or one in four people - will live with some degree of hearing loss by 2050. At least 700 million of these people will need access to hearing care and other rehabilitation services unless action is taken [1].

Hearing loss affects both children and adults and requires various treatment approaches. While hearing aids are effective for many, individuals with severe hearing loss may still struggle even with aids. Cochlear implantation has been recommended for these individuals who cannot resolve their hearing problems with hearing aids [2]. Candidates for cochlear implants are those who do not experience significant improvement in hearing after using hearing aids for three to six months. Cochlear implants work by directly stimulating the auditory nerve to provide a sense of sound, offering an alternative pathway for damaged inner ears.

Although not a cure for hearing loss, cochlear implants enable sound perception.

The signal processing block is one of the most important parts of the cochlear implant system, responsible for receiving information from the microphone, frequency division, processing, and ultimately sending it to the auditory nerve fibers [3]. Various ways and algorithms for audio signal processing for better perception have been proposed in recent years. Our study introduces a new algorithm for processing acoustic signals in cochlear implants. The algorithm is designed to be asynchronous and low-power, operating by detecting peaks and troughs in the input sound signal and stimulating the electrodes based on its frequency. By only consuming power when necessary, such as when a peak or trough is identified, our algorithm is able to significantly reduce power consumption compared to traditional algorithms that continuously operate the system.

The structure of this article is as follows. Section II offers a concise overview of the cochlear implant signal processor and outlines some of the previous strategies employed in this domain. Section III provides a detailed explanation of the proposed signal processor strategy. The results of simulations are shown in Section IV. Finally, section V presents the study's conclusion and offers concluding remarks.

II. REVIEW OF STRATEGIES FOR SIGNAL PROCESSING IN COCHLEAR IMPLANT

When broadband sounds, such as speech or music, enter the cochlea, they are filtered into a series of narrowband signals. Each of these signals has a slowly varying envelope (ENV), which represents the amplitude of the sound signal, and a rapidly oscillating carrier, known as the temporal fine structure (TFS). The timing and rate of action potentials in the auditory nerve convey the information about the ENV and TFS of the sound signal to the brain [4]. Extracting the ENV from sound is crucial for better hearing and understanding the intensity of sound. By extracting the envelope from the sound, an individual can understand more details of the sound and have a better understanding of speech and sound intensity.

979-8-3503-6020-2/23 $31.00 © 2023 IEEE

Improving the extraction of TFS information in cochlear implants (CIs) can significantly impact the ability of individuals with hearing loss to perceive sounds in noisy environments and distinguish between competing voices [5][6]. Additionally, extracting TFS information is important for music perception and tonal languages. Individuals with CIs may have difficulty perceiving and appreciating music due to the limited number of electrodes used to stimulate the auditory nerve, resulting in a loss of pitch information. By improving the extraction of TFS information in CIs, researchers aim to provide a more accurate representation of music and improve music perception for individuals with hearing loss [7][8].

Several algorithms have been developed over the years, each with different advantages and limitations. One of the most well-known and widely used algorithms is the continuous interleaved sampling (CIS) strategy. The CIS method is a simple and efficient algorithm that uses a small number of electrodes to stimulate the auditory nerve. Instead of transmitting the original signal, this algorithm extracts the envelope of the signal and delivers it to the electrodes with a fixed pulse rate [9]. While the CIS method has been successful in providing improved speech perception for many individuals with hearing loss, it has limitations in representing the TFS of sound signals.

In order to enhance the ability of a bionic ear processor to transmit not only the speech signal envelope but also the phase and TFS to the auditory nerve fibers, several speech processing techniques have been suggested for use in the Bionic Ear (BE) processor. One of the signal processing strategies used in cochlear implants is the phase-locking zero-crossing detection (PL-ZCD) strategy. PL-ZCD uses a phase-locking algorithm to detect the phase of the incoming sound waveform at the zero-crossing points, which is then used to generate stimulation pulses synchronized with the sound waveform [10]. The advantages of PL-ZCD include improved temporal coding, reduced energy consumption, and a simple algorithm. However, PL-ZCD has limitations, as it is sensitive to noise, which can cause severe performance degradation [11], and it is not always clear what the zero-crossing point represents in terms of phase [12].

The other strategy that attempted to extract phase information is named frequency amplitude modulation encoding (FAME). The FAME strategy aims to enhance ENV and TFS separately and then combine them to provide a more natural and complete representation of sound for cochlear implant users [6]. Although phase extraction plays a crucial role in enhancing sound perception within the FAME strategy, its implementation through a separate pathway independent from domain extraction poses certain drawbacks. One notable limitation is the increased demand for processing power, making it less viable for devices with limited power capabilities. Therefore, while phase extraction enhances sound quality, alternative approaches should be considered for power-constrained devices to ensure optimal performance.

Another approach involves extracting phase information by identifying peak and trough points within the signal. This technique, known as the recording peak/trough technique, allows for the conveyance of more instantaneous frequency information beyond just the fundamental frequency to the stimulation electrodes [13].

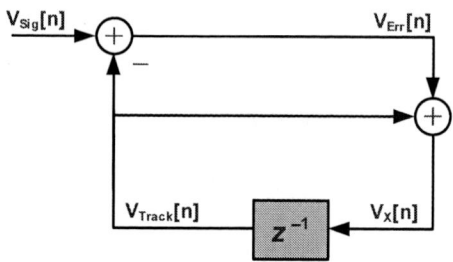

Fig. 1: Error Detection Block.

An example of this approach is phase-locked sampling (PL-PP) combined with peak instantaneous detection (PID) [14]. Additionally, Guo et al. presented an alternative implementation of this method where amplitude information is delivered to the electrodes precisely at the peaks and troughs of the signal [12]. One advantage of this method is that it enables simultaneous extraction of both phase and amplitude information by utilizing a peak-triggered sampling scheme. Since the amplitude information is embedded at the peak points, this approach offers the benefit of achieving phase and amplitude extractions simultaneously, enhancing the overall representation of the sound signal [12].

This article focuses on the peak/trough detection method that is considered the best strategy for providing superior TFS information, resulting in an enhanced perception experience. Compared to other strategies, this method has fewer drawbacks and is widely regarded as the optimal choice for peak picking. In the following section, we will delve into the details of the strategy and how it can be effectively implemented.

III. PROPOSED SIGNAL PROCESSING STRATEGY

As highlighted in the previous section, the algorithm that will be presented below places considerable importance on the peak picking strategy. This method boasts several benefits, including its ability to extract TFS in a superior manner, leading to an improved perception experience. Moreover, the strategy enables the simultaneous extraction of amplitude and phase, making it a highly efficient approach. Furthermore, the Zero Crossing block has been used as one of the simple and low-power consumption blocks in this algorithm for implementation purposes.

Before delving into the strategy and its functioning, it's important to take a closer look at Fig. 1, which is a crucial component of the proposed algorithm's circuit. Let's assume that the sound signal is sampled and recorded by a microphone with sampling frequency of 200 kHz and is then transmitted to the circuit's input as $V_{sig}[n]$. It is worth noting that the $V_{Track}[n]$ signal is identical to the input $V_{sig}[n]$ signal, but with a delay of one time cycle, as we will demonstrate below. This circuit is responsible for comparing the current signal with the previous cycle and generating the error value, $V_{Err}[n]$, between these two signals, which is then sent to the output. Additionally, $V_{Err}[n]$ is used to construct the previous cycle of the signal. In order to create the delayed signal, we can represent the mathematical equations for each node as follows. The equation at the $V_{Err}[n]$ node is as

$$V_{Err}[n] = V_{Sig}[n] - V_{Track}[n] \qquad (1)$$

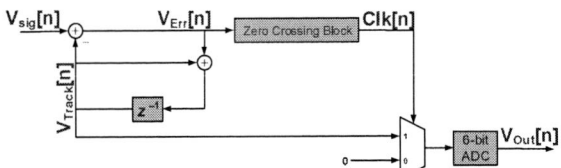

Fig. 2: Proposed Speech Processing Block Diagram.

Also, the equation at the $V_X[n]$ node can be written as

$$V_X[n] = V_{Err}[n] + V_{Track}[n] \qquad (2)$$

And the equation for the final node is as

$$V_{Track}[n] = V_X[n-1] \qquad (3)$$

By inserting (2) in (3), the following equation is obtained by

$$V_{Track}[n] = V_{Err}[n-1] + V_{Track}[n-1] \qquad (4)$$

By substituting the values from (1) into (4) and simplify it, it is denoted as

$$V_{Track}[n] = V_{Sig}[n-1] \qquad (5)$$

Equation (5) demonstrates that the $V_{Track}[n]$ is equivalent to the delayed $V_{Sig}[n]$.

To summarize the circuit thus far, it subtracts the current signal from the signal of the previous stage, and then sends the resulting difference as $V_{Err}[n]$ to the next stage. Furthermore, the same $V_{Err}[n]$ is utilized to generate the delayed signal. Fig. 2 is the block diagram of the strategy based on the Peak Picking algorithm. As depicted, the zero-crossing block is employed to alter the sign of the $V_{Err}[n]$ signal. To comprehend how the strategy functions, let us examine a sinusoidal signal as depicted in Fig. 3. As illustrated, during the ascending trend of the signal, each sample exceeds its previous sample, and conversely, during the descending trend, each sample is lower than its previous

Based on the definition of $V_{Err}[n]$, which is the difference between the current signal and the previous stage, the error rate becomes positive during the ascending trend and negative during the descending trend. Consequently, if we observe a change in the sign of $V_{Err}[n]$ from positive to negative, it indicates the point at which the input signal has reached its peak. Conversely, if the sign of $V_{Err}[n]$ changes from negative to positive, it indicates the trough point.

Based on the explanations provided regarding Fig. 3, let us revisit the circuit diagram in Fig. 2. The audio signal, $V_{Sig}[n]$, is fed into the input of the circuit, and then $V_{Err}[n]$, which is the difference between the current signal and the previous stage, is obtained from the input signal. In the subsequent step, the zero crossing circuit is employed to detect whenever the sign of $V_{Err}[n]$ changes. Thus, whenever the input reaches its peak which $V_{Err}[n]$ sign changes from positive to negative, the output of the zero crossing circuit changes from Low to High for one cycle and then returns to Low. The same process is repeated for the trough points, where the sign of $V_{Err}[n]$ changes from negative to positive. In summary, the output of the zero crossing block changes with each peak and trough detection.

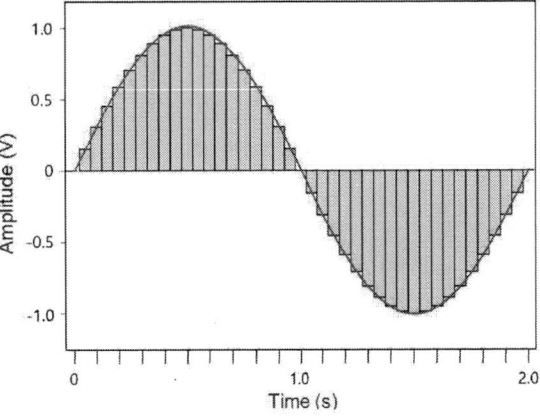

Fig. 3: Sine Function with Oversampling.

When output of the zero crossing block is High, the signal amplitude value of the previous stage, which is the maximum value in the peak and the minimum value in the trough, is transmitted to the next stage through a switch or multiplexer. Subsequently, the A/D digitizing task is initiated to process the input signal. The circuit operates in such a way that when the input signal is not at the peak or trough points, the A/D input is disabled to minimize power consumption.

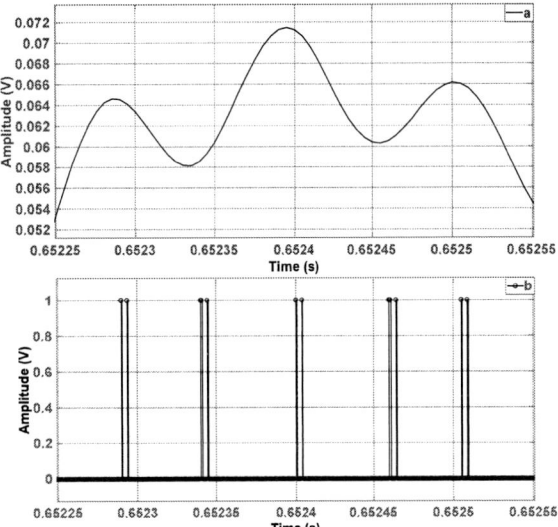

Fig. 4: (a) Part of Audio Input Signal and (b) Output of Proposed Strategy.

To summarize how this strategy works, it can be said that the circuit obtains the slope of the audio signal with a sampling frequency of 200 kHz by calculating the difference between the current signal and the previous stage. Whenever the slope of the signal changes, indicating that a peak or trough has been reached, the zero crossing block is utilized to detect the change. Subsequently, the amplitude value of the signal at the peak or trough is transmitted to the A/D input for one cycle. Until a peak or trough is reached, the circuit after the switch will not operate, which leads to a significant reduction in power consumption. As a result, the power consumption of the subsequent stages is only utilized when necessary, i.e., during the peak or trough areas. Thus, the

979-8-3503-6020-2/23 $31.00 © 2023 IEEE

power consumption varies with the input frequency, and the circuit consumes power only when required.

IV. SYSTEM SIMULATION RESAULTS

To evaluate the effectiveness of the proposed approach described in Fig. 2, the Simulink MATLAB software was used for implementation. A sound signal containing a spoken sentence was selected as the input test signal, and it was sampled at a frequency of 200 kHz.

The results of the proposed strategy are presented in Fig. 4, where part (a) displays a section of the input sound waveform that contains multiple peaks and troughs. In part (b), the output waveform of the proposed approach is shown, which accurately identifies the peaks and troughs. To further analyze the results, Fig. 5 displays an enlarged view of the first trough in Fig. 4. The amplitude of the input signal at the moment of the trough is 0.06v, which corresponds to the digital value of 1 in 6 bit. Fig. 5, part (b) has been enlarged more in the time axis to better visualize the algorithm output. The value of the digitalized signal in this magnified section is serialized and encoded as 11000001, where the first bit indicates the start of sending the amplitude, and the second bit indicates whether the strategy detects a peak (if it is 0) or a trough (if it is 1). The next 6 bits also indicate the range of the detected signal, which is 1 as it should be. The encoded value, consisting of 8 bits, is sent to the next stage in a time cycle with a frequency of 200 kHz. This value is then utilized to stimulate the corresponding electrodes in the subsequent steps.

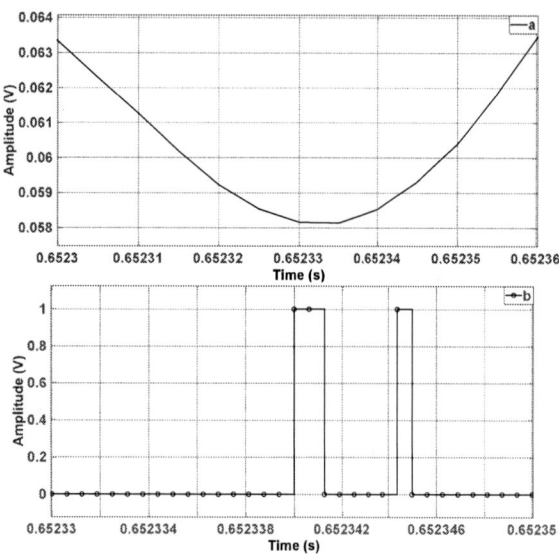

Fig. 5: (a) Magnification of the First Trough in Fig. 4 and (b) The Output of the Proposed Strategy in Selected Trough.

Guo et al. proposed an event-driven strategy based on the ADM algorithm in [12]. Guo et al. introduced an event-driven strategy based on the ADM (Amplitude Demodulation) algorithm in their publication [12]. This article represents a significant contribution to the field of signal processing in ear implant applications, as it aims to achieve optimal phase extraction with a focus on accuracy and speed. Recognized as one of the most up-to-date references in the field, the ADM strategy serves as a valuable benchmark for comparison with the strategy proposed in this conference paper.

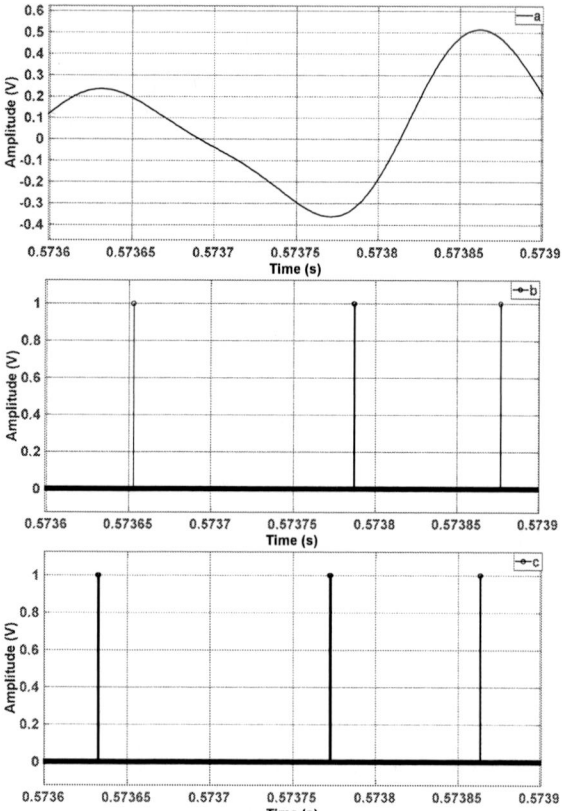

Fig. 6: (a) Part of Audio Input Signal and (b) Output of ADM Strategy and (c) Output of Proposed Strategy.

Fig. 6, part (a) displays an enlarged portion of an input sound signal. The outputs of the ADM strategy and the newly proposed strategy can be seen in parts (b) and (c), respectively. It is evident that the ADM strategy experiences significant delay compared to our strategy. Additionally, when the ADM strategy detects a peak or trough point, it sends a different amplitude value to the electrodes than the original input signal. In the ADM strategy, the phase error, known as $err_{ph}=(t_{polarity}-t_{peak}) \times f_{input}$, is defined as the error in detecting the time of the peak or trough. However, in our proposed strategy, this error is highly optimized and nearly eliminated, resulting in more accurate peak and trough detection. As a result, our strategy outperforms the ADM strategy in terms of accuracy and precision.

Fig. 7 presents another section of the input audio signal, which is different from the previous analysis. In the ADM strategy, it is known that peaks and troughs that differ by less than 1 LSB (for a 6-bit counter) cannot be detected. In part (a) of Fig. 7, the amplitude of the input audio signal changes by less than 1 LSB. Part (b) shows the output of the ADM strategy, which only detects one of the peaks and fails to identify one peak and two intermediate troughs. As a result, the ADM strategy fails to accurately perceive this section of the sound. However, our proposed strategy successfully detects all the peaks and troughs with high accuracy, and transmits them to the electrodes.

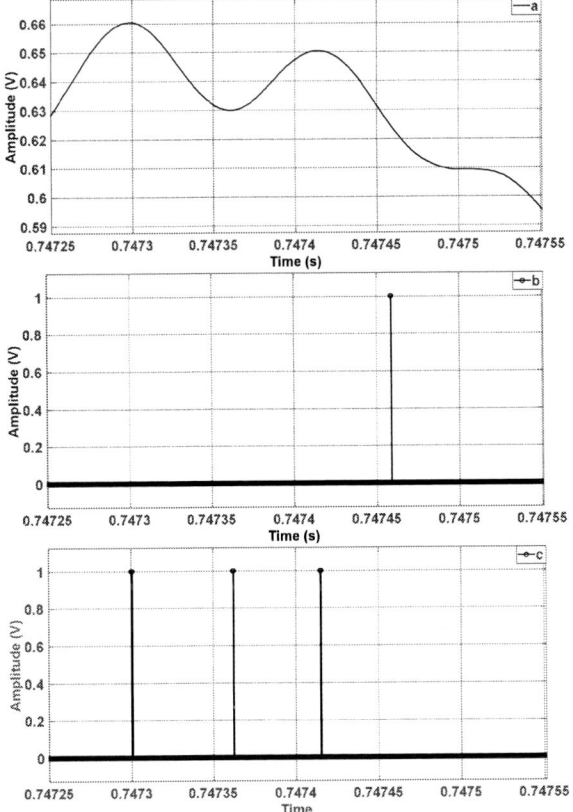

Fig. 7: (a) Shows a Section of the Audio Input Signal and (b) Output of ADM Strategy and (c) Output of Proposed Strategy.

Table I presents a qualitative comparison of various strategies. The strategies were evaluated based on their performance in phase extraction within the phase extraction line. It is worth noting that some strategies employ the amplitude envelope of the signal instead of the amplitude of the main signal. To assess the accuracy of the range, the difference between this range and the range of the original audio signal was measured and reported in the last line of the table. Our evaluation indicates that strategies employing signal push mechanisms exhibit low amplitude accuracy. Conversely, [12], which encounters a phase error, demonstrates medium accuracy. Our approach, which excels in peak and trough mining, is considered a high-accuracy strategy based on this parameter. Overall, this analysis provides valuable insights into the performance and characteristics of various signal processor strategies, emphasizing the significance of simplicity, power consumption, and accuracy in amplitude and phase extraction for optimal results.

TABLE I: COMPARISON BETWEEN SIGNAL PROCESSING STRATEGIES.

Reference	[9]	[15]	[10]	[14]	[12]	This work
complexity	Med*	High	Low	Low	Med*	Low
Sampling	Sync	Async	Sync	Sync	Async	Async
Phase Extraction	Very Low	High	Low	Med*	High	High
Power	Med*	High	Very Low	Low	Med*	Low
Amplitude Accuracy	Low	High	Low	High	Med*	High

*Med: abbreviation of Medium

V. CONCLUSION

This study proposes a novel strategy based on Peak Picking. The approach offers several advantages, including the simultaneous extraction of amplitude and phase information to better perception by recognizing the frequency of the input signal from the peaks and troughs. Additionally, the strategy utilizes simple circuits which reduces power consumption. Asynchronous timing of electrode stimulation is also implemented, which makes use of time cues of phase information. The stand-by capability of the strategy allows certain blocks, such as Analog-to-Digital Converter (ADC), to be deactivated when electrodes do not require stimulation. This effectively reduces power consumption by eliminating unnecessary operations and conserving energy. Overall, the proposed strategy offers several improvements over existing techniques and has the potential to advance the field of auditory prostheses.

REFERENCES

[1] https://www.who.int/ (Access Date: Mar. 3, 2023.)

[2] W. A. Yost, and D. W. Nielsen, *Fundamentals of hearing*: Academic Press New York, 2000.

[3] B. S. Wilson, C. C. Finley, D. T. Lawson, R. D. Wolford, D. K. Eddington, and W. M. Rabinowitz, "Better speech recognition with cochlear implants," *Nature*, vol. 352, no. 6332, pp. 236-238, Jul. 1991.

[4] B. C. J. Moore, "The roles of temporal envelope and fine structure information in auditory perception," Acoustical Science and Technology, vol. 40, no. 2, pp. 61-83, Mar. 2019

[5] F.-G. Zeng, S. Rebscher, W. Harrison, X. Sun, and H. Feng, "Cochlear implants: system design, integration, and evaluation," *IEEE reviews in biomedical engineering*, vol. 1, pp. 115-142, Nov. 2008.

[6] K. Nie, G. Stickney, and F.-G. Zeng, "Encoding frequency modulation to improve cochlear implant performance in noise," *IEEE transactions on biomedical engineering*, vol. 52, no. 1, pp. 64-73, Dec. 2004.

[7] F.-G. Zeng, "Cochlear implants in China," *Audiology*, vol. 34, no. 2, pp. 61-75, Jan. 1995.

[8] L. Bruns, D. Mürbe, and A. Hahne, "Understanding music with cochlear implants," *Scientific reports*, vol. 6, no. 1, pp. 32026, Aug. 2016.

[9] R. Sarpeshkar, C. Salthouse, J.-J. Sit, M. W. Baker, S. M. Zhak, T.-T. Lu, L. Turicchia, and S. Balster, "An ultra-low-power programmable analog bionic ear processor," *IEEE Transactions on Biomedical Engineering*, vol. 52, no. 4, pp. 711-727, Mar. 2005.

[10] R. Sarpeshkar, M. Baker, C. Salthouse, J.-J. Sit, L. Turicchia, and S. Zhak, "An analog bionic ear processor with zero-crossing detection." *Proc. IEEE Int. Solid-State Circuits Conf.*, pp. 78-79, Feb. 2005.

[11] J.-J. Sit, A. M. Simonson, A. J. Oxenham, M. A. Faltys, and R. Sarpeshkar, "A low-power asynchronous interleaved sampling algorithm for cochlear implants that encodes envelope and phase information," *IEEE Transactions on Biomedical Engineering*, vol. 54, no. 1, pp. 138-149, Dec. 2006.

[12] N. Guo, S. Wang, R. Genov, L. Wang, and D. Ho, "Asynchronous Event-driven Encoder With Simultaneous Temporal Envelope and Phase Extraction for Cochlear Implants," *IEEE Transactions on Biomedical Circuits and Systems*, vol. 14, no. 3, pp. 620-630, Apr. 2020.

[13] C. Sawigun, W. Ngamkham, and W. A. Serdijn, "Comparison of speech processing strategies for the design of an ultra low-power analog bionic ear." *Proc.IEEE EMBC*, pp. 1374-1377, Aug. 2010.

[14] C. Sawigun, W. Ngamkham, and W. A. Serdijn, "An ultra low-power peak-instant detector for a peak picking cochlear implant processor." *IEEE Conf. Biomed. Circuits Syst.*, pp. 222–225, Nov. 2010.

[15] J.-J. Sit, A. M. Simonson, A. J. Oxenham, M. A. Faltys, and R. Sarpeshkar, "A low-power asynchronous interleaved sampling algorithm for cochlear implants that encodes envelope and phase information," *IEEE Transactions on Biomedical Engineering*, vol. 54, no. 1, pp. 138-149, Dec. 2006.

The 5th Iranian International Conference on Microelectronics (IICM2023)

25 – 26 October 2023

Numerical analysis of studying the importance of choosing the right image reconstruction algorithms in tomography's accuracy and processing time

1st Maryam Ahangar Darband

Ph.D. in Electrical Engineering,
Department of Electrical
Engineering
Sahand University of Technology,
Tabriz 5331811111, Iran; E-mail:
m_ahangar94@sut.ac.ir

2nd Esmaeil Najafi Aghdam*

Professor, Electrical Engineering,
Department of Electrical Engineering,
Sahand University of Technology,
Tabriz 5331811111, Iran
najafiaghdam@sut.ac.ir

Abstract— Tomography plays an important role in imaging, especially medical imaging. One of the most important parts affecting the speed and accuracy of tomography is choosing the right image reconstruction algorithm. Photoacoustic tomography has grown significantly in recent years due to its non-invasiveness, low risk, and at the same time high optical contrast with good ultrasound resolution. In this presentation, by choosing photoacoustic tomography in diagnosing breast tumors, we studied the accuracy and speed of two image reconstruction algorithms suitable for the proposed geometry. According to the simulation results of the spatial phase-controlled algorithm and algebraic algorithm at COMSOL, the spatial phase-controlled algorithm is 24.48 seconds faster than the algebraic algorithm and has a better resolution compared to the algebraic algorithm. According to these results, the appropriate choice of the image reconstruction algorithm based on the geometry and type of tomography plays a significant role in the resolution and speed of tomography.

Keywords—Tomography, Image Reconstruction, Photoacoustic Tomography (PAT), Breast Cancer

I. INTRODUCTION

The tomography is a kind of imaging for producing a three-dimensional image of the internal structures of any solid object [1]. Tomography also plays an important role in imaging the body's internal organs. Tomography in medical topics requires extreme accuracy because failure to monitor the smallest changes in a body organ may cost the patient's or that person's life. Today, there are various methods for tomography but a highly accurate, non-invasive, low-risk, and accessible method will be always a priority. Recently, hybrid tomography has received wide attention since combining some imaging modalities takes advantage of them and improves image quality. One of the hybrid tomography imaging methods is based on the photoacoustic (PA) phenomenon. PA phenomenon was first discovered by Alexander Graham Bell in 1880 and was not used in medical

imaging before the invention of the laser [2]. Photoacoustic tomography (PAT) is an emerging imaging modality in which there is great potential for improving its different parts [3]. Different photoacoustic imaging parts have been investigated in detail and even simulated [4]. One of the most important parts of tomography is the image reconstruction algorithm (RA), which has an important effect on the accuracy and speed of imaging. Choosing and developing a suitable Image reconstruction algorithm (IRA) actually is craftwork [5]. The RA should somehow be compatible with the tomography geometry. Being surrounded by a wide range of IRAs would be a great help in choosing the right algorithm [6]. In this article, by using the software environment, we will study the effect of choosing the correct IRA. In order to have freedom of action in choosing and changing the testing geometry, low cost, and high speed, the software environment has been selected. According to the previous studies [4] and the reasons were explained in detail, we would use the COMSOL® software environment to simulate the PAI, from the sample stimulation section until capturing data from the ultrasonic sensors. Finally, the sensor data would be transferred into the MATLAB software environment and the images would be reconstructed by algorithms in order to study the results in detail. In the following, we will compare the accuracy and speed of image reconstruction based on spatial phase-controlled algorithms [8] and algebraic algorithms [9] at the same geometry. These two IRAs have the most compatibility with the presented geometry. The geometry described in this presentation is based on the model that was set up in our laboratory at Sahand University. Simple imaging system based on PA phenomenon to detect cancerous tumors in mimic breast tissue.

II. METHOD

Following this section: we will describe the basic introduction of two algorithms. Then briefly the simulation

979-8-3503-6020-2/23 $31.00 © 2023 IEEE

processes of PA phenomena in COMSOL® have been discussed. The remaining sections will be about the study of reports resulting from IRA. A flowchart description of the simulation procedure is shown in Figure 1.

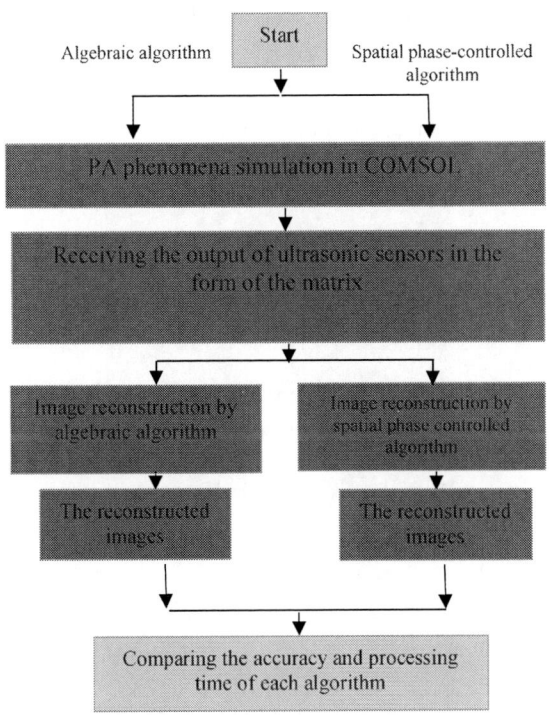

Figure 1: Flowchart description of the procedure of simulations

A. A brief description of two compared algorithms

Two major categories of image reconstruction methods are analytical reconstruction (AR) and iterative reconstruction (IR) [10]. Each RA is only available in a few special geometries [9]. To have a convenient and fast IRA based on our knowledge, we choose spatial phase-controlled algorithms and algebraic algorithms for our proposed geometry. In the following, we will show which IRA would provide accurate images with low processing time.

1. Spatial phase-controlled algorithms

For each reconstructed pixel in the image, P_m, we have the following formula [8]:

$$P_m = \sum_{k=1}^{K} D(\theta)_k P(t_{mk}) \quad (1),$$

where m is the spatial vector of the PA source, k is the position of the kth detector, and K is the total number of working detectors. $D(\theta)$ is defined as the projection intensity weight function of θ, the projection angle which is no more than the maximal acceptance angle of the array element. $P(t_{mk})$ is the signal value collected by the kth detector at position m, t_{mk} represents the time when PA pulses spread from position m to k, r_{mk} is the distance between the PA source at position m and the kth detector. v is the average velocity of the acoustic wave in tissue [4].

2. Algebraic algorithms

We can consider the algebraic reconstruction technique (ART) an iterative solver of a linear system by the following equation [9]:

$$Ax=b \quad (2).$$

$A = A_{mn}$ is the weight matrix that represents the contribution of every pixel for all different rays in the projection in which M is the total number of rays in all projections and N is the total number of pixels in the image. Conceptually, reconstructing the original image $x = (x_1, ..., x_N)^T$ from observed data $b = (b_1, ..., b_M)^T$ is the main problem in image reconstruction [11]. Mathematically, image reconstruction is posed as a system of linear equations, which must be solved to reconstruct an image [12][13][14]:

$$A_{11}x_1 + A_{12}x_2 + \cdots + A_{1N}x_N = b_1$$
$$A_{21}x_1 + A_{22}x_2 + \cdots + A_{2N}x_N = b_2$$
$$A_{31}x_1 + A_{32}x_2 + \cdots + A_{2N}x_N = b_3$$
$$.$$
$$.$$
$$.$$
$$A_{M1}x_1 + A_{M2}x_2 + \cdots + A_{MN}x_N = b_M \quad (3).$$

B. PA phenomena simulation in COMSOL

Considering the fact that the purpose of this presentation is to show the importance of choosing the right IRA, and the PA phenomenon simulation in COMSOL, described in detail in reference [4], in this section we briefly express the main and required items of simulation in COMSOL. The light radiated from high-power LED to the area of 1mm × 1mm and 1.2 W power at 1050 nm. All parameters of breast and tumor tissue that depend on the radiation wavelength were recalculated at 1050 nm radiation wavelength [4]. The simulation environment is in the form of a cylindrical chamber filled with water, for best coupling, with a diameter of 8 cm and a height of 14 cm, where eight ultrasonic sensors were placed around the circular cross-section at a height of 6 cm from the bottom of the chamber at equal distances from each other, LEDs at a distance of 5 mm is placed against the mimic breast tissue, Figure 2 and 3. The mimic breast tissue is cubic with a length of 2 cm. Due to the fact that breast cancer is angiogenesis [15], the formation of blood vessels for the supply of sufficient oxygen and nutrition for breast tumor growth, and tracking blood vessels could be an escape for the easy diagnosis of cancerous tumors of the breast tissue [16].

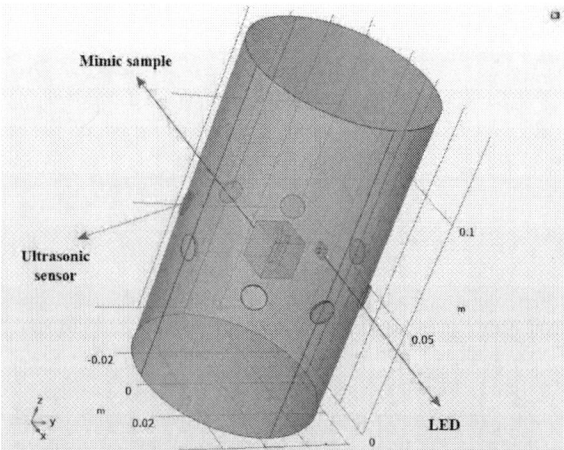

Figure 2. 3D computational domain of geometry in COMSOL Multiphysics.

In order to have precise tomography and information from different angles, the LED should have ten rotations at 27-degree intervals around the mimic breast tissue and along the Z axis in the range of -1 to +1 cm at 4 mm intervals. Finally, we will have fifty movement and fifty information matrices. There is complete synchronization between the different positions of the LED and the LED radiation. All data from eight ultrasonic sensors at the mentioned positions were captured and transferred to the MATLAB environment.

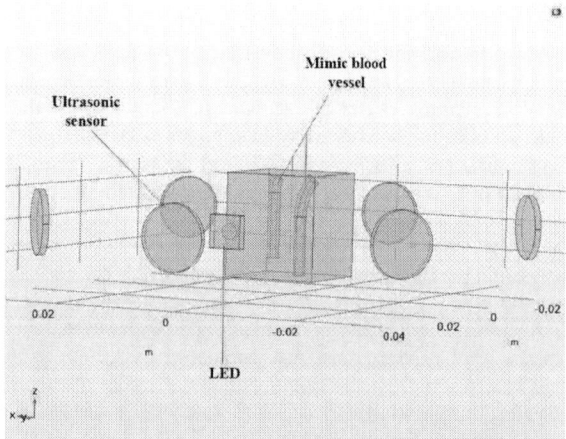

Figure 3. Schematic of the sensor's arrangement, sample, LED, and the placement of mimic blood vessels inside the sample.

C. Image reconstruction algorithms

The images will be reconstructed by two selected algorithms so that we could be able to study the accuracy and processing time of each algorithm.

III. RESULTS AND DISCUSSION

In order to clearly study the differences between the two IRAs, the two-dimensional (2D) output image, Figures 4 and 5, the data graph of eight sensors, Figures 6 and 7, and the data image of the sensors, Figure 8, were displayed separately for each IR algorithm. The difference that we were not able to show is the difference in the processing time of the two algorithms. The required time for each 2D slice image reconstruction in spatial phase-controlled algorithms and algebraic algorithms is 1 min 55.52s whereas this time for algebraic algorithms is 2 min 20 s. The algebraic algorithm is typically time-consuming, here by using a modified algebraic algorithm [9] the computation time was reduced to 2 min 20 s which is still more than the spatial phase-controlled algorithms' computation time [8].

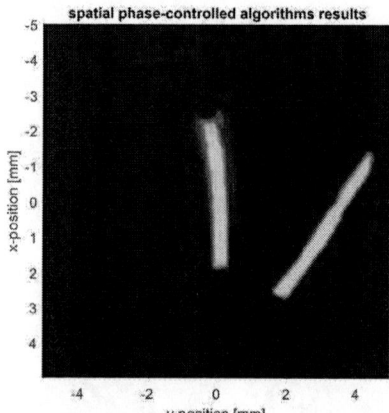

Figure 4. 2D image slice resulted from spatial phase-controlled algorithms.

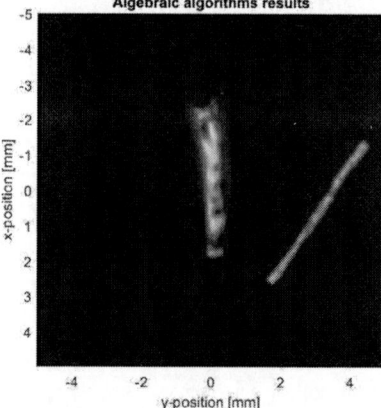

Figure 5. 2D image slice resulted from algebraic algorithms.

Figures 4 and 5 show the final results of two spatial phase-controlled algorithms and algebraic algorithms, respectively. The 2D slice images of spatial phase-controlled algorithms partially express the high resolution of this algorithm. It seems that the overlapping of PA signals in the formation of the sensitivity matrix [9], affects the resolution.

The 5th Iranian International Conference on Microelectronics (IICM2023)

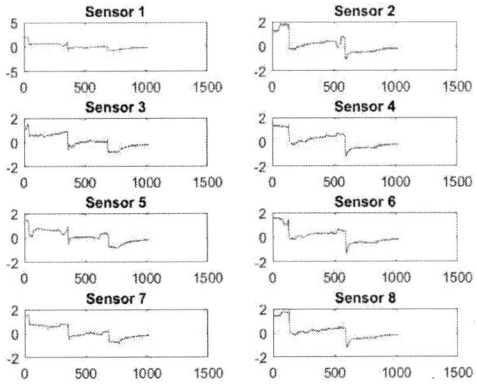

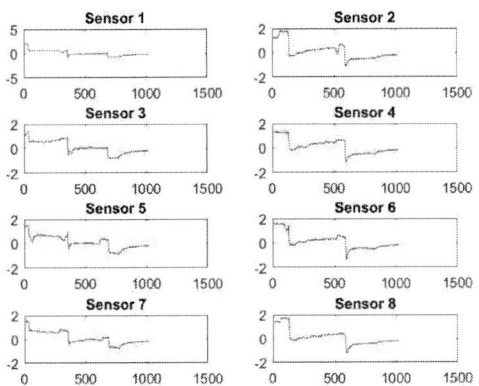

Figure 6. The output signal of eight sensors in the spatial phase-controlled algorithms.

Figure 7. The output signal of eight sensors in the algebraic algorithm.

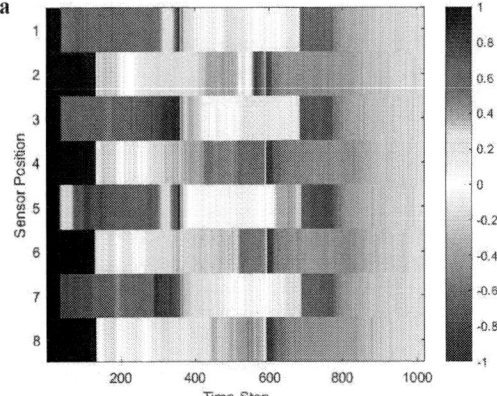

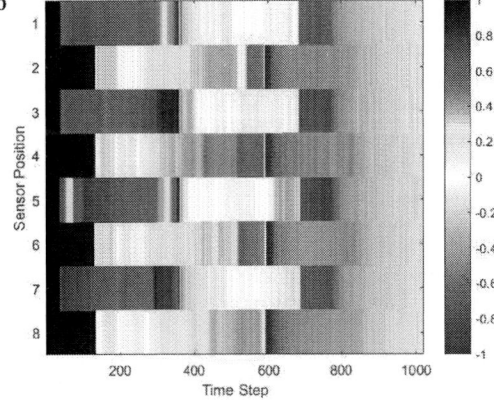

Figure 8. Showing the sensors' data as an image by using the full range of colors a) spatial phase-controlled algorithm, b) algebraic algorithm.

Figures 6 and 7 show the waveform diagram of eight ultrasonic sensors for two algorithms. The difference in the waveform of the sensors is not clearly visible in these kinds of figures. Figure 8. tried to display the sensors' data as an image by using the full range of colors, it seems the two algorithms' output difference is almost visible.

According to the presented documents, it seems that the spatial phase-controlled algorithm is preferable to the algebraic algorithm both in terms of processing and calculation time and accuracy. It seems that having sufficient knowledge of different algorithms and how they work makes it easier to choose the best and most appropriate IRA. As a result, the accuracy and speed of tomography would be clearly increased.

IV. CONCLUSION

In this study, we studied the importance of choosing the appropriate IRA in the accuracy and speed of tomography. At first, a PAI system was designed and simulated in the COMSOL environment so that the sensor output data could be used as input data for the IRA. Then, two suitable algorithms were selected based on the available knowledge, and the speed and accuracy of both were checked in detail. We came to the conclusion that the spatial phase-controlled algorithm has high accuracy and less time for the proposed geometry.

V. DATA AVAILABILITY

The data supporting this study's findings are available from the corresponding author upon reasonable request.

REFERENCES

[1] Buzug, Thorsten M., Computed tomography. Springer handbook of medical technology. Berlin, Heidelberg: Springer Berlin Heidelberg, 2011, pp. 311-342.

[2] A. G. Bell, On the production and reproduction of sound by light. American J. of Science, vol. 20, 1880, pp. 305-324.

[3] Wang, Lihong V., ed. Photoacoustic imaging and spectroscopy. CRC press, 2017.

[4] Ahangar Darband, M., Najafi Aghdam, E. and Gharibi, A., Numerical simulation of breast cancer in the early diagnosis with actual dimension and characteristics using photoacoustic tomography. Archives of Acoustics, 2023, pp.25-38.

[5] Natterer F, Wübbeling F. Mathematical methods in image reconstruction. Society for Industrial and Applied Mathematics; 2001 Jan 1.

[6] Chen L, Chu X, Zhang X, Sun J. Simple baselines for image restoration. In European Conference on Computer Vision 2022 Oct 23 (pp. 17-33). Cham: Springer Nature Switzerland.

[7] Petrou MM, Petrou C. Image processing: the fundamentals. John Wiley & Sons; 2010 May 17.

[8] Zhou, Q., Ji, X. and Xing, D., Full-field 3D photoacoustic imaging based on plane transducer array and spatial phase-controlled algorithm. *Medical physics*, *38*(3), 2011, pp.1561-1566.

[9] Ahangar Darband, M., Qorbani, O. and Najafi Aghdam, E., Modified algebraic reconstruction technique based on circular scanning geometry to improve processing time in photoacoustic tomography. Microwave and Optical Technology Letters, 65(8), 2023, pp.2456-2463.

[10] Awcock GJ, Thomas R. Applied image processing. Basingstoke, UK: Macmillan; 1995 Aug.

[11] Qu G, Wang C, Jiang M. Necessary and sufficient convergence conditions for algebraic image reconstruction algorithms. IEEE Transactions on Image Processing. 2008 Dec 12;18(2):435-40.

[12] Diaconis P, Sturmfels B. Algebraic algorithms for sampling from conditional distributions. The Annals of Statistics. 1998 Feb;26(1):363-97.

[13] Emiris IZ, Pan VY, Tsigaridas EP. Algebraic algorithms. arXiv preprint arXiv, 2013 Nov 15, pp.1311.3731.

[14] Andersen AH, Kak AC. Simultaneous algebraic reconstruction technique (SART): a superior implementation of the ART algorithm. Ultrasonic imaging. 1984 Jan;6(1):81-94.

[15] American Cancer Society. "Breast cancer facts & figures 2019–2020." Am. Cancer Soc (2019): 1-44.

[16] Giaquinto AN, Sung H, Miller KD, Kramer JL, Newman LA, Minihan A, Jemal A, Siegel RL. Breast cancer statistics, 2022. CA: a cancer journal for clinicians. 2022 Nov;72(6):524-41.

The 5th Iranian International Conference on Microelectronics (IICM 2023)

October 25-26, 2023

Design of Electrical Stimulation Circuit in 180 nm/1.8 V Standard CMOS Process

Askandar Nikzad, Mohammad Yavari, and Amir Kashi

Integrated Circuits Design Laboratory, Department of Electrical Engineering, Amirkabir University of Technology (Tehran Polytechnic), P.O. 15875-4413, Tehran 15914, Iran.

Emails: askandarnikzad@aut.ac.ir, myavari@aut.ac.ir , kashi@aut.ac.ir

Abstract— **This article presents an electrical stimulation circuit for use in implantable medical microsystems, which includes digital-to-analog converter, tuned cascade transistor structure, voltage level shifter circuit, and switches circuit for driving the stimulation current. In this research, DAC circuits and current directing switches are proposed in 180nm CMOS 1.8 V technology. Among the performed works, we can mention the increase of the current range up to 2.5 milliamperes and the stimulation voltage up to 12.6 volts to stimulate high-quality nerve tissues. In this context, a digital-to-analog converter circuit with a thermometer-coded structure has been designed and simulated with INL, DNL<0.5 LSB specifications, as well as SFDR=50.74 dB, SNDR=37.1 dB, and ENOB=5.87 Bits. Moreover the most critical parameter of the final system is the measured power consumption of 1.23 milliwatts.**

Keywords—Implantation system, electrical stimulation circuit, digital-to-analog converter (DAC), current driver, high voltage switches, low voltage technology, Monte Carlo analysis.

I. INTRODUCTION

The electrical stimulation unit consists of several subsystems, which include: a digital controller unit, digital-to-analog converter, high voltage switches, electrode, and voltage level shifter circuit [1-7]. In this project, the focus is on the optimal performance of the digital-to-analog converter and reducing the power consumption in the high-voltage switches responsible for stimulating the electrodes [8-10]. To reduce the power consumption in the electrical stimulation part, different ways have been investigated. One of these ways is how to stimulate the electrodes. In such a way that the electrodes are stimulated by a voltage applied to them, or by current injection, or by charge control [11]. Also, the way these stimulations and the shape of applied waveforms will have a significant effect on power consumption. The digital-to-analog converter circuit should be designed in such a way that it has a good resolution, and its output waveform, which is used to stimulate the electrodes, can also be easily controlled [12, 13].

The use of switches made with transistors in low voltage technology has made all the blocks of the electrical stimulation unit one-handed, and compared to high voltage technology, smaller dimensions, and lower cost are used. The use of high voltage switches in low voltage technology will bring the challenge that in order to disconnect and connect the switches, in order to protect the transistors from applying more than the permissible voltage on their drain-gate, we have to use structures to solve this problem. For this

reason, circuits designed with these features need voltage-level shifter circuits, so that they can disconnect and connect high voltage switches during electrode stimulation cycles [2, 14].

This paper proposes an integrated electrical stimulation circuit implemented in 180 nm/1.8 V standard CMOS technology. The electrical stimulation circuit is designed to apply high voltage to nerve tissues, which makes this possible by using the structure of stack transistors. Also, the current stimulation method has been used for more precise control over nerve tissue stimulation. The proposed system operates at a voltage of 12.6 V and a maximum stimulation current of 2.5 mA.

The remains of the paper are arranged as follows. The second part presents the electrical stimulation circuit architecture and principles of operation. The results of the simulations, together with the prior art comparison, are described in Section III, and finally, the conclusions are reported in Section IV.

II. PROPOSED STIMULATION CIRCUIT

The most critical challenge of the electrical stimulation structure in implanting neural prostheses is directing the stimulation current through the switches to apply high voltage to the electrode. This can be done quickly in high-voltage technology. However in low-voltage technologies, due to the existing limitations, it has its challenges. The purpose of implementing the stimulation switches in low voltage technology in this project is to integrate the whole cochlear implant structure for its implementation in a low voltage technology and the more excellent compatibility of the switches with peripheral circuits [14]. Also, the maximum amplitude of the stimulation current and its waveform are of particular importance; therefore, it has been tried to design the proposed digital-to-analog converter circuit in such a way as to produce the optimal current waveform to reduce the power consumption in stimulating the electrodes.

The electrical stimulation circuit converts the information received through the electrode into nerve impulses in three ways: voltage stimulation, load stimulation, and current stimulation. According to the investigations, the best mode of electrical stimulation is the flow mode. Because, in this case, it is possible to control the stimulation method more precisely according to the patient's condition.

Fig. 1 shows the block diagram of the proposed structure. In this structure, a digital-to-analog converter circuit with a

979-8-3503-6020-2/23 $31.00 © 2023 IEEE

coded thermometer structure, is controlled by a digital controller unit, produces an electrical stimulation current to be transferred to the electrode. This current is applied to the electrode through H-shaped bridge switches. Due to the high power required to excite the electrodes, the momentary application of this voltage to the DAC output will be problematic. To solve this problem, the adjusted cascade transistor structure is used, which provides high output impedance for DAC and prevents problems when high voltage is applied. H-bridge switches are used for biphasic stimulation of electrodes [2, 6, 15]. The advantage of this structure over the single-phase structure is that only one positive voltage source is used. The commands sent from the digital controller unit to control the switches in the form of voltage pulses, need to be transferred to higher voltage levels by an interface circuit called a voltage level shifter and then applied to the switches. However because the number of cascaded transistors is relatively large and need to be turned on and off, a bias circuit is used to turn them on.

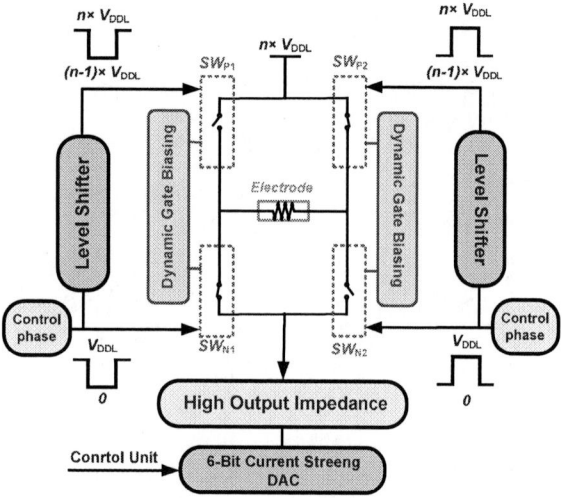

Fig. 1: Block diagram of the general structure of the electric drive circuit.

A. Thermometer-coded DAC

The work of the digital-to-analog converter is to receive a digital bit string and convert it into an analog output. In this structure, the output is the current, and the current size is the same for all cells, and when each current cell is turned on, the output current will increase by $\frac{I_{out}}{2^n}$. The advantage of this structure, in addition to uniformity, is the absence of glitches, and on the other hand, due to the tiny current of each cell, the size of the cells will be tiny. However since the number of required cells is 2^n, its total area will increase. Of course, due to the very high accuracy obtained from this structure, this limitation can be ignored to some extent.

One of the most essential advantages of this structure is that due to the simplicity of the current cells, which consist of only one NMOS transistor, the voltage changes in the source node of the transistor, which is connected to the ground is zero. There is no change in the current passing through it, which is shown in Fig. 2. Due to the measurement error, no two transistors are the same, and this effect is called mismatch. This effect in current source transistors can change their current and thus increase integral

nonlinear error (INL). The relationship between the effect of mismatch on transistor current and determining the size of INL according to mismatch [16] is given in (2) and (3), respectively:

$$\left(\frac{\sigma_I}{I}\right)^2 = \frac{1}{WL}\left(A_\beta^2 + \frac{4A_{VTH}^2}{(V_{gst})^2}\right) \tag{1}$$

$$INL = \frac{\sigma_I}{2I}\sqrt{2^N} \tag{2}$$

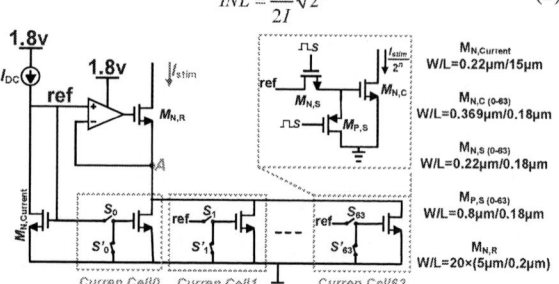

Fig. 2: Structure of Thermometer Coded DAC.

We aim is to design a digital-to-analog converter with a maximum output of 2.5 mA with a 6-bit resolution. So, the current source circuit for this structure must be able to provide the required current of each cell. Considering that the required resolution is 6 bits, the current of each cell will be calculated from the following equation:

$$I_{cell} = \frac{I_{out,max}}{2^N} \xrightarrow{N=6,\ I_{out,max}=2.5mA} I_{cell} = \frac{2.5mA}{64} \approx 39\mu A \tag{3}$$

On the other hand, for a small current to pass through the current source transistor, we try to consider its current to be about 0.01 of the current of each cell.

This structure is designed based on the gain-increasing technique to increase the output impedance of the DAC. The gain-increasing technique increases power consumption. To reduce this problem, the op-amp in the feedback loop is turned off after biphasic stimulation. The output impedance can be calculated follows:

$$R_{out} = A.g_{m(M_{N,R})}.r_{0(M_{N,R})}.r_{0(M_{N,C})} \tag{4}$$

Where A is the operational amplifier gain.

B. High voltage current driver

The electrical model of the electrode is shown in Fig. 3. Implantable electrodes are modeled as a resistor in series with a capacitor, which is placed in parallel with a a huge resistor to prevent DC current from passing into the tissue. In this model, R_S=4 kΩ, R_{ct}=10 MΩ, C_{ct}=10 nF.

This structure allows biphasic stimulation of the electrode. In this way, first, an anodic phase stimulates the electrode by turning on the SW_{P1} and SW_{N2} switches. In the next phase, which is the cathode, to balance the charge in the electrode, by turning on the SW_{N1} and SW_{P2} switches, the current in the opposite direction of the previous phase stimulates the electrode. This is done to prevent the accumulation of charge, which can lead to the production of toxic chemicals or corrosion of the electrodes. After two phases are applied, a resting phase is applied through the simultaneous closing of the lower NMOS switches, which causes the two ends of the electrode to be shorted and discharge the remaining charge in it.

The advantage of this structure over unipolar structures is that it requires only one current source. Suppose the unipolar circuits need two current sources for stimulation in the anodic and cathodic phases. Also, another advantage of the proposed circuit compared to similar structures is the reduction of the number of current directing switches.

Fig. 3 shows the proposed switches; each branch consists of a PMOS switch and an NMOS switch. To withstand the high voltage of V_{DDH} by standard transistors in 180 nm/1.8 V standard CMOS technology, a circuit with 14 transistors stacked together and able to withstand a supply voltage of $7 \times V_{DDL}$ is used. NMOS transistors are implemented in deep N-wells, while PMOS transistors have local N-wells.

In this bias circuit, we need power supplies that are an integer multiple of V_{DDL}. These voltages are provided through a circuit that changes the pump charge voltage level. To control the highest PMOS transistor in the switches, a voltage level shifter circuit is used, which brings the zero to V_{DDL} input signal to $6 \times V_{DDL}$ to $7 \times V_{DDL}$.

In this bias circuit shown in Fig. 3, M_{A1-8} transistors act like resistance dividers. Therefore, the shown bias voltages are also shown as V_{B1-5} based on this design. Also M_{B1-6} and $M_{D1,2}$ transistors act as diodes. In this way, their body is connected to their drain in all NMOS and PMOS transistors. For example, when the voltage in the M_{B1} transistor reaches lower than $5 \times V_{DDL}$, the diode turns on and returns to $5 \times V_{DDL}$. This is true for other transistors with the same conditions (transistors shown in yellow). Now, when the transistor M_1 is turned on and the transistor M_{14} is turned

off, the source of transistors M_{2-7} reaches $7 \times V_{DDL}$, and the source of transistors M_{8-14} reaches $6 \times V_{DDL}$, $5 \times V_{DDL}$, $4 \times V_{DDL}$, $3 \times V_{DDL}$, $2 \times V_{DDL}$ and V_{DDL}, respectively. Therefore, M_{A1-5} turn on and M_{A6-10} transistors turn off. This causes transistors M_{B1-4} to turn off and transistors M_{B5-8} to turn on. In the same way, the behavior of transistors is analyzed in the following classes. With these explanations, it can be seen that the voltage of the two ends of none of the transistors will exceed V_{DDL}.

The stack transistor switch has two functions: one is as a path for current to pass in the anodic and cathodic phases, connecting the current source to the electrode, and the other is as an isolation switch to protect the low-voltage digital-to-analog converter against the damage of the uniform high-voltage distribution. Will work To determine the size of the transistors in this structure, the size of each transistor that forms the output driver, considering that it is the main path of the switch, must be designed primarily, because they must pass a large current, up to 2.5 mA. Also, the dynamic bias circuit is designed and used to ensure that the voltage difference between the terminals of the switch stack transistors is maintained during the transition at 1.8 V. In addition to biasing the transistors related to the transfer switches, this circuit isolates the output of the stimulation circuit from the input signals It improves the efficiency of the circuit in this sense. Also, M_{A1-10}, M_{C1-6}, and $M_{E1,2}$ transistors should be selected so that the power of PMOS and NMOS transistors are almost the same. For this purpose, we consider the size of PMOS transistors to be

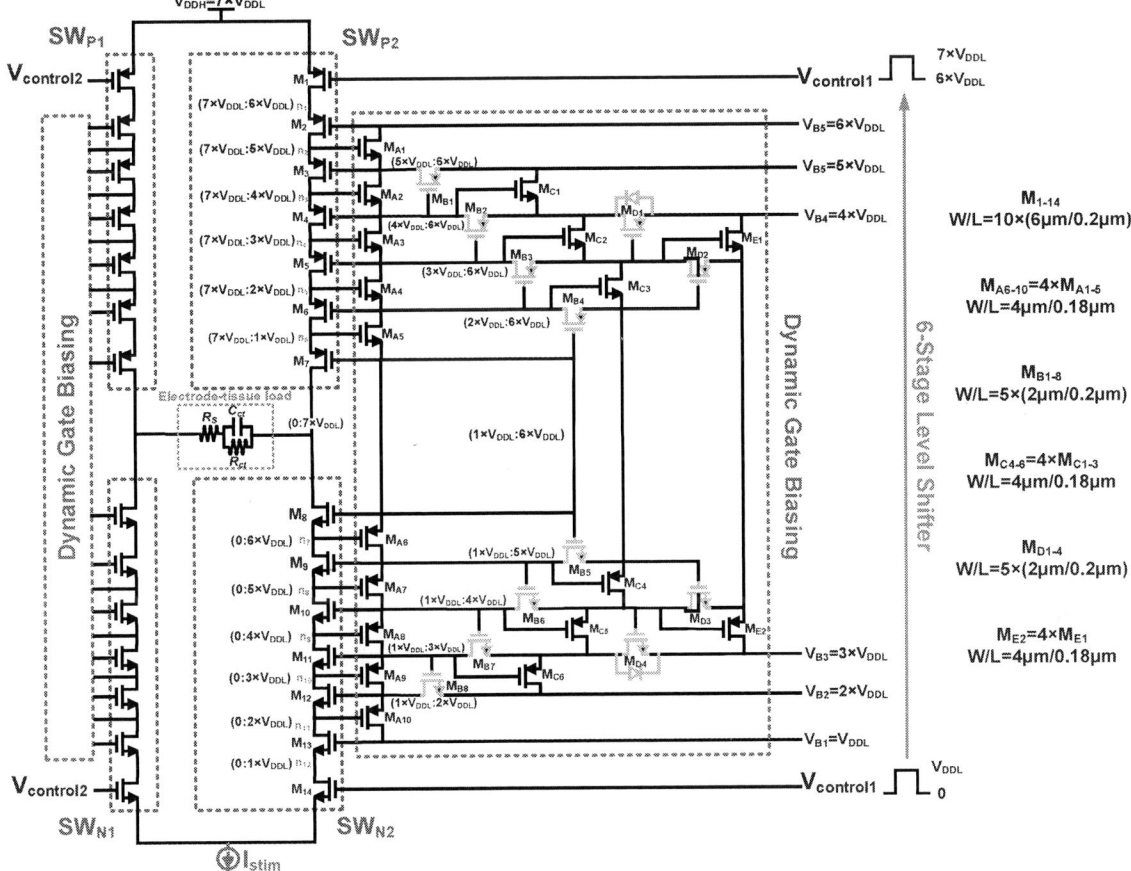

Fig. 3: Dynamic bias circuit structure of switches matrix.

about four times larger. Moreover, for M_{B1-8} and M_{D1-4} transistors, we choose relatively large sizes to increase the switching speed.

III. SIMULATION RESULTS

First, In order to check the correct operation of the DAC, by applying a digital bit string with a rate of 10 Mpulse/sec to the DAC inputs, we have drawn its output current in terms of given bits, is shown in Fig. 4(a). As expected after the design, the output current range is 39 µA. That is, the least significant bit (LSB) of the desired converter circuit is 39 µA. Differential nonlinear error (DNL) and integral nonlinear error (INL) have calculated its using MATLAB software, which were obtained at room temperature conditions of DNL=0.203 LSB, and INL=0.260 LSB for the designed converter.

To check the converter's performance, Monte Carlo statistics analysis was performed on it. In 100 simulations with different mismatches, the maximum output current was measured and its histogram was drawn. As shown in Fig. 4(b), the maximum output current has reached 2.5 mA in most tests. Which indicates the correct operation of the designed circuit.

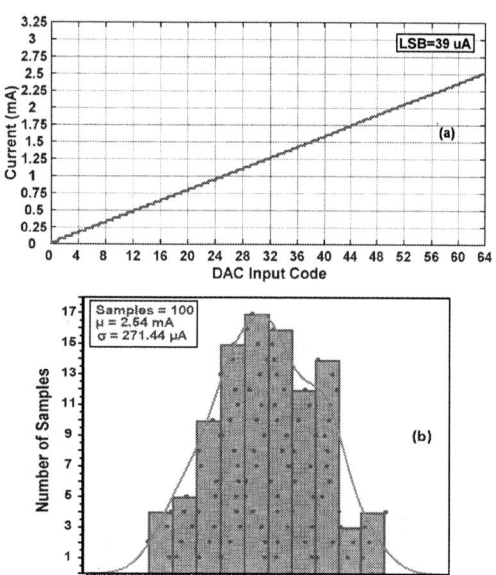

Fig. 4: (a) The anodic and cathodic currents versus DAC input code. (b) The histogram of the maximum amplitude of the output current in 100 different experiments with Monte Carlo analysis and considering mismatch.

Spurious free dynamic range (SFDR), signal-to-noise and distortion ratio (SNDR), and effective number of bits (ENOB) parameters have been calculated to investigate the dynamic behavior of the digital to analog converter. To calculate these parameters, a 6-bit ideal Flash analog-to-digital converter with thermometer-coded output is used. By applying a sinusoidal input with the input frequency f_{in}, from its output with the sampling frequency f_s and N points in the cycle number C, the sampled signals are applied to the DAC input, and the frequency response of the DAC is plotted, which is shown in Fig. 5. Also, Table I presents the simulation results in corners process.

$$f_{in} = \frac{f_{sample} \times Cycle}{N} \tag{5}$$

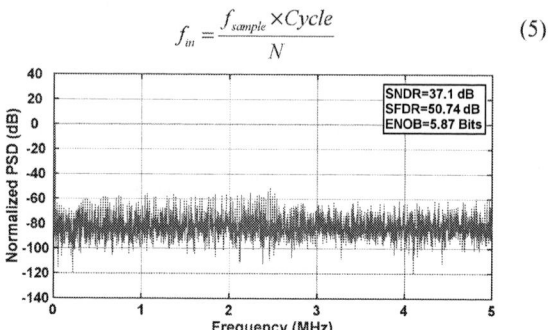

Fig. 5: SNDR and SFDR versus the input signal frequency.

TABLE I: RESULTS OF SIMULATING THE DAC IN DIFFERENT CORNERS PROCESS.

Parameters	FF(-40℃)	TT(27℃)	SS(85℃)
ENOB(Bits)	4.46	5.87	5.90
SNDR(dB)	28.63	37.1	37.2
SFDR(dB)	31.80	50.74	50.84

In the design of the high voltage electrical stimulation circuit using the accumulated transistors in the low voltage technology, it should be noted that the voltage difference between the bases of the transistors used in the circuit does not exceed the breakdown voltage of the parasitic well diode so that the transistor is not damaged. For this purpose, the terminals of the bulk transistors are connected to their sources to reduce the overdrive voltage of the stacked transistors. As mentioned before, in the structure of the high voltage electric stimulation circuit along with the dynamic gate biasing circuit, NMOS transistors are proposed in deep N-well so that their bulk can be connected to their source.

As shown in Fig. 6, the transient voltages of the drain-source, drain-gate, and gate-source terminals of M_7, M_8, M_{B6}, and M_{D3} transistors have been investigated. Some transistors have been investigated from the maximum to ensure the voltage applied to them. It can be seen that the amplitude of the applied voltage in each of the modes is lower than the breakdown voltage of the N-well diode. So it can be concluded that both the proposed bias circuit and the high voltage electrical stimulation circuit will have a good perform well in low voltage technology.

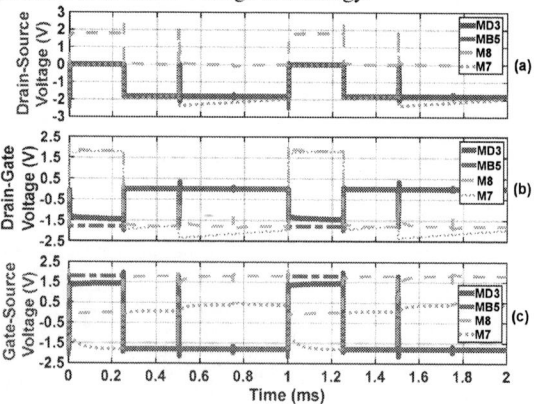

Fig. 6: (a) Drain-source voltage, (b) drain-gate voltage, and (c) gate-source voltage of transistors M_7, M_8, M_{B6}, and M_{D3}.

After the complete design of the circuit, the final simulations were performed on the electrical model of the electrode. 250 µs are considered for each of the anodic and cathodic phases, which is a suitable timing according to the frequency of nerve stimulation. As shown in Fig. 7, the voltage difference between the two electrodes and the applied current for the electric stimulation of the electrode is, clearly showing the biphasic nature of the designed system.

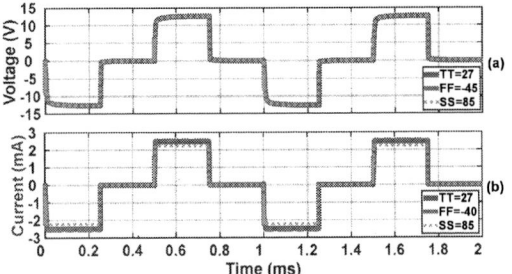

Fig. 7: (a) Differential output voltage, and (b) differential output current of the electrode in the final stimulation circuit.

Table II provides a summary and comparison of the performance between the proposed scheme and the prior art. The most important feature of the proposed design is the implementation of all circuits in 180nm CMOS 1.8V technology, which, in addition to the complete integration of the circuit, reduces the area and cost compared to high voltage technology. The excitation current range has increased, and without reducing DAC resolution, INL, and DNL nonlinearity errors have been reduced. The total power consumption has been improved according to the maximum current range and supply voltage of the excitation circuit.

TABLE II: PERFORMANCE SUMMARY AND COMPARISON WITH OTHER WORKS.

References	[2]	[10]	[17]	[18]	This Work
CMOS Process	0.18 µm High Voltage	0.18 µm High Voltage	0.18 µm High Voltage	0.18 µm low Voltage 3.3 V	0.18 µm low Voltage 1.8V
Supply Voltage (V)	9	12.8	6	12.5	12.6
Max stimulation current (mA)	0.5	0.992	1	2.08	2.5
Resolution DAC (bits)	6	5	6	5	6
INL	1.25	0.66	0.95	NA	0.260
DNL	0.16	0.2	0.5	NA	0.203
Power (mW)	0.572	NA	0.695	NA	1.23
Test	fabrication	simulation	fabrication	fabrication	simulation

IV. CONCLUSION

In this article, the general structure of the electrical stimulation system was first introduced, and its different parts were explained. These blocks included operational amplifier circuits, a digital-to-analog converter with thermometer-coded structure, and implantable electrostimulating switches. In the following, the proposed circuits for different system blocks were investigated for use in implantable microsystems in medicine. Parameters such as power consumption, the frequency response of op-amp and digital to analog converter, INL, and DNL errors and SNDR, SFDR, ENOB parameters for digital to analog converter and performance of high voltage switches in the

stimulation of implantable electrodes were thoroughly investigated. Moreover, at the end, all the required simulations were reported with Cadence and MATLAB software to ensure the accuracy of the circuit performance.

REFERENCES

[1] X. -H. Qian et al., "Design and In Vivo Verification of a CMOS Bone-Guided Cochlear Implant Microsystem, " IEEE Transactions on Biomedical Engineering, vol. 66, no. 11, pp. 3156-3167, Nov. 2019.

[2] M. Yip, R. Jin, H. H. Nakajima, K. M. Stankovic and A. P. Chandrakasan, "A Fully-Implantable Cochlear Implant SoC With Piezoelectric Middle-Ear Sensor and Arbitrary Waveform Neural Stimulation, " IEEE Journal of Solid-State Circuits, vol. 50, no. 1, pp. 214-229, Jan. 2015.

[3] B .S. Wilson, M. F. Dorman." Cochlear implants: current designs and future possibilities." J Rehabil Res Dev, Vol. 45, no. 5, pp. 695-730, 2008.

[4] Medmix, "Cochlear Implant," [Online]. Available: www.medmix.at/altern-mitallen-sinnen/#prettyPhoto. [Accessed: 13-Sep-2018].

[5] X. -H. Qian et al., "A bone-guided cochlear implant CMOS microsystem preserving acoustic hearing," 2017 Symposium on VLSI Circuits, Kyoto, Japan, pp. C46-C47, 2017,

[6] H. Uluşan, A. Muhtaroğlu and H. Külah, "A Sub-500 µ W Interface Electronics for Bionic Ears," IEEE Access, vol. 7, pp. 132140-132152, 2019.

[7] A. Abdi, and H. -K. Cha, A bidirectional neural interface CMOS analog front-end IC with embedded isolation switch for implantable devices. Microelectronics journal, pp. 70-75, Oct. 2016. 58.

[8] F. Fahimi Hanzaee; M. -M. Ahmadi. "Design and Simulation of the Digital Controller Block of a Neural Stimulation Chip for a Brain Implant". Iranian Journal of Biomedical Engineering, vol. 12, no. 2, pp. 147-159, Sep. 2018.

[9] S. -C. Liu, A. van Schaik, B. A. Minch and T. Delbruck, "Asynchronous Binaural Spatial Audition Sensor With 2 × 64 × 4 Channel Output," IEEE Transactions on Biomedical Circuits and Systems, vol. 8, no. 4, pp. 453-464, Aug. 2014.

[10] V. N. Tuan and H. -K. Cha, "A standard CMOS neural stimulator IC with high voltage compliant output current driver," 2017 International SoC Design Conference (ISOCC), Seoul, Korea (South), pp. 316-317, 2017.

[11] J. Simpson and M. Ghovanloo, "An Experimental Study of Voltage, Current, and Charge Controlled Stimulation Front-End Circuitry," 2007 IEEE International Symposium on Circuits and Systems, New Orleans, LA, USA, 2007, pp. 325-328.

[12] N. Tran et al., "A Complete 256-Electrode Retinal Prosthesis Chip." IEEE Journal of Solid-State Circuits, vol. 49, no. 3, pp. 751-765, March 2014.

[13] D. Jiang, D. Cirmirakis, and A. Demosthenous, "A vestibular prosthesis with highly-isolated parallel multichannel stimulation," IEEE transactions on biomedical circuits and systems, vol. 9, no. 1, pp. 124-137, July 2014.

[14] Z. Luo and M. -D. Ker, "A High-Voltage-Tolerant and Precise Charge-Balanced Neuro-Stimulator in Low Voltage CMOS Process," IEEE Transactions on Biomedical Circuits and Systems, vol. 10, no. 6, pp. 1087-1099, Dec. 2016.

[15] J. -J. Sit, A. M. Simonson, A. J. Oxenham, M. A. Faltys and R. Sarpeshkar, "A Low-Power Asynchronous Interleaved Sampling Algorithm for Cochlear Implants That Encodes Envelope and Phase Information," IEEE Transactions on Biomedical Engineering, vol. 54, no. 1, pp. 138-149, Jan. 2007.

[16] G. Manganaro, Advanced Data Converters, 1st Ed., New York: Cambridge University Press, 2012.

[17] H. A. Yiğit, H. Uluşan, M. Koç, M. B. Yüksel, S. Chamanian and H. Külah, "Single Supply PWM Fully Implantable Cochlear Implant Interface Circuit With Active Charge Balancing," IEEE Access, vol. 9, pp. 52642-52653, 2021.

[18] D. P. Mangut, Á. R. Vázquez and M. D. Restituto, "A Fully Integrated, Power-Efficient, 0.07-2.08 mA, High-Voltage Neural Stimulator in a Standard CMOS Process." Sensors (Basel, Switzerland) vol. 22, no. 17,Aug 2022.

Thorough Analysis of mm-Wave Broadband Planar and Vertical Transitions for Loss Reduction of Interconnects in Multilayer PCBs

Pouya Namaki
CST-Lab, School of ECE,
College of Eng, University of Tehran,
Tehran, 4396-43713 Iran
pouya.namaki@ut.ac.ir

Nasser Masoumi
CST-Lab, School of ECE,
College of Eng, University of Tehran,
Tehran, 4396-43713 Iran
nmasoumi@ut.ac.ir
CIARS, Department of ECE,
Faculty of Eng, University of Waterloo,
Waterloo, N2L3G1 Canada
nmasoumi@uwaterloo.ca

Mohammad-Reza Nezhad-Ahmadi
CIARS, Department of ECE,
Faculty of Eng, University of Waterloo,
Waterloo, N2L3G1 Canada
mrnezhad@uwaterloo.ca

Masoumeh Souri
CST-Lab, School of ECE,
College of Eng, University of Tehran,
Tehran, 4396-43713 Iran
m.souri@ut.ac.ir

Abstract—This paper presents a comprehensive investigation of mm-wave broadband planar and vertical transitions in multilayer printed circuit boards (PCBs) for high-density interconnects. The study focuses on microstrip to microstrip/stripline transitions across thin and thick substrates, addressing challenges related to chip/package attachment with small pitch sizes, the use of additional interfaces like interposers, as well as metallic losses and signal integrity. Planar transitions employ a tapered structure to connect microstrip to microstrip and stripline traces for the purpose of reducing metallic losses. Vertical transitions utilize vias to connect microstrip to stripline traces, with careful consideration of via diameter and anti-via diameter to maintain a 50 Ω characteristic impedance. Through meticulous analysis, the proposed designs exhibit excellent performance up to 60 GHz, achieving a return loss better than 10 dB. These transitions offer efficient solutions for routing high-speed signals in multilayer, high-density interconnect PCBs, ensuring minimal losses and crosstalk at mm-wave frequencies.

Keywords—*chip, interposer, multilayer printed circuit board (PCB), package, high-density interconnects, mm-wave transition, via, stripline, microstrip line.*

I. INTRODUCTION

The essential trends in modern semiconductor technology and packaging are to achieve a smaller pitch size for high-density IC/system integration, increased speed and performance, and lower cost in mass production [1]–[3]. Using thin substrate and consequently, thin traces in modern printed circuit boards (PCBs), realize the direct connection of the chip/package and PCB without any additional interface such as an interposer [4]. Nevertheless, employing a thin substrate leads to increased metallic losses in traces, which is undesirable, especially in system components involving millimeter-wave or long-length interconnects. Furthermore, numerous interconnects must be routed from the top layer to

the interlayers of the PCB to distribute chip/package pins throughout the system. As a result, at micro/mm-wave frequencies, there is a need for broadband transitions from thin to thick substrates and interlayer substrates.

So far, many transition techniques have been studied in the literature [5]–[11]. In [5] a transition from narrow to wide differential stripline and grounded coplanar waveguide (CPWG) on thin to thick substrates of low-cost PCB is presented to reduce the insertion loss of transmission line (TL) over the frequency range of DC to 40 GHz. Although differential signaling schemes offer reduced EMI production and sensitivity to induced or coupled noise in comparison to single-ended, they require two traces instead of one, or twice as much board area. Therefore, multilayer PCBs with high-density interconnects, often employing single-ended signaling schemes, are typically preferred. In [6] two broadband 3D vertical transitions and one 2D planar transition (CPW bend) on a flexible organic liquid crystal polymer (LCP) substrate were presented. Even though LCP has many advantages, it does have some drawbacks. For example, LCP is mostly considered a new material. Therefore, it is too experimental to be utilized in a developing system. In [7], another vertical transition using quasi-coaxial vias has been implemented, but it yielded limited bandwidth results. Other reported papers, such as [8]–[11], employ waveguides to microstrip/stripline transitions from one layer to another. However, these methods are constrained by limited bandwidth, bulkiness, and the necessity for waveguide-based technologies.

This paper presents a comprehensive analysis of mm-wave broadband planar and vertical transitions for multilayer PCB applications. The analysis is based on extensive 3D electromagnetic (EM) simulations of various structures, including microstrip to stripline and stripline transitions across thin and thick substrates. All the EM simulations were performed using the Ansys HFSS simulator. For the purpose of this work, the dielectric constant and dielectric loss of the

979-8-3503-6020-2/23 $31.00 © 2023 IEEE

The 5th Iranian International Conference on Microelectronics (IICM2023)

utilized multilayer PCB was set to be around 2.98 and 0.0026 for the frequencies of interest, respectively. All dimensions of the designed structures are shown in the figures, and they are based on existing commercial design rules and mechanical drilling for via formation. This paper is organized as follows. Section II investigates two transitions from thin to thick substrate, including planar microstrip to microstrip and planar and vertical stripline to stripline. A broadband microstrip to stripline transition is investigated in Section III. Additionally, the effects of physical parameters of the via in improving the transition are analyzed. Finally, brief conclusions are provided in Section VI.

II. Loss Reduction of Interconnects using Substrate with Different Thickness

Microstrip and stripline are the main forms of planar TL, frequently used in multilayer PCB designs, where high-speed signals need to be routed from one part of the system to another with minimal loss and cross talk, respectively. To increase the density of interconnects in modern PCB these transmission lines are built on thin substrates. However, it is often found that the metallic loss of traces on the thin substrate is too high at micro/mm-wave frequencies especially for long-length interconnects. In order to alleviate this problem, it is often necessary to change and increase the substrate thickness of an interconnect on its way. However, the transition from different substrate thickness represents the main discontinuity to the signal pathway, which results in reflection, loss and significant degradation of the signal if not carefully considered during the design phase. In this section, a planar and planar/vertical transition from thin to thick substrate for microstrip and stripline structures are investigated, respectively.

A. Planar transition from thin to thick substrate for microstrip line structure

Microstrip has become a critical part of RF modules due to its advantages such as low loss, low volume planar configuration, low fabrication costs, and capability to integrate with micro/mm-wave integrated circuits. Fig. 1 illustrates the structure of a mm-wave broadband transition of a microstrip line from the thin to thick substrate. The proposed structure consists of three metallic layers and two dielectrics. As Fig. 1(a) shows, on the left side of the structure a substrate with a thickness of 115 μm and a trace with a width of 240 μm is used. To increase the trace width and consequently reduce the metallic loss, on the right side of the structure the thickness of the substrate is increased to 258 μm. To keep the impedance to 50 Ω a trace with a width of 580 μm is used on the right side of the structure. At the middle point of the structure, a narrow microstrip trace is tapered to the wider trace. Ground vias in the left side of the structure helps to keep the continuity of the ground plane of the whole configuration. Fig. 1(b) and (c) show the impact of the taper length (L_{Tran}) on the insertion and return loss of the broadband transition of microstrip on thin to thick substrates, respectively. As shown in Fig. 1(b) and (c), tapered length should be kept as small as possible to improve the performance of the transition. This inference is supported by the observation that reducing the tapered length leads to a

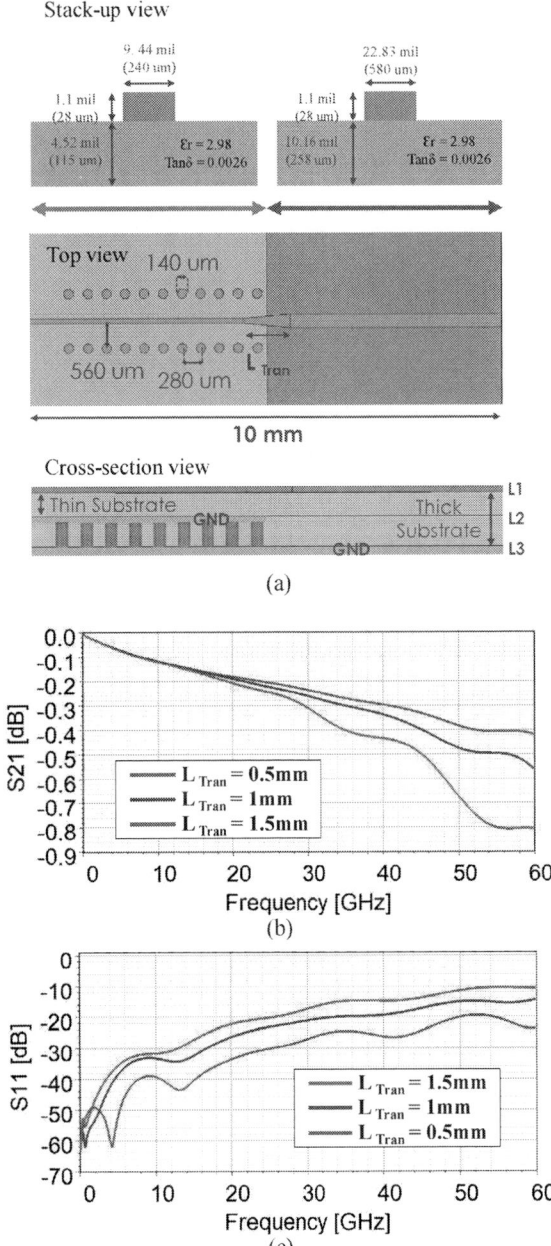

Fig. 1. Broadband transition of the microstrip on thin to the thick substrates. (a) Configuration. (b) Impact of the tapered length L_{Tran} on the insertion loss. (c) Impact of the tapered length L_{Tran} on the return loss.

decrease in the length of the narrower trace on the thick substrate, as well as a reduction in the length of the wider trace on the thin substrate. This reduction results in less signal discontinuity and degradation. In essence, minimizing the tapered length mitigates the mismatch between traces on both substrates, consequently enhancing the performance of the transition. Fig. 2 shows comparison of the insertion

979-8-3503-6020-2/23 $31.00 © 2023 IEEE   123

The 5th Iranian International Conference on Microelectronics (IICM2023)

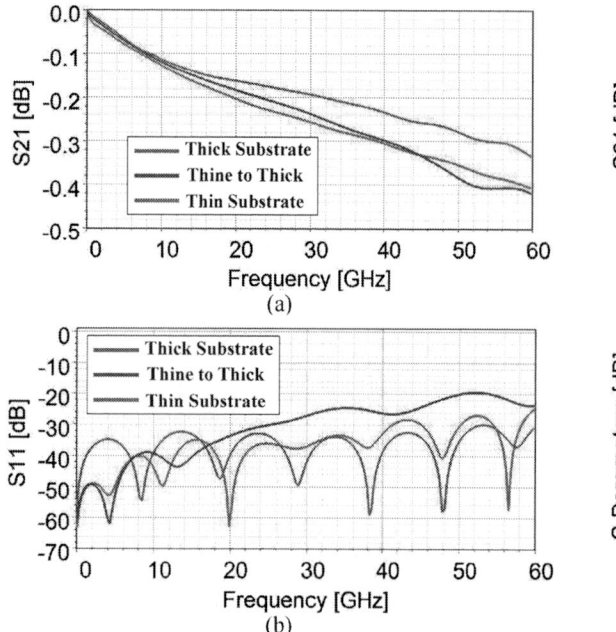

Fig. 2. Comparison of insertion loss (a) and return loss (b) for the proposed transition and narrow (red) and wide (green) microstrips on thin and thick substrates, respectively, at a total length of 10 mm.

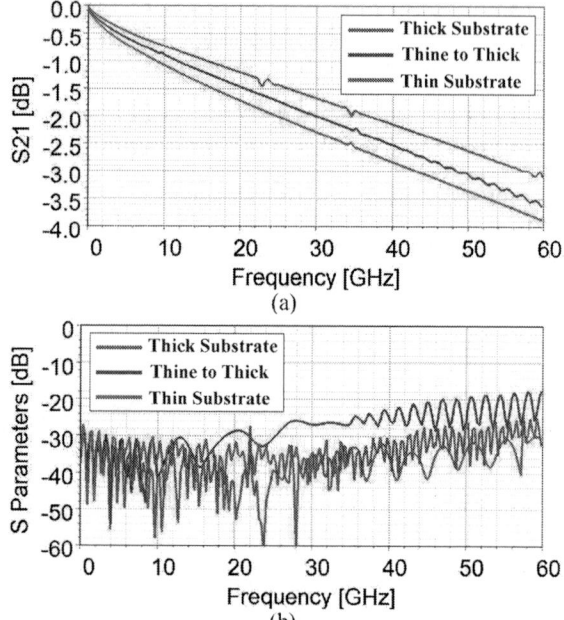

Fig. 3. Comparison of insertion loss (a) and return loss (b) for the proposed transition and narrow (red) and wide (green) microstrips on thin and thick substrates, respectively, at a total length of 100 mm.

and return loss of the proposed transition structure for a total length of 10 mm (5 mm narrow microstrip on one side, the tapered transition at the middle, and 5 mm wide microstrip on the other side of substrate) and 10 mm microstrip on the thin and thick substrate separately. Accurate simulation results show that widening the trace width and using a thicker substrate reduced insertion loss of the interconnect by 10 percent over the frequency range of DC to 60 GHz. To better show the efficiency of the proposed transition, Fig. 3 shows the simulation results for the 100 mm interconnect length. It should be noted that the loss of transition can be reduced by up to 25 percent by choosing a smaller length for the narrow microstrip line trace in comparison to the wide microstrip line trace.

B. Planar/vertical transition from thin to thick substrate for stripline structure

The capability of routing on the interlayers of a PCB makes stripline a preferred choice in a multilayer PCB design with high-density interconnects. Hence, in this section, a planar and a vertical structure for the transition of stripline through the substrate with different thicknesses are investigated. Fig. 4 shows the planar structure of the mm-wave broadband transition of the stripline from thin to thick substrate. The proposed structure consists of five metallic layers and four dielectrics. As Fig. 4(a) shows, on the left side of the structure a substrate with a thickness of 258 μm and a trace with a width of 100 μm is used. To increase the trace width and consequently reduce the metallic loss, on the right side of the structure the thickness of the substrate is

increased to 544 μm. To keep the impedance to 50 Ω a trace with a width of 260 μm is used on the right side of the structure. At the middle point of the structure, the narrow stripline trace is tapered to the wider trace. This transition requires strategically placed adjacent ground vias to suppress radiation loss and improve performance over a broad frequency range. It is important to note that the distance between the adjacent via and the trace should be at least 2 or 3 times greater than the trace width to avoid interference with microstrip propagation [12]. Fig. 4(b) and (c) show the impact of the tapered length (L_{Tran}) on the insertion and return loss of the broadband transition of stripline on thin to thick substrates, respectively. As shown in Fig. 4(b) and (c), the tapered length should be kept as small as possible to improve the performance of the transition. Fig. 5 shows in detail the vertical structure of the mm-wave broadband transition of the stripline from thin to thick substrate. Here at the middle point of the structure, one via-based vertical transition is used to connect the narrow stripline trace to the wider trace, as well as grounding vias connecting the top, bottom, and middle ground layers. Fig. 5(b) and (c) show the impact of the via diameter (D_{Via}) on the insertion and return loss of the broadband transition of stripline on thin to thick substrates, respectively. As shown in Fig. 5(b) and (c), the via diameter should be kept as small as possible to improve the performance of the transition. A 100 μm via diameter was selected. This dimension is the smallest diameter that could be mechanically drilled and metalized in our structure, thus determining the smallest via size with satisfactory mm-wave performance. For this design, another major focus was the selection of the optimum number of and spacing between

979-8-3503-6020-2/23 $31.00 © 2023 IEEE

The 5th Iranian International Conference on Microelectronics (IICM2023)

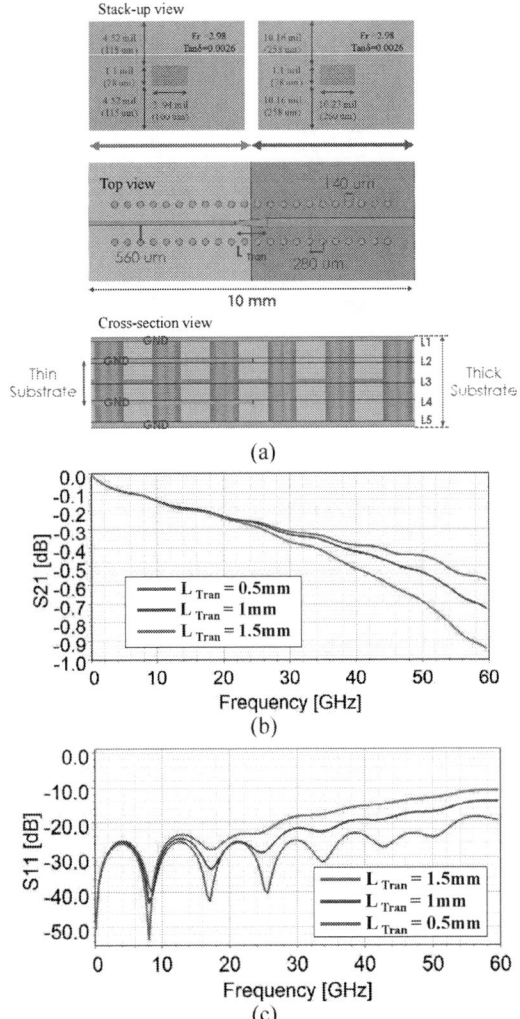

(a)

(b)

(c)

Fig. 4. Planar broadband transition of the stripline on thin to the thick substrates. (a) Configuration. (b) Impact of the tapered length L_{Tran} on the insertion loss. (c) Impact of the tapered length L_{Tran} on the return loss.

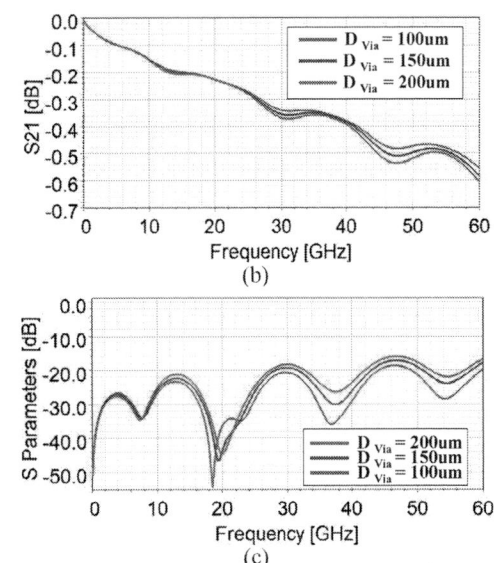

(b)

(c)

Fig. 5. Vertical broadband transition of the stripline on thin to the thick substrates. (a) Configuration. (b) Impact of the via diameter (D_{Via}) on the insertion loss. (c) Impact of the via diameter (D_{Via}) on the return loss.

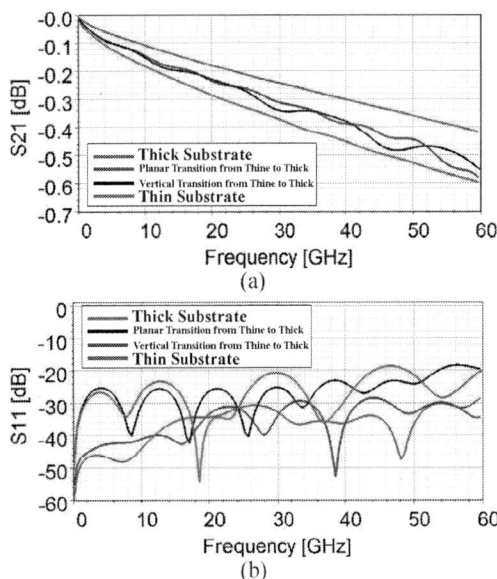

(a)

(b)

Fig. 6. Comparison of the insertion and return loss of the proposed planar/vertical transition and narrow (red) and wide (green) stripline on thin and thick substrate, respectively.

the adjacent vias. Maintaining a 560 µm and 280 µm for via–trace spacing and via–via spacing, respectively, was necessary in order to suppress the parasitic radiation due to the parallel-plate mode and achieve good performance. Fig. 6 shows the comparison of insertion and return loss of the proposed planar/vertical transition structures and stripline on the thin and thick substrate separately. Accurate simulation results show that widening the trace width and using a thicker substrate reduced insertion loss of the interconnect by 10 percent over the frequency range of the DC to 60 GHz. It

979-8-3503-6020-2/23 $31.00 © 2023 IEEE   125

should be noted that the loss of transition can be reduced by up to 33 percent by choosing a smaller length for the narrow stripline trace in comparison to the wide stripline trace.

III. MILLIMETER WAVE BROADBAND VERTICAL TRANSITION OF MICROSTRIP TO STRIPLINE

For the integration of the IC/package with a high number of I/O pins into systems many interconnects should be routed from the top layer to the interlayers of the PCB in order to spread the chip/package pins to the different parts of the system. Here, a broadband vertical transition from microstrip to stripline using via is investigated. Fig. 7 shows the structure of the mm-wave broadband transition of the microstrip line to the stripline. The proposed structure consists of eight metallic layers and seven dielectrics. As Fig. 7 shows, on the left side of the structure, a substrate with a thickness of 258 μm and a stripline trace with a width of 100 μm is used. On the right side of the structure, a microstrip trace with a width of 240 μm on the first layer is used. To connect the stripline and microstrip traces, at the middle point of the structure, a via with the diameter and length of 250 μm and 830 μm is used, respectively. Adjacent vias in the structure helps to keep the continuity of the ground plane of the whole configuration. Here, via and ant-via diameters have the main role in the performance of this transition, and they should be optimized so as to match the series inductance of the via transition and maintain a 50 Ω characteristic impedance throughout the transition, allowing for a broadband response up to 60 GHz. The effect of the chosen diameter of the via is demonstrated in the insertion and loss plots shown in Fig. 8(a) and (b), respectively. As shown in Fig. 8 (a) and (b), via diameter of 100 μm has the best performance in the terms of loss and reflection. Fig. 9 (a) and (b) show the impact of the anti-via diameter ($D_{Anti-via}$) on the insertion and return loss of the broadband transition, respectively. As shown in Fig. 9 (a) and (b), an anti-via diameter of 200 μm has the best performance in terms of loss and reflection over the frequency range of DC to 60GHz.

To further examine the performance of the proposed transition, simulated E-field distributions for the optimal configuration at 35 GHz are presented. Fig. 10(a) and (b) show the top view and cross-sectional view, respectively. The impact of strategically positioned adjacent vias on the field distribution is demonstrated in Fig. 10(a). This emphasizes the importance of maintaining specific spacing between ground array vias to effectively reduce E-field edge radiation effects. Fig. 10(b) illustrates the mode transitions occurring within our design, transitioning from TEM in the stripline to a hybrid mode in the via transition, and finally to quasi-TEM in the microstrip line. It is worth noting that the electric field distribution in our proposed vertical transition is primarily confined around the via, leading to reduced signal insertion loss and improved return loss.

Upon optimizing the via diameter, one can observe the improvement in the insertion loss up to 1.5dB at 60 GHz, while the anti-via diameter optimizes the insertion loss of 1dB at the same frequency of 60 GHz. The final design features a very good performance at frequencies up to 60 GHz with a return loss lower than 10 dB. These results are reported for the first time for frequencies up to 60 GHz and for simple to-fabricate vertical transition in multilayer PCB.

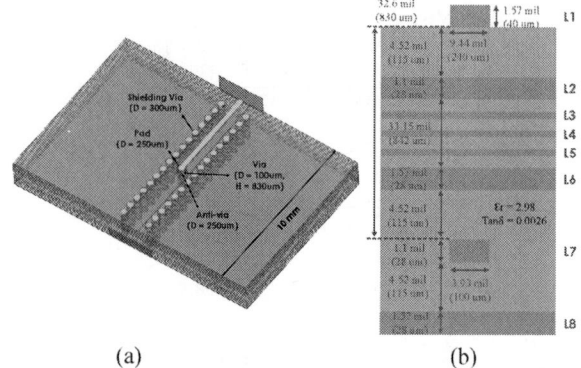

(a)　　　　　　　　　(b)

Fig. 7. Configuration (a) and stack-up (b) of the vertical broadband transition of the microstrip to stripline.

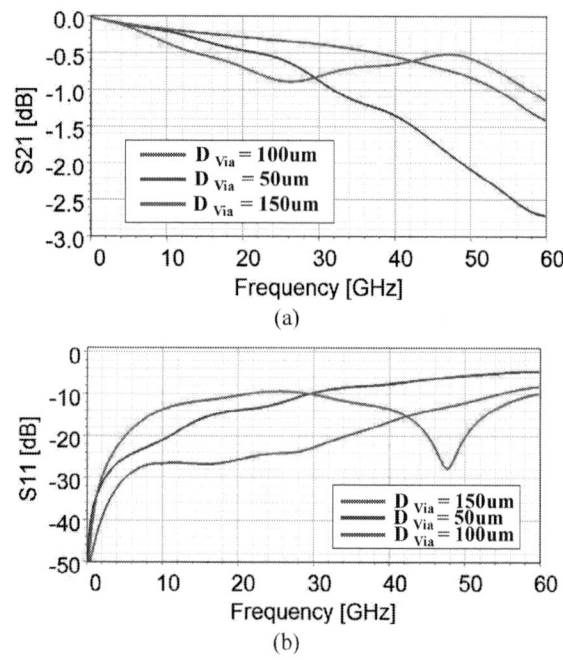

(a)

(b)

Fig. 8. Impact of the via diameter (D_{Via}) on the proposed vertical transition of the microstrip to stripline. (a) Insertion loss. (b) Return loss.

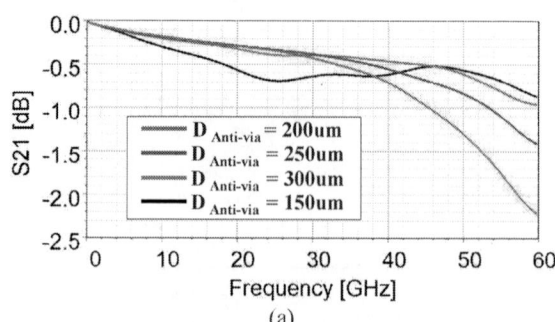

(a)

The 5th Iranian International Conference on Microelectronics (IICM2023)

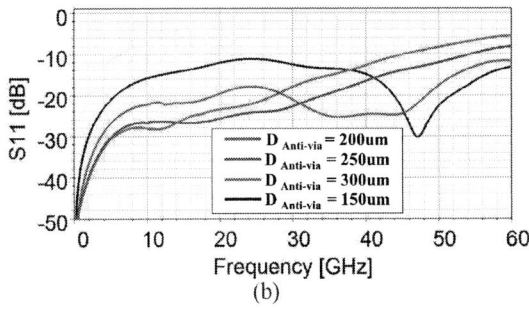

(b)

Fig. 9. Impact of the anti-via diameter ($D_{Anti-via}$) on the proposed vertical transition of the microstrip to stripline. (a) Insertion loss. (b) Return loss.

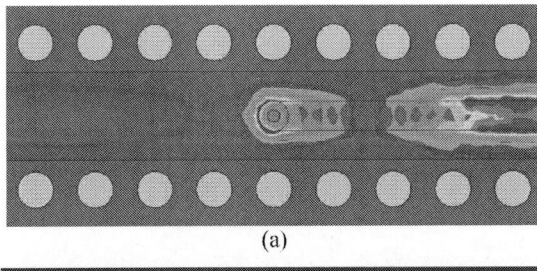

(a)

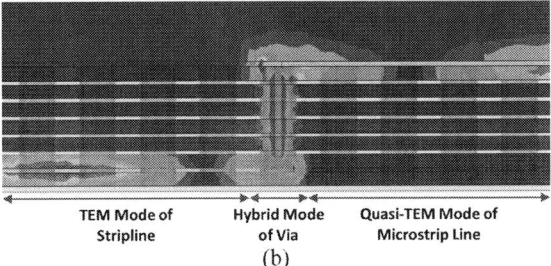

(b)

Fig. 10. The magnitude of E-field distribution for the broadband microstrip to stripline transition at 35 GHz.

This is achieved by minimizing mismatches between the lines and vertical via transition, which could be detrimental to the reflection performance of the interconnect.

IV. CONCLUSION

In this paper, a comprehensive analysis of mm-wave broadband planar and vertical transitions for loss reduction of interconnects in multilayer PCB was presented. All the proposed transitions are simple to realize and are compatible with modern multilayer PCB with high-density interconnects in which long 50 Ω microstrip and stripline traces are needed to route and spread the fine I/O pins of mounted IC/package all over the systems. The proposed designs in which tapered structure and via is used for planar and vertical transitions, respectively, achieve very good wideband mm-wave performance by placing grounding vias in appropriate locations in order to suppress parasitic modes and eliminate radiation losses. Utilizing the proposed technique reduced the microstrip/stripline interconnection loss by up to 25%.

The simulated return loss of the proposed transitions is better than 10 dB over the broadband frequency range of DC to 60 GHz.

ACKNOWLEDGMENT

The authors would like to thank the Circuits, Systems, and Test Laboratory, CST-Lab (School of Electrical and Computer Engineering, College of Eng., University of Tehran) and CIARS (Centre for Intelligent Antenna and Radio Systems, Dept. ECE, University of Waterloo) researchers, and technicians who provided insight, comments, and valuable discussions, that greatly assisted this paper.

REFERENCES

[1] J. H. Lau, *Heterogeneous Integrations*. Springer Singapore, 2019. doi: 10.1007/978-981-13-7224-7.

[2] B. Razavi, *Design of Analog CMOS Integrated Circuits*. NYUnited States: McGraw-Hill, Inc., 2016.

[3] P. Namaki, N. Masoumi, M.-R. Nezhad-Ahmadi, and S. Safavi-Naeini, "A Tunable Macro-Modeling Method for Signal Transition in mm-Wave Flip-Chip Technology," in *2021 IEEE 25th Workshop on Signal and Power Integrity (SPI)*, May 2021, pp. 1–4.

[4] P. Namaki, N. Masoumi, M. Seyedi, and M.-R. Nezhad-Ahmadi, "An Extended L–2L De-Embedding Method for Modeling and Low Return-Loss Transition of Millimeter Wave Signal Through Silicon Interposer," *IEEE Trans. Electromagn. Compat.*, pp. 1–12, 2023, doi: 10.1109/TEMC.2023.3311377.

[5] N. Ghassemi and H. Tournier, "Millimeter-wave broadband transition of stripline and CPWG on thin-to-thick substrates," in *2016 IEEE International Symposium on Electromagnetic Compatibility (EMC)*, Jul. 2016, pp. 536–541. doi: 10.1109/ISEMC.2016.7571705.

[6] A. Rida, A. Margomeno, J. S. Lee, P. Schmalenberg, S. Nikolaou, and M. M. Tentzeris, "Integrated Wideband 2-D and 3-D Transitions for Millimeter-Wave RF Front-Ends," *IEEE Antennas Wirel. Propag. Lett.*, vol. 9, pp. 1080–1083, 2010, doi: 10.1109/LAWP.2010.2091714.

[7] T. Zhang, L. Li, M. Xie, H. Xia, X. Ma, and T. J. Cui, "Low-Cost Aperture-Coupled 60-GHz-Phased Array Antenna Package With Compact Matching Network," *IEEE Trans. Antennas Propag.*, vol. 65, no. 12, pp. 6355–6362, Dec. 2017, doi: 10.1109/TAP.2017.2722867.

[8] Y. Zhu, R. Lu, C. Yu, and W. Hong, "Design and Implementation of a Wideband Antenna Subarray for Phased-Array Applications," *IEEE Trans. Antennas Propag.*, vol. 68, no. 8, pp. 6059–6068, Aug. 2020, doi: 10.1109/TAP.2020.2988946.

[9] H. A. Diawuo and Y.-B. Jung, "Waveguide-to-Stripline Transition Design in Millimeter-Wave Band for 5G Mobile Communication," *IEEE Trans. Antennas Propag.*, vol. 66, no. 10, pp. 5586–5589, Oct. 2018, doi: 10.1109/TAP.2018.2854364.

[10] X. Dai, "An Integrated Millimeter-Wave Broadband Microstrip-to-Waveguide Vertical Transition Suitable for Multilayer Planar Circuits," *IEEE Microw. Wirel. Compon. Lett.*, vol. 26, no. 11, pp. 897–899, Nov. 2016, doi: 10.1109/LMWC.2016.2614973.

[11] Y. Zhang, D. Zhao, and P. Reynaert, "A Flip-Chip Packaging Design With Waveguide Output on Single-Layer Alumina Board for E-Band Applications," *IEEE Trans. Microw. Theory Tech.*, vol. 64, no. 4, pp. 1255–1264, Apr. 2016, doi: 10.1109/TMTT.2016.2536602.

[12] S. H. HALL and H. L. HECK, *Advanced Signal Integrity for High-Speed Digital Designs*. USA: John Wiley & Sons, Inc., 2009.

Impact of Geometrical and Process Design Parameters on the Performance of Schottky Barrier Reconfigurable Field Effect Transistor

Hamid Reza Heydari
Department of Electronic
Yadegar- e- Imam Khomeini (RAH)
Shahr-e-Rey Branch, Islamic Azad
University
Tehran, Iran
weather_station2010@yahoo.com

Zahra Ahangari*
Department of Electronic
Yadegar- e- Imam Khomeini (RAH)
Shahr-e-Rey Branch, Islamic Azad
University
Tehran, Iran
Z.ahangari@gmail.com

Hamed Nematian
Department of Electronic
Yadegar- e- Imam Khomeini (RAH)
Shahr-e-Rey Branch, Islamic Azad
University
Tehran, Iran
h.nematian.n@gmail.com

Kian Ebrahim Kafoori
Department of Electronic
Yadegar- e- Imam Khomeini (RAH)
Shahr-e-Rey Branch, Islamic Azad
University
Tehran, Iran
kian_kafoori@yahoo.com

Abstract—In this study, the electrical properties of a Schottky Barrier Reconfigurable Field Effect Transistor (SBRFET) are thoroughly evaluated with the variation of important design parameters. The device proposed in this study operates in both n-type and p-type modes by exploiting electrical doping rather than additional physical doping. One of the critical design parameters is the gate work function, which significantly affects the charge density in the channel and subsequently alters the tunneling rate. The findings reveal that gate length variation has a negligible impact on the device's performance. The device achieved an on/off current ratio of 2×10^7 and a subthreshold swing of 87.5 mV/dec for n-type operation, while for p-type operation, it achieved an on/off current ratio of 1.5×10^6 with a subthreshold swing of 95.7 mV/dec. The results of this study pave the way for developing high-speed, low-power digital circuits with reprogrammable logic.

Keywords—*Reconfigurable Transistor, Schottky Barrier Transistor, Direct Tunneling, Thermionic Emission.*

I. INTRODUCTION

Next generation electronic devices and communication technologies demand low-power consumption and high-speed performance, which can be satisfied by scaling the dimensions of the conventional Metal-Oxide-Semiconductor Field Effect Transistor (MOSFET). However, MOSFET scaling below 100nm results in degraded performance due to short channel effects [1-2]. To overcome this, multi-gate structures with thin film channel thickness have been introduced as a potential solution for improving the gate control over the channel and the device electrical

performance [3-5]. Nevertheless, reducing the thickness of the source/drain increases parasitic resistance, which can degrade the on-state drive current. Thus, the Schottky barrier (metal) source/drain transistor has been introduced as a potential alternative to the conventional doped source/drain MOSFET, as it provides low parasitic source-drain resistance [6-7].

In order to surpass the limitations encountered in miniaturization and enhance the performance of integrated circuits, the Reconfigurable Field Effect Transistor (RFET) has been introduced. This technology utilizes the advantages of Schottky barrier transistors to design a single device with p-type (hole conduction) or n-type (electron conduction) without any additional physical doping. One significant advantage of RFET technology is its ability to deliver compact circuit topologies by utilizing fewer devices to realize various logic gates. Typically, reconfigurable field effect transistors are controlled by at least two separate gate electrodes. The program gate is used to select the type of charge carrier (electrons or holes), while the control gate modulates the channel conductance and therefore regulates the amount of current. The electrical characteristics of a nanowire RFET has been experimentally investigated and near-symmetric transfer characteristics has been achieved [8]. The use of multi-gate RFETs in circuit design is also explored to enable the creation of multi-stage digital circuits with two threshold voltages [9]. Furthermore, different RFET structures have been implemented using germanium and two-dimensional materials [10-11]. However, it is important to note that the critical design parameter that may significantly impact device performance has not yet been considered.

This paper centers its attention on conducting a comprehensive examination of the influence of crucial physical and structural design variables on the efficacy of the double gate Schottky barrier reconfigurable field effect transistor. In particular, achieving identical drain current for both n-type and p-type operation is crucial for designing efficient reconfigurable circuits. Any design parameter that alters the charge distribution in the channel may impact device performance. We comprehensively assess the impact of gate length, channel thickness, gate work function, and temperature to identify the optimal values for device performance. The comparison between the transfer characteristics of the experimentally conducted 3D nanowire reconfigurable field effect transistor and the simulation results is performed in order to calibrate the simulation parameters, depicted in Fig.1 [12].

II. DEVICE STRUCTURE AND SIMULATION MODELS

Figure. 2 displays the two-dimensional schematic of a Schottky Barrier Reconfigurable Field Effect Transistor (SBRFET). This device is equipped with two separate gates - the control gate and program gate. The control gate (V_{CG}) is positioned near the source region, which is responsible for activation and deactivation of the device, whereas the program gate (V_{PG}) is situated near the drain junction, which modifies the band bending for p-type or n-type mode of operation. The device differs from the conventional FET with doped source drain, as it has metal source/drain contacts, which form a Schottky barrier at the interface of the source/drain and channel region. The existing contributions in SBRFET are intricately linked to the barrier height for both electrons and holes. It is crucial that the barrier height for electrons and holes be the same, given a symmetrical current profile for both n and p carriers. This investigation employs a channel free of doping in conjunction with a mid-gap metal electrode at the source and drain, as employed in the suggested silicon device. The device features contacts composed of Nickel Silicide (NiSi$_2$), which has a low-lattice mismatch and provides an epitaxial relationship to silicon. This material also offers a Schottky barrier height of 0.59eV for electrons. By taking into account the silicon band gap value of 1.12eV, a Schottky barrier height of 0.53eV can be attained for the holes. The initial device design parameters are presented in Table.1. To evaluate the device performance, numerical simulations are conducted using the ATLAS device simulator [13] with activated models that are introduced as follows: (a) Direct Tunneling Model; The Direct Tunneling Model pertains to the Schottky contact established between the source/drain channel region and the interface, which serves as a barrier for the direct tunneling current between the sub-bands of the doping layer and the metal surface. The modulation of the tunneling rate is contingent upon the depltion region at the Schottky contact, and can be effectively controlled through the gate bias. (b) Thermionic Emission Model; The Thermionic Emission Model is based on the principle that thermally excited electrons in the metal that can cross over the Schottky barrier, generate a thermionic emission current. This current mechanism is heavily reliant on the barrier height and temperature. (c) Mobility Models; The electric field arises from the horizontal and vertical electrodes can effectively modify the carrier mobility and the carrier velocity. Therefore, field dependent mobility models are activated for calculating the drain current. In addition, the mobility is degraded through carrier-carrier scattering, which may affect the drain current. Thus, related mobility models are considered in the simulation. (d) Schottky Barrier Lowering Model ; This theoretical framework encompasses the aggregation of Image charges within the metallic electrode of a metal-semiconductor junction as carriers approach the interface between the metal and the semiconductor. This accumulation induces a potential that reduces the effective barrier height, which can greatly impact the performance of the device. Consequently, this effect is taken into consideration in the simulation. (e) Quantum Confinement Model; The Quantum confinement model primarily manifests in thin channel thicknesses and results in the formation of energy sub-bands within the channel with elevated energy level. This effect leads to an increase in the effective Schottky barrier height, which is also fundamental to the simulation. (f) Auger and Shockley-Read-Hall (SRH) recombination models; The Auger model and the Shockley-Read-Hall (SRH) model of recombination have been developed in order to provide a clear assessment of the impact of recombination and the generation of carriers on carrier transport in the presence of defects and traps.

TABLE I. INITIAL DESIGN PARAMETERS OF SBRFET

Parameter	SBRFET
Channel length -L_{ch}	40 nm
Spacer length	16 nm
Channel thickness-T_{ch}	10 nm
Control gate/ Program gate workfunction (WFG)	4.6 eV
Program gate length	12nm
Control gate length	12nm
Channel doping density	intrinsic

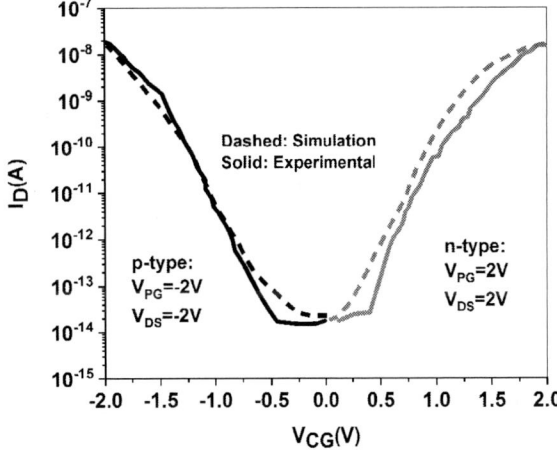

Fig. 1. Transfer characteristics of the experimental [12] and simulated 3D nanowire reconfigurable transistor.

III. RESULTS AND DISCUSSIONS

The proposed SBRFET functions in both n-type and p-type modes by manipulating bias values instead of physical doping. The activation and deactivation of the device are controlled by the bias applied to the control gate, while the program gate facilitates the movement of appropriate carriers within the channel. In n-type operation, the program gate adjusts the band bending at the drain side to prevent the transport of holes in the channel. Subsequently, by applying a positive bias to the control gate, the barrier width at the interface of the source and channel region is reduced, allowing electrons to tunnel through the channel. Conversely, p-type operation of the device can be achieved by applying an opposite bias to the control gate, program gate, and drain electrodes. The illustration of the energy band diagram for SBRFET is presented in Fig.3 (a) and (b), depicting its off-state and on-state for both n-type and p-type modes of operation, respectively. Specifically, during the off-state (i.e., $V_{CG}=0$, $V_{PG}\neq0$, $V_{DS}\neq0$), the tunneling barrier located at the interface of the source and channel region is not thin enough to facilitate the tunneling of carriers. Conversely, during the on-state state (i.e., $V_{CG}\neq0$, $V_{PG}\neq0$, $V_{DS}\neq0$), a sufficiently high bias applied to the control gate results in the accumulation of carriers in the channel, thereby reducing the width of the depletion region at the source Schottky contact and allowing for direct tunneling. Notably, the polarity of the gate bias impacts the band bending in such a way that facilitates the specific transport of either electrons or holes in the channel. Figure.4 (a) and (b) demonstrate the carrier density variation along the channel length from source to drain in both on-state and off-state operations. In Fig.4(a), we observe a significant increase in the electron density for n-type operation with an increase in the control gate bias during on-state operation. It is apparent that the accumulation of carriers in the channel decreases the depletion region width at the source Schottky contact, leading to carrier tunneling into the channel. Fig.4(b) shows the hole density for p-type mode of the SBRFET, which also exhibits a similar modulation of the hole density by the control gate bias. Figure.5 depicts the transfer characteristics of the SBRFET,

which displays reconfigurable behavior. The threshold voltage (V_{th}) denotes the control gate value at which the drain current attains 0.1A/µm. The device demonstrates uniform performance for both n-type and p-type modes of operation, with an on/off current ratio of 2×10^7 and 1.5×10^6 achieved for n-type and p-type modes of operation, respectively. Notably, the dominant current transport mechanism is direct tunneling through the Schottky barrier, owing to the high effective Schottky barrier height at the source and drain regions. The n-type and p-type operations have been able to achieve on-state currents of 1.3×10^{-6} A/µm and 1×10^{-6} A/µm, and threshold voltage of 0.53V for n-type operation, as compared to 0.78V for p-type operation. Moreover, the n-type and p-type modes have been able to achieve subthreshold swings of 87.5 mV/dec and 95.7 mV/dec, respectively. Moreover, the presence of the vertical electric field, which arises from the program gate situated at the drain side and the mid gap barrier height leads to a flat band bending at the off-state of the drain side. Consequently, the likelihood of drain leakage current, specifically the gate-induced drain leakage (GIDL) current and punch through effects, is significantly diminished. Figure.6 depicts the impact of the spacer length between the control gate and program gate on the on-state and off-state of the device for the n-type and p-type mode of operation. Our observations reveal that the drain current is impervious to the fluctuation of spacer length. Fundamentally, tunneling arises in the restricted area at the intersection of the channel region and the Schottky contact. Under these circumstances, modifying the length of control gate and polarity gate (spacer length) has no influence on the drain current value.

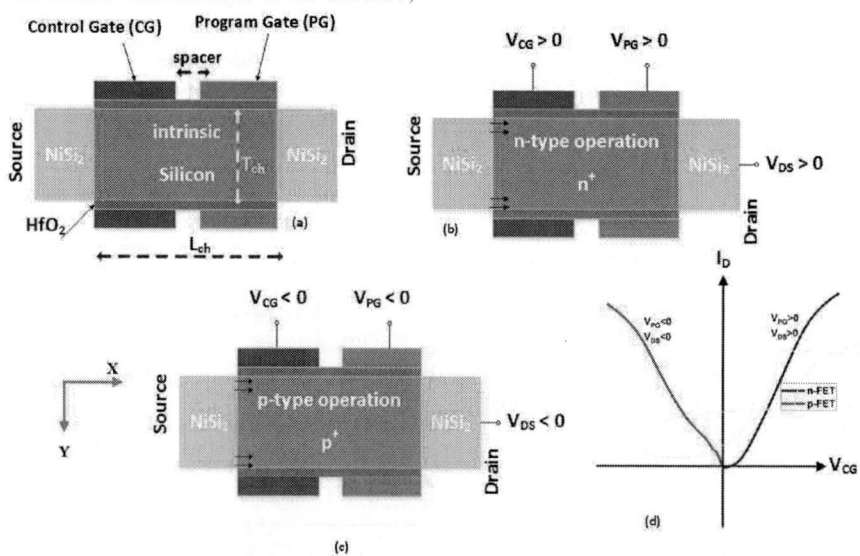

Fig. 2. Schematic of SBRFET, (a) initial structure, (b) n-type operation, (c) p-type operation and (d) transfer characteristics of reconfigurable field effect transistor. The arrows indicate the tunnelling direction at the interface of Schottky contacts and channel.

The 5th Iranian International Conference on Microelectronics (IICM2023)

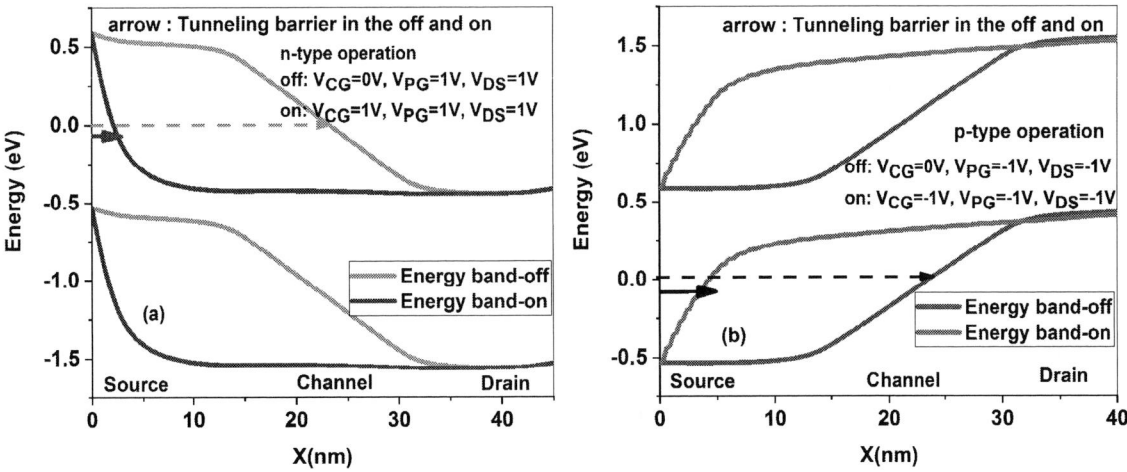

Fig. 3. Energy band profile of SBRFET in the off-state and on-state, (a) n-mode and (b) p-mode.

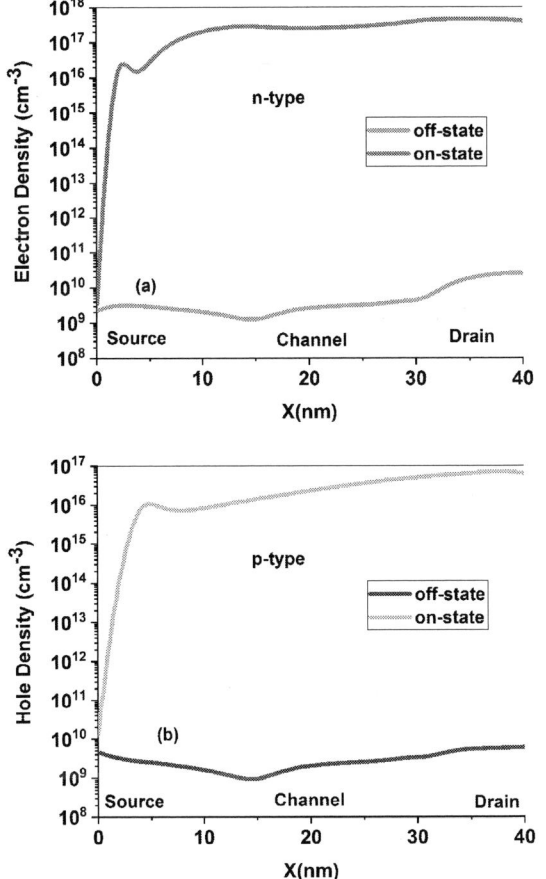

Fig. 4. Carrier density is observed along the device, extending from the source to the drain, during both the off-state and on-state for (a) n-type operation, as well as (b) p-type operation.

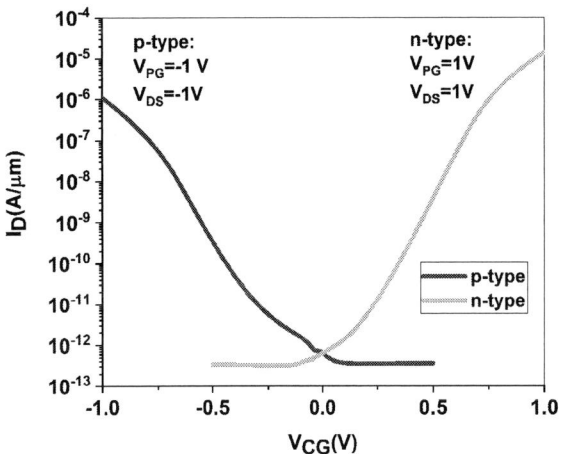

Fig. 5. Transfer characteristics of SBRFET for p-type and n-type operation.

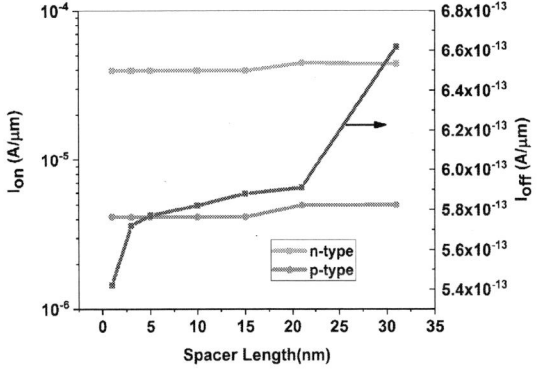

Fig. 6. Impact of spacer length on the on-state current and off-state current of SBRFET for p-type and n-type operation.

The impact of channel thickness on the electrical properties of the SBRFET is presented in Fig.7 for both n-mode and p-mode. In essence, scaling the channel thickness amplifies the gate control upon the channel and enhances the

979-8-3503-6020-2/23 $31.00 © 2023 IEEE   131

performance of the device. It is evident that decreasing the channel width from 15nm to 3nm results in a near doubling of the on-state current for both n-mode and p-mode. Nevertheless, it is crucial to note that a limited tunneling junction at the source-channel interface does not lead to a substantial improvement in the drain current when the channel width is reduced. Furthermore, a decrease in the thickness of the channel leads to an increase in the energy level within the channel, thereby leading to an increase in the effective Schottky barrier height and potentially leading to a decline in the performance of the device.

The workfunction of gate is a crucial design parameter that has the potential to influence the charge concentration in the channel, ultimately leading to modifications in the device performance. Figure.8 depicts the impact of gate workfunction on the I_D-V_{CG} curves of SBRFET for both n-type and p-type modes of operation. In essence, when the workfunction of the control gate and program gate is increased during n-type mode of operation, the electron density in the channel decreases. As a result, this phenomenon results in an elevation of the threshold voltage and a degradation in the subthreshold swing. It is noteworthy that a further increase in the gate workfunction can impact the band bending at the drain junction, potentially allowing holes to contribute to the Gate Induced Drain Leakage current (GIDL).

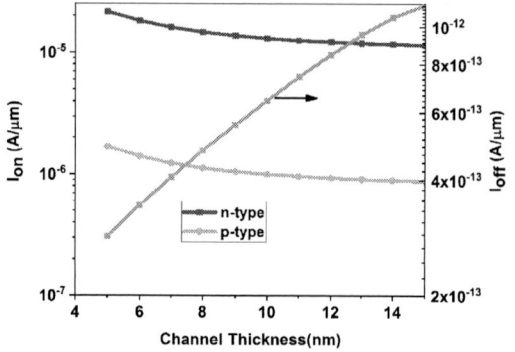

Fig. 7. Impact of channel thickness on the on-state and off-state current of SBRFET for n-type and p-type operation.

In contrast, a dissimilar trend is detected in the transfer characteristics of p-type operation when the gate workfunction is parameterized. Any increment in the gate workfunction leads to an increase in the density of holes, and consequently, a boost in the rate of tunneling and an improvement in the drain current are expected.

Figure.9 presents the impact of temperature fluctuations on the transfer characteristics of the SBRFET in both n-mode and p-mode operation. Evidently, temperature significantly affects the device's subthreshold swing and leakage current while having negligible impact on the on-state current. Essentially, temperature chiefly affects the thermionic emission component of the drain current. At low temperatures, carriers lack enough thermal energy to surmount the Schottky barrier, leading to low off-state current. With increasing temperature, some carriers obtain adequate thermal energy to traverse the barrier, causing a rise in off-state current and deterioration of the subthreshold

swing. Notably, as the gate bias increases, direct tunneling emerges as the dominant current mechanism that is impervious to temperature changes.

The thorough investigation of the variation in Schottky barrier height and its impact on the transfer characteristics of the device has yielded comprehensive results, which are visually represented in Fig. 10. The key characteristic of $NiSi_2$ is that it yields identical transfer characteristics for both n-mode and p-mode operations, owing to its mid-gap barrier height outcomes. The results clearly indicate that the effective variation in Schottky barrier height, caused by either Fermi level pinning or interface defects, fundamentally alters the transfer characteristics. Consequently, the p-type and n-type operations of the device are no longer identical.

IV. CONCLUSION

In this paper, we carry out a comprehensive assessment of the influence that different structural and physical design parameters exert on the functionality of SBRFET. One of the most fundamental design parameters that can significantly modify the carrier concentration in the channel and alter the identical behavior of n-mode and p-mode operation is the gate workfunction. Thus, it is crucial to determine the optimum value for the workfunction of both the program gate and control gate to ensure efficient operation at room temperature. Based on the obtained results, we have found that the tunneling primarily occurs at a limited region at the interface of the source and channel region near the gate oxide regions. The overall performance of the device is minimally impacted by the length of both the program gate and control gate. Furthermore, we have discovered that temperature variation may fundamentally modify the device performance due to the thermionic emission current mechanism. Hence, it is imperative to take into account the impact of temperature on the operational efficacy of the device whilst formulating effective digital logic circuits. Our study provides valuable insights into the role of design parameters on the performance of SBRFET, thus paving the way for the development of more efficient and reliable digital logic circuits.

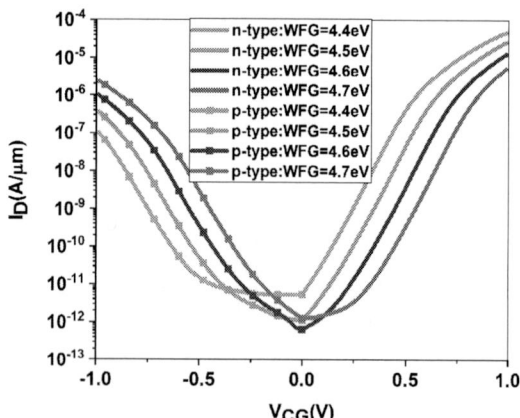

Fig. 8. I_D-V_{CG} curves of SBRFET for n-type and p-type operation as the workfunction of the program gate and control gate are varied.

The 5th Iranian International Conference on Microelectronics (IICM2023)

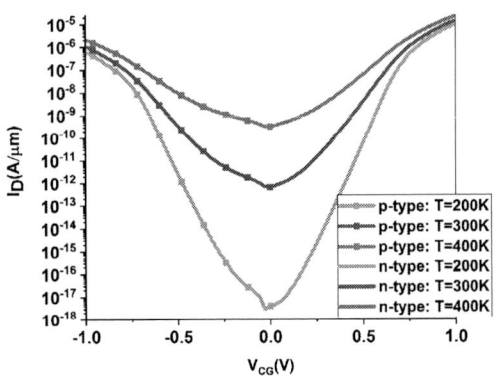

Fig. 9. Transfer characteristics of SBRFET for n-mode and p-mode operation as a function of temperature.

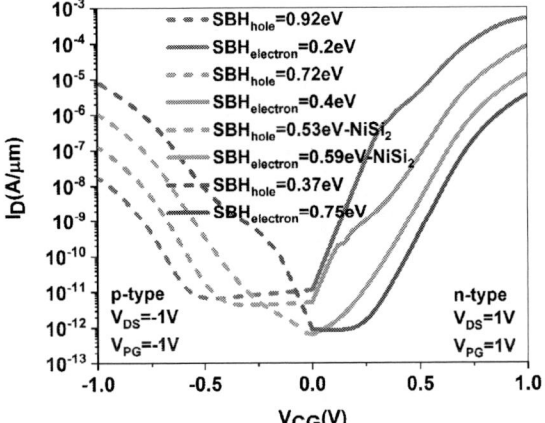

Fig. 10. I_D-V_{CG} curves of SBRFET for n-mode and p-mode operation as a function of Schottky barrier height.

REFERENCES

[1] S.L. Tripathi, P. Pathak, A. Kumar and S. Saxena, "Improved drain current with suppressed short channel effect of p+ pocket double-gate MOSFET in Sub-14 nm technology node," Silicon, vol.14, no.16, pp.10881-10891, November 2022.

[2] F.Á. Herrera, Y. Hirano, M. Miura-Mattausch, T. Iizuka, H. Kikuchihara, H.J. Mattausch, A. Ito,"Advanced short-channel-effect modeling with applicability to device optimization—Potentials and scaling," IEEE Transactions on Electron Devices, vol.66, no.9 pp. 3726-33, August 2019.

[3] Y. Sun, X. Yu, R. Zhang, B. Chen, and R. Cheng, "The past and future of multi-gate field-effect transistors: Process challenges and reliability issues," Journal of Semiconductors, vol. 42, no.2, pp.023102, February 2021.

[4] C.Yoon, S. Moon, and C. Shin, "Study of a hysteresis window of FinFET and fully-depleted silicon-on-insulator (FDSOI) MOSFET with ferroelectric capacitor," Nano Convergence, vol. 7, pp.1-7, December 2020.

[5] S. Andrić, L.O. Fhager, and L.E. Wernersson, "Millimeter-wave vertical III-V nanowire MOSFET device-to-circuit co-design." IEEE Transactions on Nanotechnology, vol. 20, pp.434-440, May 2021.

[6] Y. Wang, S. Liu, Q. Li, R. Quhe, C. Yang, Y. Guo, X. Zhang, Y. Pan, J. Li, H. Zhang, and L. Xu, "Schottky barrier heights in two-dimensional field-effect transistors: from theory to experiment. Reports on Progress in Physics," vol. 84, no.5, pp.056501, April 2021.

[7] X. Liu, A. Islam, J. Guo, and P.X.L Feng, "Controlling polarity of MoTe2 transistors for monolithic complementary logic via Schottky contact engineering" ACS nano, vol. 14, no. 2, pp.1457-1467, January 2020.

[8] T. Mikolajick, A. Heinzig, J. Trommer, T. Baldauf, and W.M. Weber,"The RFET—A reconfigurable nanowire transistor and its application to novel electronic circuits and systems," Semiconductor Science and Technology, vol. 32, no. 4, pp.043001, March 2017.

[9] J. Zhang, X. Tang, P.E. Gaillardon, and G. De Micheli, "Configurable circuits featuring dual-threshold-voltage design with three-independent-gate silicon nanowire FETs," IEEE Transactions on Circuits and Systems I: Regular Papers, vol. 61, no. 10, pp.2851-2861, July 2014.

[10] J.N. Quijada, T. Baldauf, S. Rai, A. Heinzig, A. Kumar, W.M. Weber, T. Mikolajick, and J. Trommer, "Germanium Nanowire Reconfigurable Transistor Model for Predictive Technology Evaluation," IEEE transactions on nanotechnology, vol.21, pp.728-736, November 2022.

[11] R. Ranjith, R. Jayachandran, K.J. Suja, and R.S. Komaragiri,"Two dimensional analytical model for a reconfigurable field effect transistor," Superlattices and Microstructures, vol. 114, pp.62-74, February 2018.

[12] J. Trommer, A. Heinzig, A. Heinrich, P. Jordan,M. Grube, S. Slesazeck, T. Mikolajick and W.M. Weber, " Material prospects of reconfigurable transistor (rfets)–from silicon to germanium nanowire,".MRS Online Proceedings Library (OPL), 1659, pp.225-230, November 2014.

[13] ATLAS User Manual, Santa Clara, USA: Silvaco International, 2015.

The 5th Iranian International Conference on Microelectronics (IICM2023)

25 – 26 October 2023

Modeling GaN-HEMT Electrostatic Band Diagram under full depletion approximation

Behnam Jafari Touchaei
Department of Electrical Engineering
Amirkabir University of Technology
(Tehran Polytechnique)
Tehran 15875-4413, Iran
jafari.b@aut.ac.ir

Majid Shalchian
Department of Electrical Engineering
Amirkabir University of Technology
(Tehran Polytechnique)
Tehran 15875-4413, Iran
shalchian@aut.ac.ir

Abstract— **In this work, we propose a new approach to derive the GaN-HEMT energy band diagram under full depletion conditions. The model considers the vertical electric field and vertical potential within the GaN-HEMT from gate to bulk. The effect of polarized ions due to piezoelectric and spontaneous polarization is also considered. The spike-shaped electric field within the 2D electron gas predicted by the model follows TCAD simulation results. The model is validated by TCAD and by EPFL HEMT model under the depletion condition.**

Keywords— *Energy band diagram, Gallium Nitride HEMT (GaN-HEMT), high electron mobility transistor (HEMT), modeling, piezoelectric and spontaneous polarization*

I. INTRODUCTION

GaN-HEMTs have become very popular for high-frequency and high-power switching applications [1]. GaN-HEMT illustrates attractive features distinguishable from GaAs-HEMT. Piezoelectric and spontaneous polarization effects result in the formation of an intrinsic two-dimensional electron gas (2-DEG) [2], [3]. Recently several compact models have been proposed for GaN-HEMT, including MVSG HEMT [4], ASM HEMT [5], [6] and EPFL HEMT [7], [8], [9]. The EPFL HEMT model derives the potential profile and energy band structure without calculating the electric field in the 2-DEG channel. We propose a model including a more detailed analysis that may calculate the electric field and the precise shape of the band diagram within the width of the 2-DEG region (Δ_1) under full depletion approximation.

II. DEVICE STRUCTURE AND BAND DIAGRAM

Fig. 1 shows the schematic cross-section of the GaN HEMT device used for the model and TCAD simulations. The source and drain regions are distanced from the gate by access regions. The Schottky barrier is formed between the gate and the AlGaN region. The heterostructure is formed at the interface of the AlGaN/GaN layer [3], [10]. The electrostatic solution of the Poisson equation is obtained along the vertical axis shown in Fig. 1, which also specifies the width of various layers. Device parameters are listed in Table I and used in the SILVACO simulator. The low-field mobility model, strain, and polarization effects are employed for simulation [11], [12]. There are three significant steps to derive the band diagram for this structure. First, it is mandatory to determine a width for the 2-DEG (Δ_1) and

polarization charge layer (Δ_2) to calculate the electric field from Poisson's equation. The second step refers to finding

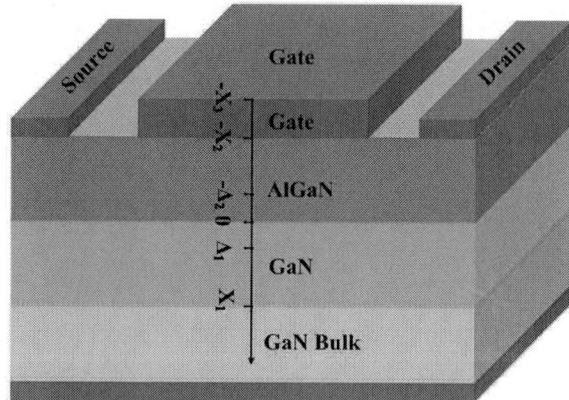

Fig. 1: Schematic of GaN-HEMT shows the structure, used by TCAD and determines all symbolic boundary parameters.

potential from the electric field in different regions, and the final step is calculating the bottom of the conduction band (E_C) and the top of the valance band (E_V) energies are shown in Fig. 2.

A. Electric field and polarization charges

Polarization charges induce a discontinuity in the electric field at the AlGaN/GaN interface. The total amount of polarization charges (ΔP) is about 0.013 C/m² taking into consideration the Aluminum mole fraction (x) of 0.15 and the detail parameters listed in Table II. The $Al_{0.15}Ga_{0.85}N$ and GaN layers are assumed to be doped with $N_D = 5 \times 10^{16}$ and $N_A = 1 \times 10^{16}$ cm⁻³, respectively. "A" and "D" indices are used to show acceptor and donor type in GaN and AlGaN regions, respectively. The proposed method applies to other doping levels as long as the full depletion approximation is valid. Although AlGaN and GaN show anisotropic behavior, several other works, including [2], consider them isotropic materials. The central equation to derive the electric field is (1). The divergence of the polarization field is zero within the GaN bulk, while the divergence of total polarization "ΔP" shows volume charge density into the Δ_2 distance from AlGaN/GaN interface.

979-8-3503-6020-2/23 $31.00 © 2023 IEEE

134

The 5th Iranian International Conference on Microelectronics (IICM2023)

$$\nabla.\vec{D} = \in \nabla.\vec{\varepsilon} + \nabla.\overrightarrow{\Delta P} \qquad (1)$$

$$\varepsilon_{2DEG}(X) = \frac{q\left[\left(X_1 - \Delta_1\right)N_A + N_S\right]}{\in_A} \qquad (6)$$

Ionized atom concentration (N_p (m^{-2})) due to spontaneous and piezoelectric polarization defined by (7).

TABLE I
DEVICE PARAMETERS [10]

Symbol	Unit	Value and Definitions
x	-	0.15 Aluminum mole fraction
L_G, L_S, L_D	m	$10\times10^{-6}, 0.1\times10^{-6}, 0.1\times10^{-6}$ length of gate, source, and drain
W, t_D	m	$10^{-6}, 20\times10^{-9}$ width of the device and the AlGaN thickness
L	m	$L_D + L_S + L_G$ Length of device
χ_D, χ_A	e.V	4.14, 4.31 AlGaN and GaN electron affinity
$\mathcal{O}_G, \mathcal{O}_S, \mathcal{O}_D$	e.V	5.04, 4.31, 4.31 gate, source, and drain electrode workfunction
N_D, N_A	cm^{-3}	5×10^{16} and 10^{16} AlGaN and GaN doping concentration
$N_{CA}, N_{C}(AlN)$	cm^{-3}	2.2×10^{18} and 4.1×10^{18} GaN and AlN electron density of state
$N_{VA}, N_{V}(AlN)$	cm^{-3}	1.16×10^{19} and 2.84×10^{20} GaN and AlN hole density of state
N_{CD}	cm^{-3}	$N_{CD} = (1-x)N_{CA} + xN_C(AlN)$ AlGaN electron density of state
N_{VD}	cm^{-3}	$N_{VD} = (1-x)N_{VA} + xN_V(AlN)$ AlGaN hole density of state
$E_{gA}, E_g(AlN)$	e.V	3.43 and 6.28 GaN and AlN electron energy gap
E_{gD}	m	$E_{gD} = E_g(AlN)x + (E_{gA} - 1.3x)(1-x)$ AlGaN energy gap
ε	V/m	electric field
$\in_0$	F/m	8.85418×10^{-12} Vacuum permittivity
$\in_A$	F/m	$9.5\times\in_0$ GaN permittivity
$\in_D$	F/m	$\in_D = (-0.5x + 9.5)\in_0$ AlGaN permittivity
a_A	m	3.189×10^{-10} GaN lattice constant
a_D	m	$a_D = (-0.077x + 3.189)\times10^{-10}$ AlGaN lattice constant
Δ_1, Δ_2	m	$12\times a_A$ Widths of 2DEG and polarization layer

The electric field in the GaN layer is deduced by considering the depletion region between Δ_1 and X_1, and the electric field in X_1 reaches zero, as shown in (2).

$$\frac{d}{dX}\varepsilon_A(X) = -\frac{qN_A}{\in_A} \qquad (2)$$

$$\varepsilon_A(X_1) = 0$$

The electric field between Δ_1 and X_1 is obtained as:

$$\varepsilon_A(X) = \frac{qN_A(X_1 - X)}{\in_A}. \qquad (3)$$

The relation between volume electron density ($n_v(X)$) and sheet electron density (N_s) in the 2-DEG channel is shown in (4).

$$\int_0^{\Delta_1}\frac{qn_v(X)}{\in_A}dX = \frac{qN_S}{\in_A} \qquad (4)$$

In the 2-DEG channel, the boundary condition is shown in (5) and is obtained by substitution Δ_1 into (3).

$$\varepsilon_{\Delta_1} = \varepsilon_A(X)\big|_{X=\Delta_1} = \frac{qN_A(X_1 - \Delta_1)}{\in_A} \qquad (5)$$

Taking into account (4) and (5), the electric field in the 2-DEG channel ($\varepsilon_{2DEG}(X)$) is obtained:

TABLE II
POLARIZATION PARAMETERS [2]

Symbol	Unit	Value and Definitions
C_{13}	Gpa	$c_{13} = (5x + 103)\times10^9$ Elastic constant wurtzite
C_{33}	Gpa	$c_{33} = (-32x + 405)\times10^9$ Elastic constant wurtzite
e_{31}	C/m^2	$e_{31} = (-0.11x - 0.49)$ Piezoelectric coefficient
e_{33}	C/m^2	$e_{33} = (0.73x + 0.73)$ Piezoelectric coefficient
P_{SpD}	C/m^2	$P_{SpD} = (-0.052x - 0.029)$ AlGaN spontaneous polarization
P_{PeD}	C/m^2	$P_{PeD} = \frac{2(a_A - a_D)}{a_D}\left(e_{31} - e_{33}\frac{C_{13}}{C_{33}}\right)$ AlGaN piezoelectric polarization
P_D	C/m^2	$P_D = P_{SpD} + P_{PeD}$ Total AlGaN polarization
P_A	C/m^2	-0.029 Total GaN polarization
ΔP	C/m^2	$\Delta P = |P_D| - |P_A|$ Total polarization at AlGaN/GaN interface

$$N_p = \frac{\Delta P}{q} \qquad (7)$$

A closed Gaussian surface is assumed between Δ_1 to Δ_2, and a relation between the electric field in Δ_2 and Δ_1 is concluded by (8).

$$\in_A \varepsilon_{\Delta_1} - \in_D \varepsilon_{\Delta_2} = q(N_P - N_S) \qquad (8)$$

The electric field in Δ_2 is isolated from (8) and given by (9).

$$\varepsilon_{\Delta_2} = \frac{q\left[\left(X_1 - \Delta_1\right)N_A + N_S - N_P\right]}{\in_D} \qquad (9)$$

To calculate the electric field at $-X_2$ (ε_{X_2}), we assume a closed Gaussian surface around $-X_2$. The electric field at $-X_2$ is a boundary condition for the differential equation in the AlGaN region between $-\Delta_2$ to $-X_2$, shown in (10).

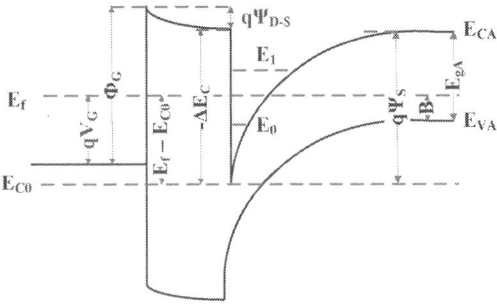

Fig. 2 The energy band diagram of GaN-HEMT.

979-8-3503-6020-2/23 $31.00 © 2023 IEEE   135

The 5th Iranian International Conference on Microelectronics (IICM2023)

$$\frac{d}{dX}\varepsilon_D(X)=\frac{qN_D}{\in_D} \tag{10}$$

$$\varepsilon_{X_2}=\frac{qN_g}{\in_D}$$

The electric field between $-\Delta_2$ and $-X_2$ into the AlGaN region is solved based on (10) and is shown in (11). The AlGaN region is considered fully depleted.

$$\varepsilon_D(X)=\frac{q\left[(X+X_2)N_D+N_g\right]}{\in_D} \tag{11}$$

The electric field around $-\Delta_2$ is defined in (12) and calculated by introducing $-\Delta_2$ into (11):

$$\varepsilon_{\Delta_2}=\varepsilon_D(X)\big|_{X=-\Delta_2}=\frac{q\left[(X_2-\Delta_2)N_D+N_g\right]}{\in_D} \tag{12}$$

The charge neutrality relation between N_g, depletion charges, 2DEG, and polarization charges is written as:

$$N_g=N_A(X_1-\Delta_1)+N_S-N_P-N_D(X_2-\Delta_2) \tag{13}$$

resulted from (12) and (9). Here we conclude the relations for the electric field under depletion approximation at different regions in (14):

$$\varepsilon(X)=\begin{cases}\varepsilon_G=0 & -X_3\le X<-X_2 \quad (14)\\[2mm]\varepsilon_D(X)=\dfrac{q\left[(X+X_2)N_D+N_g\right]}{\in_D} & -X_2\le X<-\Delta_2\\[2mm]\varepsilon_{\Delta_2}=\dfrac{q\left[(X_1-\Delta_1)N_A+N_S-N_P\right]}{\in_D} & -\Delta_2\le X<0\\[2mm]\varepsilon_{2DEG}(X)=\dfrac{q\left[(X_1-\Delta_1)N_A+N_S\right]}{\in_A} & 0\le X<\Delta_1\\[2mm]\varepsilon_A(X)=\dfrac{qN_A(X_1-X)}{\in_A} & \Delta_1\le X\le X_1\end{cases}$$

The electric field from the gate to the GaN region is depicted in Fig. 3. (a). The electric field in the gate electrode is zero, and the electric field is negative in the AlGaN region. There is a big spike of about 1.5 MV/cm around the 2-DEG region at $V_G=0.5$ V, and the electric field decreases sharply in AlGaN/GaN interface around Δ_1 and smoothly goes to zero towards the substrate electrode. The electric field calculated by the model follows TCAD simulation.

B. Potential, E_C, and E_V

The potential is calculated by integrating the electric field (15):

$$\nabla\Psi=-\varepsilon \tag{15}$$

The electric field in the GaN region near the substrate is zero. This is the first boundary condition for the potential:

$$\Psi_A(X_1)=0 \tag{16}$$

Potential at Δ_1 is denoted as a boundary condition for a 2-DEG potential as presented in (17).

$$\Psi_{\Delta_1}=\frac{qN_A(\Delta_1-X_1)^2}{2\in_A} \tag{17}$$

The potential in the 2-DEG layer between Δ_1 and "0" is obtained as (18).

$$\Psi_{2DEG}(X)=-\frac{q\left[N_A(X_1-\Delta_1)\left(X-\dfrac{X_1}{2}-\dfrac{\Delta_1}{2}\right)+N_S(X-\Delta_1)\right]}{\in_A} \tag{18}$$

Ψ_S, the surface potential at $X=0$, is obtained in (19).

$$\Psi_S=\frac{q\left[N_A(X_1^2-\Delta_1^2)+2\Delta_1 N_S\right]}{2\in_A} \tag{19}$$

Next, the potential between "0" to $-\Delta_2$ is obtained as:

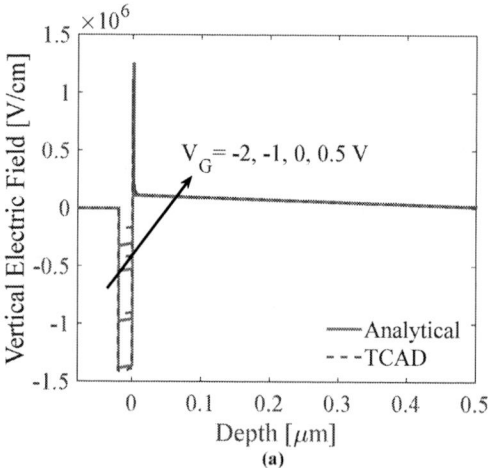

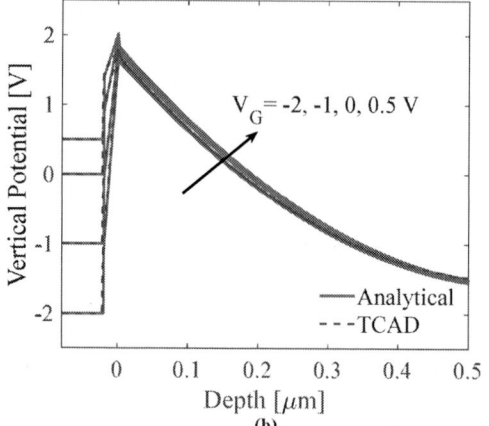

Fig. 3: (a) Vertical electric field and (b) potential from gate to bulk

$$\Psi_{\Delta_2}(X)=-\frac{qX\left[N_A(X_1-\Delta_1)+N_S-N_P\right]}{\in_D} \tag{20}$$

$$+\frac{q\left[N_A(X_1^2-\Delta_1^2)+2\Delta_1 N_S\right]}{2\in_A}$$

This potential is independent of Δ_2. Finally, the potential between $-\Delta_2$ to $-X_2$ is deduced according to (21).

$$\Psi_D(X)=\Psi_{D-S}(X)+\Psi_S \tag{21}$$

$$\Psi_{D-S}(X)=\frac{qN_D\Delta_2^2}{2\in_D}-\frac{q\Delta_2\left[X_2N_D-(X_1-\Delta_1)N_A+N_g-N_S+N_P\right]}{\in_D}$$

$$-\frac{qX\left[N_D(X+2X_2)+2N_g\right]}{2\in_D}$$

979-8-3503-6020-2/23 $31.00 © 2023 IEEE   136

The minimum of conduction band energy: $E_C(x)$, follows vacuum-level energy $E_{Vac}(x)$ by a fixed distance called electron affinity χ. The band bending inside the GaN layer is obtained as $E_C(X) - E_C(X_1) = -q(\Psi_A(X) - \Psi_A(X_1))$. Therefore using (16) we obtain.

$$E_C(X) = -q\Psi_A(X) + E_{Vac}(X_1) - \chi \quad \text{(22)}$$

This relation might be used to obtain the band diagram at the GaN layer under the depletion approximation. In the AlGaN region, the electron affinity is different from GaN, and the difference is shown as $\Delta\chi = \chi_A - \chi_D$. Taking into consideration, the continuity of the vacuum level, E_C at the boundary of the AlGaN layer.

$$E_C\big|_{AlGaN}(X=0) = E_C\big|_{GaN}(X=0) - \Delta\chi \quad \text{(23)}$$

Again, we may assume that the conduction band and valance band bending inside the AlGaN layer are obtained from the solution of the Poisson equation:

$$E_C\big|_{AlGaN}(X) = E_C\big|_{AlGaN}(X=0) - q\Psi_D(X) \quad \text{(24)}$$

Assuming ideal Schottky barrier contact without interface traps, as demonstrated in Fig 2. the Fermi level energy in the gate region is equal to $E_f\big|_{gate} = E_C(-X_2)\big|_{AlGaN} - \Phi_G$, where Φ_G is defined as a metal-semiconductor Schottky barrier.

Fig 2. shows the band diagram obtained from this analytical model. The gate potential and bulk Fermi level energy are defined by $\Psi_G = \Psi(-X_2)$ and $E_f = E_C(X_1) - E_{gA} + B$, respectively, where $B = U_T \ln(N_{VA}/N_A)$. The bulk Fermi level is simplified to $E_f = E_{Vac}(X_1) - \chi_A - E_{gA} + B$. The difference in Fermi level between bulk and gate is related to the gate voltage, shown in (25). The Ψ_G is derived from (25) and given in (26).

$$E_f\big|_{gate} - E_f = -qV_G \quad \text{(25)}$$

$$q\Psi_G = qV_G - \Phi_G + E_{gA} - B \quad \text{(26)}$$

This relation suggests that the electrostatic potential at the gate boundary with reference to the substrate depends on gate Schottky barrier height (Φ_G) and GaN layer band gap (E_{gA}). The electrostatic potential at different regions of the device along the X direction is listed in (27). To draw the band diagram, we set the Fermi level at zero ($E_f = 0$). The edges of the conduction band and valance band are obtained according to (28) and (29) respectively.

$$\Psi(X) = \begin{cases} \Psi_G(X) & -X_3 \leq X < -X_2 \\ \Psi_D(X) & -X_2 \leq X < -\Delta_2 \\ \Psi_{\Delta_2}(X) & -\Delta_2 \leq X < 0 \\ \Psi_{2DEG}(X) & 0 \leq X < \Delta_1 \\ \Psi_A(X) & \Delta_1 \leq X \leq X_1 \end{cases} \quad \text{(27)}$$

$$E_C\big|_{AlGaN(GaN)} = -q\Psi(X)\big|_{AlGaN(GaN)} + E_{gA} - B \quad \text{(28)}$$

$$E_V\big|_{AlGaN(GaN)} = E_C\big|_{AlGaN(GaN)} - E_{gD(A)} \quad \text{(29)}$$

Fig. 3(b) shows the potential plotted along the X-axes. It clearly illustrates a sharp peak around $X=0$ within the Δ_1 interval where 2-DEG is formed. Although this feature of the potential has been concluded by other papers [3], Fig. 3 (b) depicts how the proposed model facilitates and how the potential varies in response to the 2-DEG formation.

Fig 4. compares the edges of the conduction band and the valance band calculated by (23) and (24) with the TCAD simulation result. It indicates perfect agreement between the proposed analytical model and the TCAD simulation. To calibrate the model with experimental data, we used "Δ_1" and "Δ_2" as

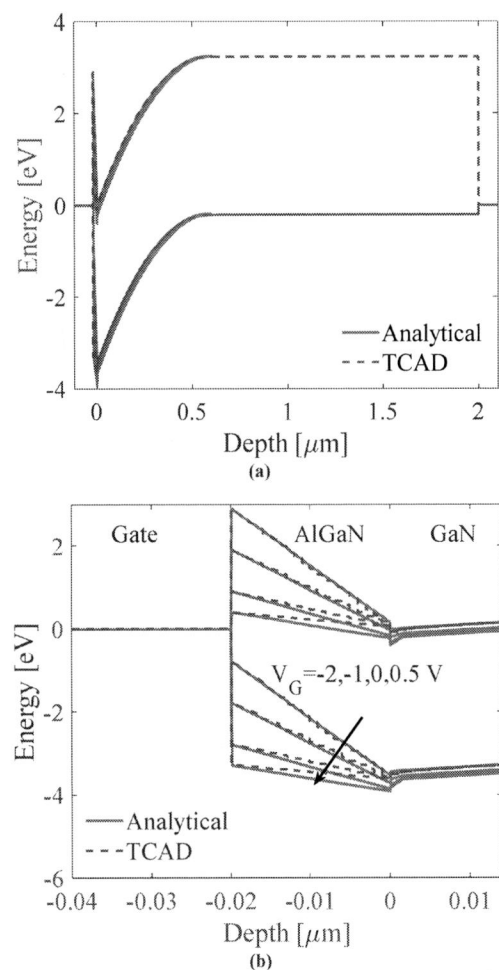

Fig. 4: (a) Energy band diagram of GaN HEMT device at zero bias from the gate to the bulk (b) magnified band diagram around the 2-DEG layer.

fitting parameters and we obtained the best fitting values for both parameters as $12 \times a_A \approx 3.8$ nm, where a_A is the GaN lattice constant listed in Table I. The results have been validated for the range of fraction mole from 0.15 to 0.3, where AlGaN/GaN interface is smooth enough to avoid dislocations.

The proposed model uses the thickness of 2DEG (Δ_1) and polarization layer (Δ_2) to derive the electric field and potential. The proposed model reverts to the EPFL HEMT model if we neglect the thickness of those layers and assume the 2DEG and polarization layers as charge sheets at the GaN/ AlGaN interface.

I. CONCLUSION

We proposed a model for the electrostatic solution of GaN HEMT under the depletion approximation. The widths of the 2DEG layer and polarized ions at the AlGaN/GaN interface significantly influence the electric field distribution. The spike-shaped electric field in the 2-DEG channel, with a maximum of about 1.5 MV/cm at $V_G = 0.5V$, is obtained and validated with TCAD. The potential and band diagram are also derived from the electric field and validated with TCAD. The proposed model converts to the EPFL HEMT model under full depletion, if we neglect the thickness of 2DEG and polarization layers.

REFERENCES

[1] M. N. Yoder, "Wide bandgap semiconductor materials and devices," in IEEE Transactions on Electron Devices, vol. 43, no. 10, pp. 1633-1636, Oct. 1996, doi: 10.1109/16.536807.

[2] O. Ambacher et al., "Two-dimensional electron gases induced by spontaneous and piezoelectric polarization charges in N- and Gaface AlGaN/GaN heterostructures," *J. Appl. Phys.*, vol. 85, no. 6, pp. 3222–3233, Mar. 1999, doi: 10.1063/1.369664.

[3] M. Delagebeaudeuf and N. T. Linh, "Metal-(n) AlGaAs-GaAs twodimensional electron gas FET," IEEE Trans. Electron Devices, vol. ED- 29, no. 6, pp. 955–960, Jun. 1982, doi: 10.1109/T-ED.1982.20813.

[4] U. Radhakrishna, T. Imada, T. Palacios, and D. Antoniadis, "MIT virtual source GaNFET-high voltage (MVSG-HV) model: A physics based compact model for HV-GaN HEMTs." physica status solidi (c) 11, no. 3-4 (2014): 848-852. doi: 10.1002/pssc.201300392

[5] A. Dasgupta, S. Ghosh, Y. S. Chauhan, and S. Khandelwal, "ASMHEMT: compact model for GaN HEMTs." In 2015 IEEE International Conference on Electron Devices and Solid-State Circuits (EDSSC), pp. 495-498. IEEE, 2015. doi: 10.1109/EDSSC.2015.7285159

[6] S. A. Ahsan, S. Ghosh, S. Khandelwal, and Y. S. Chauhan, "Physics-based multi-bias RF large-signal GaN HEMT modeling and parameter extraction flow." IEEE Journal of the Electron Devices Society 5.5 (2017): 310-319. doi: 10.1109/JEDS.2017.2724839

[7] F. Jazaeri and J.-M. Sallese, "Charge-based EPFL HEMT model," IEEE Trans. Electron Devices, vol. 66, no. 3, pp. 1218–1229, Mar. 2019, doi: 10.1109/TED.2019.2893302.

[8] F. Jazaeri, M. Shalchian, and J.-M. Sallese, "Transcapacitances in EPFL HEMT model," IEEE Trans. Electron Devices, vol. 67, no. 2, pp. 758–762, Feb. 2020, doi: 10.1109/TED.2019.2958180.

[9] M. Allaei, M. Shalchian, and F. Jazaeri, "Modeling of short-channel effects in GaN HEMTs," *IEEE Trans. Electron Devices*, vol. 67, no. 8, pp. 3088–3094, Aug. 2020, doi: 10.1109/TED.2020.3005122.

[10] B. J. Touchaei, and M. Shalchian. "Non-Quasi-Static Intrinsic GaN-HEMT Model." IEEE Transactions on Electron Devices (2022). doi: 10.1109/TED.2022.3211485

[11] A. Bykhovski, B. Gelmont, and M. Shur, "The influence of the strain-induced electric field on the charge distribution in GaN-AlN-GaN structure," J. Appl. Phys., vol. 74, no. 11, pp. 6734–6739, Dec. 1993, doi: 10.1063/1.355070.

[12] J. D. Albrecht, R. P. Wang, P. P. Ruden, M. Farahmand, and K. F. Brennan, "Electron transport characteristics of GaN for high temperature device modeling," J. Appl. Phys., vol. 83, no. 9, pp. 4777–4781, May 1998, doi: 10.1063/1.367269.

The 5th Iranian International Conference on Microelectronics (IICM2023)

25 – 26 October 2023

Magnetic Properties of Permalloy (Co-Ni-Fe) Electroplated Film on Graphene-Oxide (GO) Thin Film Based on Copper Substrate

Ali Rezaei
Iran University of Science and
Technology (IUST)
Tehran, Iran

ali.rezaei@ieee.org

Abstract—In this paper, properties of a thin film of permalloy, an alloy of Fe and Ni accompanied by Co, electroplated on graphene-oxide (GO) film on a copper substrate have been studied at ambient temperature. Firstly, electrophoresis method was utilized to deposit GO on the copper substrate. Afterwards, electroplating has been carried out using proper aqueous solution of chlorides and sulfates of aforementioned metallic elements. Surface morphological properties of the thin film were studied by field emission scanning electron microscopy (FESEM) technique. X-ray diffraction analysis (XRD) and vibrating sample magnetometer (VSM) were applied to study the structural and magnetic properties of the sample thereby extracting coercivity (H_c), remanence (M_r) and saturation magnetization (M_s) of it. The extracted s-shaped narrow width hysteresis curve demonstrates the soft magnetic characteristics on the surface of the film. In order to study the existence of desired elements including magnetic element on the deposited films, EDS (Energy Dispersive Spectroscopy) analysis was performed and the elemental atomic and weight percentages along with dispersion map of elements on the surface were extracted.

Keywords— Electroplating, electrophoresis (EP), graphene-oxide (GO), thin films, ferromagnetism

I. INTRODUCTION

Thin films by and large have been in use in many industrial applications like sensors [1], solar cells [2], media recording devices [3] and anti-corrosion coatings [4] for decades. Graphene as an extraordinary allotrope of carbon in two dimensions and its derivatives like GO have been pervasively in the literature in recent years. Its splendid optical, mechanical and electrical properties have enhanced prospect of emerging novel electronic devices in the future. In this regard, some properties like high electrical and thermal conductivity, high electron mobility, very high elasticity modulus and optical transparency can be named but a few [5]. These extraordinary properties that can hardly be seen in other materials, can make it a forerunner material in electronic applications. Graphene and its derivatives found their way to futuristic applications like energy storage devices [6], solar cells [7, 8], electronic sensors [9], biomedical applications like drug delivery [10] and even food industry as toxin detector [11].

Electrophoretic (EP) deposition is a simple and straightforward technique to move charged particles which are dispersed in liquid, by exerting an electric field and deposit on a target electrode with the contrary electrical charge [12]. EP process is at large cost-effective, fast, highly controllable and also can be utilized on any conductive surface. Contrary to graphene, GO has hydrophilic characteristics and can be dispersed in water by conventional methods like ultrasonication. Therefore, GO dispersion in water can be utilized for electrophoresis process. After successful deposition of GO film on the substrate (the anode of the electrophoresis bath) and reducing it using conventional methods, graphene film will be achieved [13-17].

Permalloy an alloy of two prominent ferromagnetic metals namely iron and nickel, has been in use for several decades because of its significant magnetic characteristics like high saturation flux density (B_s), low coercivity (H_c), low hysteresis and low magnetostriction [18]. Permalloy [19] has dramatically found its way to magnetic applications like magnetic MEMS devices [20, 21], permanent storage devices read/write head [22], magnetoresistive random access memories (MRAM) [23] and micro actuators [24, 25] due to its low cost, fast deposition rate and precise controllability by changing the bath conditions like temperature, pH, current magnitude and shape and so on. In this technique the reduction of metallic cations in the electroplating bath leads to precipitation of metallic elements on the cathode which forms a uniform film on the target substrate [26].

In this brief experimental study, morphological, structural, magnetic and elemental characteristics of electroplated permalloy film along with Co (henceforth referred to as CoNiFe) deposited on the previously deposited GO film using EP technique on a copper substrate were studied using FESEM, XRD, VSM and EDS and the empirical results were concisely discussed.

II. EXPERMINTAL SETUP

A. EPD of GO on copper sheet

GO dispersion in de-ionized (DI) water was readily bought from a specialized company in carbon products to take away the burden of making GO flakes by conventional

979-8-3503-6020-2/23 $31.00 © 2023 IEEE   139

The 5th Iranian International Conference on Microelectronics (IICM2023)

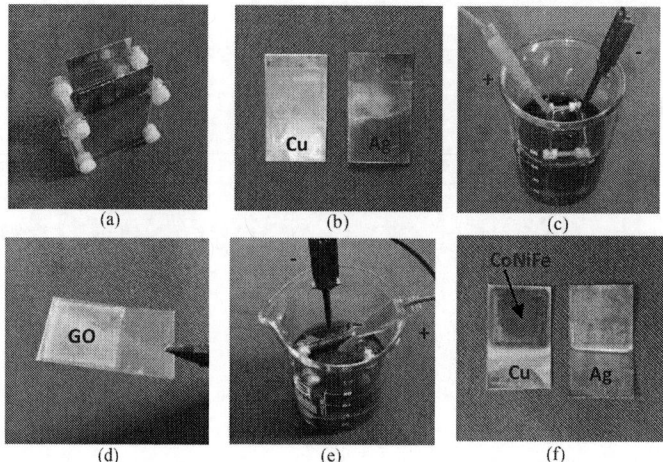

Fig. 1. (a) electrodes holder totally made from polymer (b) polished copper sheet (left) and pure silver sheet (right) (c) EP bath including the electrodes and the holder (d) GO thin film on the copper substrate after EPD process (e) electroplating bath including the electrodes and the holder and (f) CoNiFe electroplated on copper substrate (left) and corroded silver electrode (right)

methods namely Tour [27, 28] and Hummer method [29].

The EP bath was formed by adding 1 cm3 of GO aqueous dispersion (concentration=5 grams/liter) to 60 cm3 DI water. Silver (Ag) sheet cathode (thickness=0.1 mm) and target anode electrode from copper (Cu) sheet (thickness=0.1 mm) were cut in 20x30 mm pieces and mounted in parallel in a laser cut polymer holder holding the electrodes at the distance of nearly 1.6 cm which can be observed in Fig. 1 (a).

Before immersing the polymer structure along with the electrodes in the EP bath and exerting the voltage to silver and copper electrodes, target copper electrode was polished by polish paste and a handheld polish machine to ensure a shiny surface which is devoid of any oxide. Afterwards, it was immersed and rinsed by acetone and isopropyl alcohol to remove any smattering of fat or polish paste from the

TABLE I. CHEMICALS USED TO FORM THE ELECTROPLATING BATH

Compound	Wt %
$NiCl_2.6H_2O$	1.0 g
$NiSO_4.6H_2O$	1.0 g
$CoCl_2.6H_2O$	1.0 g
$CoSO_4.7H_2O$	1.0 g
$FeCl_3.6H_2O$	1.0 g
$FeCl_2.4H_2O$	1.0 g
$FeSO_4.7H_2O$	1.0 g
$NaC_{12}H_{25}SO_4$ (Sodium Lauryl Sulfate)	50 mg
$KNaC_4H_4O_6 \cdot 4H_2O$ (Potassium Sodium Tartrate)	180 mg
H_3BO_3 (Boric acid)	160 mg

surface. Fig. 1 (b) demonstrates the cleaned Cu electrode on the left and Ag electrode on the right.

Fig. 1 (c) depicts the electrophoresis bath including the electrodes and the GO murky dispersion. The process lasted for 11 minutes by applying 7.5 volts DC between electrodes. After finishing the process, the deposited Cu substrate (anode) was rinsed with DI water and left to become desiccated. It is worth mentioning that no visible harm to Ag electrode was observed after accomplishment of EPD process as no chemical reaction took place during the deposition process. Fig. 1 (d) demonstrates the deposited GO film which is observable as a matte thin film on the Cu substrate exactly as the same size of the holder opening.

B. CoNiFe electroplating on the GO film

In this step, the cations of nickel, iron and cobalt were introduced to 100 cm3 of DI water forming the electroplating bath by adding sulfates and chlorides of the aforementioned metallic elements. Table. 1 demonstrates the salts and other compounds inside the bath including their relevant weight percentage. It is worth mentioning that two iron chlorides with both II and III valences were utilized inside the bath. After dissolving the compounds in the DI water, the solution was filtered by an ashless filter paper (Grade 41) to remove any undissolved compounds from the bath thereby forming a translucent yellowish solution. The GO coated copper substrate from the previous stage, was mounted in the same holder again so that the GO film faced the Ag electrode. After a role reversal, the Ag was chosen as the anode electrode and the target GO-deposited Cu substrate was used playing the role of the cathode as the negative polarity electrode absorbs cations in case of exerting enough electric field inside the bath.

The aqueous solution of aforementioned compounds was poured into a 50 cm3 beaker until the two electrodes became totally immersed in the solution. Fig.1 (e) depicts the electroplating bath containing a lucid solution. The Ag electrode was connected to the positive terminal of a lab DC power supply and the Cu electrode to the negative terminal, to reduce metallic cations thereby depositing them on the GO film.

979-8-3503-6020-2/23 $31.00 © 2023 IEEE

The 5th Iranian International Conference on Microelectronics (IICM2023)

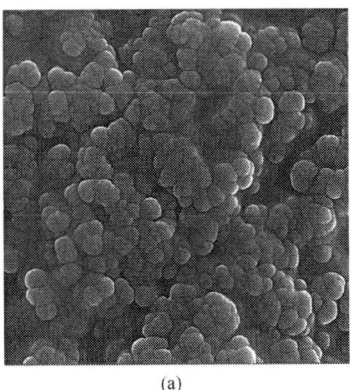  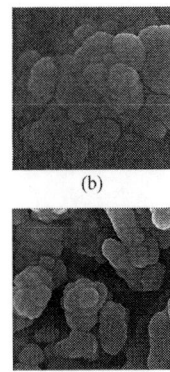

Fig. 2. FESEM micrograph from surface of the electrophoresed GO film on the Cu substrate (a) magnification=30kx and (b) 100kx and the electroplated CoNiFe on the GO film on the Cu substrate (c) magnification=30kx and (d) 100kx

The voltage was chosen to be 3.0 volts DC and electrical current passed through the bath for about a single minute. A number of bubbles were clearly appeared after passing the current from the bath and a dark film on the Cu substrate was appeared which pertained to successful CoNiFe electrodeposition. The generation of bubbles during the process made the solution fizzy. After finishing the process, it was observed that the Ag electrode was corroded dramatically which was a consequence of electrochemical side reactions that is not of interest in this study.

Afterwards, surface film residues were wiped off by flowing DI water and the sample left to be desiccated in free air. Fig.1 (f) depicts the final deposited CoNiFe film on the Cu substrate on the left in addition to the partly corroded Ag electrode on the right. The sample was cut into four equal pieces around 1x1cm and each was used to be characterized in the next step.

III. CHARACTERISATION

In order to study the morphological properties of both the GO and the CoNiFe film on Cu substrate, FESEM was employed. A number of micrographs with different zoom magnitudes were extracted using MIRA III Tescan microscope. Fig. 2 (a) and (b) depict the extracted micrographs from the surface of the GO thin film at magnitude equal to 30kx and 100kx, respectively whereas the Fig. 2 (c) and (d) depict the CoNiFe film on the GO film

at the same magnitudes of zoom, respectively. As the GO film micrograph suggests, the island-like structure on the surface validates the existence of the GO which conforms to the related literature. On the other hand, it can be clearly observed from the CoNiFe thin film that there is a granular structure on the surface which is somewhat uniformly distributed. The surfaces seem to be free from any microcracks and no remarkable defects were observed in different magnifications which prove the existence of the uniform CoNiFe film on the surface.

In order to study the structural and crystallinity of the GO film on Cu substrate and the CoNiFe film on GO, normal XRD analysis was performed ($10<2\theta<90$ degrees) and in order to maximize the dynamic range and avoid unimportant data representation, just patterns in the range of 40 to 76 degrees were depicted in Fig. 3. After matching the extracted patterns using Highscore software using its build-in library, the peaks diffraction locations to a high degree match to Cu. The peaks' locations were observed at 43.4°, 45.4°, 48.2°, 50.7° and 74.4° for the GO film on Cu sheet before and after CoNiFe electroplating. In Fig. 3 (b), although metal atoms were introduced on the GO film surface, no extra remarkable peak was observed. It might be attributed to low mass of the deposited CoNiFe film which is beyond the sensitivity of the XRD machine. The only detected difference in Fig. 3 patterns corresponds to the ratio of S4 or the significant peak

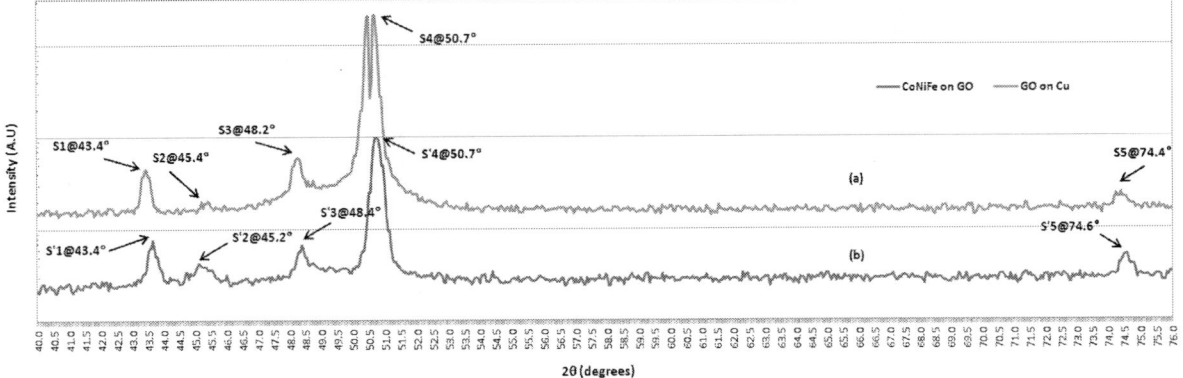

Fig. 3. XRD scattering patterns for (a) the GO film and (b) the electroplated CoNiFe film on the Cu substrate

979-8-3503-6020-2/23 $31.00 © 2023 IEEE   141

The 5th Iranian International Conference on Microelectronics (IICM2023)

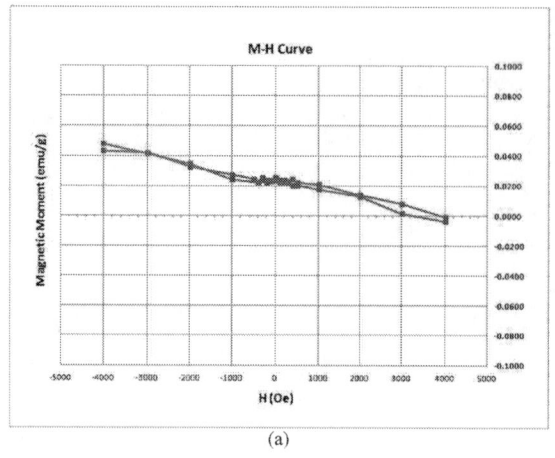
(a)

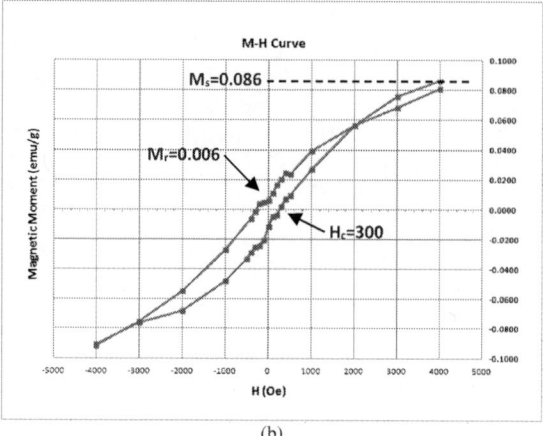
(b)

Fig. 4. Thin film magnetization curve extracted by VSM for (a) GO on Cu substrate and (b) CoNiFe film on GO

in Fig. 3 (a), located at 50.7°, to other side peaks namely S1, S2, S3 and S5 which have greater values in contrast to the same ratios calculated in Fig. 3 (b). Based on the empirical patterns of XRD analysis, there is no remarkable peaks' shift before and after CoNiFe electroplating on the Cu substrate.

VSM was utilized to extract the magnetic curve which incorporates the hysteresis, coercivity, remanence flux density and saturation flux density of the deposited ferromagnetic film. Fig. 4 (a) and (b) depict the magnetic curve of the GO film on the Cu substrate and the CoNiFe film on the GO film, respectively. As the Fig. 4 (a) suggests, no conventional characteristics associated with a ferromagnetic material can be observed on the GO film as expected. Coercivity (H_c), remanence magnetization (M_r),

and saturation magnetization (M_s) were extracted from the Fig. 4 (b) equal to 300 Oe, 0.006 emu/g and 0.086 emu/g, respectively. The low value of coercivity and the narrow width hysteresis are enough evidences to regard the under study CoNiFe film as a soft magnetic material. Furthermore, a nearly symmetrical curve of the Fig. 4 (b) denotes isotropic magnetic properties of the CoNiFe thin film.

As the last analysis, EDS was carried out on both the GO and the CoNiFe films to gain an understanding of the surface elemental distribution. Fig. 5 (a) depicts the GO film micrograph (zoom magnitude=3kx) which exhibits a nearly uniform surface morphology. On the other hand, Fig. 5 (b) – (e) depict the C, O, Cu and cumulative elemental distribution on the surface, respectively.

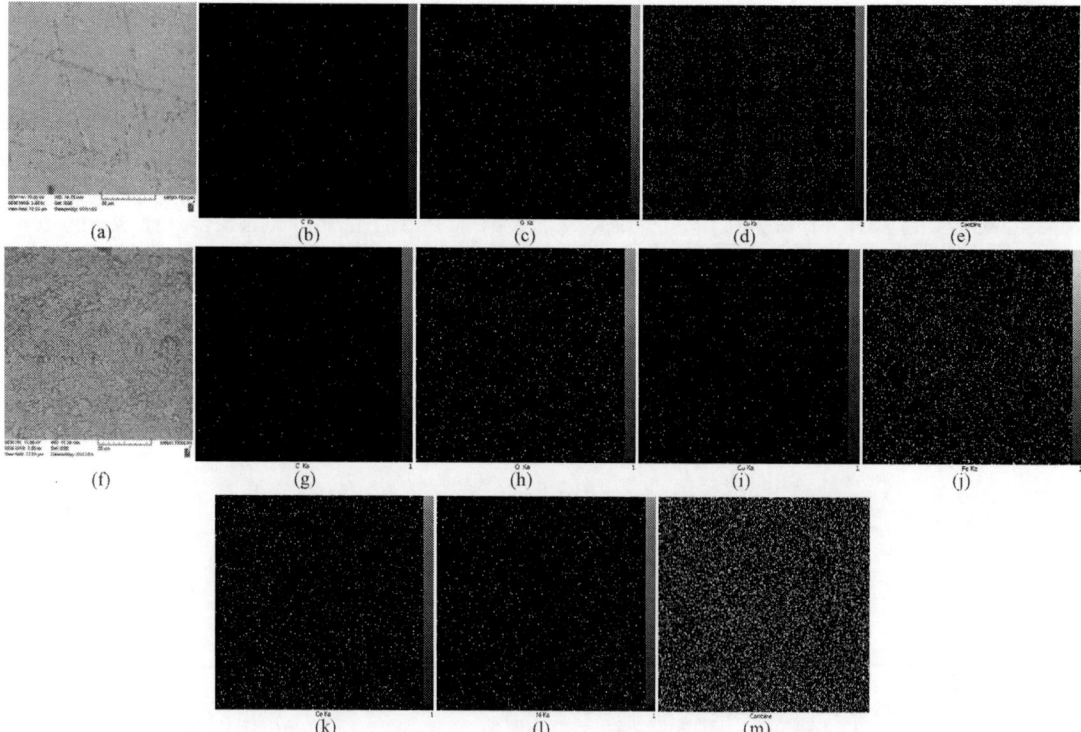

Fig. 5. (a) EDS micrograph from the GO film (b) C map (c) O map (d) Cu map on the GO thin film (e) cumulative distribution of C, O and Cu on the surface (f) EDS micrograph from the CoNiFe film (g) C map (h) O map (i) Cu map (j) Fe map (k) Co map (m) Ni map on the CoNiFe film and (n) cumulative distribution of C, O, Cu, Fe, Co and Ni on the surface

979-8-3503-6020-2/23 $31.00 © 2023 IEEE   142

The 5th Iranian International Conference on Microelectronics (IICM2023)

TABLE II.   WEIGHT AND ATOMIC PERCENTAGES OF ELEMENTS ON THE SURFACE (A) FOR THE GO FILM ON THE COPPER SUBSTRATE AND (B) FOR THE CONIFE FILM ON THE GO FILM

(a)

Element	Wt %	Atomic %
C	7.79	28.73
O	3.37	9.33
Cu	88.84	61.94
Cumulative	100.00	100.00

(b)

Element	Wt %	Atomic %
C	12.94	30.54
O	11.15	18.76
Cu	25.89	24.33
Fe	28.37	10.45
Co	13.41	6.63
Ni	8.24	9.29
Cumulative	100.00	100.00

As for the GO film and as the distribution mappings qualitatively demonstrate, the most atoms detected at the surface belong to Cu atoms (Fig. 5 (d)). Moreover, the extracted mappings exhibit nearly homogeneous distribution of elements on the thin films which make their utilization proper enough in real applications. As for the CoNiFe film, Fig. 5 (f) depicts the micrograph of the CoNiFe film on the GO (zoom magnitude=3kx). Fig. 5 (g) – (m) correspond to C, O, Cu, Fe, Co, Ni atoms and their cumulative distribution on the surface of the CoNiFe film, respectively.

In order to have a quantitative knowledge of the surface elements, both in terms of weight and atomic percentage, the numerical results have been provided in Table. II (a) and (b) for the GO and the CoNiFe thin films, respectively. As the results in Table. II (a) suggest and expected, Cu atoms account for the majority of weight and atomic percentage on the surface. That is because the GO film thickness is too thin in contrast to the substrate. Oxygen atoms are available on the surface in addition, as the GO functional groups incorporate oxygen as one of the main elements. Carbon and oxygen atoms are relatively light atoms, and the empirical results validate the fact that their atomic percentages of them are greater than their weight percentages.

As for the CoNiFe thin film, the desired magnetic elements on the surface were studied in addition to O, C and Cu atoms, and the results are provided in Table. II (b). Likewise, in this table the atomic percentages for O and C atoms are greater than their weight percentage. Furthermore, C atoms outnumber other atoms and account for almost 30.5% of the surface atoms whereas the Fe atoms outnumber in terms of weight by approximately 28.3%. Cu atoms almost account for a quarter of surface atoms both in terms of atomic and weight percentages. As for the Fe and Co, which are heavier atoms than O and C, their weight percentages are much greater than their atomic percentages at least by a factor of two whereas Ni atoms weight and atomic percentages are nearly the same. The latter fact arises further investigation and more thorough research on how this happened. In view of the empirical atomic percentages extracted by EDS analysis, an experimental formula of $Co_{25}Ni_{35}Fe_{40}$ can be suggested for the CoNiFe thin film on the surface.

IV. CONCLUSION

In this work, EP was utilized as an economical and ubiquitous technique to deposit the GO from its aqueous dispersion in DI water on the Cu substrate. Afterwards, CoNiFe was electroplated on the GO film. Conventional characterization methods namely FESEM, XRD, VSM and EDS were utilized to study the films' morphology, crystallinity, magnetic and elemental characteristics,

respectively. After comparing the FESEM micrographs, it was deduced that these images highly correspond to the previous research reports. The XRD patterns do not reveal a noticeable difference before and after the CoNiFe electroplating which might be attributed to the low mass of the film which is beyond the sensitivity of the XRD device. VSM analysis validates the existence of some kind of soft magnetic alloy on the substrate after the CoNiFe electrodeposition. In the end, EDS elemental mapping patterns on the surface reveal a uniform and homogeneous distribution of atoms and its corresponding quantitative results bring a reasonable understanding of the atoms on the surface of the Cu substrate.

REFERENCES

[1] P. C. Moura, T. P. Pivetta, V. Vassilenko, P. A. Ribeiro, and M. Raposo, "Graphene Oxide Thin Films for Detection and Quantification of Industrially Relevant Alcohols and Acetic Acid," *Sensors,* 23, 2023].

[2] I. Gharibshahian, A. A. Orouji, and S. Sharbati, "An Analytical Model for Sb₂Se₃ Thin-Film Solar Cells by Considering Current-Voltage Distortion," *Advanced Theory and Simulations,* vol. 6, no. 1, pp. 2200438, 2023/01/01, 2023.

[3] M. Li, Z. Tian, X. Yu, D. Yu, L. Ren, and Y. Fu, "Two-bit Storage Recording Head Based on the Magnetic Thin Film Device." pp. 1-3.

[4] W.-T. Ting, K.-S. Chen, and M.-J. Wang, "Dense and anti-corrosion thin films prepared by plasma polymerization of hexamethyldisilazane for applications in metallic implants," *Surface and Coatings Technology,* vol. 410, pp. 126932, 2021/03/25/, 2021.

[5] Ö. Güler, and N. Bağcı, "A short review on mechanical properties of graphene reinforced metal matrix composites," *Journal of Materials Research and Technology,* vol. 9, no. 3, pp. 6808-6833, 2020/05/01/, 2020.

[6] L. Grande, V. T. Chundi, D. Wei, C. Bower, P. Andrew, and T. Ryhänen, "Graphene for energy harvesting/storage devices and printed electronics," *Particuology,* vol. 10, no. 1, pp. 1-8, 2012/02/01/, 2012.

[7] X. Kong, L. Zhang, B. Liu, H. Gao, Y. Zhang, H. Yan, and X. Song, "Graphene/Si Schottky solar cells: a review of recent advances and prospects," *RSC Advances,* vol. 9, no. 2, pp. 863-877, 2019.

[8] F. Shahnooshi, A. A. Orouji, and A. Abbasi, "Enhanced performance of Graphene/AlGaAs/GaAs heterostructure Schottky solar cell using AlGaAs drainage," *Journal of Materials Science: Materials in*

Electronics, vol. 33, no. 7, pp. 4617-4627, 2022/03/01, 2022.

[9] J. Liu, S. Bao, and X. Wang, "Applications of Graphene-Based Materials in Sensors: A Review," *Micromachines,* 13, https://mdpi-res.com/d_attachment/micromachines/micromachines-13-00184/article_deploy/micromachines-13-00184.pdf?version=1643186752, 2022].

[10] K. Tadyszak, J. K. Wychowaniec, and J. Litowczenko, "Biomedical Applications of Graphene-Based Structures," *Nanomaterials,* 8, 2018].

[11] S. Srivastava, V. Kumar, M. A. Ali, P. R. Solanki, A. Srivastava, G. Sumana, P. S. Saxena, A. G. Joshi, and B. D. Malhotra, "Electrophoretically deposited reduced graphene oxide platform for food toxin detection," *Nanoscale,* vol. 5, no. 7, pp. 3043-3051, 2013.

[12] Y. Ma, J. Han, M. Wang, X. Chen, and S. Jia, "Electrophoretic deposition of graphene-based materials: A review of materials and their applications," *Journal of Materiomics,* vol. 4, no. 2, pp. 108-120, 2018/06/01/, 2018.

[13] D. Perumal, E. L. Albert, and C. A. Abdullah, "Green Reduction of Graphene Oxide Involving Extracts of Plants from Different Taxonomy Groups," *Journal of Composites Science,* 6, 2022].

[14] V. Agarwal, and P. B. Zetterlund, "Strategies for reduction of graphene oxide – A comprehensive review," *Chemical Engineering Journal,* vol. 405, pp. 127018, 2021/02/01/, 2021.

[15] Y. Lin, G. Hou, S. Bi, X. Su, and H. Li, "Synthesis of reduced graphene oxide paper for EMI shielding by a multi-step process," *Functional Materials Letters,* vol. 13, no. 05, pp. 2051024, 2020/07/01, 2020.

[16] S. W. Lee, C. Mattevi, M. Chhowalla, and R. M. Sankaran, "Plasma-Assisted Reduction of Graphene Oxide at Low Temperature and Atmospheric Pressure for Flexible Conductor Applications," *The Journal of Physical Chemistry Letters,* vol. 3, no. 6, pp. 772-777, 2012/03/15, 2012.

[17] F. Sun, H. Ghosh, J. Wang, Z. Tan, and S. Sivoththaman, "A Two-Step Process for Reduced Graphene Oxide Films With Work Function Tunability," *IEEE Transactions on Nanotechnology,* vol. 21, pp. 481-488, 2022.

[18] N. S. A. Rani, S. N. Ibrahim, M. Zahangir, and N. A. Malik, "Characterization of Electroplated Permalloy Film on Microstructure for Bio-MEMS Application." pp. 184-187.

[19] J.-M. Quemper, S. Nicolas, J. P. Gilles, J. P. Grandchamp, A. Bosseboeuf, T. Bourouina, and E. Dufour-Gergam, "Permalloy electroplating through photoresist molds," *Sensors and Actuators A: Physical,* vol. 74, no. 1, pp. 1-4, 1999/04/20/, 1999.

[20] P.-C. Yeh, H. Duan, and T.-K. Chung, "A Novel Three-Axial Magnetic-Piezoelectric MEMS AC Magnetic Field Sensor," *Micromachines,* 10, 2019].

[21] D. Niarchos, "Magnetic MEMS: key issues and some applications," *Sensors and Actuators A: Physical,* vol. 106, no. 1, pp. 255-262, 2003/09/15/, 2003.

[22] E. I. Cooper, C. Bonhote, J. Heidmann, Y. Hsu, P. Kern, J. W. Lam, M. Ramasubramanian, N. Robertson, L. T. Romankiw, and H. Xu, "Recent developments in high-moment electroplated materials for recording heads," *IBM Journal of Research and Development,* vol. 49, no. 1, pp. 103-126, 2005.

[23] J. Wang, X. Zhang, X. Lu, J. Zhang, Y. Yan, H. Ling, J. Wu, Y. Zhou, and Y. Xu, "Magnetic domain wall engineering in a nanoscale permalloy junction," *Applied Physics Letters,* vol. 111, no. 7, pp. 072401, 2017.

[24] L. Chang, and Y. W. Yi, "Micromachined magnetic actuators using electroplated Permalloy," *IEEE Transactions on Magnetics,* vol. 35, no. 3, pp. 1976-1985, 1999.

[25] U. Abidin, B. Y. Majlis, and J. Yunas, "$Ni_{80}Fe_{20}$ V-shaped magnetic core for high performance MEMS sensors and actuators." pp. 66-69.

[26] M. Theis, S. Ediger, M. T. Schmitt, J.-E. Hoffmann, and M. Saumer, "Nanocrystalline electroplated NiFe-based alloys for integrated magnetic microsensors," *physica status solidi (a),* vol. 210, no. 5, pp. 853-858, 2013/05/01, 2013.

[27] V. O. Kotsyubynsky, V. M. Boychuk, I. M. Budzulyak, B. I. Rachiy, M. A. Hodlevska, A. I. Kachmar, and M. A. Hodlevsky, "Graphene oxide synthesis using modified Tour method," *Advances in Natural Sciences: Nanoscience and Nanotechnology,* vol. 12, no. 3, pp. 035006, 2021/09/02, 2021.

[28] D. C. Marcano, D. V. Kosynkin, J. M. Berlin, A. Sinitskii, Z. Sun, A. Slesarev, L. B. Alemany, W. Lu, and J. M. Tour, "Improved Synthesis of Graphene Oxide," *ACS Nano,* vol. 4, no. 8, pp. 4806-4814, 2010/08/24, 2010.

[29] H. Yu, B. Zhang, C. Bulin, R. Li, and R. Xing, "High-efficient Synthesis of Graphene Oxide Based on Improved Hummers Method," *Scientific Reports,* vol. 6, no. 1, pp. 36143, 2016/11/03, 2016.

Analysis of electrostatic interaction between a charge trap and a quantum dot based single electron transistor

Fatemeh Hamedvasighi
Amirkabir University of Technology
(Tehran Polytechnique)
Tehran, Iran
fatemeh_hv@aut.ac.ir

Majid Shalchian
Amirkabir University of Technology
(Tehran Polytechnique)
Tehran, Iran
shalchian@aut.ac.ir

Abstract— In this article, we studied the effect of a single charge trap on the charge stability diagram of a quantum dot-based single electron transistor. The anomalies in the Coulomb characteristic diagram of the quantum dot, system energy, occupation probabilities, and conductivity of the quantum dot have been investigated. These anomalies are caused by the electrostatic interaction of the quantum dot with the charge trap. The trap is capacitively connected to the quantum dot and can be empty or occupied by an electron; considering a few quantum states of the system, we solved the master equation using Fermi's golden rule to obtain the tunneling rates and the matrix of tunneling coefficients. Then, calculating the inverse of the matrix of coefficients, the probability of each quantum state has been obtained. Validation of the analysis is performed by comparing the simulation results with the experimental data. Finally, we have demonstrated that our study provides a tool to detect the charge presence in a trap near a quantum dot, which can potentially be used for the readout of quantum gates.

Keywords— trap, quantum dot, single electron transistor, quantum gates, silicon semiconductor

I. INTRODUCTION

Transport through single electron systems, such as quantum dots, dopant atoms, and single electron transistors, is a powerful tool in condensed matter physics. Furthermore, standard nanofabrication techniques have enabled the routine fabrication of complicated arrays of these single-electron systems, and they are being actively pursued as a potential architecture for quantum computation [1, 2].

To read single-electron charges or spins [3] as logic bits, either quantum point contacts (QPC) or single electron transistors (SETs) are used [4, 5]. QPCs are constrictions or narrow channels through which electrons can flow, and they are designed to exhibit quantized conductance due to the quantization of electron energy levels in one or more dimensions. The presence or absence of individual electrons passing through the QPC can be detected by carefully engineering the QPC and its surrounding environment. When the QPC captures an electron, it modifies the conductance of the QPC, leading to measurable changes in the electrical current. These changes can be detected and used as a signal to read out the charge state of the single electron [6, 7]. Using a QPC as a charge detector, the distribution function of current

fluctuations in the QD can be directly measured by counting electrons.

SETs have been used as very sensitive electrometers for the charge on a second quantum dot [8, 9]. Single electron transistors (SETs) based on metallic tunnel junctions and gate-defined sensors quantum dots (SQD), conceptually equivalent to SETs, have also been widely used as proximal sensors and provide similar sensitivity and bandwidth [10-13]. In [14], mutual charge sensing between electron and hole quantum dots using a single electron transistor (SET) and a single-hole transistor (SHT) is reported. Both quantum dots sense charge displacements in the other quantum dots simultaneously. Moreover, in [12], a real-time observation of individual electron tunneling occurrences within a quantum dot is presented, achieved through an integrated radio-frequency single-electron transistor (SET). Electron counting is employed to directly assess the quantum dot's tunneling frequency and the likelihood of its charge states being occupied.

The single-electron transistor (SET) surpasses the quantum point contact (QPC) in charge sensing due to its unparalleled sensitivity to single-electron charge changes, low noise characteristics, and quantized response, making it ideal for high-precision applications.

A system consisting of a trap connected to a quantum dot is used for various purposes, such as qubit initialization, information storage, and manipulation. The trap may be intentionally placed in the system to limit and control the charge state of the quantum dot, which allows precise manipulation and measurement. Or it might be located due to unavoidable factors such as lattice defects and deep impurities inside the structure and in the vicinity of the main quantum dot. In both cases, there is a need to evaluate and analyze the effect of the trap and whether the trap is full or empty on the charge stability diagram of the quantum dot. By coupling the trap to the quantum dot, it is possible to control the energy levels and interactions of the quantum dot system, which promises applications in quantum information processing and quantum computing.

In this work, we focus on a charge trap created by an isolated impurity locality capacitively coupled to the main dot. We have considered the electrostatic interaction between the trap and the quantum dot. In quantum dot operation, we have used Coulomb confinement theory and single-electron tunneling and calculated the tunneling equations of individual quantum states in the system using the method of original equations and Fermi's golden rule. Finally, we have simulated the anomalies in the characteristics of the quantum dot, which represents the charge in the trap.

II. DEVICE STRUCTURE AND PARAMETERS

Fig. 1 shows schematically the electrical model of the structure used in our simulation [15]. The main dot is between the source and drain leads with two tunnel junctions. The gate electrode is coupled only capacitively (with zero tunneling probability) to the main dot (C_{mg}) and the trap (C_{tg}). G_L and G_R specify the tunneling rates between the main dot to/from the left and the right leads.

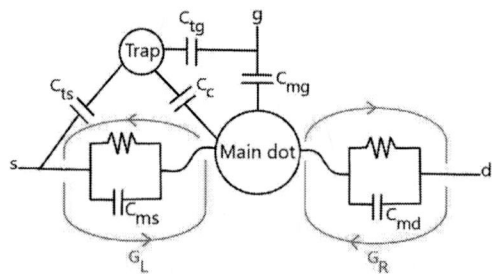

Fig. 1. Electrical model of the quantum dot in the vicinity of the trap [15]

The trap is capacitively coupled to the source (Cts) and the quantum dot (Cc). Table I lists the values of coupling capacitors specified in Fig. 1, which are obtained for an experimental structure [11]. The gate electrode controls the number of electrons in the quantum dot and the trap. The source and drain electrodes bias the device and establish a current through the source and drain tunnel junctions.

TABLE I: PHYSICAL PARAMETERS OF THE SYSTEM USED IN THE SIMULATIONS

Parameter	Value
Cmg	10×10-18 (F)
Cmd	11×10-18 (F)
Cms	11×10-18 (F)
Ctg	0.007×10-18 (F)
Cts	0.3×10-18 (F)
Ctd	0 (F)
Cc	0.15×10-18 (F)

III. SIMULATION OF THE ELECTROSTATIC ENERGY DIAGRAM OF THE SYSTEM

Fig 2(a) obtained from [11] shows the electrostatic energy of the system (W) in different charge states (E0 for zero electrons in the trap and E1 for 1 electron in the trap) as a function of gate voltage (Vg).

To calculate the electrostatic energy, we assume that the number of electrons inside the quantum dot (n_m) can vary from 0 to 9, and the number of electrons inside the trap (n_t) can be 0 or 1. Therefore, there are ten different charge states as (0,0), (1,0), (2,0), ..., (9,0) in which the trap is empty and ten other charge states as (0,1), (1,1), (2,1), ..., (9,1) in which the trap is occupied by one electron and in general the system can have twenty different charge states.

Then, the energy in different charge states has been calculated using (1), which expresses the electrostatic energy of the system as a function of the charge in the quantum dot (Q_m) and the charge in the trap (Q_t).

$$W(Q_m, Q_t) = \frac{(Q_m + \beta_t Q_t + X)^2}{2C} + \frac{(Q_t + X_t)^2}{2(Ct + Cc)} \quad (1)$$

The parameters of the above relation are defined as relations (2-8) in which ($i = m$ (for main dot) or t (for trap)) and q_0 is the background charge in the dot, which is a fitting parameter.

$$Q_m(n_m) = -n_m q + q_0 \quad (2)$$

$$Q_t(n_t) = -n_t q \quad (3)$$

$$Ci = Cis + Cid + Cig \quad (4)$$

$$\beta_t = \frac{Cc}{Ct + Cc} \quad (5)$$

$$C = Cm + \beta_t Ct \quad (6)$$

$$Xi = Cis\, V_s + Cid\, V_d + Cig\, V_g \quad (7)$$

$$X = X_m + \beta_t X_t \quad (8)$$

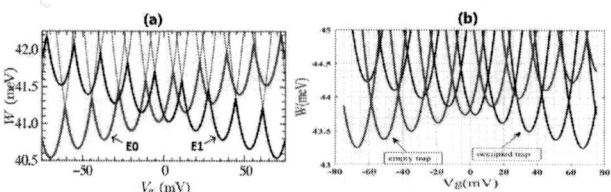

Fig. 2. Electrostatic energy of the system in different charge states at zero bias as a function of Vg (a) results from [15] (b) result of this work simulation in MATLAB

As can be seen in (Fig 2. b), the number of electrons inside the quantum dot increases with the increase of the gate voltage. Also, by increasing the gate voltage, the system's energy increases when the trap is empty, blue parabolas. Conversely, the energy of the system decreases when the trap is occupied with an electron, black parabolas, which is due to the repulsion that exists between the trap electron and the quantum dot electrons.

979-8-3503-6020-2/23 $31.00 © 2023 IEEE

The 5th Iranian International Conference on Microelectronics (IICM2023)

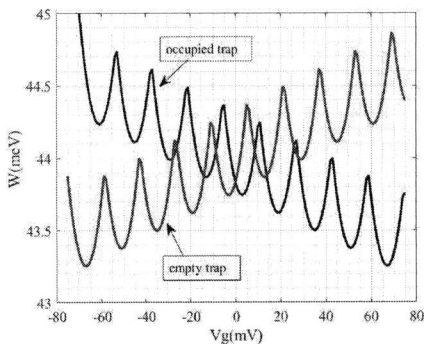

Fig. 3. The ground state energies of the main dot for the empty and occupied trap

According to Fig. 3, as the gate voltage increases, the number of quantum dot electrons (n_m) increases, and when the energy of the system is equal in both consecutive states, an electron is added to the quantum dot to decrease the energy of the system and the system remains in its ground states.

IV. SIMULATION OF TRAP OCCUPATION AND CHARGE STABILITY DIAGRAM

A common technique used to simulate transport is the master equation approach [16, 17]. The aim is to determine the probability that a system occupying a given charge state in a steady state [18].

The system can have twenty different charge states, and the electrostatic energy of each state is described by (1). To calculate the tunneling rates, we must calculate all possible transitions for each of these states to neighboring states. Fig. 4 shows all possible transitions among twenty different charge states in the system. The paths that are red and blue represent the transitions between the quantum dot and the source or drain leads, respectively, when the trap is empty and occupied, the paths in green represent the transitions between the leads and the trap, and the paths in Black color represents transitions between quantum dot and trap.

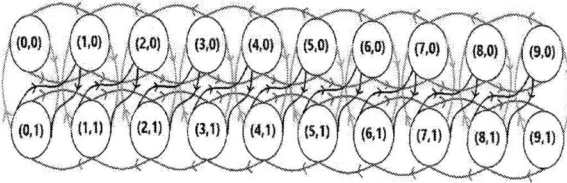

Fig. 4. All possible transitions among twenty different charge states of the system

Fermi–Dirac distribution describes the occupation of the electron state. In this work, we have used the Fermi distribution function (9) to calculate the tunneling rates between the quantum dot and the trap and the tunneling rates between the trap and the leads. We have also used Fermi-Dirac auto-convolution (10) to calculate the tunneling rates between the quantum dot and the leads due to the continuous nature of energy states on both sides of the transition [19].

$$f(x) = \frac{1}{e^x + 1} \tag{9}$$

$$f^*(x) = (f * f)(x) = \frac{x}{e^x - 1} \tag{10}$$

For example, we assume the system is initially in state (2,0). By adding an electron from the right or left leads (source or drain) to the quantum dot, the system moves to the (3,0) state, and conversely, by removing an electron from the quantum dot to the left or right leads, the system returns to state (2,0), in which the tunneling rates are according to relations (11) and (12) respectively.

$$G_{L,R(3,0),(2,0)} = T_1 f^* \left(\frac{W_{(3,0)} - W_{(2,0)} + \mu_{L,R}}{K_B T} \right) \tag{11}$$

$$G_{L,R(2,0),(3,0)} = T_1 f^* \left(\frac{W_{(2,0)} - W_{(3,0)} - \mu_{L,R}}{K_B T} \right) \tag{12}$$

In the above relations, the indices L and R represent the left and right electrodes; K_B is Boltzmann's constant, T1 is a constant coefficient, which is 1000 times larger than T2, considering that the capacitors connecting the quantum dot to the source and drain are much larger than the capacitors connecting the trap to the source and quantum dot and $\mu_{L,R}$ is the variation of the electrostatic energy of the system due to removing/adding a single electron from/to the left/right lead specified as (13):

$$\mu_{L,R} = -eV_{L,R} \tag{13}$$

In another case, an electron can be added to the trap from the left or right electrode, and as a result, the system moves from the (2,0) state to the (2,1) state, or vice versa, the electron inside the trap moves to the left or right electrodes, and the system returns from the state (2,1) to state (2,0), which are described in relations (14) and (15) respectively.

$$G_{L,R(2,1),(2,0)} = T_2 f \left(\frac{W_{(2,1)} - W_{(2,0)} + \mu_{L,R}}{K_B T} \right) \tag{14}$$

$$G_{L,R(2,0),(2,1)} = T_2 f \left(\frac{W_{(2,0)} - W_{(2,1)} - \mu_{L,R}}{K_B T} \right) \tag{15}$$

Finally, in the (2,0) state, an electron can tunnel from the quantum dot to the left or right electrode, and the system will move to the (1,0) state, or an electron can tunnel from the dot to the trap and the system will move from the (2,0) state to the (1,1) state. Both of these tunnelings are also possible in reverse, that is, from the state (1,0) to (2,0) by adding an electron from the left or right electrode to the quantum dot and also from the state (1,1) to (2,0) by tunneling an electron from the trap to the quantum dot, all these tunnelings are shown in relations (16-19) respectively.

$$G_{L,R(1,0),(2,0)} = T_1 f^* \left(\frac{W_{(1,0)} - W_{(2,0)} - \mu_{L,R}}{K_B T} \right) \tag{16}$$

979-8-3503-6020-2/23 $31.00 © 2023 IEEE

$$G_{(1,1),(2,0)} = T_2 \, f\left(\frac{W_{(1,1)} - W_{(2,0)}}{K_B T}\right) \qquad (17)$$

$$G_{L,R(2,0),(1,0)} = T_1 \, f^*\left(\frac{W_{(2,0)} - W_{(1,0)} + \mu_{L,R}}{K_B T}\right) \qquad (18)$$

$$G_{(2,0),(1,1)} = T_2 \, f\left(\frac{W_{(2,0)} - W_{(1,1)}}{K_B T}\right) \qquad (19)$$

The net tunneling rate of the (2,0) is obtained from the steady-state solution of the master equation. In other words, the difference between all the tunneling events leaving the (2,0) state and all the tunneling events entering the (2,0) state as indicated in (20).

$$
\begin{aligned}
\Gamma_{(2,0)} &= G_{L,R(2,0),(1,0)}\,P_{(1,0)} + G_{(2,0),(1,1)}\,P_{(1,1)} \\
&+ G_{L,R(2,0),(2,1)}\,P_{(2,1)} + G_{L,R(2,0),(3,0)}\,P_{(3,0)} \\
&- P_{(2,0)}\big(G_{L,R(1,0),(2,0)} + G_{(1,1),(2,0)} \\
&+ G_{L,R(2,1),(2,0)} + G_{L,R(3,0),(2,0)}\big)
\end{aligned}
\qquad (20)
$$

In the above relationship, P(i,j) is the probability of charge state (i,j). After writing all the net tunneling rates for each state, the tunneling equations are written in matrix form using (21).

$$C = P A \qquad (21)$$

In relation (21), P is the vector of probabilities in the form of $(P_{(0,0)}, P_{(1,0)}, P_{(2,0)}, \ldots, P_{(9,0)}, P_{(0,1)}, P_{(1,1)}, P_{(2,1)}, \ldots, P_{(9,1)})$, A is the matrix of tunneling rates, which is obtained by taking into account the exchange of all states as we briefly demonstrated for (2,0) state, and C is a vector in the form of $(c_1=0, c_2=0, c_3=0, \ldots, c_{20}=1)$. The last equation indicates that the sum of probabilities of all charge states of the system should be unity.

Fig. 5 shows the collective probability for all states with one electron occupying the trap, including {(0,1),(1,1), …(9,1)} obtained from our calculations as a function of gate voltage and drain voltage. This figure shows a close correlation with Fig. 3. which suggests that the ground state of the system is "Empty trap (E0)" at $V_g < -50$ mV and it switches to "Occupied Trap" at $V_g > 50$ mV, while the ground state oscillates between E_0 and E_1 as V_g varies in this range and therefore, we expect the probability of trap occupation is also varies from 0 to 1 and shows significant variations between 0 and 1 as we increase V_g.

The tunneling current of the system is calculated by (22)

$$
\begin{aligned}
I = -q\sum_{n_m,n_t} \big(& G_{L(n_m+1,n_t)(n_m,n_t)}P_{(n_m,n_t)} \\
&+ G_{L(n_m,n_t+1)(n_m,n_t)}P_{(n_m,n_t)} \\
&- G_{L(n_m-1,n_t)(n_m,n_t)}P_{(n_m,n_t)} - G_{L(n_m,n_t-1)(n_m,n_t)}P_{(n_m,n_t)} \big)
\end{aligned}
\qquad (22)
$$

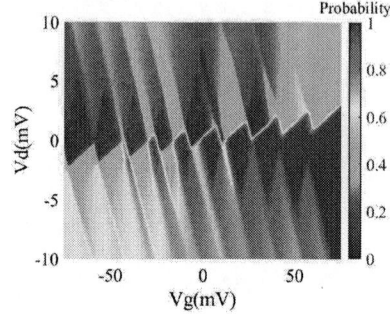

Fig. 5. Characteristics of trap occupancy probability as a function of gate and drain voltage

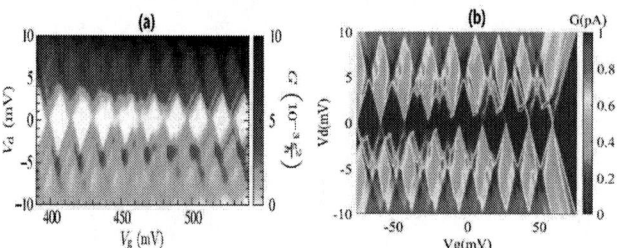

Fig. 6. Conductance characteristic of a quantum dot in the vicinity of an occupied trap (a) Experimental results [11] (b) result of this work simulation in MATLAB

Fig. 6 shows the tunneling conductance (a derivative of the tunneling current) with respect to Vd. The conductance curve is usually referred to as the charge stability diagram, and our simulation results (Fig 7. b) show excellent agreement with the experimental data (Fig 7. a [11]). The anomalies of the charge stability diagram are observed if we compare Fig. 7 with the charge stability diagram of a simple single electron transistor and the footprint of the charge trap.

V. SIMULATION OF COULOMB BLOCKADE CHARACTERISTIC

Fig. 7 shows the boundaries of neighboring charge states of the main dot in V_g and V_d plane for two scenarios: a) empty trap and b) occupied trap.

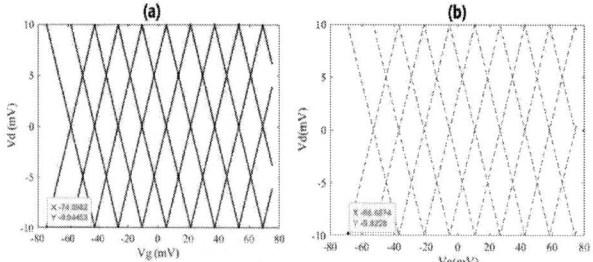

Fig. 7. Coulomb characteristic diagram of a quantum dot in the vicinity of a trap (a) When the trap is empty (b) When the trap is occupied

As can be seen, there is a horizontal shift of 5.4mV in Fig 7. b compared to Fig 7. a. This shift can be attributed to the coulombic repulsion between the charged trap and the electrons in the main dot, which leads to additional energy (gate voltage) required to add a single electron to the main dot. By increasing the capacitor between the trap and the gate (Ctg), the horizontal shift decreases because by increasing the coupling between the trap and the gate and by increasing the gate voltage, the effect of the gate on the presence of electrons in the trap rises, and as a result, the impact of electron repulsion at the main point with the trap decreases, and less shift is created.

VI. CONCLUSION

In this work, the effect of a charge trap in the vicinity of the quantum dot (SET) on the charge stability diagram of the quantum dot has been investigated by adding a trap that can only have 0 or 1 electron; new charge states were created in the system. Tunneling rates were calculated using the master equations for all permitted transitions with neighboring states, and the matrix of tunneling coefficients was calculated. The current diagram, conduction, and Coulomb blockade characteristics were simulated in MATLAB for both empty and occupied traps, and the simulation results are in good agreement with the experimental results. It is demonstrated that occupation of the trap induces a shift in the order of 5.4 mV in the charge state transition boundaries. This value can be used to estimate the location of the trap and its charge state with respect to the main dot. This analysis might be used for intentionally positioning a trap near a qubit for manipulation and readout of the charge state of the main dot in quantum computing applications.

ACKNOWLEDGMENT

The authors would like to thank Dr. Farzan Jazaeri from EPFL for technical discussions and Dr. Amin Rassekh for sharing his work on double quantum dots.

REFERENCES

[1]   M. Veldhorst, H. Eenink, C.-H. Yang, and A. S. Dzurak, "Silicon CMOS architecture for a spin-based quantum computer," Nature communications, vol. 8, no. 1, p. 1766, 2017.

[2]   C. H. Yang et al., "Operation of a silicon quantum processor unit cell above one kelvin," Nature, vol. 580, no. 7803, pp. 350-354, 2020.

[3]   A. Rassekh, M. Shalchian, J.-M. Sallese, and F. Jazaeri, "Design space of quantum dot spin qubits," Physica B: Condensed Matter, p. 415133, 2023.

[4]   B. E. Kane, "A silicon-based nuclear spin quantum computer," nature, vol. 393, no. 6681, pp. 133-137, 1998.

[5]   D. Loss and D. P. DiVincenzo, "Quantum computation with quantum dots," Physical Review A, vol. 57, no. 1, p. 120, 1998.

[6]   S. Gustavsson et al., "Counting statistics of single electron transport in a quantum dot," Physical review letters, vol. 96, no. 7, p. 076605, 2006.

[7]   Y. Yin, "Emission rate of electron transport through a quantum point contact," Journal of Physics: Condensed Matter, vol. 35, no. 35, p. 355301, 2023.

[8]   P. Lafarge, H. Pothier, E. R. Williams, D. Esteve, C. Urbina, and M. H. Devoret, "Direct observation of macroscopic charge quantization," Zeitschrift für Physik B Condensed Matter, vol. 85, pp. 327-332, 1991.

[9]   L. Molenkamp, K. Flensberg, and M. Kemerink, "Scaling of the Coulomb energy due to quantum fluctuations in the charge on a quantum dot," Physical review letters, vol. 75, no. 23, p. 4282, 1995.

[10]   T. Buehler et al., "Single-shot readout with the radio-frequency single-electron transistor in the presence of charge noise," Applied Physics Letters, vol. 86, no. 14, 2005.

[11]   T. Fujisawa, T. Hayashi, Y. Hirayama, H. Cheong, and Y. Jeong, "Electron counting of single-electron tunneling current," Applied physics letters, vol. 84, no. 13, pp. 2343-2345, 2004.

[12]   W. Lu, Z. Ji, L. Pfeiffer, K. West, and A. Rimberg, "Real-time detection of electron tunnelling in a quantum dot," Nature, vol. 423, no. 6938, pp. 422-425, 2003.

[13]   R. Schoelkopf, P. Wahlgren, A. Kozhevnikov, P. Delsing, and D. Prober, "The radio-frequency single-electron transistor (RF-SET): A fast and ultrasensitive electrometer," science, vol. 280, no. 5367, pp. 1238-1242, 1998.

[14]   A. S. de Almeida et al., "Ambipolar charge sensing of few-charge quantum dots," Physical Review B, vol. 101, no. 20, p. 201301, 2020.

[15]   M. Hofheinz, X. Jehl, M. Sanquer, G. Molas, M. Vinet, and S. Deleonibus, "Individual charge traps in silicon nanowires: Measurements of location, spin and occupation number by Coulomb blockade spectroscopy," The European Physical Journal B-Condensed Matter and Complex Systems, vol. 54, pp. 299-307, 2006.

[16]   S. Datta, Quantum transport: atom to transistor. Cambridge university press, 2005.

[17]   A. Rassekh, M. Shalchian, J.-M. Sallese, and F. Jazaeri, "Tunneling Current Through a Double Quantum Dots System," Ieee Access, vol. 10, pp. 75245-75256, 2022.

[18]   R. A. Bush, E. D. Ochoa, and J. K. Perron, "Transport through quantum dots: An introduction via master equation simulations," American Journal of Physics, vol. 89, no. 3, pp. 300-306, 2021.

[19]   M. Hofheinz, "Coulomb blockade in silicon nanowire MOSFETs," Université Joseph-Fourier-Grenoble I, 2006.

A Low-Power Differential Ring VCO Using An Active Inductor For Wireless Applications

Mahdi Alijani
School of Electerical Engineering, Iran university of Science and Technology, Tehran, Iran
mahdi_alijani@elec.iust.ac.ir

Mohammadmahdi Javanmardi
School of Electerical Engineering, Iran university of Science and Technology, Tehran, Iran
m_javanmardi@elec.iust.ac.ir

Adib Abrishamifar
School of Electerical Engineering, Iran university of Science and Technology, Tehran, Iran
abrishamifar@iust.ac.ir

Abstract— In this paper, the design of a Voltage-Controlled Oscillator (VCO), which is considered one of the essential blocks in modern wireless communication systems, is the focus. The general structure of the proposed VCO consists of a four-stage differential dual-delay-path cell interconnected in a ring topology. The output signal frequency is controlled by altering the value of the control voltage. The circuit has profited from Lin-Payne active inductor and pre-discharged path at the output of each delay circuit to achieve an efficient power consumption, wide tuning range, and higher frequency. This work is designed and simulated in TSMC 0.18μm technology with a 1.8 V supply voltage (V_{DD}). The design could achieve the frequency tuning range from 2.27 to 3.6 GHz with the control voltage varied from 2 to 0.8 V. By consuming average power of only 1.4 mW, -90.74 dBc/Hz at 1MHz phase noise achieved, which could gain a figure of merit (FoM) equals to -158.6 dBc/Hz.

Keywords—Ring oscillator, Differential delay-cell, Active inductor, VCO.

I. INTRODUCTION

Over the past few years, with technological developments, there has been a growing demand for portable devices with superior performance. Increasing the speed and accuracy of communication requires the use of high-frequency circuits. In order to achieve this goal, a high-frequency oscillator with adequate phase noise is necessary to ensure the speed and accuracy of communication systems such as phase-locked-loop (PLL), on-chip clock generators, frequency synthesizers, and data recovery circuits [1].

Radio frequency integrated oscillators can be constructed using an LC tank or a CMOS ring configuration. The LC type VCOs have advantages of better phase noise and higher output frequency and are widely used in wireless transceiver systems. They have limitations on tuning range and relatively large silicon area consumed because of the essential passive component, inductor, and capacitor [2], [3]. In contrast, the ring oscillator circuit is inductor-free, constructed using only delay cells without an external resonant circuit. Ring topologies can provide a wider tuning range than LC VCOs. They also have more efficient power consumption, less complexity in design compared to LC structures, and smaller layout area because of using CMOS technology delay cells. Ring topology suffers from a lower-quality phase noise performance due to its low-quality factor, and the maximum frequency rate is limited to the minimum delay of its inverter cell [2]–[4]. In order to decrease the minimum delay of a cell, the dual-delay

configuration has been proposed to compensate for the delay in turning on and off in transistors [5]–[11]. Many delay cells with a straightforward design have been used in a ring oscillator proposed in recent works to achieve the desired performance in communication systems [2], [6]. Differential structure in a ring oscillator, which gives an advantage of flexibility in the number of delay cells and better performance in phase noise, has been used in [6], [10]. A ring oscillator designed in a four-stage dual-delay configuration has been presented in [7], but it suffers from high power consumption and poor performance in terms of phase noise. In [5], A high-speed fully-differential ring VCO with the inductive shunt peaking technique using an active inductor configuration has been suggested to enhance the output frequency by sacrificing a wide tuning range and more power efficiency. Despite applying the inductive shunt peaking technique and acceptable power dissipated in [9], the frequency tuning range is not quiet wide. The differential delay cell with an active load and a MOS varactor topology has been reported in [12], [13] to achieve better performance on the power dissipated. However, the tuning range and phase noise have been affected. Using an active inductor circuit in a specific biasing condition could lead to a wide tuning range (almost 100%) with acceptable power and silicon area usage performance, according to [4]. A dual-path differential delay has been presented in [6] where the Hara active inductor has been used to solve the tradeoff between the power and tuning range of the VCO. However, the power efficiency has been disregarded. In [1], the feed-forward phase noise cancellation technique has been applied to minimize the phase noise of the VCO. A four-stage differential delay cell VCO to improve the overall performance of the output frequency and phase noise using a pre-discharged technique has been presented in [7].

In this paper, a four-stage dual-delay-path topology using a pre-discharged path, MOS varactor, and inductive peaking technique, using Lin-Payne active inductor in a differential ring VCO configuration is proposed to improve the tradeoff and achieve a desired performance in phase noise, power dissipation, tuning range, and output frequency. The remaining sections of the document are arranged in the following manner. Section II introduces the details of the proposed active inductor. Section III describes the schematic of the proposed cell and its detailed operation. Section IV presents the simulation results and table of comparison. Finally, Section V contains the overall conclusion of the entire work.

II. ACTIVE INDUCTOR

The limitations of bulky passive spiral on-chip inductors in digital circuits have sparked an interest in using small-sized active inductors instead. CMOS active inductor can be introduced as an active network exhibiting inductor-like characteristics under certain conditions. This type of inductor offers several advantages, such as tunable and self-resonant frequency, inductance and high quality factor. Additionally, because CMOS technology is used, the network consumes a lower silicon area and is compatible with digital systems [4]. Typically, designing an active inductor can be achieved by using an op-amp, which consumes much space, or a gyrator-C, which has high linearity, consumes less area, and is possible to be used in the sub-gigahertz to the gigahertz frequency range [6]. In this design, the Lin-Payne active inductor [14], a gyrator-C type, is utilized, as shown in Fig. 1. M1 and M2 perform the inductive shunt peaking technique in Fig. 1(a). The equivalent small signal model of the Lin-Payne active inductor is presented in Fig. 1(b). The input impedance of the proposed active inductor can be determined as follows:

$$V_{sg1} = -V_{in} \tag{1}$$

$$V_{gs2} = -g_{m1}V_{sg1} \times (\frac{1}{S \times C_{gs2}}) = \frac{g_{m1}V_{in}}{S \times C_{gs2}} \tag{2}$$

By writing a KCL in the input node:

$$g_{m1}V_{sg1} + (\frac{V_{in}}{\frac{1}{S \times C_{gs1}}}) = I_{in} + g_{m2}V_{gs2} \tag{3}$$

By applying (1) and (2) in (3):

$$-g_{m1}V_{in} + (S.C_{gs1}V_{in}) = I_{in} + \frac{g_{m1}g_{m2}V_{in}}{S.C_{gs2}} \tag{4}$$

Equivalent input inductance and resistance of the proposed circuit are carried out by making a decent estimation as follows:

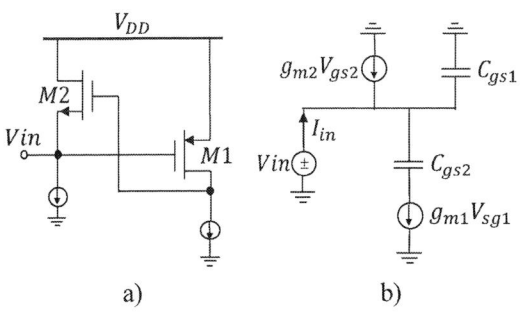

Fig. 1. (a) Schematic of Lin-Payne active inductor, (b) Small-signal model of Lin-Payne active inductor.

$$| L(\omega) | = \omega \frac{C_{gs2}}{g_{m1}g_{m2}} \tag{5}$$

$$| R_p | = \frac{1}{g_{m1}} \tag{6}$$

III. ARCHITECTURE OF THE PROPOSED DUAL-DELAY CELL

The differential architecture can be utilized in both even and odd numbers of stages, but in communication systems, two, three, and four stages are commonly employed. The three-stage topology offers the advantage of higher frequency owing to the lower number of stages, but the quadrature phases (90° phase difference) can be generated in a four-stage topology. The ability to generate precise quadrature signals is important in various applications where phase-sensitive operations are required. Fig. 2 demonstrates the block diagram of the proposed four-stage ring VCO. In a ring oscillator, assuming the supply voltage is constant, the output frequency can be calculated as follows:

$$f_o = \frac{1}{2 \times t_d \times N} \tag{7}$$

The delay of a single cell, which comprises fall time and rise time, is represented by t_d, and the variable N represents the number of delay cells that could be even or odd in differential structure. Based on (7), two parameters are able to control the ring frequency. In order to increase the

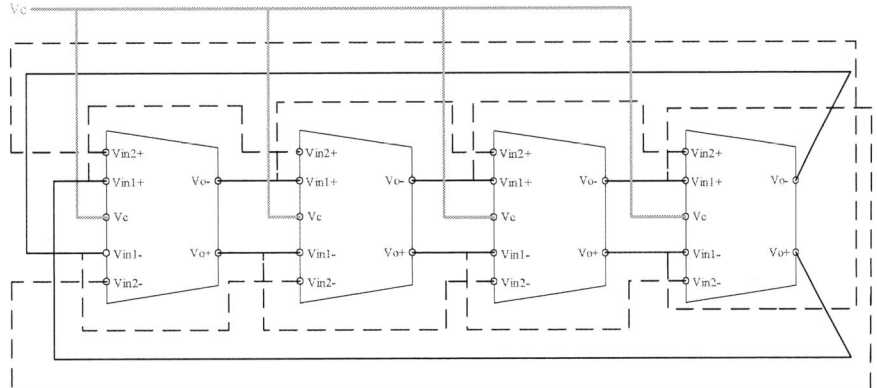

Fig. 2. Block diagram of four-stage ring VCO.

979-8-3503-6020-2/23 $31.00 © 2023 IEEE

frequency, the delay time of each cell should be reduced, which can be achieved by using the dual-delay-path technique, as shown in Fig. 3. The dual-delay-path technique allows the transistors in the subsequent stages to operate more quickly, resulting in a reduction in t_d and a rise in the output frequency.

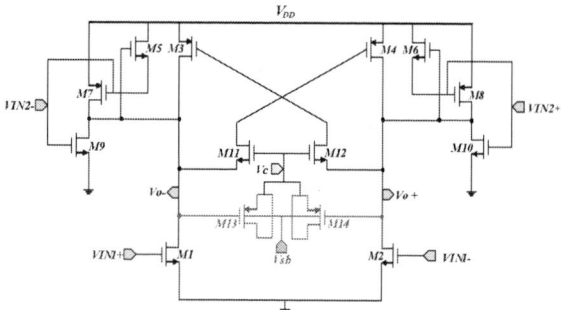

Fig. 3. Schematic of the proposed differential delay cell.

In Fig. 2, the primary inputs of the proposed cell are VIN1+ and VIN1-, and the secondary inputs, 45° leading in phase from the primary inputs, are VIN2+ and VIN2. The differential outputs of the cell are Vo+ and Vo-, which are connected to Vin- and Vin+ of the next stage, respectively. The proposed dual-delay-path cell in a differential configuration is presented in Fig. 3. Primary delay paths are applied to the gates of M1, M2, and M6, M8. At the same time, M5 and M7 are two active inductors, used to perform the inductive shunt peaking technique. The secondary inputs are connected to the gates of M7, and M8, which compensates for the delay of PMOS transistors. M9 and M10 can pre-discharge the output nodes, reducing the fall time. This reduction in fall time results in a decline in the delay of a single cell, thereby causing the frequency of the ring to rise. M3 and M4 are used as an active load of M12 and M11 in a cross-coupled topology in the cell circuit, operating in the linear region to provide enough voltage for M3 and M4 to work as a strong latch. M13 and M14 are used in a varactor structure to control the delay of cells, so the capacitance can be regulated by varying the control voltage (V_C). By increasing the control voltage (V_C), the capacitance of varactors increases, the delay of the cell increases, and the frequency decreases, and vice versa.

Phase noise, as one of the most important parameters in oscillators, can be calculated by (8), which has been defined in [15] as follows:

$$L_{\min}(\Delta f) = \frac{8}{3n} N \frac{KT}{P} \left(\frac{V_{DD}}{V_{CR}} + \frac{V_{DD}}{R_L I_{Tail}} \right) \frac{f_o^2}{\Delta f^2} \qquad (8)$$

Where K, T, and n are Boltzmann, temperature in Kelvin, and proportional constant, respectively. N is the number of delay cells. V_{CR}, f_o and Δf are the characteristics voltage of the device, oscillation frequency, and the offset frequency of carrier, respectively. One of the trade-offs is between phase noise and power consumption. The higher the frequency, the more the phase noise is expected; so in other words, as the frequency increases, more power is needed to keep the phase noise constant.

IV. SIMULATION RESULTS

The presented four-stage differential ring VCO was designed and simulated in TSMC 0.18 μm CMOS process with a 1.8 V supply voltage. Relevant analyses were conducted to extract power consumption, tuning range, output swing, phase noise, and transient output oscillation signal of the proposed VCO. The transient response for the control voltage (VC) of 1.3 V and varactors with a width (W) of 5μm is presented in Fig. 4, which shows the output peak-to-peak swing is equal to 1.5 V. A diagram that represents the variation of frequency versus control voltage (V_C) from 2 to 0.8 V at three different values for width of varactor, is shown in Fig. 5. Fig. 6 presents the variation of dissipated power versus control voltage variation in three different supply voltages. Fig. 7 demonstrates variation of frequency over control voltage at three different source voltages. Fig. 8 presents the frequency variation over voltage control at five different temperatures. The average power consumption by the proposed VCO is 1.4 mW, and the equation to calculate the tuning range is mentioned in (9). The phase noise of the designed VCO, which is illustrated in Fig. 9, is -90.74 dBc/Hz at 1 MHz offset frequency. The equation for the FoM of the ring VCO is calculated using (10), resulting in a value of -158.6 dBc/Hz.

$$Tuning\_Range(TR\%) = \left(\frac{f_{\max} - f_{\min}}{\dfrac{f_{\max} + f_{\min}}{2}} \right) \times 100 \qquad (9)$$

$$FoM = L\{\Delta f\} - 20\log(\tfrac{f_o}{\Delta f}) + 10\log(\tfrac{P_{diss}}{1_{mW}}) \qquad (10)$$

Where $L\{\Delta f\}$ represents the phase noise at Δf offset frequency, f_o is the oscillation frequency, $P_{diss.}$ is the dissipated power of the ring VCO in mW and f_{max}, f_{min} are maximum and minimum oscillation frequency. As the frequency increases, it is expected that the phase noise also rises, so achieving a better FoM becomes more challenging, and the key is to solve the trade-offs between power, frequency, and phase noise.

The layout area of the proposed ring VCO, considering the layout factors, is 40μm × 20μm which is shown in Fig. 10. A comparison in performance parameter of the proposed ring VCO with previously reported designs in terms of CMOS technology, supply voltage, number of stages, oscillation frequency, tuning range, power consumption, phase noise and range of control voltage, are demonstrated in Table 1. The results of the proposed design are extracted from schematic simulation analysis.

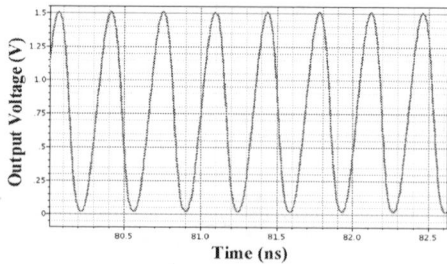

Fig. 4. Transient response of proposed ring VCO in V_{DD} = 1.8 V, V_C = 1.3 V and W of varactors = 5 μm.

The 5th Iranian International Conference on Microelectronics (IICM2023)

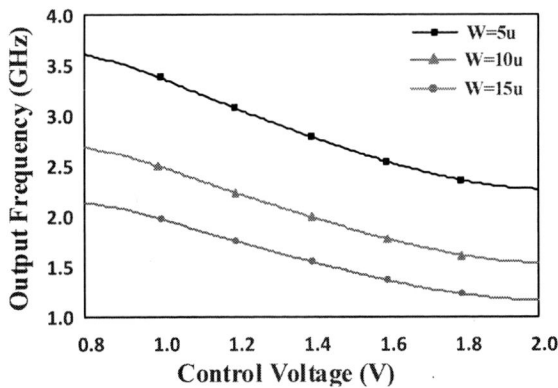

Fig. 5. Frequency variation over the range of control voltage from 0.8 V to 2 V at different varactor widths.

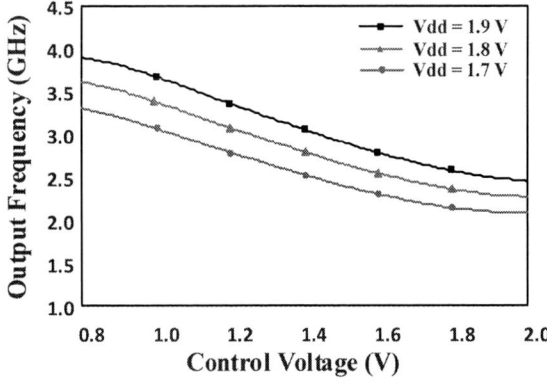

Fig. 7. Frequency variation over the range of control voltage from 0.8 V to 2 V at different supply voltages.

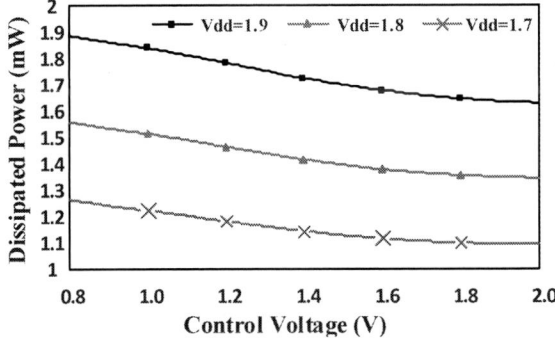

Fig. 6. Power dissipation variation over the range of control voltage from 0.8 V to 2 V at different supply voltage.

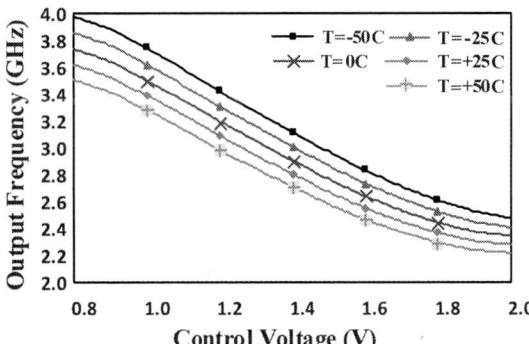

Fig. 8. Frequency variation over the range of control voltage from 0.8 V to 2 V at different temperatures.

Table 1. Performance comparison.

Ref.	CMOS Process (nm)	Supply Voltage (V)	Number of Stage (N)	Oscillation Frequency (GHz)	Tuning Range* (%)	Power Dissipation (mW)	Phase Noise (dBc/Hz) at 1 MHz	FoM (dBc/Hz)	Control Voltage (V)
[6]	180	1.8	4	5.2-7.75	39.38	6.18-10.8	-72.57	149.33	0.1-1
[7]	180	1.8	4	1.61-3.71	78.94	10.54	-89	-150.1	0.1-1
[5]	180	1.8	4	4.9-5.9	18.51	8.1	-86.7	-149.7	0-1.8
[13]	180	1.8	3	4.782-5.516	14.25	2.22	-98.63	-168.41	1-2
	180	1.8	3	4.656-5.333	13.55	2.53	-102.37	-172.86	1-2
[8]	180	1.8	4	4.02-6.12	41.42	4.47	-89.7	-155.9	1-2
[3]	65	1	4	0.485-1.11	78.36	10	-110.8	-157	0-0.8
[12]	180	1.8	3	6.395-6.687	4.46	1.84-2.20	-76.20	-149.43	1-2
	180	1.8	3	5.726-6.142	7.0	1.97-2.17	-73.42	-145.58	-0.5 - +0.5
	180	1.8	3	3.796-4.499	16.94	2.06	-75.63	-145.17	1-2
[11]	65	1.2	4	4.25-21.31	133.5	12.36	-90.47	-184.72	0-1.2 0.6-1.2**
[10]	180	1.8	4	0.72-2.81	118.41	4.62	-97.35	-145.1	0-1.0
[9]	180	1.8	4	3.25-4.2	25.5	7	-92.3	-154.8	0-1.8
This Work	180	1.8	4	2.27-3.6	45.31	1.4	-90.74	-158.6	0.8-2

* Calculated by (9) ** Using two control voltage

979-8-3503-6020-2/23 $31.00 © 2023 IEEE   153

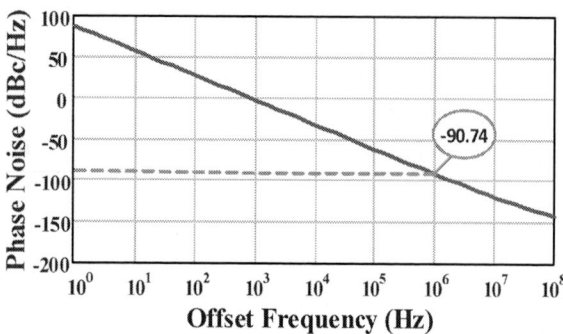

Fig. 9. Phase noise over offset frequency.

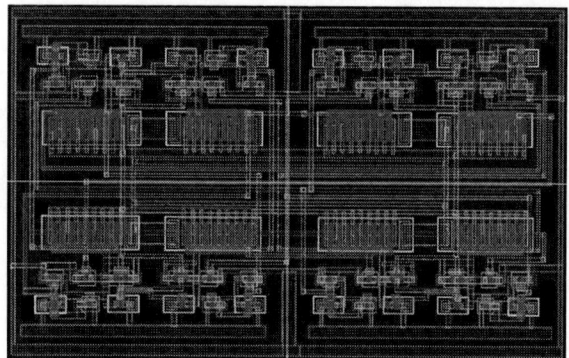

Fig. 10. Layout of proposed four-stage differential ring VCO.

V. CONCLUSION

In this paper, a new design of a dual-delay-path cell in a four-stage ring VCO architecture is presented. In this design, the Lin-Payne active inductor, pre-discharge path, and varactor are used to improve the overall performance of the ring VCO, resulting in a noticeable decline in power consumption. The tuning range of the VCO varied between 2.27 and 3.6 GHz by varying the value of the control voltage. The average power dissipated in the proposed ring VCO is 1.4 mW. The measured phase noise is -90.74 at 1 MHz offset, and the estimated FoM of the designed circuit is -158.6 dBc/Hz.

REFERENCES

[1] P. Cancellation, S. Min, T. Copani, A. Member, and S. Kiaei, "A 90-nm CMOS 5-GHz Ring-Oscillator PLL With Delay-Discriminator-Based Active," *IEEE J. Solid-State Circuits*, vol. 48, no. 5, pp. 1–10, 2013.

[2] M. Moghavvemi and A. Attaran, "Recent advances in delay cell VCOs [application notes]," *IEEE Microwave Magazine*, vol. 12, no. 5, pp. 110–118, 2011.

[3] J. Kim *et al.*, "A Low-Noise Four-Stage Voltage-Controlled Ring Oscillator in Deep-Submicrometer CMOS Technology," *IEEE Transactions on Circuits and Systems I: Regular Papers*, vol. 60, no. 2, pp. 71-75, 2013.

[4] F. Yuan, "CMOS Active Inductors and Transformers: Principle, Implementation, and Applications," *Springer*, Jun. 2008.

[5] C. Zhang, Z. Li, J. Fang, J. Zhao, Y. Guo, and J. Chen, "A novel high-speed CMOS fully-differentical ring VCO," *12th IEEE International Conference on Solid-State and Integrated Circuit Technology (ICSICT)*, Guilin, China, pp. 1-3, 2014.

[6] M. M. Kiloo, V. Singh, M. Kumar, N. Kumar, N. Tripathi, and A. Bhardwaj, "Active inductor based cross coupled differential ring voltage controlled oscillator for UWB applications," *International Journal of Information Technology (Singapore)*, vol. 15, no. 4, pp. 1895–1900, 2023.

[7] N. Kumar and M. Kumar, "Design of CMOS-based low-power high-frequency differential ring VCO," *International Journal of Electronics Letters*, vol. 7, no. 2, pp. 143–153, 2019.

[8] N. Kumar and M. Kumar, "Low Power, Ring VCO with Pre-Charge and Pre-Discharge Circuit for 4 GHz-6.1 GHz Applications in 0.18 µ m CMOS," *Journal of Circuits, Systems and Computers*, vol. 28, no. 11, 2019.

[9] C. Zhang, "A CMOS fully differential ring VCO with active inductors and I / Q outputs," *Microwave and Optical Technology Letters*, vol. 61, no. 10, pp. 2407-2412, Oct. 2019.

[10] Y. Kumar, A. Raman, R. Ranjan, S. Deep, and R. K. Sarin, "Differential RVCO with low power, low phase no ise and wider tuning range for PLL application," *International Journal of Electronics*, vol. 00, no. 00, pp. 1–14, 2023.

[11] S. Askari, M. Saneei, and S. Salem, "Design and Analysis of Wide Tuning Range Ring VCO in 65nm CMOS Technology," *Radioelectronics and Communications Systems*, vol. 62, no. 5, pp. 232–240, 2019.

[12] V. Jangra and M. Kumar, "New low power differential VCO circuit designs with active load and IMOS varactor," *AEU - International Journal of Electronics and Communications*, vol. 119, p. 153191, 2020.

[13] V. Jangra and M. Kumar, "Low power active load and IMOS varactor based VCO designs using differential delay stages in 0 . 18 µ m technology," *Microelectronics Journal*, vol. 98, no. December 2019, p. 104728, 2020.

[14] T. Y. K. Lin and A. J. Payne, "Design of a low-voltage, low-power, wide-tuning integrated oscillator," *2000 IEEE International Symposium on Circuits and Systems (ISCAS)*, Geneva, Switzerland, vol.5, pp. 629-632, 2000.

[15] A. Hajimiri, S. Limotyrakis, and T. H. Lee, "Jitter and phase noise in ring oscillators," *IEEE Journal of Solid-State Circuits*, vol. 34, no. 6, pp. 790–804, 1999.

The 5th Iranian International Conference on Microelectronics (IICM2023)

25 – 26 October 2023

A Low-Noise Amplifier with Bandwidth Extension and Noise Cancellation for 5G Receivers

Pardis Javanbakht
Department of Electrical and Computer Engineering
Urmia University
Urmia, Iran
pardis.javanbakht@aut.ac.ir

Mortaza Mojarad
Department of Electrical and Computer Engineering
Urmia University
Urmia, Iran
m.mojarad@urmia.ac.ir

Abstract—This paper proposes a wideband Low-Noise amplifier (LNA) with low power consumption for 5G receiver frontends. The presented LNA is a two-stage amplifier to obtain sufficient gain over the wide frequency range from 1 GHz to 6 GHz. The proposed LNA has been simulated in a standard TSMC 0.13 μm CMOS process using ADS. It achieves a minimum Noise Figure (NF) of 2.61 dB by using a novel noise-cancellation scheme. Also, a new technique has been proposed to improve the input matching and the maximum S_{11} is almost -8 dB over the specified frequency range. The LNA dissipates total power of 6.93 mW from a single 1.2-V supply voltage. Moreover, a new structure and a novel method have been proposed to implement the LNA which result in wide bandwidth for S_{21}. The simulation results prove that the LNA has a relatively flat gain with the maximum S_{21} of 15.8 dB. The two-tone test shows that the IIP3 is approximately equal to 1.35 dBm.

Keywords—Low-noise amplifier, LNA, noise cancellation, linearity, input matching, bandwidth

I. INTRODUCTION

A Low-Noise Amplifier (LNA) is often known as the first active building block in a wireless receiver, which amplifies the received low-power signal and guarantees a sufficiently high signal-to-noise ratio (SNR) for the communication link. There are a lot of specifications for an LNA in wireless receivers which need to be satisfied such as high gain, low noise figure (NF), high linearity, and maximum RF power transfer. In order to acquire LNAs with low NF, noise canceling schemes have been widely investigated in the literature [1]-[4]. However, these structures suffer from the excess noise of the auxiliary circuits and elevated power consumption.

To accomplish noise and distortion cancellation simultaneously, a noise cancelling structure has been presented in [5] that reduces noise and power consumption considerably by utilizing a positive feedback. In [6] a local feedback between the CG and CS stages boosts the transconductance of a CG stage to alleviate the noise figure. The drawback of this method is a considerably increased quiescent current. In [7]-[13], it is demonstrated that using common gate (CG) and cascode topologies can lead to wideband operation. However, the aforementioned topologies suffer from high NF and power dissipation. Moreover, the maximum attainable bandwidth for the S_{21} for

these LNAs is limited and is not adequate for 5G frontends. Therefore, to implement wideband LNAs for the next-generation communication standards, short-channel CMOS processes are exploited which increases NF and cost. Hence, implementing a wideband LNA with low noise figure remains a big challenge. As mentioned in [14], [15], for lowering the supply voltage while having a low NF, common source (CS) LNAs can be utilized. However, conventional CS LNAs achieve limited bandwidths and there is a need to use close-loop topologies for achieving the desired bandwidth in CS LNAs which leads to gain reduction. Furthermore, the conventional matching networks for CS stages include degeneration inductors which increases the die area and cost [16]. Amplifiers with shunt feedback have also been extensively discussed in the previous works which are often accompanied by elevated noise and gain reduction [17].

In this paper a wideband low-noise amplifier has been proposed with a bandwidth from 1 GHz to 6 GHz, while preserving a low NF and small power consumption. The remaining of this article is organized as follows. The proposed structure and circuit realization of the new LNA is presented in Section II. In Section III simulation results are reported, and Section IV concludes the paper.

II. THE STRUCTURE AND THE CIRCUIT IMPLEMENTATION OF THE PROPOSED LNA

The schematic of the proposed LNA has been depicted in Fig. 1. It is a two-stage amplifier with the primary goal to achieve adequate impedance matching and sufficient gain. AC analysis neglecting the channel length modulation demonstrates an approximate value for input impedance which is equal to

$$R_{in,amp} = \frac{1}{g_{m1}} \parallel \frac{1}{g_{m2}} = \frac{1}{g_{m1} + g_{m2}}. \quad (1)$$

The impedance in (1) implies that appropriate and wideband input matching can be achieved. It is worth mentioning that the drawback of a conventional common-gate is that the accurate input resistance calculated from its source is almost twice the input impedance expressed in (1). The first stage of the proposed LNA circuit can be considered as the combination of two common-gate amplifiers. One common-gate stage is made up of an NMOS transistor M_1, and the second one is made by the PMOS transistor M_2. The first

979-8-3503-6020-2/23 $31.00 © 2023 IEEE   155

The 5th Iranian International Conference on Microelectronics (IICM2023)

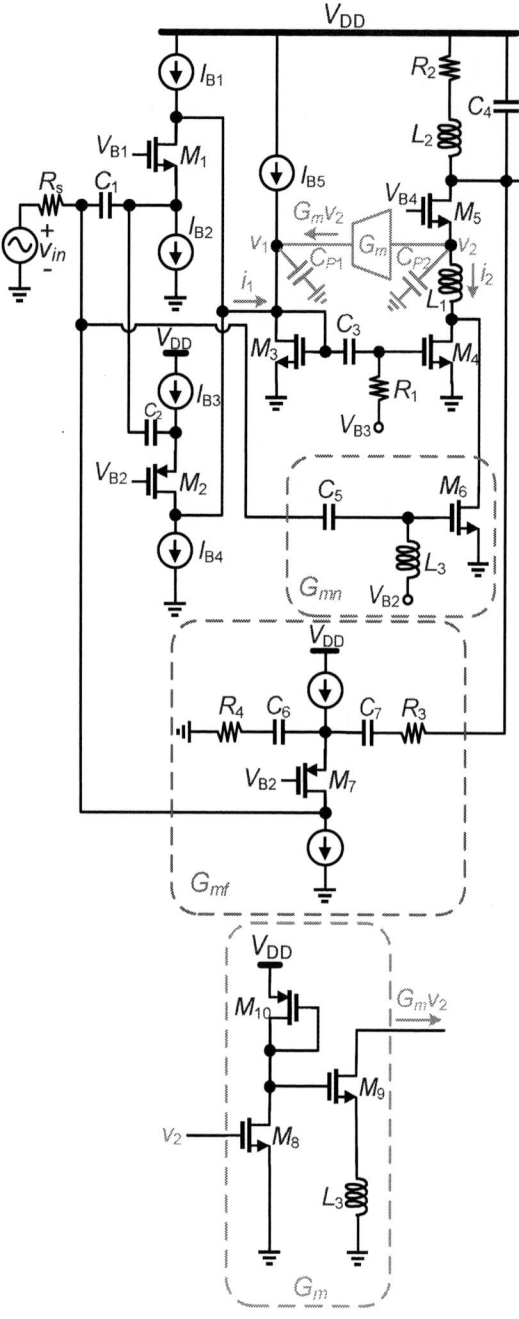

Fig. 1   Circuit implementation of the proposed LNA

stage is in fact a transconductance amplifier and its output signal is current which can be derived as:

$$i_1 = (g_{m1} + g_{m2})v_{in} \tag{2}$$

The output current of the first stage i_1, flows through the current mirror comprised of transistors M_3 and M_4. The aspect ratio of M_4 is n times larger than M_3 and therefore i_1 is amplified and $i_2 \approx n \times i_1$, assuming R_1 to be a large resistance.

The second stage is another common-gate amplifier including transistor M_5, the inductors L_1 and L_2, the capacitor C_4, and the resistor R_2. The main purpose of this stage is to enhance the overall gain of the LNA and perform an I/V conversion. The central frequency is adjusted using a tank circuit made of L_2 and C_4. Herein, in order to increase the bandwidth, a new technique has been proposed which consists of a feedback transconductance amplifier G_m, as depicted in Fig. 1. It is very well known that the reason for bandwidth shrinkage in an amplifier is the parasitic capacitances and low-frequency poles in the transfer function of the amplifier. In the new LNA, the lumped parasitic capacitance at node v_1 which is denoted by C_{P1}, creates a low frequency pole in the transfer function of i_2/v_{in}. The frequency of this pole can be approximately calculated as follows:

$$\omega_p = \frac{1}{\frac{1}{g_{m3}} \times C_{P1}} \tag{3}$$

In this work, the feedback path G_m, has been used in the proposed topology to increases the bandwidth in order to accommodate all of channels of the 5G standard. By conducting frequency analysis it is shown that the exploited techniques has been proven to be effective in increasing the bandwidth by generating a left-half-plane zero. Moreover, the pole associated with node v_1 is pushed to higher frequencies using this feedback transconductaor. The frequency of the new pole and zero frequencies can be given by

$$\omega_z = \frac{1}{\frac{1}{g_{m5}} \times C_{P2}} \tag{4}$$

$$\omega_p = \frac{1}{\frac{g_{m5}}{n \times G_m \times g_{m3}} \times C_{P1}} \tag{5}$$

In order to obtain a relatively flat gain for the LNA, the zero in (4) and the pole in (5) are required to be approximately equal. Therefore, a design constraint can be obtained as follows:

$$G_m = \frac{g_{m5}^2}{n \times g_{m3}}. \tag{6}$$

By satisfying (6), the bandwidth extension has been achieved. As mentioned before, in the proposed structure, the LNA is comprised of two cascaded wideband band-pass amplifiers with different center frequencies. As the first stage undergoes a decreasing gain as frequency increases, the second stage makes up the decrement in gain at high frequencies, to preserve constant gain at the designated frequency range.

The noise contribution of transistors M_1 and M_2 in the first stage of the amplifier in the depicted schematic of Fig. 1, has been calculated. The power spectral density due to noise of these two transistors is obtained as

$$\overline{V_{n,out}^2} = (g_{m1} + g_{m2})(\frac{n}{2} \times R_2)^2 (4KT\gamma) \tag{7}$$

Based on the above discussions, in order to cut back on the noise in the presented LNA, a new method has been proposed. In this method, the main effort is to mitigate the noise contribution of the transistors of the first stage M_1 and M_2. This has been accomplished by using a transconductance amplifier (G_{mn}) as shown in Fig. 2. The transconductance amplifier (G_{mn}) can provide the noise current i_{n3} to neutralize the noise current i_{n4}. Eventually the noise current signal of i_{n4} which is the sum of two noise currents i_{n2} and i_{n3} will be equal to zero and no noise due to M_1 and M_2 will appear at the output of LNA. By conducting a noise analysis according to Fig. 2, the noise current i_{n1} can be derived as follows :

$$i_{n1} = \frac{1}{2}(i_{n,M2} - i_{n,M1}) \tag{8}$$

The current mirror consisting of transistors M_3 and M_4 multiply the current by n and i_{n2} can be given by

$$i_{n2} = \frac{n}{2}(i_{n,M2} - i_{n,M1}) \tag{9}$$

Herein, the transconductance amplifier G_{mn} has been implemented for noise cancellation. By inspecting the equivalent circuit in Fig. 2, it can be deduced that the voltage at the source of transistors M_1 and M_2 can be calculated as

$$v_{n,s} = \frac{R_s}{2}(i_{n,M1} - i_{n,M2}) \tag{10}$$

The output current of the auxiliary transconductance amplifier G_{mn} is thus equal to

$$i_{n3} = G_{mn}\frac{R_s}{2}(i_{n,M1} - i_{n,M2}) \tag{11}$$

According to Fig. 2 and from (11), it is apparent that the noise current i_{n4} which is the noise current that flows to the output node of the LNA can be given by

$$i_{n4} = \frac{n}{2}(i_{n,M2} - i_{n,M1}) + G_{mn}\frac{R_s}{2}(i_{n,M1} - i_{n,M2}) \tag{12}$$

From (12), it can be concluded that i_{n4} will be equal to zero if the constraint in (13) is satisfied.

$$G_{mn} = \frac{n}{R_s} \tag{13}$$

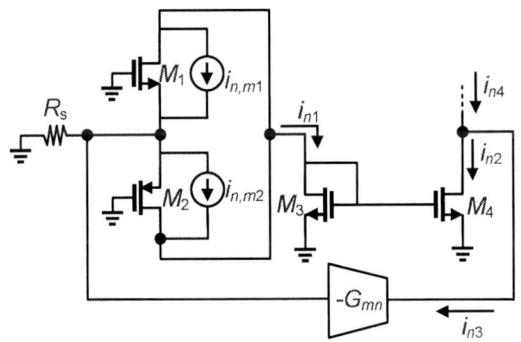

Fig. 2. AC equivalent circuit for noise analysis

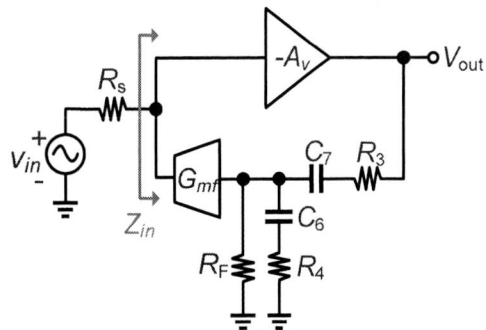

Fig. 3. The calculation of the input impedance

The other advantage of the proposed structure is the enhanced linearity. In the first stage, as it is obvious, the drain currents of the large sized transistors M_1 and M_2 consist of nonlinear terms which by adjusting the biasing currents and the bias voltages V_{B1} and V_{B2} as depicted in Fig. 1, non-linear terms cancel out at the output, and the third harmonic distortion can be eliminated. This procedure causes a significant improvement in the overall linearity of the LNA.

As mentioned earlier, wideband input matching cannot be accomplished using the conventional and basic amplifier topologies. In this work, the matching has been improved by exploiting a new technique which is creating an active feedback path consisting of the frequency-dependent transconductance amplifier G_{mf}, capacitors C_6 and C_7, and resistors R_3, R_4, and R_F as shown in Fig. 3. The input impedance Z_{in} has been calculated by carrying out a small-signal analysis and is given by

$$Z_{in} = \frac{R_{in,amp}}{1 + G_{mf}R_{in,amp}A_v \dfrac{Z_2}{Z_1 + Z_2}} \tag{14}$$

where the impedance Z_1 is the series combination of R_3 and C_7 and the impedance Z_2 represents the series equivalent impedance of C_6 and R_4 which is connected in parallel to R_F. Also, $-A_v$ is the total gain of the LNA. By assuming $Z_2 \gg Z_1$ and choosing $G_{mf}R_{in,amp}A_v \gg 1$, the input resistance of the LNA is divided by a large amount and consequently the presented method will significantly cause a reduction in the input impedance. This facilitates proper input matching consuming low DC power. The main feature of the proposed technique is ensuring input matching over a wide frequency range. The important reason for the limited bandwidth of the input matching is that the input impedance decreases because the impedances of the parasitic capacitances also decrease for higher frequencies. In Fig. 3, C_7 is chosen to be much larger than C_6. Therefore, at low frequencies the coefficient $\alpha = Z_2/(Z_1 + Z_2)$ is almost equal to unity. Also, R_F is much smaller than R_3 which makes α smaller than unity for higher frequencies. As a result, as frequency increases, the reduction in α compensates the reduction of $R_{in,amp}$, by increasing the input impedance Z_{in}.

III. SIMULATION RESULTS

The proposed LNA has been designed and simulated in a standard 0.13 μm CMOS process using ADS. In Fig. 4 simulation result for S_{11} is depicted. The maximum S_{11} is almost -8 dB proving a good input matching.

979-8-3503-6020-2/23 $31.00 © 2023 IEEE

The simulation results for S_{21} and the noise figure are shown in Fig. 5 and Fig. 6, respectively. The simulations have also been carried out for different process corners and for different temperatures. The result from a two-tone test has been depicted in Fig. 7 which proves +1.35 dBm IIP3.

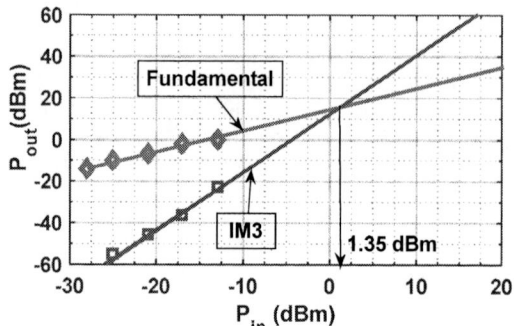

Fig. 7 The two-tone test result

A summary of the performance of the proposed LNA has been given in Table 1. Moreover, the performance has been compared to the previously reported works which demonstrates that the proposed LNA achieves low noise figure due to employing the noise cancellation technique. It also has a better overall gain, higher bandwidth due to employing a novel structure to implement the LNA and auxiliary circuits to further boost the bandwidth. A new technique has also been employed to improve the input matching over the entire frequency band while consuming low quiescent current. It is worth noting that the attained bandwidth is even larger than the previous works some of which utilize short-channel CMOS processes. The results for the figure of merit has been compared in Table 1 using the following FoM.

$$FoM = 20\log\frac{IIP3(mW)\times Gain\times BW(GHz)}{P_{DC}(mW)\times(NF-1)} \quad (15)$$

The Monte Carlo simulation for IIP3 has been conducted with 250 iterations and the result is shown in Fig. 8. The μ stability test shown in Fig. 9 proves that the LNA is unconditionally stable over the specified frequency range.

IV. Conclusion

The presented paper exhibits a new noise cancelling LNA with a novel bandwidth extension method. Also the input matching has been improved over a wide frequency range by using a new technique. It achieves small amount of power consumption and a superior overall performance in comparison with the prior art .The new LNA is appropriate to be used in 5G Receivers frontends.

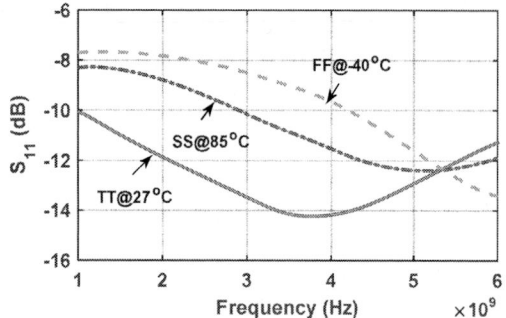

Fig. 4. Simulation results for S_{11}

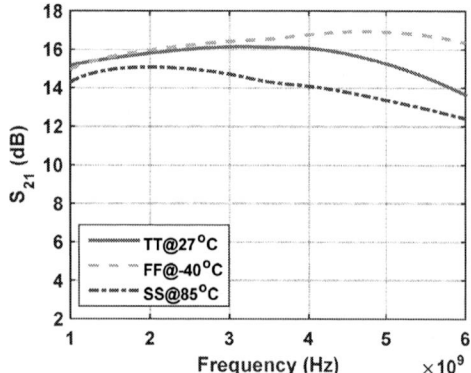

Fig. 5 Simulated S_{21}

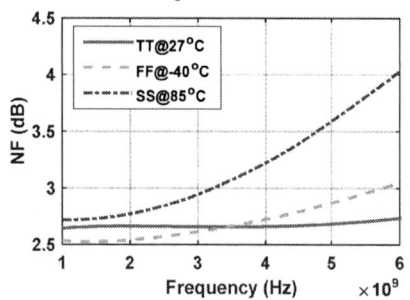

Fig. 6 Simulation results for the noise figure

Table 1. Comparison against prior Art

	Tech.	Frequency (GHz)	BW(GHz)	S_{21}(dB)	NF(dB)	IIP3(dBm)	Power(mW)	FOM
This work*	130nm	1~6	5	15.8	2.61~2.8	1.35	6.93	33
[18]	65nm	0.4~2.2	1.8	16.4	2~2.5	-5	29	12.9
[19]	180nm	0.1~2	1.9	17.5	2.9~3.5	10.4	21.3	18.27
[20]	90nm	0.1~20	19.9	12.7	3.3~5.5	-2.5	20.4	6.4
[21]	130nm	0.8~10.6	9.8	16	3.4~5.6	1.6	14.4	14.4
[22]	180nm	0.5~1.3	0.8	10	2.9~3.2	7.5	18	1.6
[23]	65nm	0.1~4	3.9	18	2~4.4	-5.5	12	1.9
[24]*	65nm	0.47~3.3	2.83	22	2.57~3.5	2.81	12.5	34.8
[25]*	180nm	0.9~1.8	0.9	19	4	5	10.8	22.83

*Simulation results

The 5th Iranian International Conference on Microelectronics (IICM2023)

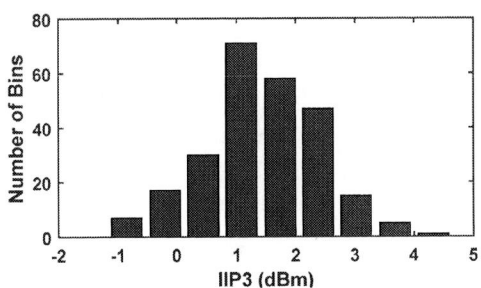

Fig. 8 Monte Carlo simulation results for IIP3 with 250 iterations

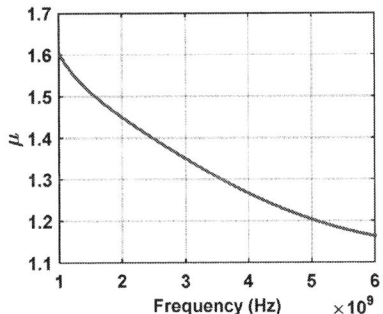

Fig. 9 μ stability test

REFERENCES

[1] F. Bruccoleri, E. A. M. Klumperink and B. Nauta, "Wide-band CMOS low-noise amplifier exploiting thermal noise canceling," in *IEEE Journal of Solid-State Circuits*, vol. 39, no. 2, pp. 275-282, Feb. 2004, doi: 10.1109/JSSC.2003.821786.

[2] D. Im, I. Nam, H. -T. Kim and K. Lee, "A Wideband CMOS Low Noise Amplifier Employing Noise and IM2 Distortion Cancellation for a Digital TV Tuner," in IEEE Journal of Solid-State Circuits, vol. 44, no. 3, pp. 686-698, March 2009, doi: 10.1109/JSSC.2008.2010804.

[3] Z. Liu, C. C. Boon, X. Yu, C. Li, K. Yang and Y. Liang, "A 0.061-mm² 1–11-GHz Noise-Canceling Low-Noise Amplifier Employing Active Feedforward With Simultaneous Current and Noise Reduction," in *IEEE Transactions on Microwave Theory and Techniques*, vol. 69, no. 6, pp. 3093-3106, June 2021, doi: 10.1109/TMTT.2021.3061290.

[4] A. Bozorg and R. B. Staszewski, "A 0.02–4.5-GHz LN(T)A in 28-nm CMOS for 5G Exploiting Noise Reduction and Current Reuse," in *IEEE Journal of Solid-State Circuits*, vol. 56, no. 2, pp. 404-415, Feb. 2021, doi: 10.1109/JSSC.2020.3018680.

[5] Z. Liu, C. C. Boon, C. Li, K. Yang, Y. Dong and T. Guo, "A 0.0078mm2 3.4mW Wideband Positive-feedback-Based Noise-Cancelling LNA in 28nm CMOS Exploiting **G**m Boosting," *2022 IEEE International Solid- State Circuits Conference (ISSCC)*, San Francisco, CA, USA, 2022, pp. 1-3, doi: 10.1109/ISSCC42614.2022.9731719.

[6] H. Wang, L. Zhang, and Z. Yu, "A Wideband inductorless LNA with local feedback and noise cancelling for low-power low-voltage applications," *IEEE Trans. Circuits Syst. I, Reg. Papers*, vol. 57, no. 8, pp. 1993–2005, Aug. 2010.

[7] H. Zhang, X. Fan, and E. S. Sinencio, "A low-power, linearized, ultra-wideband LNA design technique," *IEEE J. Solid-State Circuits*, vol. 44, no. 2, pp. 320–330, Feb. 2009.

[8] Y. Shim, C.-W. Kim, J. Lee, and S.-G. Lee, "Design of full band UWB common-gate LNA," IEEE Microw. Wireless Compon. Lett., vol. 17, no. 10, pp. 721–723, Oct. 2007.

[9] Y.-T. Lo and J.-F. Kiang, "Design of wideband LNAs using parallel-to-series resonant matching network between common-gate and common-source stages," IEEE Trans. Microw. Theory Techn., vol. 59, no. 9, pp. 2285–2294, Sep. 2011.

[10] H. Gao et al., "A 48–61 GHz LNA in 40-nm CMOS with 3.6 dB minimum NF employing a metal slotting method," in Proc. IEEE Radio Freq. Integr. Circuits Symp. (RFIC), May 2016, pp. 154–157.

[11] M. Elkholy, S. Shakib, J. Dunworth, V. Aparin, and K. Entesari, "A wideband variable gain LNA with high OIP3 for 5G using 40-nm bulk CMOS," IEEE Microw. Wireless Compon. Lett., vol. 28, no. 1, pp. 64–66, Jan. 2018.

[12] S. Kong, H.-D. Lee, S. Jang, J. Park, K.-S. Kim, and K.-C. Lee, "A 28-GHz CMOS LNA with stability-enhanced G m -boosting technique using transformers," in Proc. IEEE Radio Freq. Integr. Circuits Symp. (RFIC), Jun. 2019, pp. 7–10.

[13] L. Gao and G. M. Rebeiz, "A 24–43 GHz LNA with 3.1–3.7 dB noise figure and embedded 3-pole elliptic high-pass response for 5G applications in 22 nm FDSOI," in Proc. IEEE Radio Freq. Integr. Circuits Symp. (RFIC), Jun. 2019, pp. 239–242.

[14] S.-C. Shin, M.-D. Tsai, R.-C. Liu, K.-Y. Lin, and H. Wang, "A 24 GHz 3.9-dB NF low-noise amplifier using 0.18 μm CMOS technology," IEEE Microw. Wireless Compon. Lett., vol. 15, no. 7, pp. 448–450, Jul. 2005.

[15] Y. Ding, S. Vehring, and G. Boeck, "Design of 24 GHz high-linear high-gain low-noise amplifiers using neutralization techniques," in IEEE MTT-S Int. Microw. Symp. Dig., Jun. 2019, pp. 944–947.

[16] D. Kim and D. Im, "A Reconfigurable Balun-LNA and Tunable Filter With Frequency-Optimized Harmonic Rejection for Sub-GHz and 2.4 GHz IoT Receivers," in *IEEE Transactions on Circuits and Systems I: Regular Papers*, vol. 69, no. 8, pp. 3164-3176, Aug. 2022, doi: 10.1109/TCSI.2022.3169364.

[17] E. V. P. Anjos, D. Schreurs, G. A. E. Vandenbosch and M. Geurts, "A Compact 26.5–29.5-GHz LNA-Phase-Shifter Combo With 360° Continuous Phase Tuning Based on All-Pass Networks for Millimeter-Wave 5G," in *IEEE Transactions on Circuits and Systems I: Regular Papers*, vol. 68, no. 9, pp. 3927-3940, Sept. 2021, doi: 10.1109/TCSI.2021.3089630.

[18] S. S. Regulagadda, B. D. Sahoo, A. Dutta, K. Y. Varma, and V. S. Rao, "A packaged noise-canceling high-gain wideband low noise amplifier," IEEE Trans. Circuits Syst. II, Exp. Briefs, vol. 66, no. 1, pp. 11–15, Jan. 2019.

[19] B. Guo, J. Chen, L. Li, H. Jin, and G. Yang, "A wideband noisecanceling CMOS LNA with enhanced linearity by using complementary nMOS and pMOS configurations," IEEE J. Solid-State Circuits, vol. 52, no. 5, pp. 1331–1344, May 2017.

[20] M. Chen and J. Lin, "A 0.1–20 GHz low-power self-biased resistivefeedback LNA in 90 nm digital CMOS," IEEE Microw. Wireless Compon. Lett., vol. 19, no. 5, pp. 323–325, May 2009

[21] S. Lou and H. C. Luong, "A 0.8 GHz–10.6 GHz SDR low-noise amplifier in 0.13-μm CMOS," in Proc. IEEE Custom Integr. Circuits Conf., Sep. 2008, pp. 65–68.

[22] D. Im, "A +9dbm output P1dB active feedback CMOS wideband LNA for SAW-less receivers," IEEE Trans. Circuits Syst. II, Exp. Briefs, vol. 60, no. 7, pp. 377–381, Jul. 2013.

[23] X. Wang, W. Aichholzer, and J. Sturm, "A resistive feedback LNA with feedforward noise and distortion cancellation," in Proc. Eur. Solid-State Circuits Conf. (ESSCIRC), Sep. 2010, pp. 406–409.

[24] B. Shirmohammadi and M. Yavari, "A Linear Wideband CMOS Balun-LNA With Balanced Loads," *IEEE Transactions on Circuits and Systems II: Express Briefs*, vol. 69, no. 3, pp. 754-758, Mar. 2022.

[25] D. Pathak, S. Vardhan, A. Kumar and A. Dutta, "Reconfigurable Concurrent Dual-Band Low Noise Amplifier with Dynamic Output Load Network for Software Defined Radio," *IEEE International Symposium on Circuits and Systems* (ISCAS), Seville, Spain, 2020, pp. 1-5.

979-8-3503-6020-2/23 $31.00 © 2023 IEEE

The 5th Iranian International Conference on Microelectronics (IICM 2023)

October 25-26, 2023

A Low-Power Inductor-Less Linear Wideband CMOS Balun-LNA Using Current Reuse And Linearity Techniques

Soroush Hashemi Bani and Mohammad Yavari

Integrated Circuits Design Laboratory, Department of Electrical Engineering, Amirkabir University of Technology (Tehran Polytechnic), P.O. 15875-4413, Tehran 15914, Iran.

Emails: soroushhashemi@aut.ac.ir, myavari@aut.ac.ir

Abstract—This brief presents an inductor-less low-power balun-LNA employing negative feedback and post-distortion technique for the purpose of acquiring third-order non-linearity elimination without disturbing second-order intermodulation cancellation. The common gate (CG)-common source (CS) complementary push-pull topology, partially declines the common gate (CG) transistors thermal noise and the balun technique noise cancellation leads to total noise figure (NF) reduction. The post-distortion technique employs auxiliary cross-coupled transistors in weak-inversion region, which not only improves third-order linearity but also boosts voltage gain. The proposed balun-LNA structure designed in 65nm CMOS technology, reveals a noise figure (NF) of 3.25 dB, a S_{11} of less than -11 dB, and a S_{21} of 16.1 dB covers the -1 dB frequency range of 0.3-5 GHz. It achieves mean second-order intercept point (IIP2) and third-order intercept point (IIP3) of +27 dBm and +5 dBm, respectively. The circuit operates at a nominal supply voltage of 1.2 V with a bias current of 2.48 mA, which leads to low-power consumption.

Keywords—*CMOS balun-LNA, inductor-less, linearity, noise cancellation, post-distortion, IIP2, IIP3, low-power.*

I. INTRODUCTION

Low noise amplifier (LNA) role as the first active block in wireless receivers that improves signal to noise ratio (SNR) of the receivers which happens by noise cancellation techniques and simultaneously high amount of gain. The performance of LNAs evaluates by several factors and parameters compatible with the required applications. Some of these factors are sufficient impedance matching, a flat good amount of gain, high linearity, low noise figure (NF) over the entire bandwidth, and low gain and phase mismatch. Over the past decade, balun-LNAs received more attention since demanding differential outputs and provision of lower second-order distortion without using passive baluns that use a lot of area and have a loss.

Recently, linearity factor and noise mitigation techniques have been bold in circuit designers' eyes [1]. creating two different paths from the input to the output with the same gains weakens the noise of the input transistors to a good extent and at the same time increases the gain of the circuit [1, 4, 8, 10, 13]. Various linearization methods have attracted the attention of designers, and according to the improvement of the linearity type, each method can optimize the desired circuit structure. One of the linearization methods used is the derivative superposition technique, which by biasing the auxiliary transistors in the weak-inversion region, the third-order current coefficients of these transistors are similar to those of the main transistors, and IIP3 is significantly improved [1, 4, 3]. The only problem with this method is the degradation in input impedance matching. Another method used is the post-distortion method, which will once again enhance the linearity of the circuit by removing the third-order coefficients [1, 2, 7, 9, 3]. This method suffers from high power loss due to the bias of auxiliary transistors in the strong-inversion region. Symmetrical adoption of the output stage and load resistors is one of the punch lines that must be observed to maintain gain and phase mismatch [5, 6, 7, 11, 14, 17]. from this perspective, the use of balloon structure in designing low noise amplifier circuits shows itself more. utilization of passive balloons is not recommended due to their high cost and losses [15, 16, 18]. LNAs with an inherent balloon are the only choice for cost reduction and optimal design. The most common balloon structure is the mixture of CS and CG transistors at the input, which brings challenges. Matching the input impedance is highly dependent on the CG transistor, and satisfying this action will require choosing a lower transconductance of the CG transistor than the CS, which abolishes the symmetry of the circuit. One of the proposed solutions to prevent this phenomenon is auxiliary transistors in the role of current bleeding, which solves this problem by injecting additional current into the CS transistor, nevertheless causing an increase in power consumption [5, 7, 8]. One of the keys to facing this limitation is to use positive feedback, which adjusts the input impedance without shrinking the CG transistor and establishes symmetry in the circuit [12, 17].

This brief proposed a low-power inductor-less wideband balun-LNA with balanced loads and linearity techniques such as negative feedback and a mixture of derivative-superposition, and post-distortion that improves LNA's IIP3 by consuming negligible additional power. The complimentary CG-CS topology increases IIP2 and affords partial noise cancellation to provide a degree of freedom to design LNA for better linearity conditions. The future sections of the paper are organized as follows. Section II presents the structure of the proposed balun-LNA in detail. Section III provides the simulation results, and finally, the conclusion is given in Section IV.

979-8-3503-6020-2/23 $31.00 © 2023 IEEE

The 5th Iranian International Conference on Microelectronics (IICM 2023)

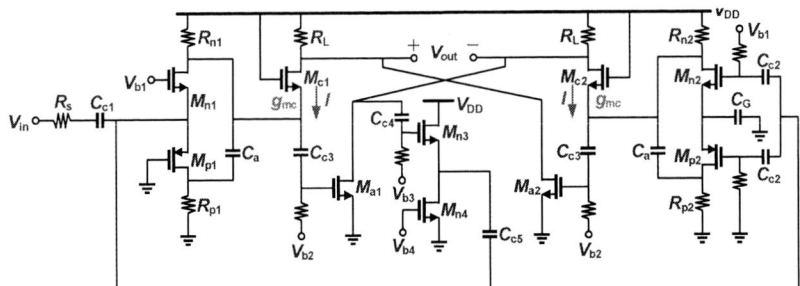

Component Name	Value (M×F×W/L)
$M_{n1,2}$	1×25×1.2 μm /60 nm
$M_{p1,2}$	4×20×1.2 μm /60 nm
$M_{c1,2}$	1×20×1.2 μm /60 nm
$M_{a1,2}$	1×1×600 nm /60 nm
M_{n3} , M_{n4}	1×1×600 nm /60 nm

Component Name	Value
$R_{n1,2}$, $R_{p1,2}$, R_L (Ω)	2.25 k, 200, 200
C_{c1} , C_G , C_a , C_{c2} ,	20 p, 20 p, 10 p, 10 p
C_{c3} , C_{c4} , C_{c5} (F)	5 p, 1 p, 1 p

Fig. 1. Proposed CMOS Balun-LNA.

II. STRUCTURE OF THE PROPOSED BALUN-LNA

The proposed Balun-LNA circuit illustrates in Fig. 1. The input stage of the LNA is made of the complimentary CG-CS structure, which in addition to removing the thermal noise of the M_{n1} and M_{p1} establishes the second-order linearity to a convenient extent. M_{n2} and M_{p2} are used to create the inherent balloon on the other side of the circuit, which, in addition to eliminating the noise of CG transistors, has the ability to eliminate phase and gain mismatch. In the following, M_{c1} and M_{c2} as cascode transistors are used to establish the symmetry of the circuit, whose bias is established by adjusting the resistance in their source. The total current of the cascode transistors and the bias resistor passes through the input transistors reused, which prevents current bleeding and significantly reduces power consumption. In the next step, to increase the linearity of the circuit, M_{a1} and M_{a2} as auxiliary transistors are used in the weak-inversion region in parallel with the cascode transistors. Due to the auxiliary transistors in the weak-inversion region, they pass a small amount of current and consume a small amount of power. CS transistors intensify the effects of third-order nonlinearity, therefore, to improve IIP3, negative feedback is considered for the transistors. utilizing such a technique, the attenuation loop of the third-order coefficients is formed and creates this degree of freedom that can eliminate third-order intermodulation. the input impedance matching circumstances, voltage gain calculation, noise figure analysis, and linearity of the proposed balun-LNA are represented in the following.

A. Input Impedance Matching

According to the structure of the circuit in Fig. 1, the input of LNA consists of complementary CG-CS, which generally CG transistors are responsible for matching the input impedance. According to these points, the following equation represents the matching of input impedance.

$$R_S = \frac{1}{g_{mn1} + g_{mp1}} \quad (1)$$

In the above relationship, R_S is the resistance seen from the source, g_{mn1} and g_{mp1} are the transconductance of CG transistors. It should be noted that to adjust the bandwidth of the circuit, capacitors C_G and C_{c1} play a key role in flattening the gain, NF, and S_{11}.

B. Voltage Gain

Without considering the drain-source resistance of the

transistors and the effect of cross-coupled transistors, taking into account the negative feedback to establish the linearity of the CS transistors, the proposed LNA gain is obtained according to the following equation:

$$A_v = \frac{2(g_{mn1} + g_{mp1})g_{mc1}R_{n1}R_{p1}R_L}{[R_S(g_{mn1} + g_{mp1}) + 1][R_{n1}R_{p1}g_{mc1} + R_{n1} + R_{p1}]} \quad (2)$$

Due to symmetry, the half-circuit gain calculation depicted in (2), where g_{mn1} and g_{mp1} are transconductance of CG transistors, g_{mc1} transconductance of cascode transistors, R_{n1} and R_{p1} drain resistances of CG transistors, R_S source resistance, and R_L load resistance. By adding auxiliary transistors to increase the linearity, the overall gain of the circuit is increased by 1 dB according to (3), and in addition, it improves the phase mismatch.

$$A_v = \frac{2(g_{mn1} + g_{mp1})(g_{mc1} + g_{ma2})R_{n1}R_{p1}R_L}{[R_S(g_{mn1} + g_{mp1}) + 1][R_{n1}R_{p1}g_{mc1} + R_{n1} + R_{p1}]} \quad (3)$$

where g_{ma2} is the transconductance of the auxiliary transistor.

C. Noise Analysis

According to the operation of noise cancellation methods, if the amplification of the two noise paths leading to the output is the same, the noise of the desired elements can be eliminated. Considering the selection of the right transconductances for CG transistors to establish input impedance matching, the optimal dimensions of CS transistors are directly effective in eliminating the noise of CG transistors. Considering that the cascode transistors and the output resistors are designed in the same way due to establishing symmetry in the output, to create the same gain in the CS branch, it is necessary to adjust the resistors R_{p2} and R_{n2}. The noise figure (NF) of the balun-LNA authorizing input impedance matching using equation (1), is calculated with simplification, according to the following equation:

$$NF = 1 + \frac{\gamma}{2}\left[\frac{1 - R_S^2(g_{mn2} + g_{mp2})^2}{(1 + g_{mp1}R_S)}\right]^2 + \frac{2\gamma(g_{mn2} + g_{mp2})}{R_S^3}$$
$$+ \frac{4\gamma g_{mc1}}{R_S^3}\left[1 + \frac{R_{n1} + R_{p1}}{g_{mc1}R_{n1}R_{p1}}\right]^2 + \frac{4R_S(R_{n1} + R_{p1})}{R_{n1}R_{p1}g_{mc1}^2} \quad (4)$$
$$+ 4R_S R_L\left[1 + \frac{R_{n1} + R_{p1}}{g_{mc1}R_{n1}R_{p1}}\right]^2$$

979-8-3503-6020-2/23 $31.00 © 2023 IEEE   161

The second and third term of (4) stands for the thermal noise of M_{n1}, M_{p1}, M_{n2}, and M_{p2} respectively. The fourth term depicted the M_{c1} and M_{c2} thermal noise and the rest of the equation shows the thermal noise of the resistors.

As mentioned before, noise cancellation technique must implement on CS transistors. the condition to satisfy the criteria shows in (5) which improves NF by 2 dB.

$$1 - R_S(g_{mn2} + g_{mp2}) = 0 \qquad (5)$$

D. Linearity Analysis

The linearization technique used in this article is a mixture of derivative-superposition and post-distortion. In the conventional derivative-superposition technique, the auxiliary transistors are located in the weak-inversion region and are parallel to the main transistors, which will have a significant effect in matching the input impedance, and also in the conventional post-distortion technique, the auxiliary transistors are parallel to the output cascode transistors, which are They will not directly affect the matching of the input impedance, but due to being in the strong inversion region, they consume high power. According to the above points, in the proposed technique, auxiliary transistors are used in parallel with cascode transistors in the weak inversion region, which will not only have no effect on input impedance matching, but their power consumption will be negligible. In addition, the negative feedback technique has been used to create a loop to eliminate the third-order coefficients of CS transistors in order to increase the linearity of CS transistors. The CS stage linearity of the balun-LNA calculated according to the following equations:

$$V_{in} - V_1 = R_S(g_{mn1} + g_{mp1})V_1 \qquad (6)$$

$$I_{out-} = I_{mc2} - I_{ma1} \qquad (7)$$

$$-I_{mc2} = I_{mn2} + I_{mp2} \,\square\, (g_{mn2} + g_{mp2})V_1$$
$$+ (g'_{mn2} + g'_{mp2})V_1^2 + (g''_{mn2} + g''_{mp2})V_1^3 \qquad (8)$$

Equations (6) and (7) attains by Kirchhoff's voltage and current law, and putting Volterra series to use results relation (8) to accomplish fundamental, second-order, and third-order coefficients of output current an follows.

$$I_{out-, fund} = -\frac{(g_{mn2} + g_{mp2})(g_{mc2} + g_{ma1})R_{n2}R_{p2}}{2(g_{mc2}R_{n2}R_{p2} + R_{n2} + R_{p2})}V_{in} \qquad (9)$$

$$I_{out-, 2nd} = -(g'_{mn2} + g'_{mp2})(g_{mc1}R_{n1}R_{p1} + R_{n1} + R_{p1})^2$$
$$\times \frac{g'_{ma1}[(g_{mn2} + g_{mp2})R_{n2}R_{p2}]^2}{4(g_{mc2}R_{n2}R_{p2} + R_{n2} + R_{p2})^2}V_{in}^2 \qquad (10)$$

$$I_{out-, 3rd} = -(g''_{mn2} + g''_{mp2})(g_{mc1}R_{n1}R_{p1} + R_{n1} + R_{p1})^3$$
$$\times \frac{g''_{ma1}[(g_{mn2} + g_{mp2})R_{n2}R_{p2}]^3}{8(g_{mc2}R_{n2}R_{p2} + R_{n2} + R_{p2})^3}V_{in}^3 \qquad (11)$$

The fundamental relation depicts voltage gain of the balun-LNA which gives in equation (9). From (10) and (11), the second and third non-linearity factors appear that leads to calculation of the non-linearity attenuation by reaching

The following relations at the same time.

$$(g'_{mn2} + g'_{mp2})g_{mc1}^2 + g'_{ma1}(g_{mn2} + g_{mp2})^2 = 0 \qquad (12)$$

$$(g''_{mn2} + g''_{mp2})g_{mc1}^3 + g''_{ma1}(g_{mn2} + g_{mp2})^3 = 0 \qquad (13)$$

According to the relation (13), by biasing auxiliary transistors in weak-inversion region, g''_{ma1} will be positive while g''_{mn2} and g''_{mp2} are negative. As a result, the IIP3 improvement is achievable due to reducing third-order intermodulation terms.

As equation (9) shows, the fundamental relation stands for balun-LNA gain, and it shows that the auxiliary transistors increase the gain by 1 dB. According to the simulation results, which are presented in the next section, the proposed linearization techniques improve IIP3 of the proposed LNA by 7 dB with only 130 μA additional current consumption.

III. SIMULATION RESULTS

The proposed balun-LNA circuit is designed for wideband applications at the operating frequency of 0.3-5 GHz. The aspect ratios of the LNA transistors are shown in Fig. 1. The circuit was designed in Spectre-RF using 65 nm RF-CMOS technology with a 1.2 V power supply and metal-insulator-metal (MIM) capacitors and poly resistors are used for the circuit structure. The circuit structure provides this capability to establish input impedance matching without the need for passive inductors. The main transistors of the circuit use the complementary CG-CS structure to provide local noise cancellation for the CG transistors in addition to creating second-order linearity. By choosing the optimal dimensions for the CG transistors, the input impedance matching, as well as suitable S_{11} less than -11 dB over the entire bandwidth is obtained shown in Fig. 2.

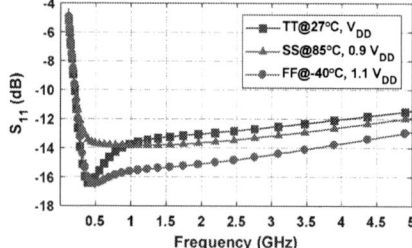

Fig. 2. Simulated S_{11} of the proposed LNA over -1dB gain bandwidth.

In addition to establishing an inherent balun, the path of CS transistors can eliminate the noise of CG transistors by 2 dB, and optimal NF is obtained by choosing their dimensions optimally shows in Fig. 3.

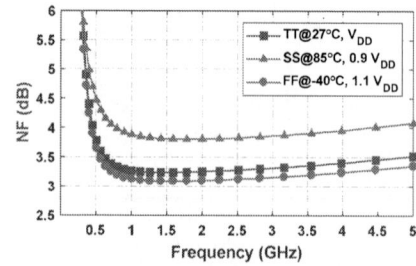

Fig. 3. Simulated NF of the proposed LNA over -1dB gain bandwidth.

The auxiliary transistors in the weak-inversion region are biased with the gate-source voltage to have little power consumption which improves IIP3 by 4 dB. in addition to weakening the third-order intermodulation, the negative feedback path is from the CS transistors to the input capacitor improves IIP3 by 3 dB, which has the task of establishing the third-order coefficient attenuation loop of the CS transistors, which have adopted small dimensions so that, in addition to the low power consumption, they do not disturb the matching of the input impedance. The IIP2 and IIP3 of the proposed LNA illustrate in Fig. 4 and Fig. 5 respectively at typical corner case. The overall gain of the balun-LNA shown in Fig.6. Wide-swing cascode constant-gm structure drives the LNA shown in Fig. 7.

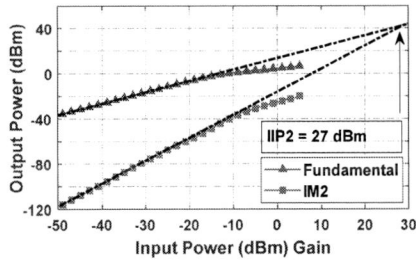

Fig. 4. Simulated IIP2 considering elements mismatches in typical corner.

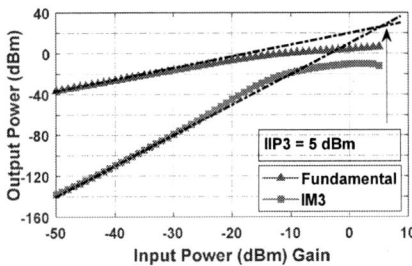

Fig. 5. Simulated IIP3 considering in typical corner.

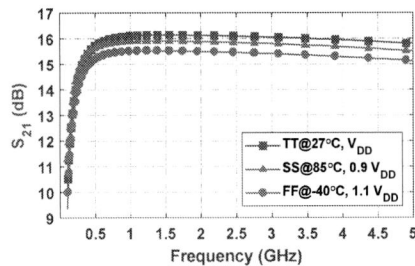

Fig. 6. Simulated S_{21} of the proposed LNA over -1dB gain bandwidth.

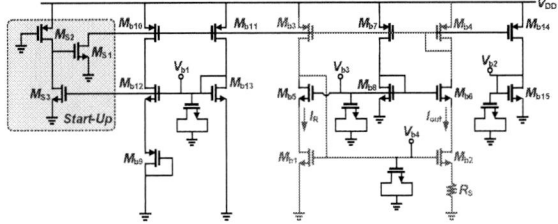

Fig. 7. Proposed constant-Gm bias circuit.

TABLE I: PVT SIMULATION RESULTS

Parameter	TT @ 27°C, V_{DD}	SS @ 85°C, 0.9 V_{DD}	FF @-40°C, 1.1V_{DD}
BW$_{-3dB}$ (GHz)	0.15-10	0.16-9.5	0.14-12
S$_{21}$ (dB)	16.2	15.8	15.5
NF$_{min}$ (dB)	3.32	3.80	3.13
IIP3 (dBm)	+5	+2	+1
IIP2 (dBm)	28	24	30
Power (mW)	2.97	2.5	3.6

Table I summarized the simulation results of the proposed balun-LNA in PVT conditions and it shows the circuit is well designed against PVT variations. To test the stability of the circuit from the Rollet stability factor (K-factor) given in (14). Considering that, for stability of the circuit, K-factor should be greater than 1, and Δ smaller than 1 in the operating frequency of the circuit, the simulation of these parameters has been done up to the frequency of 20 GHz, which indicates the unconditional stability of the circuit shows in Fig. 8.

$$K = \frac{1 + |\Delta|^2 - |S_{11}|^2 - |S_{22}|^2}{2|S_{21}||S_{12}|} \quad (14)$$

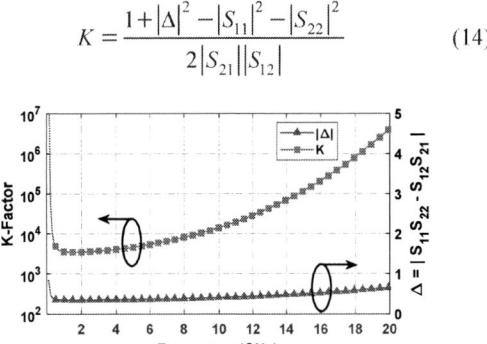

Fig. 8. Simulated the Rollet stability factor (K-factor) and Δ.

Fig. 9(a), Fig. 9(b) Illustrates the Monte Carlo simulation results of IIP3 before and after the linearization technique implementation with the main RF tone at 2 GHz and 10 MHz spacing. The figures indicate the proposed idea improves the IIP3 by the amount of 7 dB. The IIP3 versus the input signal frequency with 10 MHz spacing and also the Monte Carlo simulation results of IIP2 shows in Fig. 9(c) and Fig. 9(d) respectively.

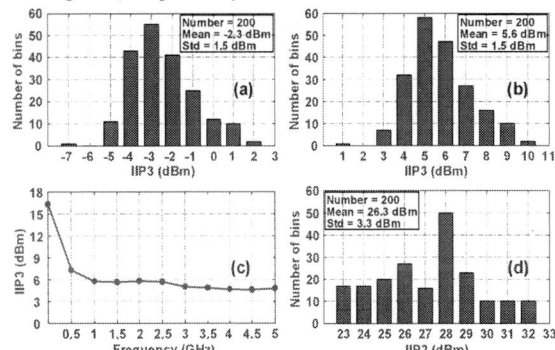

Fig. 9. IIP3 Monte Carlo simulation (a) before and (b) after using the linearity enhancement technique, (c) IIP3 versus the input signal frequency with 10 MHz spacing, (d) IIP2 Monte Carlo simulation results.

Table II represents a comparison between the proposed balun-LNA and some of recent wideband balun-LNAs. As it seems, the proposed structure reaches higher linearity with lower power consumption and yet low NF with extensive

The 5th Iranian International Conference on Microelectronics (IICM 2023)

TABLE II: PERFORMACE COMPARISON WITH SEVERAL PREVIOUS WIDEBAND CMOS BALUN-LNA

Reference	CMOS Process	BW (GHz)	S11 (dB)	S21 (dB)	NF (dB)	IIP3 (dBm)	IIP2 (dBm)	Power (mW)	V_{DD} (V)	FoM (dB)
Mejo'15 [2][b]	90 nm	2.4-10.4	<-11.2	9.5	3.5	+13.1	+42.8	14.8	1.2	18.99
TCAS-I'19 [5][#]	65 nm	0.05-1	<-10	24-30	2.3-3.3	-1.7	20	19.8	2.2	16.67
ICEE'20 [7][b]	65 nm	0.65-8.5	<-13	17	2.75-3.5	-1.358	-	7.4	1	16.43
AICSP'20 [8][#]	180 nm	0.1-6	<-10	15.5	3-4	+1.5	-	14	1.8	13.27
TCAS-I'20 [11][#]	65 nm	0.05-2.1	-	24-27.5	2.3-3	-2.2	+19.6	5.7	1	22.42
TCAS-I'21 [12][#]	180 nm	0.1-3.1	-	7.2	2.5-3.4	+17.8	-	23.9	1.8	17.08
TCAS-II'21 [13][#]	65 nm	0.7-2.2	-	21.9-26.8	2.9-3.8	+11.4	-	15.1	1	28.4
TMTT'22 [14][b]	180 nm	0.13-0.93	<-10	16.6-19.6	3.6-5	-8.5	+12	3	1.8	4.25
TCAS-I'22 [15][#]	130 nm	0.05-0.7	-	26.5	<4.25	-11.5	-	33.6	1.2	-3.93
TR'22 [16][#]	180 nm	0.065-0.84	<-10	12.5-15.5	3.2-5	+8.3	+43	6.4	1.8	14.29
TCAS-II'22 [17][b]	65 nm	0.47-3.3	<-10	19.45-22	2.57-3.5	+2.81	29.27	12.5	1.5	19.29
TCAS-II'23 [18][#]	22 nm	2.1-5.2	<-10	12.7-15.7	1.46-1.96	-	-	16	1	-
This Work *[b]	65 nm	0.3-5	<-11	15.4-16.1	3.25-3.7	+5	+28	2.97	1.2	22.89

*-1 dB gain bandwidth   [b]Simulation Results   [#]Measurement results

bandwidth. The following figure of merit (FoM) defines in [17] represents a better comparison which used in Table II.

$$FoM(dB) = 10 \times \log \frac{S_{21}(abs) \times IIP3(mW) \times BW(GHz)}{(F-1)(abs) \times P_{DC}(mW)} \quad (15)$$

In above equation S_{21} is the maximum magnitude of the power gain, BW is the bandwidth of the LNA, F is the magnitude of the minimum NF and P_{DC} is the LNA's power consumption.

IV. CONCLUSION

In this brief, new structure of a low-power balun-LNA employing linearity techniques introduced in 65 nm CMOS technology. The linearization technique not only improves IIP3, but also increase voltage gain with consuming negligible amount of power. Simulation results shows that the circuit operates in 0.3-5 GHz -1 dB bandwidth with, a maximum S_{21} of 16.1 dB, minimum NF of 3.25 dB, IIP3 of +5 dBm, IIP2 of +27 dBm, and a S_{21} less than -11 dB over the entire bandwidth.

REFERENCES

[1] H. Zhang and E. Sánchez-Sinencio, "Linearization Techniques for CMOS Low Noise Amplifiers: A Tutorial," *IEEE Transactions on Circuits and Systems I: Regular Papers*, vol. 58, no. 1, pp. 22-36, Jan. 2011.

[2] B. Mazhabjafari and M. Yavari, "A UWB CMOS Low-Noise Amplifier with Noise Reduction and Linearity Improvement Techniques," *Microelectronics Journal*, vol.46, no.2, pp. 198-206, Feb. 2015.

[3] M. Asghari and M. Yavari, "Using the Gate–Bulk Interaction and a Fundamental Current Injection to Attenuate IM3 and IM2 Currents in RF Transconductors," *IEEE Transactions on Very Large Scale Integration (VLSI) Systems*, vol. 24, no. 1, pp. 223-232, Jan. 2016.

[4] P. Solati and M. Yavari, "A Wideband High Linearity and Low-Noise CMOS Active Mixer Using the Derivative Superposition and Noise Cancellation Techniques," *Circuits Syst and Signal Process*,38, pp. 2910-2930, Jan. 2019.

[5] S. Kim and K. Kwon, "A 50-MHz–1-GHz 2.3-dB NF Noise-Cancelling Balun-LNA Employing a Modified Current-Bleeding Technique and Balanced Loads," *IEEE Transactions on Circuits and Systems I: Regular Papers*, vol. 66, no. 2, pp. 546-554, Feb. 2019.

[6] M. Yaghoobi, M. Yavari, M. H. Kashani, H. Ghafoorifard and S. Mirabbasi, "A 55–64-GHz Low-Power Small-Area LNA in 65-nm CMOS With 3.8-dB Average NF and ~12.8-dB Power Gain," *IEEE Microwave and Wireless Components Letters*, vol. 29, no. 2, pp. 128-130, Feb. 2019.

[7] B. Shirmohammadi and M. Yavari, "A Low Power Wideband Balun-LNA Employing Local Feedback, Modified Current-Bleeding Technique, and Balanced Loads," *2020 28th Iranian Conference on Electrical Engineering (ICEE)*, Tabriz, Aug. 2020.

[8] S Han, Tingting, Zhiqun Li, and Mi Tian. "An inductor-less CMOS broadband balun gm-boosting LNA exploiting noise cancellation techniques," *Analog Integrated Circuits and Signal Processing*, vol 104, pp.121-129. Jun. 2020.

[9] B. Guo, D. Prevedelli, R. Castello and D. Manstretta, "A 0.08mm2 1-6.2 GHz Receiver Front-End with Inverter-Based Shunt-Feedback Balun-LNA," *2020 IEEE Radio Frequency Integrated Circuits Symposium (RFIC)*, Los Angeles, CA, USA, pp. 379-382. Aug. 2020.

[10] M. Yaghoobi, M. H. Kashani, M. Yavari and S. Mirabbasi, "A 56-to-66 GHz CMOS Low-Power Phased-Array Receiver Front-End With Hybrid Phase Shifting Scheme," *IEEE Transactions on Circuits and Systems I: Regular Papers*, vol. 67, no. 11, pp. 4002-4014, Nov. 2020.

[11] S. Kim and K. Kwon, "Broadband Balun-LNA Employing Local Feedback gm-Boosting Technique and Balanced Loads for Low-Power Low-Voltage Applications," *IEEE Transactions on Circuits and Systems I: Regular Papers*, vol. 67, no. 12, pp. 4631-4640, Dec. 2020.

[12] B. Guo, J. Gong and Y. Wang, "A Wideband Differential Linear Low-Noise Transconductance Amplifier With Active-Combiner Feedback in Complementary MGTR Configurations," *IEEE Transactions on Circuits and Systems I: Regular Papers*, vol. 68, no. 1, pp. 224-237, Jan. 2021.

[13] S. Tiwari and J. Mukherjee, "An Inductorless Wideband Gm-Boosted Balun LNA With nMOS-pMOS Configuration and Capacitively Coupled Loads for Sub-GHz IoT Applications," *IEEE Transactions on Circuits and Systems II: Express Briefs*, vol. 68, no. 10, pp. 3204-3208, Oct. 2021.

[14] D. Shin, K. Lee and K. Kwon, "A Blocker-Tolerant Receiver Front End Employing Dual-Band N-Path Balun-LNA for 5G New Radio Cellular Applications," *IEEE Transactions on Microwave Theory and Techniques*, vol. 70, no. 3, pp. 1715-1724, March. 2022.

[15] D. Kim and D. Im, "A Reconfigurable Balun-LNA and Tunable Filter With Frequency-Optimized Harmonic Rejection for Sub-GHz and 2.4 GHz IoT Receivers," *IEEE Transactions on Circuits and Systems I: Regular Papers*, vol. 69, no. 8, pp. 3164-3176, Aug. 2022.

[16] S. Tiwari and J. Mukherjee, "An Inductorless Low Power Wideband Push-Pull Balun LNA with Dual Stage Noise Cancelation andIIP2/IIP3 Linearity Modes for Sub-GHz Wireless Applications," TechRxiv, May. 2022.

[17] B. Shirmohammadi and M. Yavari, "A Linear Wideband CMOS Balun-LNA With Balanced Loads," *IEEE Transactions on Circuits and Systems II: Express Briefs*, vol. 69, no. 3, pp. 754-758, March 2022.

[18] X. Wang, Z. Li and Z. Li, "A 1.46-1.96dB-NF 2.1-5.2-GHz Wideband Passive Balun LNA in 22-nm CMOS," *IEEE Transactions on Circuits and Systems II: Express Briefs*. May. 2023.

Optimization of 6.5 GHz CMOS Low Noise Amplifier Applying Multi-objective Firefly Algorithm

Maryam Babasafari
Department of Electrical Engineering
University of Zanjan
Zanjan, Iran
Mbabasafari@znu.ac.ir

Mostafa Yargholi
Department of Electrical Engineering
University of Zanjan
Zanjan, Iran
Yargholi@znu.ac.ir

Mohammad Mostafavi
Department of Electrical Engineering
University of Zanjan
Zanjan, Iran
mmostafavi@znu.ac.ir

Abstract— **This paper demonstrates an optimal low-noise amplifier (LNA) design with applying Firefly Algorithm (FA). To optimize the noise figure (NF) and gain (similar to the input and output matching and linearity) the FA is charity, although validating entirely the design constraints. Subsequently, there are five objectives for optimizing; they may treat as multi-objective optimization. Weighted sum access is accustomed to alter these purposes to a single objective function. The Weights are set consistent through the objective function precedence. The recommended LNA has a cascode manufacture with an inductive source degeneration circuit for usage at 6.5 GHz frequency; it is exploited to implement in ADS software at the UMC 0.18 μm CMOS technology through a 1.8 V supply voltage. The considered LNA has simulated assessments: Input reflection coefficient (S_{11}) of -21dB, Output reflection coefficient (S_{22}) of -14 dB, Voltage gain (S_{21}) of 19.74 dB, noise figure (NF) of 1.42 dB, and IIP3 of -3 dBm at 6.5 GHz. The optimized LNA parameter assessment utilizing FA, at what time simulated in MATLAB situation, is instituted to be -26.46 dB, -18.41 dB, 20.51 dB, 1.29 dB and 0.325 dBm for S_{11}, S_{22}, S_{21}, NF, and IIP3, respectively. The FA execution in optimizing the LNA parameters is also associated through the execution of comparable fashionable algorithms similar particle swarm optimization (PSO) and cuckoo search algorithm (CSA). The consequences completely verify the FA's superiority ended additional approaches in expressions of its computational consistency and proficiency.**

Keywords— *Firefly Algorithm (FA), Low Noise Amplifier, Multi-objective Function*

I. INTRODUCTION

Low-noise amplifier (LNA) productions a crucial function in radio frequency (RF) course intend. It is the main constituent that is employed next to the radio receiver front-end accustomed to strengthen the weak signals taken through the antenna [1]. This process sets a convinced gain necessity toward the LNA. The predictable signal must have a influenced signal to noise ratio (SNR) to let appropriate revealing. Afterward, the further noise thru the LNA circuit must be diminished to the similar extent possible [2]. A bulky interfering signal or blocker may take place at the LNA input. The units must be properly linear to have a logical signal response. For transportable uses, reasonable power dissipation is an unconventional restraint. Up till now, published CMOS-based wideband amplifiers have been employed in some topologies [3]. It is required to preserve the accurate balance among some parameters similar to Gain,

NF, power consumption, and linearity [4]. From this time, the substance of circuit demonstration and sizing become a supposed bring out the issue for the researchers in the current time. So the LNA demonstration and sizing have come to be research an active area to the extent that the analog VLSI technology development is concerned [5]. In the existing technology, space restriction and saving subjects are the researcher's chief worries. So as to encounter these necessities, a number of useful trials are accomplished via creators utilizing simulations and computer project software. Tunable parameters guide tuning is a time-consuming occupation, consequently, an knowledgeable project engineer is needed to organize the equivalent. Observance these in attention, in the past few periods, some assistants have attempted to improve the LNA parameters by applying assorted optimization methods. Many classical optimization methods, for instance, gradient search methods, quadratic programming, linear programming [6], Newton's method, and Nonlinear programming [7], are employed to resolve compound engineering difficulties. These algorithms accustomed the gradient information toward searching the solutions nearby key points to have a greater propensity to trick toward the confined optima inside. Besides, these algorithms are accomplished to resolve merely constant difficulties. The alternative algorithm established upon magnetic field [8] is suggested that electrical and magnetic forces are considered through the set appreciated information aim for the optimization procedure. To the extent that the engineering difficulties are anxious, the greatest of them are greatly non-smooth, non-convex, non-linear, or discrete difficulties. Consequently as an unconventional to this, researchers have fulfilled the difficulties of meta-heuristic optimization methods. These algorithms styles importance are that they do not necessity the copied of the objective purposes and restrictions related through them. Overhead that, they put away less time to converge and make available a capable explanation. Genetic algorithm [9], Local search [10], particle swarm optimization (PSO) [11], human behavior particle swarm optimization (HB-PSO) [12], cuckoo search algorithm (CSA) [13], ant colony algorithm (ACA) [14], backtracking search optimization [15], simulated annealing (SA) [16], gravitational search algorithm (GSA) [17], bacterial foraging [18] are various of the samples of frequently accustomed meta-heuristic algorithms which are mainly employed to resolve dissimilar engineering difficulties. In current years, assistants have been tested to optimize LNA parameters through nature-inspired algorithms. To improvement the converging procedure to a universal maximum, a multi-unit algorithm is announced [19] according to the direct universal optimization method. Optimization challenging of LNA parameters is the greatest regularly confronted optimization difficulty in electrical and as fine electronics engineering. It is a multi-objective optimization tricky, where some inconsistent objectives (Gain, NF, and linearity) are to be improved concurrently. Multi-objective optimization arranges for several achievable explanations; consequently, a profound experience side by side is requisite to choice the greatest transaction. In this paper, a favorable algorithm as of the meta-heuristic family named Firefly Algorithm (FA) is practical to resolve numerous objectives related through the LNA project. Subsequently quite a lot of objectives, for instance NF, Gain, linearity, and input and output matching [20-21], are to be optimized concurrently, the tricky may be preserved as a multi-objective optimization difficulty. At this time, the weighted sum method is accustomed to alter multiple-objective purposes toward the inside of a single objective purpose. According to the skill, the greatest solution is selected and stated as the give-in solution. The suggested effort is simulated in MATLAB version 2014b and ADS Spectre, UMC, 0.18 µm technology. This manuscript residue is prearranged as follows: Section II defines the suggested inductive source degeneration LNA. Section III describes the firefly procedure. In Section IV, consequences established from LNA utilizing the optimization methods with PSO, CSA, and FA which converse along with the assessment of the three optimization methods. In Section V, several consequences get from considered LNA simulated at 6.5GHz frequency, and as a final point, the conclusion of paper is set in Section VI.

II. CMOS LNA SCHEME

For optimization obtainable of the circuit in Fig. 1 is the considered LNA which working frequency is 6.5 GHz, where L_g, L_s, and L_d are gate, source, and drain inductors, respectively [22]. They are entirely on-chip spiral inductors. Inductors usage in the considered LNA is set as follows: (a). The input capacitance values are regulated out utilizing L_g. (b). Input matching is obtained through applying L_s. (c). L_d is operated to have output resonance by output capacitance, to have an LNA with significant gain [23]. M_1 is an input device, and M_2 is the cascoding device accustomed to arrange for isolation among regulated output and regulated input. Moreover, M_2 decreases the consequence of C_{gd} of M_1, which may be an actual problem [24]. In designing any LNA, the first phase is to allocate the assessments of M_1 and C_{gs}, in view of both noise toward the inside account and bias current [23-25]. The wanted input resistance may be became by the right choice the assessments of L_g and L_s. M_3 and R_1 form a bias circuit somewhere M_3 fundamentally forms a current mirror through M_1 [26], and its width is prepared nearby one-tenth of the M_1 width to minimize the bias circuit power dissipation. Capacitors C_{in} and C_{out} are accustomed to block DC voltages. In this paper, we optimized inductive source degeneration LNA's constructions in terms of the gain, NF, linearity, and corresponding with conveying the appropriate weights. For this LNA development, the M_1 size is 245 µm and the M_2 size is selected a reduced amount of than the M_1 (229 µm) to have the finest isolation.

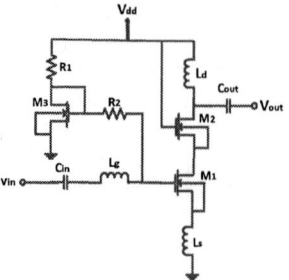

Fig. 1. General schematic of the LNA

III. FIREFLY ALGORITHM

FA is advanced via Xin-She Yang that is established upon the flashing activities and the food searching fireflies' progression through the night. Afterward meticulous fireflies-investigation, Yang detected that throughout the food looking for a course, a collection of fireflies guide themselves headed for the firefly, which has supplementary bright. This course aids in abbreviation the expanse flanked by two fireflies. This was the main detection understood through Xin-She Yang, that aids him in evolving the precise typical of informing the firefly locations . The flashing light released via firefly is the means of passing along among them. There are approximately two thousands firefly species accessible, and the flash arrangement is dissimilar for the distinct species. They mostly release two kinds of light, the first to represent of the obtainability of food source and the second for the coupling aim. FA has extended abundant admiration as a result of its simple engagement course and parameter capability. The real demonstration of the manners of any expected species in the typical mathematical direction is a hard duty; therefore, for simplicity, Yang prototypes three main orders for the fireflies' development [27]. 1. Altogether, fireflies are supposed to be unisexual. 2. Brightness and appeal are completed straightly relational. From now, a minimum of bright fireflies will be enchanted to the brighter ones. Correspondingly, distance and brightness are inside out relational. Firefly will transfer casually if none of them set up brighter. 3. A firefly brightness is the objective purpose suitability; consequently, for a maximization difficult, the brightness may merely be relational to the objective purpose and reverse assessment. Exploiting the above-stated instructions, the grade of appeal β of the FA can be stated as:

$$\beta = \beta' e^{-\gamma r^2} \qquad (1)$$

So that r denotes the distance among two fireflies, and γ is the light preoccupation coefficient. At $r = 0$, the appeal grade is determined by β'. A brighter firefly incomes the firefly has more intensity (I) than the less bright firefly, from now for any maximum optimization difficulties [19], A firefly brightness I at a specific position x can be selected as $I(x) \alpha F(x)$. It can correspondingly be supposed that the appeal grade acquires deviated through the preoccupation from its source, and at the similar time, some light is captivated through the media. In an easy procedure, it can be definite that light intensity $I(r)$ differs from the inverse-square instruction, and can be denoted as:

$$I(r) = \frac{I'}{r^2} \qquad (2)$$

So that I' signifies the intensity by the source side. When an average has a secure light preoccupation coefficient γ, then the intensity of light (I) deviates through distance (r), which is stated as:

$$I = I'e^{-\gamma r} \tag{3}$$

The distance within the two fireflies i and j with their locations X_i and X_j, respectively, may be exemplified as:

$$r_{ij} = |X_i - X_j| = \sqrt{\sum_{k=1}^{d}(X_{i,k} - X_{j,k})^2} \tag{4}$$

In (4), $X_{i,k}$ and $X_{j,k}$ are the kth constituents of X_i and X_j matches of ith and jth fireflies, respectively. In FA, the more fascinating or brighter firefly (j) has the power to appeal to the gesture of alternative firefly (i), and is distinct by:

$$X_i = X_i + \beta_0 e^{-\gamma r_{ij}^2}(X_j - Xi) + \alpha\left(rand - \frac{1}{2}\right) \tag{5}$$

So that the current assessment is indicated with X_i. The term $\beta_0 e^{-\gamma r_{ij}^2}(X_j - Xi)$ is because of the fascinating, whereas the last term $\alpha\left(rand - \frac{1}{2}\right)$ happens as a result of the random choice of a trial from a people through α engaged as a randomization adjustable. At this time, rand is accidental figure of originators which are uniformly circulated inside [0, 1]. Mostly, in this execution, β_0 is occupied as 1 and $\alpha \in [0, 1]$. An essential graphical demonstration of the FA is shown in Fig. 2. Actual optimizing difficulties majority are multi-objective in countryside. That revenues that several objectives necessity be calculated with the equal time. There are two manners to resolve multi-objective optimizing difficulties. The primary choice is that the multi-objective are composed, and then allocated weights to them to alter the difficulty toward a mono-objective one. In this situation, some non-negative weights will be allocated to every objective, and objectives through upper precedence will be specified with advanced weight [28]. An alternative way is expending non-dominated method. In non-dominated method, every objective is optimized independently to get a collection of non-dominated solutions named Pareto optimal explanation [29]. Multi-objective FA usages a way particular kind to optimize multi-objective difficulties. The optimization complicated shape of multi-objective is defined as [29]:

$$Minimize(f_1(x), f_2(x), f_3(x), ..., f_i(x))$$
$$Subject$$
$$to \tag{6}$$
$$h_k(x) \le 0, k = 1, ..., l$$
$$where,$$
$$x = (x_1, x_2, x_3, ..., x_n) \in x,$$
$$where,$$
$$x \subset \Re^n$$
$$f_n(x_i)\Re^n \to \Re$$
$$n = 1, ..., P(P \ge 2)$$
$$h_k(x) \le 0, k = 1, ..., l$$

So that $x \subset \Re^n$ is the choice space for the variables, $f_n(x_i)\Re^n \to \Re$, $n = 1, ..., P(P \ge 2)$ are objective functions, and $h_k(x) \le 0, k = 1, ..., l$ are act restraints. There are different methods to optimize multi-objective difficulties, and scalarization way is one of them. Scalarization methods are techniques that scalarize a multi-objective difficult to a single-objective difficult. In this paper, the weighted sum technique is accustomed to scalarization. The weighted sum technique allocates a positive weight to every objective, and typically the weights sum up to one.

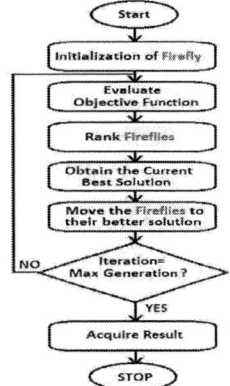

Fig. 2. The FA illustration [30]

The weighted sum technique mathematical description is shown in (7):

$$Minimize$$
$$F(x) = \sum_{m=1}^{M}\omega_m f_m(x) \tag{7}$$
$$Subject$$
$$to$$
$$h_k(x) \le 0, k = 1, ..., l$$
$$where$$
$$\sum_{m=1}^{M}w_m = 1$$

So that, M is the whole objective function number. The whole objective function in the weighted sum method may be definite as:

$$\tag{8}$$
$$F_{obj} = w_1 * F_1 + w_2 * F_2 + w_3 * F_3 + ... + w_M * F_M$$

where, F_{obj} is the net objective function to be optimized. $F_1, F_2, ...,$ and F_M are the separate objective functions to be optimized concurrently. $w_1, w_2, ...,$ and w_M are the altered weights related through the objective functions.

TABLE I.   OPTIMIZATION PARAMETER SETTINGS

Factors	Alpha (α)	Beta (β)	Gama (γ)	Pop Size	Max iteration
Assessment accustomed	0.8	0.2	1	50	500

IV. RESULTS AND CONVERSATION

To optimize numerous LNA parameters, optimization algorithms are applied well. In this paper, LNA module with five objectives is optimized all together via exchanging them in the single objective function direction. The weighted sum method is accustomed to exchanging the different objectives in a single objective direction. Consequences are moreover associated with famous state-of-the-art approaches as PSO and CSA. Good input and output matching, high gain, low NF, and high linearity are the most important LNA project objectives. In other words, the gain should be high, and the NF must be minimum, so that there is adequate gain to decrease the noise from the following sequence of constituents. Also the LNA must have the corresponding network, thus it has insignificant reflections among first and last phase [29]. So, it is required to have a minimum of $S_{11}(dB)$ and $S_{22}(dB)$. In this work, study contains of five assignment variables, and the aim is the $S_{11}(dB)$, $S_{22}(dB)$, $S_{21}(dB)$, NF, and IIP3 optimization.

TABLE II. AMONG ASSOCIATION THROUGH ALGORITHM RESULTS FROM PSO, CSA AND FA

Aim	PSO Optimized assessment	CSA Optimized assessment	FA Optimized assessment
S_{11}	-19.10 dB	-21 dB	-26.46 dB
S_{22}	-17.70 dB	-18 dB	-18.41 dB
S_{21}	19.80 dB	20.37 dB	20.51 dB
NF	1.54 dB	1.4 dB	1.29 dB
IIP3	0.128 dBm	0.298 dBm	0.325 dBm

TABLE III. ASSOCIATION AND THE SIMULATED CONSEQUENCES IN ADS

Aim	Optimized consequences in MATLAB by FA	Simulated consequences in ADS by FA
S_{11}	-26.46 dB	-21 dB
S_{22}	-18.41 dB	-14 dB
S_{21}	20.51 dB	19.74 dB
NF	1.29 dB	1.42 dB
IIP3	0.325 dBm	-3 dBm

The preferred limitations are L_s and L_g, which may be got utilizing the appearance for good input and output matching:

$$L_s = \frac{R_s C_{gs}}{g_m} \tag{9}$$

$$L_g = \frac{1}{\omega_0^2 C_{gs}} - L_s \tag{10}$$

So that, C_{gs}, g_m and ω_0 are the gate to source capacitance, the transconductance of the transistor M_1 and the frequency definition as $\omega_0 = 2\pi f_0$ where $f_0 = 6.5\,GHz$, respectively. As mentioned in (7), the entire weighted sum that may be assumed to be the any considered system objective functions is one. Utilizing this information, considered LNA objective functions have set some weight depending upon the importance. The optimization design of LNA parameters is shown in Table I. Table II shows among association with results achieved from PSO, CSA, and FA; besides it is established that FA exceeds in performance of PSO and CSA in calculative proficiency and strength terms. The greatest assessment presented in Table II from every one procedure is the consequence of 50 independent trial runs. Since these algorithms are established upon random initialization, variation in consequences is a clear topic. Consequences got from the FA and consequences simulated in ADS are illustrated in Table III.

V. SIMULATED RESULTS ACHIEVED BY ADS

In this section, we define the considered LNA simulated consequences for wideband implementation and simulated results in a standard through UMC 0.18 μm technology by employing the optimized project variable found from FA. The LNA through an output buffer put away 8.5 mW from a 1.8 V supply voltage for the cascode topology. Fig. 3 shows the distinction of S_{11} through concerning to frequency when simulated on the ADS tool. In this scheme, S_{11} is found to be -21 dB, which demonstrations good input matching. The output reflection coefficient is displayed in Fig. 4 where S_{22} is -14 dB, which may be considered a respectable output matching. Fig. 5 shows S_{21} of the considered LNA as 19.74

dB; and in Fig. 6, S_{12} is found to be -40.09 dB. Fig. 7 shows the NF. The NF assessment is become as 1.42 dB. As of the Fig. 8, it is displayed that the achieved IIP3 in this project is -3 dBm. Table III shows the simulated consequences of optimized LNA at 6.5 GHz frequency.

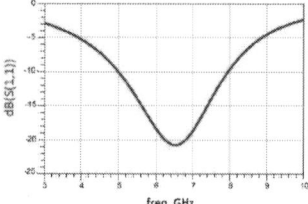

Fig. 3. The simulation result of S_{11} by FA

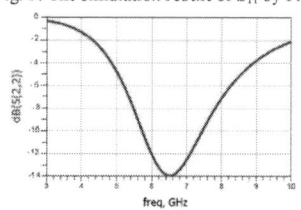

Fig. 4. The simulation result of S_{22} by FA

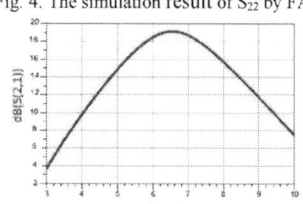

Fig. 5. The simulation result of S_{21} by FA

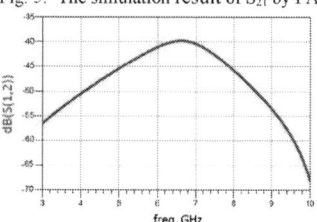

Fig. 6. The simulation result of S_{12} by FA

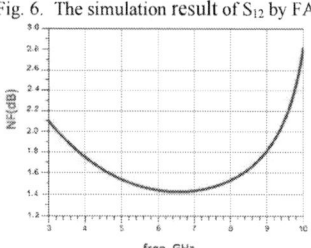

Fig. 7. The simulation result of NF by FA

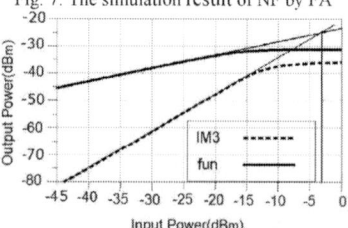

Fig. 8. The simulation result of IIP3

In Table IV gain variation for dissimilar approaches in 50 autonomous prosecutions are shown.

The 5th Iranian International Conference on Microelectronics (IICM2023)

TABLE IV.   GAIN VARIABLE WITH DIFFERENT PROSECUTIONS RUNS

	Gain in dB				Gain in dB		
	PSO	CSA	FA		PSO	CSA	FA
1	19.8	19.65	20.51	26	19.8	20.37	20.51
2	19.73	20.37	20.51	27	19.8	20.37	20.51
3	19.6	19.8	20.51	28	18.5	20.37	20.51
4	19.37	20.37	20.51	29	19.8	20.25	20.51
5	19.8	20.37	20.51	30	19.8	19.71	19.48
6	19.8	20.37	20.41	31	19.8	20.37	20.51
7	19.8	20.26	20.51	32	19.5	20.37	20.51
8	19.03	20.37	20.51	33	19.8	20.37	20.51
9	19.8	20.37	20.51	34	19.8	20.37	20.51
10	19.23	20	20.51	35	19.8	20.37	20.51
11	19.8	19.28	19.95	36	18.9	19.86	19.37
12	18.57	19.4	20.51	37	19.8	20.37	20.51
13	19.8	20.37	20.51	38	19.8	20.37	20.51
14	19.61	20.37	20.51	39	18.5	20.37	20.14
15	19.8	20.37	20.51	40	19.8	20.37	20.51
16	19.8	20.37	20.15	41	19.8	19.84	20.51
17	19.4	20.37	20.51	42	19.2	20.37	20.24
18	19.8	19.46	20.51	43	19.8	20.06	20.51
19	19.33	20.37	20.51	44	19.8	20.37	20.51
20	19.8	20.37	20.92	45	19.8	20.37	20.51
21	19.44	19.15	20.51	46	18.8	20.37	19.98
22	19.8	19.51	20.51	47	19.8	19.81	20.51
23	19.8	20.37	20.51	48	19.8	20.37	20.51
24	19.8	20.14	20.51	49	19.6	20.37	20.51
25	19.51	20.37	19.81	50	19.8	20.37	20.47

TABLE V.   ASSOCIATION OF THE SUGGESTED METHOD BY OTHER PUBLISHED CMOS UWB LNA

Ref.	[22][a]	[22][b]	[30][a]	[30][b]	[31][c]	[32][d]	[33][e]	[34][e]	Our Work [d]
Technology (µm)	0.18	0.18	0.18	0.18	0.18	0.18	0.18	0.18	0.18
Frequency (GHz)	5.5	5.5	6.5	6.5	5.5	5.5	4.4	2.14	6.5
S_{11} (dB)	-12.75	-26.35	-10.3	-25	-26.1	-36.65	-37	-11.07	-21
S_{22} (dB)	-23.46	-12	-15.82	-17.5	-23.31	-22.62	-42	-7.37	-14
S_{21} (dB)	22.37	22.84	19.371	19.84	17.19	22.15	12.52	26.01	19.74
NF (dB)	0.76	0.87	1.776	1.886	1.24	1.16	1.8	0.32	1.42
IIP_3 (dBm)	-7.2	-6.9	-8	-7.6	-	-2.6	-	-	-3
Power dissipation (mW)	10.8	10.8	10	10	10.8	2.43	-	-	8.5

[a] Case 1: Minimization of NF in Single- Objective FA Algorithm   [b] Case 2: Maximization of Gain in Single- Objective FA Algorithm
[c] Firefly Algorithm   [d] Multi-Objective FA Algorithm   [e] Particle Swarm Optimization Algorithm

As of the Table IV, it may be observed that the FA hits 39 times out of 50 prosecutions, although PSO and CSA hit 32 and 34 times to their particular finest assessment. This analysis approves the FA reliability completed PSO and CSA. The experimental run conspiracy for FA is shown in Fig. 9, and the experimental run conspiracy for different algorithms are shown in Fig. 10. To approve the schematics presentations of the considered LNAs and their ability for employment, the comprehensive suggested LNAs layouts are done utilizing of Cadence tool which are established in Fig. 11. It reside in 350.20 µm×370.27 µm silicon die area for the LNA topology. As established in Fig. 12, the peak gain Post-layout importance of is 19.85 dB. Fig. 13 shows NF simulation and Post-layout consequences. The NF of post-layout consequences is 1.57 dB. Figs 12 and 13 exhibit the good matching among simulation and Post-layout significances. Table V, sum up the act of recently printed CMOS UWB LNAs. Table VI shows the ADS simulation settings.

Fig. 9. Prosecutions run plot for the gain of FA

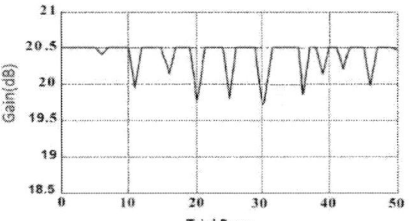

Fig. 10. Prosecutions run plot for gains of algorithms of PSO, CSA, and FA

TABLE VI.   ADS SIMULATION SETTING

Technology (μm)	VDD (V)	P_{diss} (mW)
0.18	1.8	8.5

979-8-3503-6020-2/23 $31.00 © 2023 IEEE

The 5th Iranian International Conference on Microelectronics (IICM2023)

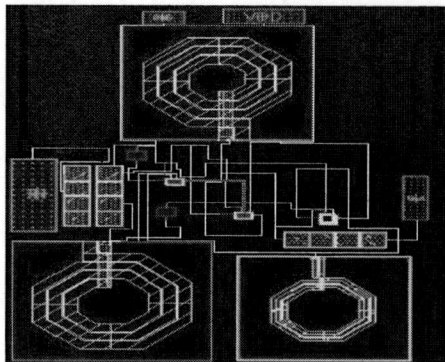

Fig. 11. The layout of the LNA by FA

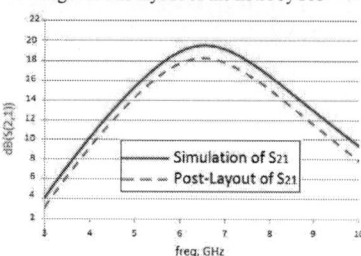

Fig. 12. Simulation and Post-layout of S_{21} by FA

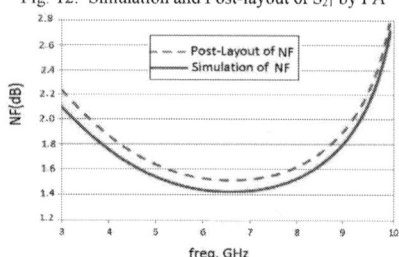

Fig. 13. Simulation and Post-layout of NF by FA

VI. CONCLUSION

In this paper, multi-objective FA has been planned and simulated for the UWB LNA for 6.5 GHz frequency to optimize the LNA parameters based on the 0.18 μm CMOS technology. Some objectives associated through LNA are optimized by applying the multi-objective optimization weighted sum method. The enhanced LNA when simulated utilizing of ADS tool, exhibits -21 dB, -14 dB, 19.74 dB, 1.42 dB, and -3 dBm for, S_{11}, S_{22}, S_{21}, NF, and IIP3, respectively. While simulated on MATLAB the consequences found are -26.46 dB, -18.41dB, 20.51 dB, 1.29 dB and 0.325 dBm for S_{11}, S_{22}, S_{21}, NF and IIP3, respectively. Results of the FA are superior than the consequences achieved through PSO and CSA. prosecutions run plot and numerical examination display in looking for the global optima that the FA persistence is superior than PSO and CSA.

REFERENCES

[1] TH . Lee, "5-GHz CMOS wireless LANs", IEEE Trans Microw Teory Tech.vol.50, pp. 268–280, 2002.

[2] M. Babasafari, M. Yargholi, and M. Mostafavi, "Design of CMOS LNAs with Low Power and Out of Band Rejection Capabilities Using the Improvement of Noise and Linearity in the UWB System, " ISC J. of Electronic Industries , vol.10, pp. 39-50, 2019.

[3] M. Babasafari, and M. Yargholi, "A low power CMOS UWB LNA with dual-band notch filter using forward body biasing, " IETE J. of research, Taylor & Francis, 2020.

[4] M. Kotti, A. Sallem, M. Bougharriou, M. Fakhfakh, and M. Loulou, "Optimizing CMOS LNA circuits through multi-objective meta heuristics, In: XIth international workshop on symbolic and numerical methods modeling and applications to circuit design, " 2010.

[5] M. Fakhfakh, Y. Cooren, A. Sallem, M. Loulou, and P. Siarry, "Analog circuit design optimization through the particle swarm optimization technique," Analog Integr Circuits Signal Process 63, pp. 71–82, 2010.

[6] DR. Jones, CD. Perttunen, and BE. Stuckman Lipschitzian, "optimization without the Lipschitz constant," J Optim Theory Appl 79, pp.157–181, 1993 .

[7] A. Kaveh, and S. Talatahari, "A novel heuristic optimization method: charged system search, " Acta Mech 213, pp. 267–289, 2010.

[8] A. Kaveh, A. Mohammad, M. Share, and M. Moslehi, " Magnetic charged system search: a new meta-heuristic algorithm for optimization," Acta Mech 224, pp. 85–107, 2013.

[9] M. Taherzadeh, R. Lotfi, H. Zare, and O. Shoaei, "Design optimization of analog integrated circuits using simulation-based genetic algorithm," Proc IEEE Int Symp Signals Circuits 1, pp.73–76, 2003.

[10] E. Aarts, and K. Lenstra, "Local search in combinatorial optimization, " Princeton University Press, Princeton, 2003.

[11] J. Kennedy, and RC. Eberhart, "Particle swarm optimization, " Proc IEEE Int Conf Neural Netw 4, pp. 1942–1948, 1995.

[12] L. Hao, X. Gang, G. Ding, and Y-B Sun, "Human Behavior-Based particle swarm optimization, " Sci World J. pp. 1–14, 2014.

[13] XS.Yang, and S. Deb, Cuckoo Search via Levy Flights, In: World congress on nature and biologically inspired computing, pp. 210-214.

[14] M. Dorigo, V. Maniezzo, and A. Colorni, "The ant system: optimization by colony of cooperating agents, " IEEE Trans Syst Man Cybern Part B26, pp. 29–41, 1996.

[15] E. Gonza´lez, O. A ´ lvarez, Y. Dı´az, C. Parra, and C. Bustacara, "BSA: a complete coverage algorithm," In: Proceedings of the IEEE international conference on robotics and automation, pp. 2040–2044, 2005.

[16] S. Kirkpatric, CD. Gelatt, and MP. Vecchi, "Optimization by simulated annealing, " J Sci 220, pp. 671–680, 1983.

[17] E. Rashedi, H. Nezamabadi-pour, S. Saryazdi, "GSA: a gravitational search algorithm, " Inform Sci 179, pp. 2232–2248, 2009.

[18] CS. Mohanty, PS. Khuntia, and D. Mitra, "A modified bacterial foraging optimized PID controller for time-delay systems, " Int J Adv Intell Paradig 6, vol. 4, pp. 255–271, 2014.

[19] SZ. Khong, D. Nesic, C. Manzie, and Y. Tan, "Multidimensional global extremum seeking via the DIRECT optimisation algorithm, " Automatica,vol. 49, pp. 1970–1978, 2013.

[20] AP Tarighat, and M Yargholi, "Wideband Input Matching CMOS Low-Noise Amplifier with Noise and Distortion Cancellation," Journal of Circuits, Systems, and Computers, , vol. 29, No. 4, 2020.

[21] AP Tarighat, and M Yargholi, "Design of Low-Noise Two-Path Amplifier for 6–16 GHz Applications," Journal of Circuits, Systems, and Computers, vol. 30, No. 8, 2021.

[22] R. Kumar, A. Rajan, FA. Talukdar, N. Dey, V. Santhi, and VE. Balas, "Optimization of 5.5-GHz CMOS LNA parameters using firefly algorithm, " Neural Comput & Applic, 2016.

[23] B. Razavi, RF microelectronics. Prentice-Hall PTR, Upper Saddle River, 1997.

[24] M. Babasafari, M. Yargholi and M. Mostafavi, "A CMOS Low Noise Amplifier for 5.6GHz with Employing Current-Reused Topology and Modified Derivative Superposition Technique, "ICEE. of Electronic 28th Conf., 2020.

[25] AP Tarighat, M Yargholi, "Low power active shunt feedback CMOS low noise amplifier for wideband wireless systems," integration, Springer, pp. 189-197, 2020.

[26] F. Ellinger, Radio frequency integrated circuits and technologies, 2nd edn. Springer, New York, 2007.

[27] XS. Yang, Firefly algorithm, Nature-inspired metaheuristic algorithms, 20. Wiley Online, Library, pp.79–90, 2008.

[28] K. Deb, " Single and multi-objective optimization using evolutionar algorithms, " In: Presented at KanGAL Report February 2004.

[29] E. Zitzler, Evolutionary algorithms for multi-objective optimization: methods and applications. PhD thesis, Swiss Federal Institute of Technology, Zurich, Switzerland, 1999.

[30] M. Babasafari, M. Yargholi, and M. Mostafavi, "Optimal Design of a 6.5 GHz Low Noise Amplifier By Single Objective Firefly Algorithm, " ISC J. of Electronic Industries , 2022.

[31] R. Kumar, A. Rajan, FA. Talukdar, N. Dey, and V. Santhi, "Quality Factor optimization of Spiral Inductor using Firefly Algorithm and its Application in Amplifier, " International Journal of Advanced Intelligence Paradigms, February 2016.

[32] R. Kumar, F.A. Talukdar, A. Rajan, A. Devi, R. Raja, " Parameter Optimization of 5.5-GHz low noise amplifier using multi-objective firefly algorithm, " Neural Comput & Applic, Springer, 2018.

[33] A. Parhizkar Tarighat, M. Yargholi, "Optimization of Current-Reused LNA with PSO Algorithm," Journal of Telecommunication, Electronic and Computer Engineering ,vol. 11, no. 2,. June 2019.

[34] M. M. Karkhanehchia, S.Naderib, F.Jafaria, S.Majidifarc, "Design and Optimization of a Very Low Noise Amplifier using Particle Swarm Optimization Technique, " International Journal of Engineering & Technology Sciences (IJETS) pp., 122-131, 2014.

Design of a High-Efficiency Deep Bias Class-AB Power Amplifier With 70% PAE at P_{1dB}

Fazel Ziraksaz
Faculty of Electrical Engineering
Shahid Beheshti University
Tehran, Iran
f_ziraksaz@sbu.ac.ir

Alireza Hassanzadeh
Faculty of Electrical Engineering
Shahid Beheshti University
Tehran, Iran
a_hassanzadeh@sbu.ac.ir

Abstract—This paper presents a high-efficiency class-AB gallium nitride (GaN) power amplifier (PA) in the frequency range of 800-900MHz. The proposed class-AB PA is deeply biased to achieve high efficiency. The suggested PA is minimized the overlap between the output current and the output voltage waveforms to reduce power loss. Also, the proposed structure is reduced radiation and interference in other bands by using series stubs instead of short circuited and open circuited stubs in the input and output matching networks. The result indicates that the adjacent channel power ratio (ACPR) is around 37dBc. At the gate and drain, $\frac{\lambda}{4}$ lines have been used instead of inductors. Simulation results of the proposed PA show that 17 dB Gain, 42.5dBm output power, 41.86dBm P_{1dB}, and 70% power-added efficiency (PAE).

Keywords— Power Amplifier (PA), Deep Bias Class-AB, Gallium Nitride (GaN), High Power-Added Efficiency (PAE), Harmonic Termination Network (HTN).

I. INTRODUCTION

Power amplifier (PA) is one of the most power-hungry building blocks in the transceivers and there is a permanent trade-off between efficiency and linearity in PA [1-5]. High power added efficiency (PAE) and high 1-dB compression point power (P_{1dB}) are important in PAs. P_{1dB} shows the range of the input power that the PA is in the linear region. High P_{1dB} is one of the key requirements of linearity. Moreover, in recent years, wireless telecommunications networks have been growing dramatically [6-9]. The design of PAs is typically complicated by the stringent linearity requirements of these new telecommunication networks.

Switching PAs have high efficiency but suffer from low bandwidth and poor linearity. One of the methods to minimize the overlap of voltage and current waveforms is to manipulate harmonics to increase efficiency. References [10-12] use harmonic manipulation in Class-F to achieve high efficiency. Harmonic termination is a simple and effective method to enhance efficiency and minimize the overlap between the current and voltage waveforms [13].

On the other hand, although the linearity and bandwidth of the linear PAs are good, efficiency in this type of PA faces challenges. The Class-B PAs, despite having better efficiency than other linear classes [5], suffer from zero-crossing distortion. The Class-C PAs are also limited by the conduction angle and do not have proper linearity and are suitable for applications such as an auxiliary amplifier in the

Doherty structure. The Class-AB PA, subject to the appropriate choice of bias, matching networks, and design, can have proper efficiency and linearity along with solving the problem of zero-crossing distortion.

Input and output matching networks play a significant role in the efficiency of the PA [3]. The open circuited and short circuited stubs may lead to radiation. Moreover, using passive elements on the board and soldering them adds parasitic elements to the circuit, which increases the losses, but the microstrip lines are made of board and have better performance.

In this paper, a deep bias class-AB power amplifier is designed with 17dB Gain, 42.5dBm output power, 41.86dBm P_{1dB}, 70.1% PAE at P_{1dB} and 73.6% maximum PAE. This work uses series stubs. In addition, this design employs a harmonic termination network to minimize the overlap between voltage and current waveforms.

The following sections are organized as follows. Section II discusses design considerations including quiescent point, harmonic termination network, matching network and stability. Section III presents the proposed power amplifier and the simulation results. Finally, the conclusion is at the end.

II. DESIGN PROCEDURE

The most desirable state for any PA is to achieve the highest efficiency at the highest P_{1dB}. In other words, the most desirable state is to achieve the highest efficiency in the linear region so that the PA can have high efficiency along with linearity.

In this work, the goal is to design a high-efficiency power amplifier at P_{1dB}. This PA must be stable in the frequency band of 800-900MHz. In this PA, a harmonic termination network (HTN) is provided to minimize the overlap between voltage and current waveforms. To achieve high efficiency, the class-AB PA is deeply biased. To prevent radiation and the addition of parasitic elements, the series stubs have been used in matching networks.

In the following sections, the design procedure is presented in detail.

979-8-3503-6020-2/23 $31.00 © 2023 IEEE

A. $\frac{\lambda}{4}$ Lines at the Gate and the Drain

This design employs $\frac{\lambda}{4}$ lines at the gate and the drain of the main PA. These lines are used to apply the bias voltage to the gate and the supply voltage to the drain of the transistor. In this design, the supply voltage is 50 (V). These lines must be able to withstand the current passing through them. Here, the width of the drain transmission line through which the supply voltage is applied is 1mm, and the width of the gate transmission line through which the bias voltage is applied is considered to be 0.6mm. The thickness of the transmission line is 18μm. The drain and gate transmission lines can withstand up to 1.5 A and 1 A, respectively. For the central frequency of 850 MHz, the lengths of the drain and gate transmission lines are 54 mm and 55 mm, respectively. Fig. 8 shows the gate and drain transmission lines.

B. Quiescent Point

In this design, the goal is to bias the PA in the linear class. Since the efficiency of the linear classes is lower than switching classes, the choice of quiescent point plays a key role in the efficiency. In Fig.1, Class-A and Class-B are shown. Class-AB is the interval between Class-A and Class-B. In this design, to achieve high efficiency while also solving the problem of zero-crossing distortion, the class-AB PA is deeply biased. In this, the conduction angle will be greater than class B and the problem of zero crossing distortion will not occur. For this purpose, the quiescent point must be considered slightly higher than class B. Considering that the bias of class B is at V_{GS} = -2.4 (V), for the deep bias of class-AB, V_{GS} = -2.2 (V) has been considered.

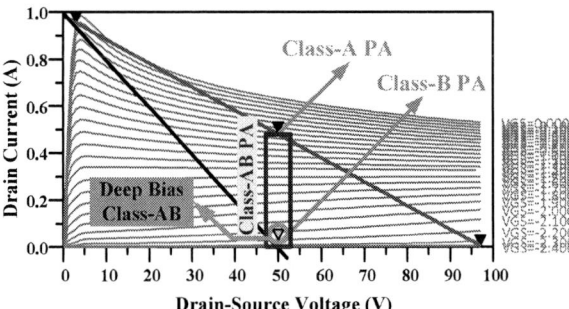

Fig. 1. Bias of the PA

C. Harmonic Termination Network (HTN)

To minimize the overlap between voltage and current waveforms, a second and fourth harmonic termination network (HTN) is used as shown in Fig.2. Therefore, power losses reduce. Fig.3 shows the current and voltage waveforms at the output.

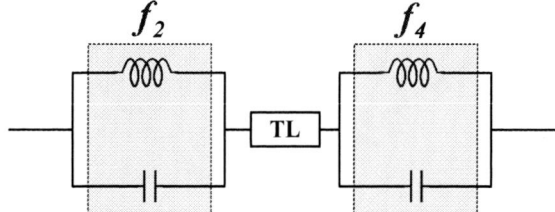

Fig. 2. Harmonic Termination Network (HTN)

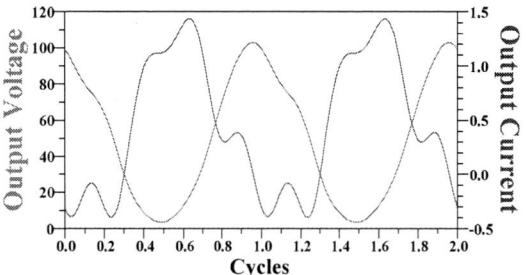

Fig. 3. Voltage and Current waveforms at the output

D. Matching Networks

Input and output matching networks play a key role in the efficiency of the PA. In this design, to prevent radiation and the use of passive elements, series stubs have been used in input and output matching networks instead of using the open circuited and short circuited stubs or inductors and capacitors. Fig.4 shows input and output matching networks. Fig.5 and Fig.6 present input matching and output matching results, respectively. The results show proper matching in the desired frequency band.

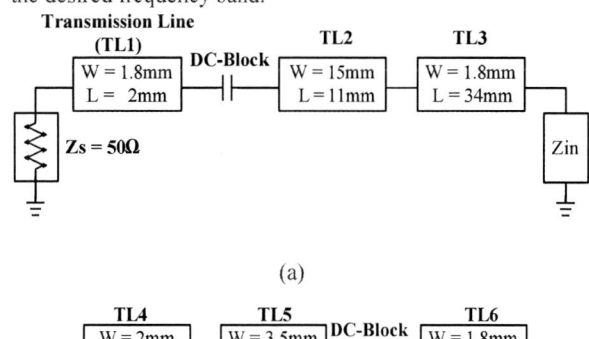

(a)

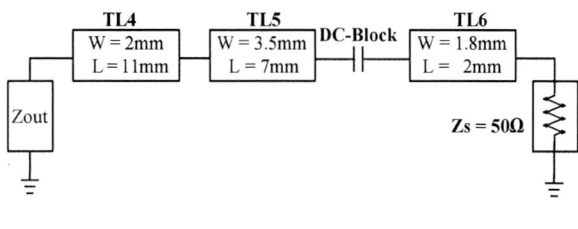

(b)

Fig. 4. Matching Network a) Input b) Output of the PA

The 5th Iranian International Conference on Microelectronics (IICM2023)

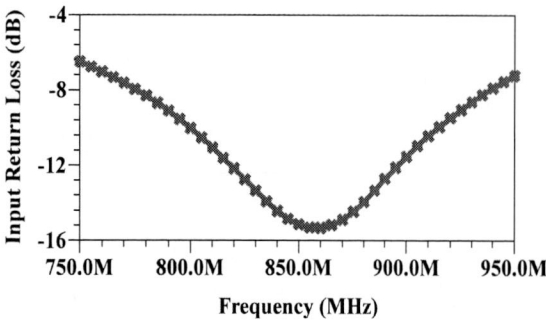

Fig. 5. Input Matching

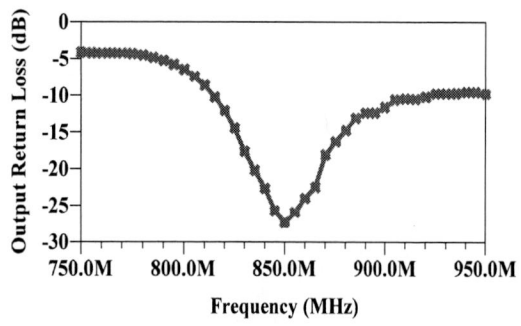

Fig. 6. Output Matching

E. Stability

The proposed PA must be stable in a wide frequency range. For this purpose, coefficient k must be greater than 1 (K > 1) and coefficient β must be greater than zero (β > 0). An inductor and a paralleled capacitor are used in the gate of the transistor for stabilization, shown in Fig.8. The suggested PA has met these conditions in a very wide range. As a result, the proposed PA is stable. Fig.7 shows K and β factors of the PA, respectively.

III. THE PROPOSED POWER AMPLIFIER AND SIMULATION RESULTS

Fig.8 shows the schematic of the proposed power amplifier.

C_1 and C_2 are DC-block capacitors. R_g and C_g are used as a Stabilization network.

In this design, the supply voltage and bias voltage are 50(V) and -2.2 (V), respectively.

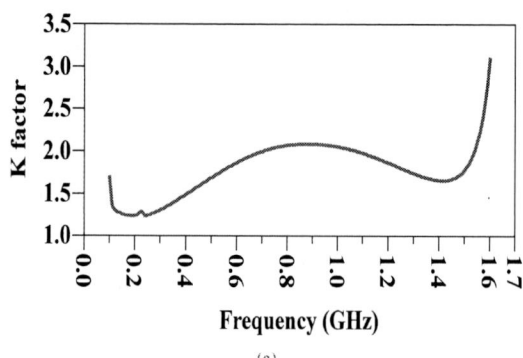

(a)

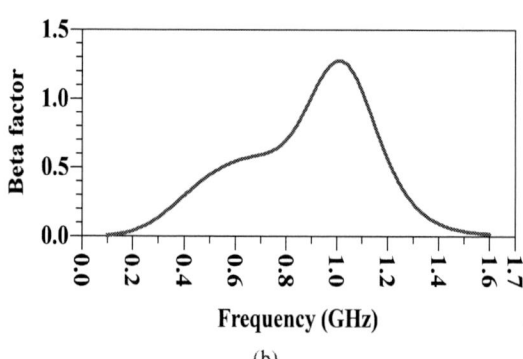

(b)

Fig. 7. Stability a) K factor b)β factor

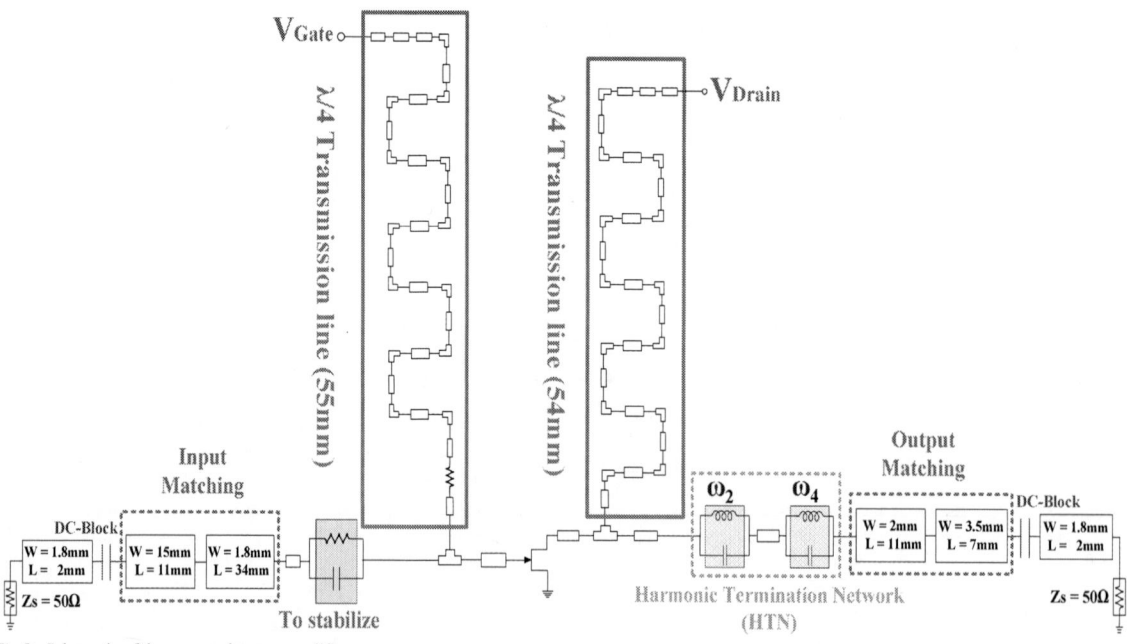

Fig 8. Schematic of the proposed power amplifier

The 5th Iranian International Conference on Microelectronics (IICM2023)

Fig.9 presents the layout of the proposed structure. The Area of the circuit is 108.6mm × 40.6mm. She substrate is Rogers_R04003 and transistor is CLF1G0060.

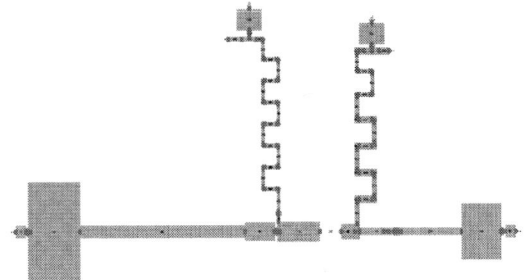

Fig.9. Layout of the proposed structure

Fig.10 shows the EM simulation results of the power gain of the proposed PA. The power gain of the PA in the frequency range of 800-900MHz is more than 16.7dB. The result indicates that the gain changes in the mentioned frequency range in Fig.10 less than 1dB.

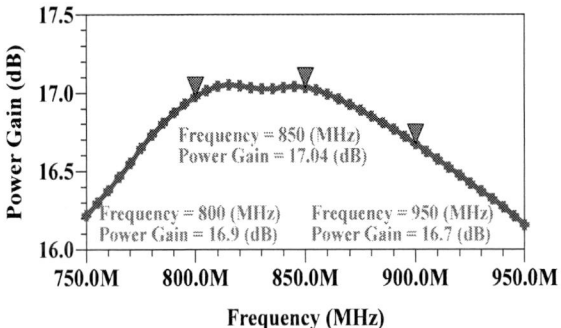

Fig.10. EM simulation results of the power gain

Fig.11 presents the EM simulation results of the output power of the suggested PA. The output power of the PA in the desired frequency band is more than 41.8dB.

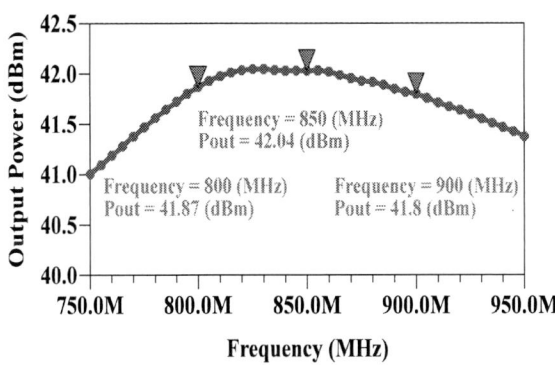

Fig.11. EM simulation results of the output power

Power-added efficiency of the PA is 70.7% at 850MHz. PAE of the suggested PA is more than 66.9% in the desired frequency band as shown in Fig.12.

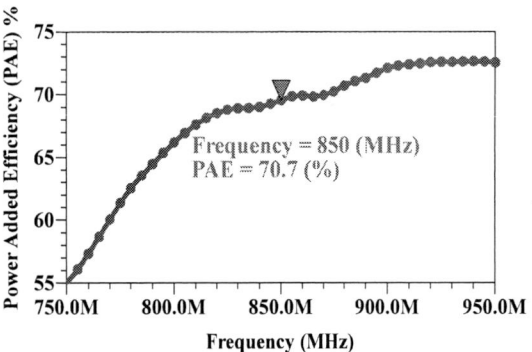

Fig.12. EM simulation results of the Power-Added Efficiency (PAE)

Fig.13 shows the output power versus input power. P_{1dB} of the PA is 41.86dBm.

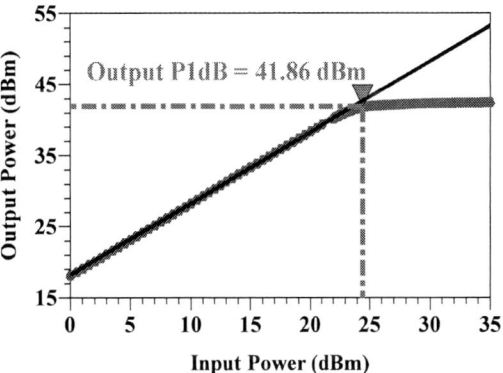

Fig.13 Output power versus input power

Fig.14 shows the power gain and the gain compression versus input power.

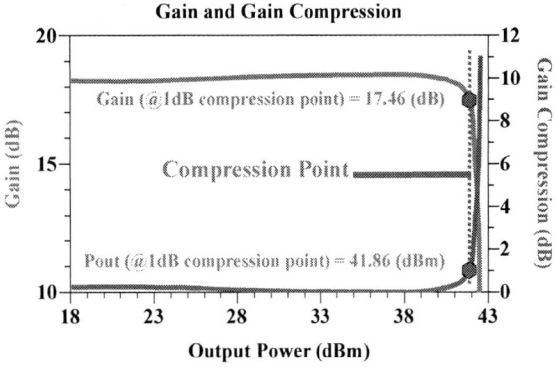

Fig.14 power gain and gain compression versus input power

Fig. 15 shows the power spectra response of the proposed PA to a 10-MHz QAM signal at input power=20dBm. The result indicates the adjacent channel power ratio (ACPR) is around 37dBc. PAE at P_{1dB} and the maximum PAE are 70.1% and 73.6%, respectively. Fig.16 shows PAE versus input power.

The 5th Iranian International Conference on Microelectronics (IICM2023)

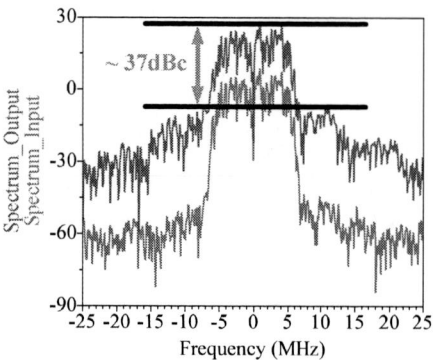

Fig. 15 Output spectra of the proposed PA

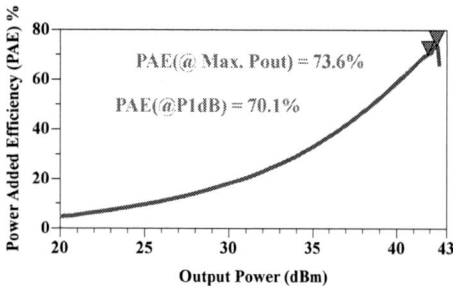

Fig. 16 PAE versus input power

Table 1 presents a comparison between this work and other works. The proposed PA has the highest output power. Moreover, this work has reached the highest P_{1dB}. P_{1dB} shows the range of the input power (P_{in}) that the PA is in the linear region. This work has the highest PAE at the center frequency of 850MHz.

TABLE 1. COMPARSION BETWEEN THIS WORK AND OTHER SIMILAR WORKS

Ref.	Frequency (GHz)	P_{out} (dBm)	P_{1dB} (dBm)	PAE (@ P_{1dB}) (%)
14	0.8	40.8	----	33.2
15	0.85	37.8	----	----
16	2.45	40	~39*	~70 (η_D)**
17	2.3	41.5	~40.5*	~80 (η_D)**
This Work	0.85	42.5	41.86	70.1

* According to the curves, this value is estimated.

** These works report drain efficiency (η_D) that is greater than PAE.

IV. CONCLUSION

This paper presents a high-efficiency deep bias class-AB power amplifier (PA) in the frequency band of 800-900MHz. The deep bias leads to high efficiency. In addition, it also overcomes the problem of the zero-crossing distortion. In this work, the PAE of the proposed PA at P_{1dB} is 70.1%. The maximum PAE is 73.6%. This design employs the harmonic termination network to minimize the overlap between the output voltage and the output current waveforms. Moreover, this design uses series stubs instead of short circuited and

open circuited stubs in the input and output matching networks to reduce radiation and interference in other bands. The ACPR of the proposed PA is around 37dBc. The simulation results of the proposed PA indicate that Gain, output power, and P_{1dB} are 17dB, 42.5dBm and 41.86dBm, respectively.

REFERENCES

[1] F. Ziraksaz and A. Hassanzadeh. "An enhanced ultra-wideband single phase hybrid supply envelope tracking modulator for modern wireless communications" *AEU-International Journal of Electronics and Communications* 156 (2022): 154393.

[2] B. Razavi, "Microelectronics RF. New York", NY, USA: Prentice-Hall; 2012, pp. 751.

[3] F. Ziraksaz and A. Hassanzadeh, "A 23.4-31.9 GHz Tunable RF-MEMS Impedance Matching Network for 5G Power Amplifier," *2021 29th Iranian Conference on Electrical Engineering (ICEE)*, 2021, pp. 69-73.

[4] X. Zhou, W. S. Chan, T. Sharma, J. Xia, S. Chen and W. Feng, "A Doherty Power Amplifier With Extended High-Efficiency Range Using Three-Port Harmonic Injection Network," in *IEEE Transactions on Circuits and Systems I: Regular Papers*, vol. 69, no. 7, pp. 2756-2766, July 2022.

[5] F. Ziraksaz and A. Nabavi, "Design of a Linear Class AB Amplifier with 55dB Gain, 890MHz Bandwidth and Low Output Impedance for Envelope Tracking Supply Modulator," *2019 27th Iranian Conference on Electrical Engineering (ICEE)*, 2019, pp. 253-257.

[6] S. K. Noor *et al.*, "A Review of Orbital Angular Momentum Vortex Waves for the Next Generation Wireless Communications," in *IEEE Access*, vol. 10, pp. 89465-89484, 2022.

[7] F. Ziraksaz and A. Hassanzadeh, "Design of a Tunable Wideband Differential Amplifier Based on RF-MEMS Switches," *2021 Iranian International Conference on Microelectronics (IICM)*, 2021, pp. 1-5.

[8] S. Hashima *et al.*, "On Softwarization of Intelligence in 6G Networks for Ultra-Fast Optimal Policy Selection: Challenges and Opportunities," in *IEEE Network*, vol. 37, no. 2, pp. 190-197, March/April 2023.

[9] F. Ziraksaz and A. Hassanzadeh, "Design of a Tunable Impedance Matching Network Based on MEMS Cantilever Switches," *2022 Iranian International Conference on Microelectronics (IICM)*, Tehran, Iran, Islamic Republic of, 2022, pp. 39-42.

[10] M. G. Sadeque, Z. Yusoff, S. J. Hashim, A. S. M. Marzuki, J. Lees and D. FitzPatrick, "Design of Wideband Continuous Class-F Power Amplifier Using Low Pass Matching Technique and Harmonic Tuning Network," in *IEEE Access*, vol. 10, pp. 92571-92582, 2022.

[11] S. K. Dhar *et al.*, "Modeling of Input Nonlinearity and Waveform Engineered High-Efficiency Class-F Power Amplifiers," in *IEEE Transactions on Microwave Theory and Techniques*, vol. 68, no. 10, pp. 4216-4228, Oct. 2020.

[12] G. Nikandish, E. Babakrpur and A. Medi, "A Harmonic Termination Technique for Single- and Multi-Band High-Efficiency Class-F MMIC Power Amplifiers," in *IEEE Transactions on Microwave Theory and Techniques*, vol. 62, no. 5, pp. 1212-1220, May 2014.

[13] M. Seo *et al.*, "High-Efficiency Power Amplifier Using an Active Second-Harmonic Injection Technique Under Optimized Third-Harmonic Termination," in *IEEE Transactions on Circuits and Systems II: Express Briefs*, vol. 61, no. 8, pp. 549-553, Aug. 2014.

[14] Y. Hu and S. Boumaiza, "Doherty power amplifer distortion correctionusing an RF linearization amplifier," IEEE Trans. Microw. Theory Techn.,vol. 66, no. 5, pp. 2246-2257, May 2018.

[15] Y. Hu and S. Boumaiza, "Power-scalable wideband linearization ofpower amplifers," IEEE Trans. Microw. Theory Techn., vol. 64, no. 5, pp. 1456-1464, May 2016.

[16] A. Dani, M. Roberg, and Z. Popovic, "PA effciency and linearity enhance-ment using external harmonic injection," IEEE Trans. Microw. TheoryTechn., vol. 60, no. 12, p. 4097-4106, Dec. 2012.

[17] S. Rahimizadeh, T. Cappello, and Z. Popoviḉ, "An effcient linear power amplifer with 2nd harmonic injection," in Proc. IEEE Topical Conf. RF/Microw. Power Modeling Radio Wireless Appl. (PAWR), Jan. 2019, pp. 14.

The 5th Iranian International Conference on Microelectronics (IICM2023)

25 – 26 October 2023

Piezoresistive pressure sensor based on flexo photosensitive Resin Plate

Ferdos Akrami
Department of Electronics and Electrical Engineering, Shiraz University of Technology,
Shiraz, Iran
fs.akrami@sutech.ac.ir

Samaneh Hamedi[*]
Department of Electronics and Electrical Engineering, Shiraz University of Technology,
Shiraz, Iran
Hamedi@sutech.ac.ir

Abstract— **In this paper, we propose and fabricate a new piezoresistive pressure sensor by combining micro-structured flexo photosensitive Resin Plate/Ag and Ag interdigital electrodes through a low-cost process. These plates are made from light-sensitive polymers that are hardened by UV light. The proposed material as a flexible substrate has advantages like Excellent rebound resilience enables high-quality printing with sharp images, high resolution, Cost-effective, etc. our proposed sensor shows sensitivity (18.14 kPa^{-1} and 458.2 kPa^{-1} in the pressure range of 0 Pa to 1.2 kPa and 1.2-11 kPa), low detection limit (40 Pa), signal response (3.51 s) and recovery response (0.81 s). these characteristics allow the device to work in applications in intelligent robots, human-machine interfaces, and so forth.**

Keywords— pressure sensor, piezoresistive, lithography

I. INTRODUCTION

The rapid advancement of technology and the growing demand to improve the interaction between humans and machines have led to remarkable progress in wearable electronics. Wearable sensors, in particular, have emerged as a key component in bridging the gap between humans and machines, enabling seamless and more natural interactions. These sensors possess unique characteristics, such as the ability to withstand bending, stretching, and compression deformations, which make them highly adaptable for various applications [1].

Among the various factors, pressure stands out as a significant variable, representing the ratio of applied force to the surface area. The emergence of flexible pressure sensors has garnered significant attention compared to their rigid counterparts, primarily due to their unique properties. These sensors can withstand pressure, bending, stretching, and twisting, making them highly adaptable and suitable for use on non-smooth surfaces. Moreover, their formability adds to their versatility [1]. Flexible pressure sensors play a crucial role in converting mechanical deformation into electrical output, making them an integral component of flexible electronics [2]. Similar to the sensory receptors in the human

skin, these sensors effectively convert pressure signals into electrical signals [3], [4].

The operating mechanism of piezoresistive sensors is based on the principle of resistance change. When pressure is applied to a flexible structure, it leads to an alteration in resistance, resulting in a change in current under a bias voltage. This pressure-induced change increases the conductive paths between the structure and the flexible electrodes, causing the sensor to respond to the applied pressure [5], [6].

Piezoresistive sensors are typically constructed by coating a layer of sensitive material on a flexible substrate or filling an elastomeric structure with conductive materials. Various conductive materials, such as carbon-based materials (carbon nanotubes, graphene, carbon black, etc.), metal nanoparticles, and polymers, are utilized to enhance charge transfer paths [1],[2]. One of the ongoing challenges in pressure sensor development is improving their performance, which involves enhancing sensor parameters, including sensitivity and detection range [3]. An effective approach to enhance sensor performance is the utilization of microfabrication technology, which plays a key role in functional integration and miniaturization of electronics by incorporating micro and nanoscale structures [4]. This enables an increased number of conductive paths, leading to improved alignment between sensitivity and linear measurement range [2], [6]. Solid-sheet photopolymer plates are the most popular material used in the modern flexographic printing industry, designed to meet constant high-quality requirements. In our study, we fabricated micro-structured flexo photosensitive Resin Plate/Ag and Ag interdigital electrodes using traditional lithography with UV-A, UV-C, and DC-sputtering methods. Subsequently, we compared the performance of our sensor to that of others in the field.

979-8-3503-6020-2/23 $31.00 © 2023 IEEE

II. EXPERIMENTAL PROCEDURE

A. Materials and Fabrication

As depicted in Fig. 1, the flexo photosensitive Resin plate consists of a cover sheet that protects the virgin plate surface from contamination, a top-release protective film placed upon the block copolymer layer, a curable layer at the heart of the photo-polymerization process, an adhesive layer between the support and the curable layer, and a dimensionally-stable substrate, which serves as the support for the photopolymer layer [7]. First, we carefully cut 1 piece of flexo photo-sensitive resin plate measuring 1.5cm x 1.5 cm with a cutter. standard lithography process and experimental parameters of pattern for printing and transferring microstructure onto the flexo photo-sensitive resin plate were listed as follows, mainly including back exposure, main exposition, washout, drying, and post-exposure, Followed by the back exposure to harden the base of the plate (FIG.2. 1), followed by main exposition for 15 min (correspondence with the diameter of the top contact area of 314µm) to harden the plate in the image areas, in which the mask (150mm * 150mm * 0.1 mm), including circular array pattern with a diameter of 400µm, and centroid distance of 400µm, was adopted and aligned on the face the plate carefully (FIG.2. 2), followed by removing the protective cover film and washed out in the development unit (octane+ Polyethylene perchlorate) to remove the unhardened photopolymer from the non-image areas and create the desired design (FIG.2. 3), followed by air-drying at about 60°C (FIG.2. 4). Finally, the post-exposure flexo photo-sensitive resin plate was also treated with UV light again to cure the whole plate (FIG.2. 5). The plate is then stored in air at room temperature and light finishing for about 24 hours for stabilization. In the end, silver with a thickness of 70nm was deposited on the patterned sides of the flexo photo-sensitive resin plates by DC-sputtering (FIG.2. 6), respectively. To Prepare the Polyethylene terephthalate (PET)/ Ag interdigital electrode, First, the commercially available PET (Polyethylene terephthalate) substrate with 1 mm thickness was cut into dimensions of 1.7 cm × 2.7 cm. After cleaning by ultrasonication in DI water, acetone, ethanol, and isopropanol, for 10 min, the Ag was deposited on the substrate (dimensions of 1.5 cm × 2.5 cm *200 nm) by using DC sputtering method via a patterned shadow mask (FIG.2. 7). In the end, The flexible pressure sensor was assembled with an Ag/ flexo photo-sensitive resin plate and interdigital electrode on PET followed by encapsulation with surgical semipermeable polyurethane film [8].

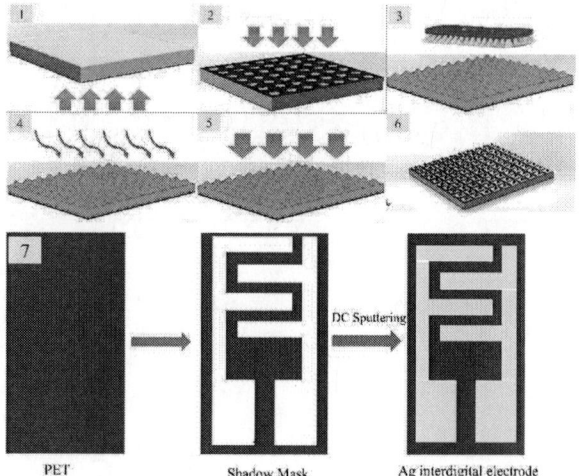

FIG.2. Schematic illustration of the fabrication process of the Ag/ FPSRP microstructure and Ag interdigital electrode

B. Measurements

The current-voltage (I-V) characteristics were measured at 27°C using a Keithley (2450) instrument. The measurements involved applying a sweep voltage and bias voltage to the two points of the interdigital electrodes, and the corresponding currents were recorded.

III. RESULTS AND DISCUSSION

FIG. 3 displays both the optical and SEM images of the micro-structured flexo photosensitive Resin Plate/Ag and Ag interdigital electrode.

FIG.3. optical image and SEM image of the flexo photo-sensitive resin plate/Ag and (PET)/ Ag interdigital electrode

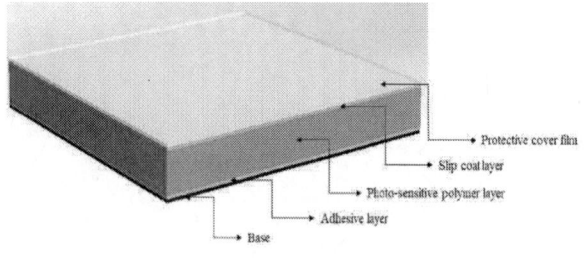

FIG. 1. Schematic of the flexo photosensitive Resin Plate

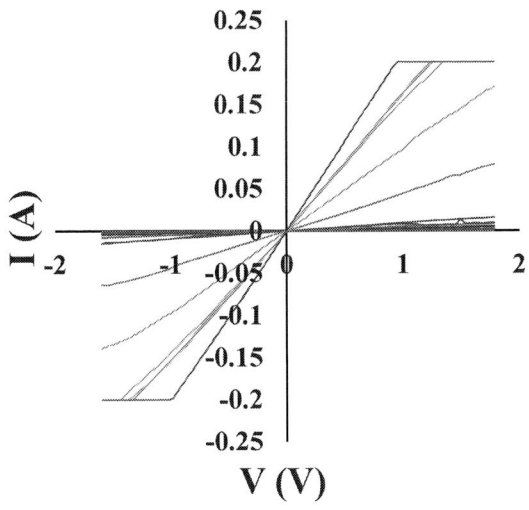

FIG.4. I-V curves with different loading pressures

In FIG. 4, the I-V curves for various pressures are presented. As the applied pressure increases, the slope of the I-V curves also increases. Furthermore, the entire curves demonstrate linear characteristics and ohmic conductive behavior. It is worth noting that the current measurement limitation was set at 0.2A.

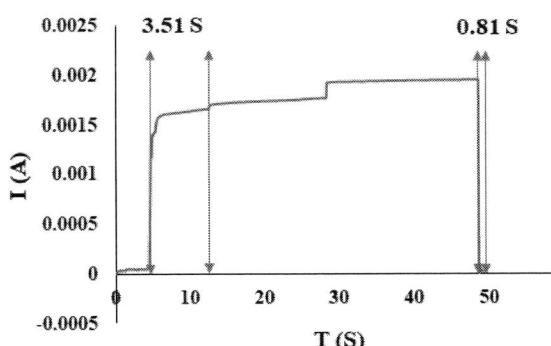

FIG.5. Response time and Recovery time of sensor in Pressure 484 Pa

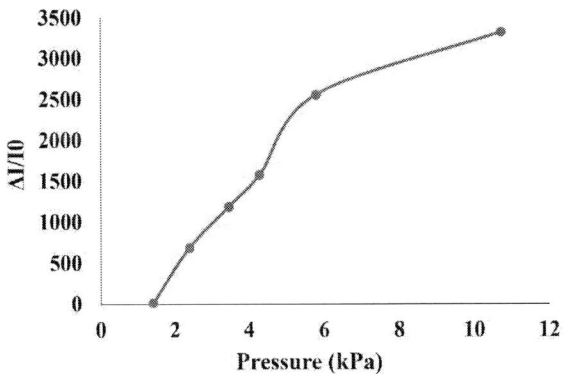

FIG.6. The relative current variations ΔI/I0 under different pressures

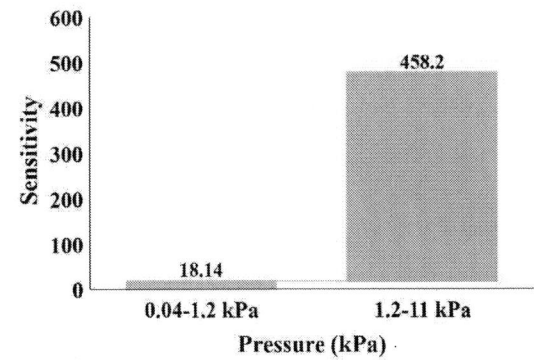

Fig.7. sensitivity of pressure sensor in ranges of 0-1.2 kPa and 1.2-11 kPa

In FIG.5. the pressure sensor responded to a pressure of 484 Pa, which corresponds to the pressure of a mass of 8345 mg onto an area of 1.5 cm^2 with a delay of 3.51 s under a supply voltage of 1 V, using Keithley (2405) when considering the points taken one-hundredth of a second. In FIG. 6, the relative current variations ($\Delta I/I_0$) under different pressures are shown. The sensitivity (Y) of the proposed sensor was evaluated using the following equation:

$$Y=(\Delta I/I_0)/\Delta P \qquad (1)$$

where I_0 represents the initial current, and $\Delta I = (I - I_0)$ indicates the current variation under the applied pressure (ΔP). In FIG. 7, the pressure-sensing performance is illustrated, showing a two-step sensing behavior throughout the entire pressure range. The sensitivity in both linear regimes was determined by extracting the slope of the fitting curves, resulting in a sensitivity of 18.14 in the pressure range of 0.04-1.2 kPa, and a high sensitivity of 458.2 in the pressure range of 1.2-11 kPa.

TABLE I. SENSING PERFORMANCE COMPARISON OF PRESSURE SENSORS IN PREVIOUS RESEARCH STUDIES

MATERIALS	METHOD	SENSITIVITY (kPa^{-1})	YEAR
PDMS/AG NW	Peel-off/spin coating/heating	3.81 (0.8-2.1 kPa) 2.73 (0.4-12 kPa)	2019[10]
PDMS/PEDOT: PSS/PUD & CNTs	3D-Printing	$3.54*10^{-3}$ (0-22.2 kPa)	2021[9]
(PAAm)/poly (vinyl alcohol) (PVA)	ultraviolet polymerization/ freeze–thaw treatment	2.27 (0-0.5 kPa)	2023[11]
photosensitive Resin Plate/Ag	Lithography/DC-Sputtering	18.14 (0.04-1.2 kPa) 458.2(1.2-11 kPa)	This work

According to Table I, Yongyun et al. reported a flexible pressure sensor using two sheets of polydimethylsiloxane (PDMS) pillar arrays coated with silver nanowires (AgNWs), which were assembled face-to-face. Their proposed pressure sensor exhibited sensitivities of 3.81 kPa^{-1} and 2.73 kPa^{-1} within the linearity ranges of 0.8 to 2.1 kPa and 0.4 to 12 kPa. Yiwei et al. achieved a piezoresistive pressure sensor by utilizing a 3D printing method with PDMS and PEDOT: PSS/PUD&CNTs composite film. The sensor demonstrated a sensitivity of 3.54×10^{-3} kPa^{-1} over the range of 0 to 22.2 kPa. Yun et al. fabricated a piezoresistive pressure sensor by decorating pyramid microarrays on the surface of a tough and ionically conductive polyacrylamide (PAAm)/poly(vinyl alcohol) (PVA) hydrogel, sandwiched between two flat Aluminum electrodes. Their proposed sensor achieved a sensitivity of 2.27 over the range of 0 to 0.5 kPa. In comparison, our proposed sensor exhibited significantly higher sensitivity, with values of 18.14 kPa^{-1} and 458.2 kPa^{-1}, within the pressure ranges of 0.04-1.2 kPa and 1.2-11 kPa, respectively.

IV. CONCLUSION

We fabricated a new piezoresistive pressure sensor based on micro-structured flexo photosensitive Resin Plate/Ag and Ag interdigital electrode over a low-cost method and measured the I-V characteristics. We found out that the pressure sensor shows ultra-high sensitivity of 458.2 in the low pressure regime. Our proposed pressure sensor has a long time life and the fabrication process does not need complex methods and materials.

REFERENCES

[1] C. Cui, Q. Fu, L. Meng, S. Hao, R. Dai, and J. Yang, "Recent Progress in Natural Biopolymers Conductive Hydrogels for Flexible Wearable Sensors and Energy Devices: Materials, Structures, and Performance," *ACS Applied Bio Materials*, vol. 4, no. 1. American Chemical Society, pp. 85–121, Jan. 18, 2021.

[2] Y. Duan, S. He, J. Wu, B. Su, and Y. Wang, "Recent Progress in Flexible Pressure Sensor Arrays," *Nanomaterials*, vol. 12, no. 14. MDPI, Jul. 01, 2022.

[3] Y. Ding, T. Xu, O. Onyilagha, H. Fong, and Z. Zhu, "Recent advances in flexible and wearable pressure sensors based on piezoresistive 3D monolithic conductive sponges," *ACS Applied Materials and Interfaces*, vol. 11, no. 7. American Chemical Society, pp. 6685–6704, Feb. 20, 2019.

[4] H. S. Oh, C. H. Lee, N. K. Kim, T. An, and G. H. Kim, "Review: Sensors for biosignal/health monitoring in electronic skin," *Polymers*, vol. 13, no. 15. MDPI AG, Aug. 01, 2021.

[5] T. Tsuda, S. Chae, M. Al-Hussein, P. Formanek, and A. Fery, "Flexible Pressure Sensors Based on the Controlled Buckling of Doped Semiconducting Polymer Nanopillars," *ACS Appl Mater Interfaces*, vol. 13, no. 31, pp. 37445–37454, Aug. 2021.

[6] M. Y. Liu et al., "Advance on flexible pressure sensors based on metal and carbonaceous nanomaterial," *Nano Energy*, vol. 87. Elsevier Ltd, Sep. 01, 2021.

[7] X. Liu and J. T. Guthrie, "A review of flexographic printing plate development."

[8] Q. Du et al., "High-Performance Flexible Pressure Sensor Based on Controllable Hierarchical Microstructures by Laser Scribing for Wearable Electronics," *Adv Mater Technol*, vol. 6, no. 9, Sep. 2021.

[9] Y. Shao et al., "Flexible pressure sensor with micro-structure arrays based on pdms and pedot:Pss/pud&cnts composite film with 3d printing," *Materials*, vol. 14, no. 21, Nov. 2021.

[10] Y. Mao, B. Ji, G. Chen, C. Hao, B. Zhou, and Y. Tian, "Robust and Wearable Pressure Sensor Assembled from AgNW-Coated PDMS Micropillar Sheets with High Sensitivity and Wide Detection Range," *ACS Appl. Nano Mater.*, vol. 2, no. 5, pp. 3196–3205, 2019.

[11] Y. Xia Zhang et al., "Sensitive piezoresistive pressure sensor based on micropyramid patterned tough hydrogel," *Appl. Surf. Sci.*, vol. 615, no. December 2022, 2023.

The 5th Iranian International Conference on Microelectronics (IICM2023)

25 – 26 October 2023

A MEMS Resonant Pressure Sensor Based on 2D Graphene Material

Amir Noroolahi
Department of Electrical Engineering
Yadegar-e-Imam Khomeini (RAH),
Shahr-e-Rey Branch, Islamic Azad
University, Tehran, Iran
noroolahi@yahoo.com

Farshad Babazadeh
Department of Electrical Engineering
Yadegar-e-Imam Khomeini (RAH),
Shahr-e-Rey Branch, Islamic Azad
University, Tehran, Iran
babazadeh@um.ac.ir

Abstract— In this manuscript, we offer a comprehensive investigation of the operational principle, performance assessment, and prospects of resonant pressure sensors that rely on 2D material membrane. Specifically, the membrane in question is a Graphene-based one, possessing a radius of 2500 nm and a thickness of 25 nm. Our simulation setup consists of a multi-layer Graphene membrane that suspended on a Si substrate and stimulated by an electrostatic excitation. When acting as a resonator, the membrane displays a frequency shift of roughly 57 MHz in response to pressures ranging from zero to 1 bar applied pressure. Our sensor exhibits a sensitivity of 56 kHz/mbar, which is 200 times higher than that of pressure sensors based on Silicon, despite employing a membrane area that is 1200 times smaller. Our proposed pressure sensor exploits the mechanical and electrical properties of Graphene to enable exceptionally accurate pressure measurements.

Keywords—2D material, Resonant Frequency, Pressure Sensor, Nanoelectromechanical systems (NEMS), MEMS

I. INTRODUCTION

Resonant pressure sensors operate based on of the principle of mechanical resonance. They consist of a sensing element, typically a diaphragm or vibrating structure, and a means to measure the resulting resonant frequency changes. When subjected to pressure, the mechanical properties of the sensing element, such as stiffness and mass, are altered, leading to a shift in the resonant frequency. This frequency shift is measured to determine the applied pressure. Resonant pressure sensors offer numerous advantages, including high sensitivity, wide dynamic range, and fast response time [1-2].

In recent years, there has been significant progress in of the development of resonant pressure sensors through the integration of two-dimensional (2D) materials, with Graphene emerging as a prominent candidate. Graphene is a remarkable 2D material consisting of a single layer of carbon atoms arranged in a hexagonal lattice. It possesses exceptional mechanical, electrical, and thermal properties, making it an attractive material for various applications [3-4]. One of the key properties of Graphene that enhances the performance of resonant pressure sensors is its exceptional mechanical strength. Graphene is one of the strongest materials known, with a tensile strength that exceeds that of Silicon. The related mechanical properties of the proposed materials are summarized in *TABEL. I*. This high mechanical strength enables Graphene-based sensors to exhibit exceptional sensitivity to pressure-induced deformations.

Even minute changes in pressure can be detected accurately, expanding the range of applications where precise pressure measurement is required[5]

TABLE I.   MATERIALS MECHANICAL PROPERTIES(GRAPHENE AND SILICON) [6]

Materials	Mechanical Properties			
	Young's Modulus	Poisson's Ratio	Density	Tensile Strength
Graphene	1 TPa	0.15	2267	130 GPa
Silicon	170 Mpa	0.28	2329	165 MPa

Graphene also possesses high electrical conductivity, allowing for reliable measurement of resonant frequency changes in the sensor. By monitoring the changes in the resonant frequency, pressure variations can be accurately determined. The combination of Graphene's mechanical and electrical properties enables the development of highly sensitive and accurate resonance pressure sensors [7]. Furthermore, Graphene's atomic thickness contributes to its exceptional properties and enhances the performance of resonant pressure sensors. The atomic thickness of Graphene results in low mass, leading to a faster response time of the sensor. This is advantageous in applications where real-time pressure monitoring and control are crucial. The low mass of Graphene-based sensors enables them to rapidly respond to changes in pressure, making them suitable for dynamic pressure measurement applications.

The integration of Graphene in resonance pressure sensors offers several potential benefits and advancements. Firstly, the exceptional mechanical properties of Graphene allow for improved sensitivity in pressure detection. Graphene-based sensors can detect even minute pressure variations, expanding the range of applications where precise pressure measurement is required. This includes applications in industries such as aerospace, automotive, and medical devices, where accurate pressure monitoring is critical for optimal performance and safety. Graphene's atomic thickness enables the development of miniaturized sensors. The compact size and reduced dimensions of Graphene-based sensors allow for integration into smaller and more portable devices[8]. This opens up possibilities for the integration of pressure sensing capabilities in a wide range of applications, including wearable devices, Internet of Things (IoT) devices, and biomedical implants[9].

979-8-3503-6020-2/23 $31.00 © 2023 IEEE

The 5th Iranian International Conference on Microelectronics (IICM2023)

Additionally, Graphene exhibits excellent biocompatibility, making it suitable for biomedical and healthcare applications. Graphene-based pressure sensors can be integrated into implantable devices or used in diagnostics, contributing to advancements in personalized healthcare. The combination of Graphene's mechanical and electrical properties, along with its biocompatibility, makes it a promising material for the development of next-generation pressure sensing technologies in the healthcare field[10]. Another advantage of Graphene is its surface functionalization capability. The surface of Graphene can be functionalized with specific molecules or nanoparticles to enhance selectivity in pressure sensing. This allows for tailored sensing capabilities for specific gases or liquids, expanding the range of applications for resonance pressure sensors. Functionalization of Graphene surfaces enables the development of highly selective and sensitive sensors for environmental monitoring, gas detection, and chemical sensing applications [6-11].

The main objective of this paper is to explore the integration of Graphene, an exceptional two-dimensional material, into resonant pressure sensors. The sensitivity of the proposed pressure sensor is thoroughly analyzed for various pressure ranges and membrane thicknesses to attain an optimal design. The persistent research and development endeavors in the domain of two-dimensional materials are consistently enhancing the performance and potential of resonance pressure sensors, propelling innovation in pressure measurement technology.

II. STRUCTURE AND RINCIPLE OF OPERATION

A. Structure

The pressure sensor structure consists of a Graphene membrane with radius R=2500 nm and a thickness of t=25nm. The Graphene membrane serves as the primary sensing element in the device, as show in Fig.1. Its small size allows for high sensitivity to pressure variations. The membrane is securely anchored within the sensor structure, ensuring its stability and accurate pressure measurement. Under an applied pressure, the Graphene membrane undergoes mechanical deformation, such as bending or stretching, which result in changes in its resonance characteristics. The precise response of the membrane to pressure is influenced by its dimensions and thickness. The low mass of the Graphene membrane contributes to its high sensitivity, as it allows for rapid response to pressure variations. To ensure the stability and accurate measurement of pressure, the Graphene membrane is securely anchored within the sensor structure. The support structure surrounding the membrane is designed to minimize unwanted vibrations and external interferences, thereby enabling precise pressure measurements.

By applying alternating voltage to the electrode and measuring its feedback, the resonant frequency is determined. It is clear that the resonant frequency changes due to the deformation of the membrane is depended on different pressures [12].

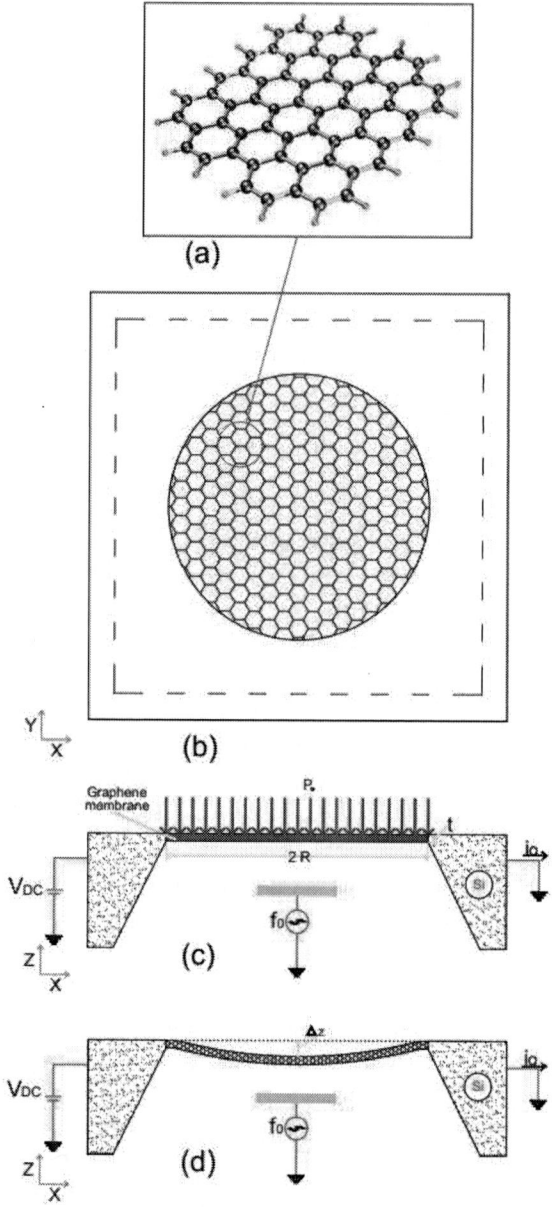

Fig. 1. Schematic structure of the pressure sensor with electrostatic excitation, (a) schematic of a Graphene sheet, (b) top view, (c) before applied pressure, (d) after applied pressure.

The resonance frequency of the membrane is calculated from Equation (1) [13].

$$f = \frac{10.4}{\pi} \sqrt{\frac{E t^2}{\rho \pi R^4 (1 - v^2)}} \qquad (1)$$

Where f is mechanical resonance frequency, E is the Young's modulus, t is the thickness of the membrane, ρ is the mass density of the membrane, R is the radius of membrane, and v is the Poisson's ratio.

979-8-3503-6020-2/23 $31.00 © 2023 IEEE

B. Simulation

This paper focuses on the utilization of COMSOL Multi-physics, for simulating resonant pressure sensors and exploring their behavior under different operating conditions. Through the implementation of structural modeling and the application of varying degrees of pressure across a range of radii spanning from 1250 to 3500 nm, the resulting frequency response can be observed in Fig. 2.

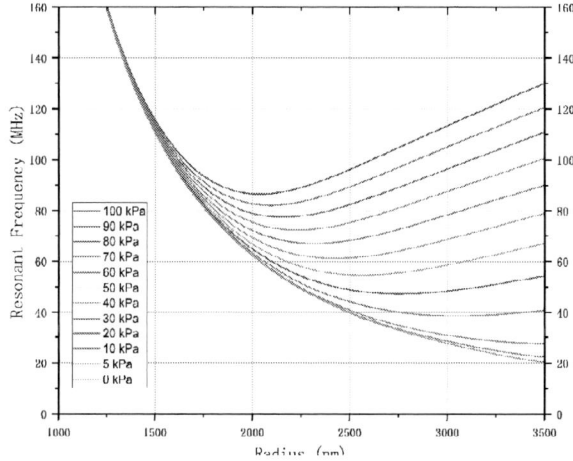

Fig. 2. Resonant Frequency versus different radius of membrane (thickness 25 nm).

As evidenced by the diagram, the resonant frequency experiences an initial decrease as the radius of the membrane expands, followed by a subsequent increase. Determining the appropriate value of the membrane radius based on the lack of deformation of the membrane at zero pressure and the desired accuracy, as summarized in TABLE. II. In this sensor, considering the manufacturing conditions, we considered the thickness of the membrane to be 25 nm and its radius to be 2500 nm.

TABLE II. PRESSURE SENSOR CHARACTERISTICS(GRAPHENE).

Parameter	Explanation	Value	Units
R	Membrane radius	2500	nm
t	Membrane thickness	25	nm
f	Membrane Resonance Frequency	39.746	MHz

The shape of the membrane is altered by the application of pressure, and this alteration results in a modification of the natural resonance frequency, as illustrated in Fig. 3. The resonance frequency is dependent on the edge boundary conditions of the membrane in relation to the Si substrate.

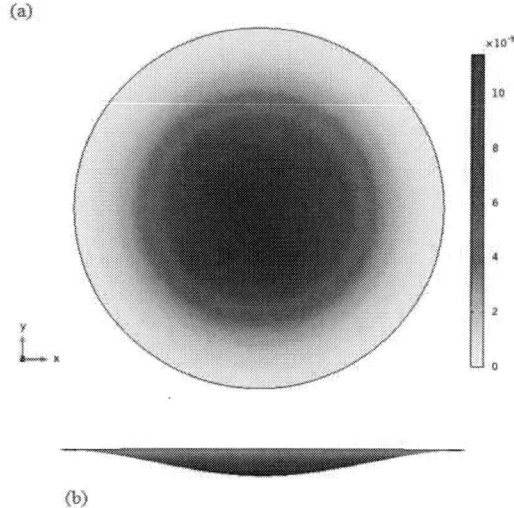

Fig. 3. Deformation of the membrane (m) (a) top view and (b) side view

C. Eigenfrequncy

Eigenfrequencies play a fundamental role in resonant based pressure sensors, resonant pressure sensors operate based on the principle of mechanical resonance, where the frequency response of the sensor is maximized at specific Eigenfrequencies. These Eigenfrequencies are determine by the mechanical properties of the sensing element, including its stiffness, mass, and geometry.

The geometric factors of the sensing element also play a significant role in determining the Eigenfrequencies in resonant pressure sensors:

a. Diaphragm Thickness: Thinner diaphragms exhibit higher Eigenfrequencies due to reduced mass and increased stiffness, resulting in enhanced sensor sensitivity.

b. Diaphragm Size and Shape: The size and shape of the diaphragm influence its effective stiffness and Eigenfrequencies. Different geometries can be explored to optimize the sensor's response and sensitivity.

c. Boundary Conditions: The boundary conditions of the sensing element, such as clamping or support locations, affect the Eigenfrequencies. Proper consideration of boundary conditions is necessary for accurate modeling and analysis of the sensor.

At a pressure of zero atmospheric pressure, the resonance frequency is observed to be 40 MHz A gradual increase in pressure by increments of 10 kPa up to 100 kPa results in a linear change in the resonance frequency, as depicted in Fig. 4, with a final frequency of 96 MHz The proposed structure is advantageous due to the linearity of the frequency response in the range of 20 to 100 kPa.

The 5th Iranian International Conference on Microelectronics (IICM2023)

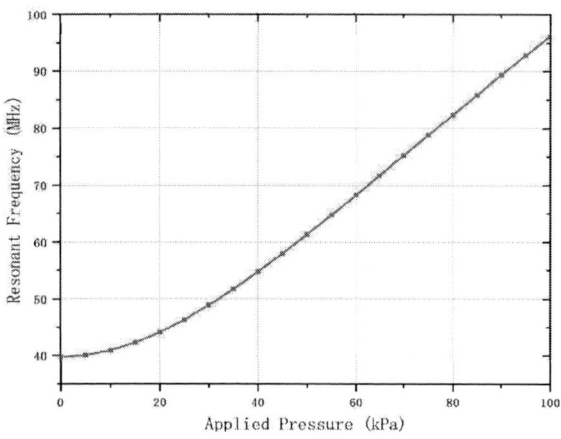

Fig. 4. Resonant Frequency of the Graphene membrane versus applied pressure

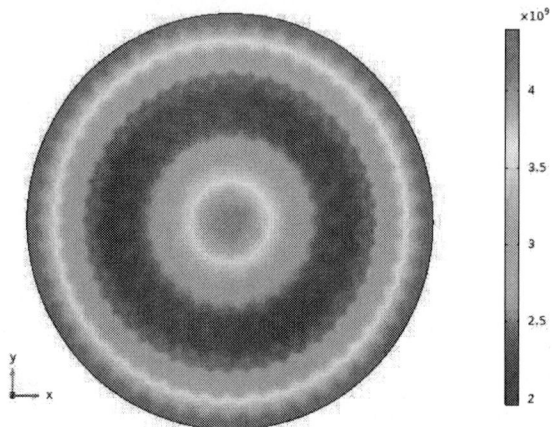

Fig. 6. Stress distribution in the Graphene membrane (Pa)

The resonant frequency at 40 MHz, which results in the most significant displacement in the membrane as per the first mode shape of the structure, is depicted in Fig.5,

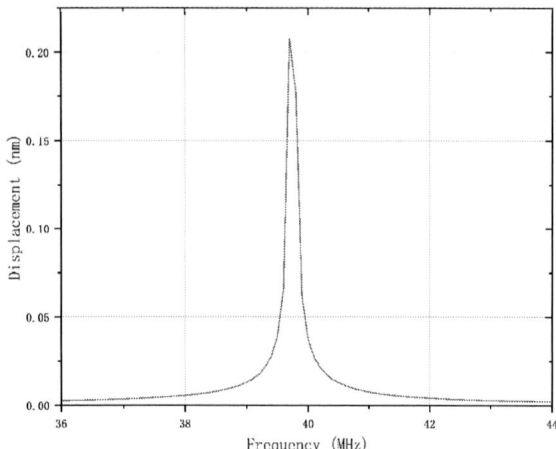

Fig. 5. Frequency response of the Graphene membrane.

When pressure is applied, the membrane is displaced, resulting in the creation of tension stress in the structure, as shown in Fig. 6, The maximum intensity is observed at the fixed edges, but it is still below the Graphene's yield point.

D. Sensivity

Sensitivity is a critical performance parameter for resonant pressure sensors, as it determines their ability to detect and accurately measure small changes in pressure. This paper focuses on the sensitivity of resonant pressure sensors and explores various techniques for enhancing their sensitivity. Sensitivity refers to the ability of a pressure sensor to detect and measure small changes in pressure. In resonance pressure sensors, sensitivity is typically express as the change in resonant frequency per unit pressure. A high sensitivity allows for precise pressure measurements, enabling the sensor to detect subtle variations and provide accurate readings [14-15].

Several factors affect the sensitivity of resonance pressure sensors:

a. Mechanical Properties: The mechanical properties of the sensing element, such as stiffness, mass, and damping, play a crucial role in determining the sensor's sensitivity. Materials with higher stiffness and lower mass exhibit higher sensitivity.

b. Diaphragm Design: The diaphragm's geometry and dimensions influence its effective stiffness and sensitivity. Optimizing the diaphragm design can enhance sensitivity by increasing its responsiveness to pressure changes.

c. Sensor Excitation: The excitation method used to stimulate the sensor's mechanical response affects its sensitivity. Employing an excitation mechanism that matches the sensor's resonant frequency can improve sensitivity.

d. Signal Detection: The detection and measurement technique employed to monitor the sensor's resonant frequency changes can affect sensitivity. Using high-resolution and accurate measurement systems enhances the sensor's sensitivity.

The demonstrated data portrays that the designed structure showcases a sensitivity of 56 kPa / mbar over a pressure range of 0 to 100 kPa. This value is considerably superior when compared to other designed structures, and thus, is considered as one of the advantageous features of the designed structure index.

III. CONCLUSION

In this paper, we have investigated the progress and uses of resonance pressure sensors that are constructed using two-dimensional (2D) material membranes, with a specific focus on Graphene. The proposed pressure sensor indicates a frequency shift of approximately 57 MHz in response to an applied pressure of 0 to 1000 mbar, and a sensitivity of 56 kHz/mbar. Additionally, the pressure sensor boasts a membrane area that is 1200 times smaller and a sensitivity that is 200 times higher than that of a Silicon base pressure sensor. To conclude, Graphene-based resonant pressure sensors exhibit significant potential to revolutionize the field of pressure measurement technology. The extraordinary properties of Graphene, combined with continuous advances in MEMS technology, pave the way for the development of highly sensitive, dependable, and adaptable pressure sensors with applications across various industries, including healthcare, environmental monitoring, industrial processes, and consumer electronics.

IV. REFRENCE

[1] Z. Li, L. Zhao, Z. Ye, H. Wang, Y. Zhao, and Z. Jiang, "Resonant frequency analysis on an electrostatically actuated microplate under uniform hydrostatic pressure," *Journal of Physics D*, vol. 46, no. 19, p. 195108, Apr. 2013, doi: 10.1088/0022-3727/46/19/195108.

[2] S. Kohli and A. Saini, "MEMS BASED PRESSURE SENSOR SIMULATION FOR HEALTHCARE AND BIOMEDICAL APPLICATIONS," *International Journal of Engineering Sciences & Emerging Technologies*, vol. 6, no. 3, pp. 308–315, 2013, Accessed: Sep. 15, 2023. [Online]. Available: https://www.ijeset.com/media/0002/4N13-IJESET0603132-v6-iss3-308-315.pdf

[3] M. Šiškins *et al.*, "Sensitive capacitive pressure sensors based on graphene membrane arrays," *Microsystems & Nanoengineering*, vol. 6, no. 1, Nov. 2020, doi: 10.1038/s41378-020-00212-3.

[4] Y. Dai *et al.*, "Suspended 2D Materials: A Short Review," *Crystals*, vol. 13, no. 9, p. 1337, Sep. 2023, doi: 10.3390/cryst13091337.

[5] Jahan Zeb Hassan, A. Raza, Zaheer, Usman Qumar, Ngeywo Tolbert Kaner, and A. Cassinese, "2D material-based sensing devices: an update," *Journal of materials chemistry. A, Materials for energy and sustainability*, vol. 11, no. 12, pp. 6016–6063, Jan. 2023, doi: 10.1039/d2ta07653e.

[6] D. Akinwande *et al.*, "A review on mechanics and mechanical properties of 2D materials—Graphene and beyond," *Extreme Mechanics Letters*, vol. 13, pp. 42–77, May 2017, doi: 10.1016/j.eml.2017.01.008.

[7] A. Gupta, S. Thangavel, and S. Seal, "Recent development in 2D materials beyond graphene," *Progress in Materials Science*, vol. 73, pp. 44–126, Aug. 2015, doi: 10.1016/j.pmatsci.2015.02.002.

[8] Manuel Vázquez Sulleiro, A. Dominguez-Alfaro, Nuria Alegret, A. Silvestri, and I. Jénnifer Gómez, "2D Materials towards sensing technology: From fundamentals to applications," *Sensing and bio-sensing research*, vol. 38, pp. 100540–100540, Dec. 2022, doi: 10.1016/j.sbsr.2022.100540.

[9] Vicente Orts Mercadillo *et al.*, "Electrically Conductive 2D Material Coatings for Flexible and Stretchable Electronics: A Comparative Review of Graphenes and MXenes," *Advanced Functional Materials*, vol. 32, no. 38, pp. 2204772–2204772, Jul. 2022, doi: 10.1002/adfm.202204772.

[10] J. Liu, S. Bao, and X. Wang, "Applications of Graphene-Based Materials in Sensors: A Review," *Micromachines*, vol. 13, no. 2, p. 184, Jan. 2022, doi:10.3390/mi13020184.

[11] Y. Xiao, F. Luo, Y. Zhang, F. Hu, M. Zhu, and S. Qin, "A Review on Graphene-Based Nano-Electromechanical Resonators: Fabrication, Performance, and Applications," *Micromachines*, vol. 13, no. 2, p. 215, Jan. 2022, doi:10.3390/mi13020215.

[12] G. W. Vogl and A. H. Nayfeh, "A Reduced-Order Model for Electrically Actuated Clamped Circular Plates," *Proceedings of the ASME 2003 International Design Engineering Technical Conferences and Computers and Information in Engineering Conference*, vol. 5, Jan. 2003, doi: 10.1115/detc2003/vib-48530.

[13] M. Das and A. Bhushan, "Investigation of an electrostatically actuated imperfect circular microplate under transverse pressure for pressure sensor applications," *Engineering Research Express*, vol. 3, no. 4, Nov. 2021, doi: 10.1088/2631-8695/ac3771.

[14] R. J. Dolleman, D. Davidovikj, S. J. Cartamil-Bueno, H. S. J. Van Der Zant, and P. G. Steeneken, "Graphene Squeeze-Film Pressure Sensors," *Nano Letters*, vol. 16, no. 1, pp. 568–571, Dec. 2015, doi: 10.1021/acs.nanolett.5b04251.

[15] G. Baglioni *et al.*, "Ultra-sensitive graphene membranes for microphone applications," *Nanoscale*, vol. 15, no. 13, pp. 6343–6352, Jan. 2023, doi:10.1039/d2nr05147h.

ZnO-Based Surface Acoustic Wave Droplet Sensor

Farzaneh Soleimanpour[+]
Electrical and Computer Engineering Department
Tarbiat Modares University
Tehran, Iran
f.soleimanpour@modares.ac.ir

Behdad Barahimi[+]
Electrical and Computer Engineering Department
Tarbiat Modares University
Tehran, Iran
b.barahimi@modares.ac.ir

Sara Darbari*
Electrical and Computer Engineering Department
Tarbiat Modares University
Tehran, Iran
s.darbari@modares.ac.ir

Mohammad Kazem Moravvej-Farshi*
Electrical and Computer Engineering Department
Tarbiat Modares University
Tehran, Iran
moravvej@modares.ac.ir

Abstract— In this work we designed and fabricated a ZnO-based SAW sensor with focused aluminum IDTs for DI water droplet sensing. We employed ZnO/SiO₂/Si as the piezoelectric substrate in this study. A minimum measurable DI water droplet of 6 mgr was obtained in the presented SAW sensor.

Keywords—Surface Acoustic Wave (SAW), Sensor, Piezoelectric, SAW Device, Zinc Oxide (ZnO).

I. INTRODUCTION

Acoustic sensors and biosensors provide sensitive, label-free, fast, and selective detection of analytes for gas and liquid samples at low costs [1]. Different acoustic sensors include bulk acoustic waves (BAW) and surface acoustic waves (SAW) sensors.

SAW sensors are used in the frequency range from a few MHz to a few GHz frequencies [1]. Interdigital transducers (IDTs) are responsible for both exciting and receiving the surface wave on the piezoelectric material. The delay line, defined as the distance between the input and output IDTs, results in phase and amplitude shifts between input and output signals due to the time delay [1].

SAW devices generate surface acoustic waves by altering the applied RF electrical energy into mechanical displacement and energy. The electrical element produces the mechanical energy as acoustic waves within the piezoelectric substrate caused by the dynamic variation of the electrical input signal due to the electro-mechanical coupling. The generated SAW travels along the surface of the piezoelectric substrate and reaches the other interdigitated electrodes. Through the mechanoelectrical coupling of the piezoelectric substrate, the receiver IDT produces electrical signals and senses the surface waves via the reverse mechanism [2].

Changes in mass on the surface of the device directly impact both the acoustic wave propagation velocity and the signal response of the sensor [1].

In SAW devices, energy is concentrated within a range of one to two acoustic wavelength thicknesses from the surface of the piezoelectric substrate. As a result, the acoustic amplitude diminishes along the thickness direction. [3].

SAW devices find utility in numerous domains, including applications in electrical, mechanical, biological, chemical, and physical science fields, thanks to their advantageous electro-mechanical transduction capabilities [2].

II. WORKING PRINCIPLE

A. SAW configuration

Figure 1 shows the conventional SAW configuration, consisting of two IDTs pairs on a piezoelectric substrate.

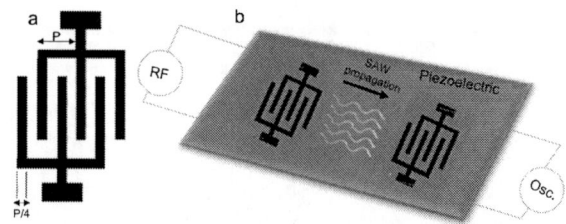

Figure 1. a) IDT parameters b) SAW configuration

At the applied RF frequency, f, the mechanical waves generated exhibit constructive interference when the pitch of the IDTs, represented by p, matches the acoustic wavelength λ. The resonant frequency of the device corresponds to the frequency given by [2]:

+ Equal contribution
* Corresponding authors

$$f = \frac{v}{p} \qquad (1)$$

where v is the velocity of the SAW on the piezoelectric substrate. Piezoelectric substrates, such as quartz, $LiTaO_3$, and $LiNbO_3$, and thin films, such as Aluminum Nitride (AlN), Lead zirconate titanate (PZT), and zinc oxide (ZnO), are used in SAW devices [3].

B. SAW design

We designed a $ZnO/SiO_2/Si$ SAW device with aluminum-focused IDTs (FIDTs) with pitch p=100 μm. This structure is a type of surface acoustic wave sensor that is used in the detection of various biomarkers and proteins; it can provide high sensitivity and selectivity and low detection limit [4] and can be fully integrated with CMOS chips, enabling the development of portable real-time detection devices [4]. The commercial Si/SiO_2 wafer utilized here, is a single-side polished 500 μm P-type prime silicon wafer covered with 300 nm SiO_2, with about 0.01~0.001 Ω cm resistivity.

ZnO, as an inorganic II–VI binary compound semiconductor, has unique functional and nano morphological properties [5]. ZnO is the surface acoustic wave conducting layer, and SiO_2 works as the insulation and protection layer of the Si substrate surface [5, 6].

Kharusi et al. presented FIDT for the first time in 1972 [7]. FIDT is obtained by changing the electrode design in the SAW device from straight electrodes to focally curved electrodes. The FIDT acoustic waves concentrate toward a localized zone and intensify the acoustic force gradient [7]. This leads to higher amplitude acoustic wave generation and could provide rapid acoustic streaming.

Another advantage of FIDT is the concentration of SAW energy at the center of the IDT with higher intensity, resulting in a high beam-width compression ratio within a small, localized area. [4]. The FIDT can achieve the same mixing efficiency with significantly lower SAW power compared to the conventional IDT configuration, known as standard IDT (SIDT) [7].

The best ohmic contacts for the ZnO thin films, are Al contacts [8]. These issues can be attributed to the electron affinity of ZnO, which is 4.1 eV, and the work function of Al, which is 4.24 and closely matches the electron affinity of ZnO. [8].

III. FABRICATION AND EXPERIMENTAL SETUP

A. Fabrication

Initially, the Si/SiO_2 wafer was cleaned with RCA cleaning protocol. Afterward, a thin layer of 500 nm ZnO was RF sputtered on the SiO_2, followed by a thermal annealing process at 600°C for an hour to regularize the crystal lattices, Figure 2- 1.

The SAW sensor chips were made by patterning Al IDTs through a lithographic lift-off process (Figure 2) on a ZnO thin layer. The bi-layer lift-off process, conventionally used to finely pattern or remove metals that are hard to etch [9], is hired here to avoid ZnO thin layer removal in the direct Al layer etching. In this case, first, a layer of Shipley-1318 positive photoresist was spin-coated and prebaked in the oven at 75°C for 10 minutes (Figure 2- 2). Subsequently, total lithography was done (Figure 2- 3). Then, the second photoresist layer was spin-coated and prebaked at the same condition as the first layer (Figure 2- 4). This time, IDT patterns were lithographed (Figure 2- 5) and developed by NaOH solution (Figure 2- 6). Afterward, a thin film Al layer (about 30 nm) was deposited through the Physical Vapor Deposition (PVD) process (Figure 2- 7), which was subsequently washed in acetone to lift-off unwanted sections (Figure 2- 8), resulting in IDTs (Figure 3).

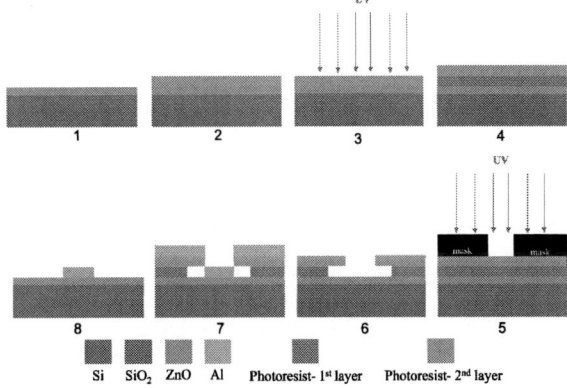

Figure 2. Schematic of the lift-off process

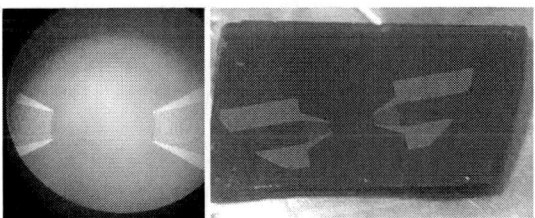

Figure 3. Lithography and patterned IDTs

The sensor chip was assembled on the designed printed circuit board (PCB) that has SMA connectors for input/output connections, Figure 4- a. Aluminum ingot for environmental noise reduction was used beneath the PCB as the ground and copper wires and silver paste were used for chip connection to the PCB (Figure 4- b and c).

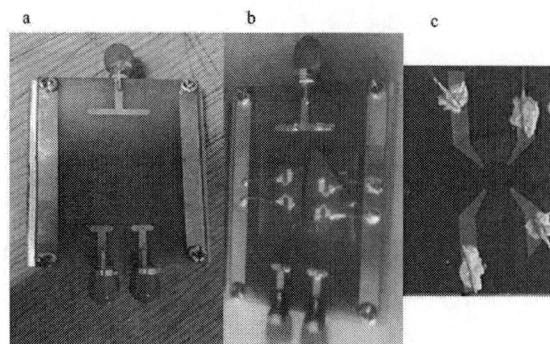

Figure 4. a) PCB b) final sensor configuration c) silver paste connections

B. Experimental Setup

The experimental setup used in our experiments is shown in Figure 5. A signal generator and power amplifier were used to apply a sinusoidal input signal and oscilloscope to read out the output signal of the SAW sensor Figure 5.

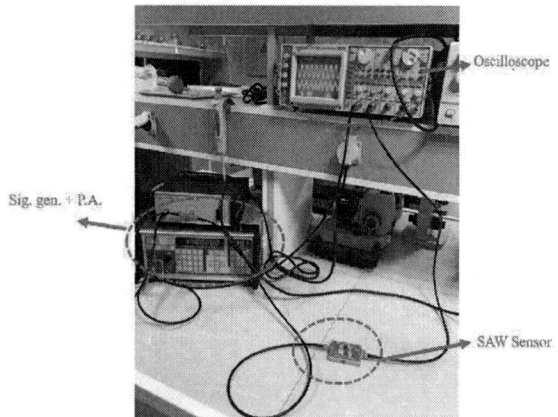

Figure 5 Experimental setup

IV. RESULTS AND CONCLUSION

We aim to evaluate our ZnO-based SAW sensor functionality for sensing different DI water droplet volumes placed on the delay line of the FIDTs (Figure 6).

Figure 6. Droplet test

To obtain the resonance frequency of the device, first, we applied a sinusoidal signal in the 44- 94 MHz frequency range and read out the voltage amplitude of the output FIDT (Figure 7). Plotting the output (v_{out})/input (v_{in}) signals ratio at each frequency when no droplet was placed on the sensing area of the sensor resulted in the resonance frequency at which this ratio is maximized. For our sensor, the resonance frequency was around 68 to 72 MHz.

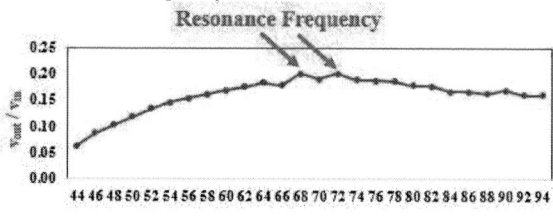

Figure 7. Frequency response of the device

Around the resonance frequency, we introduced a 6, 7, 8, 9, 10, 15, and 20 µL DI water droplet on the sensing area (delay line) and measured the output for 3 repetitions (Figure 8).

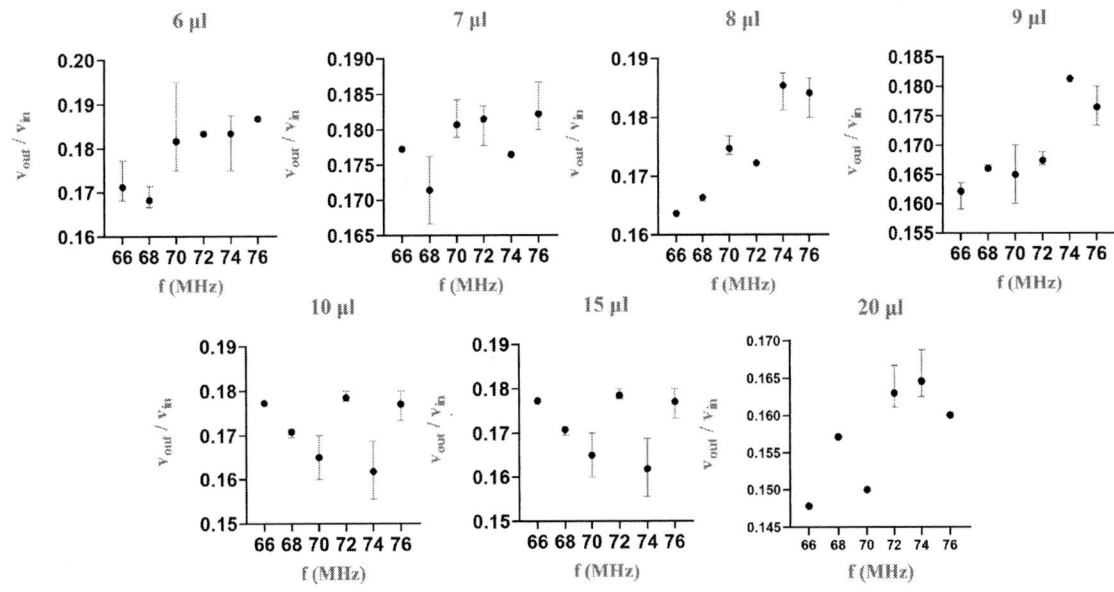

Figure 8. Result of testing different droplet volumes on the device

Figure 9 shows the mass effect on the v_{out}/v_{in} ratio at $f = 70$ MHz, indicating a decreased v_{out}/v_{in} ratio as the droplet volume (equivalently the droplet mass) increases.

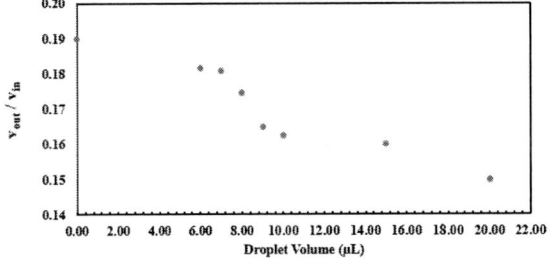

Figure 9. The mass effect shows a decreasing v_{out}/v_{in} ratio by increasing droplet volume

V. REFERENCES

[1] K. Länge, "Bulk and surface acoustic wave sensor arrays for multi-analyte detection: A review," *Sensors,* vol. 19, no. 24, p. 5382, 2019.

[2] D. Mandal and S. Banerjee, "Surface acoustic wave (SAW) sensors: Physics, materials, and applications," *Sensors,* vol. 22, no. 3, p. 820, 2022.

[3] M. P. Nair, A. J. Teo, and K. H. H. Li, "Acoustic biosensors and microfluidic devices in the decennium: Principles and applications," *Micromachines,* vol. 13, no. 1, p. 24, 2021.

[4] A. T. Le, M. Ahmadipour, and S.-Y. Pung, "A review on ZnO-based piezoelectric nanogenerators: Synthesis, characterization techniques, performance enhancement, and applications," *Journal of Alloys and Compounds,* vol. 844, p. 156172, 2020.

[5] S. Abbasi, B. Barahimi, S. Darbari, M. K. Moravvej-Farshi, and M. Zabetian, "ZnO-based Acoustofluidics: Droplet-based Particle Manipulation," in *2022 30th International Conference on Electrical Engineering (ICEE),* 2022: IEEE, pp. 331-335.

[6] S. Krishnamoorthy, "Development of a ZnO/SiOsub2/Si High Sensitivity Interleukin-6 Biosensor," 2007.

[7] M. B. Mazalan, A. M. Noor, Y. Wahab, S. Yahud, and W. S. W. K. Zaman, "Current development in interdigital transducer (IDT) surface acoustic wave devices for live cell in vitro studies: A review," *Micromachines,* vol. 13, no. 1, p. 30, 2021.

[8] S. J. Ikhmayies, N. M. A. El-Haija, and R. N. Ahmad-Bitar, "A comparison between different ohmic contacts for ZnO thin films," *Journal of Semiconductors,* vol. 36, no. 3, p. 033005, 2015.

[9] J. Liang, F. Kohsaka, T. Matsuo, X. Li, and T. Ueda, "Improved bi-layer lift-off process for MEMS applications," *Microelectronic Engineering,* vol. 85, no. 5-6, pp. 1000-1003, 2008.

The 5th Iranian International Conference on Microelectronics (IICM2023)

25 – 26 October 2023

Design and Fabrication of Carbon Nanoparticles-Based Sensor by Arc Discharge Method

Golsa Taghizadeh Afshari
Electrical and Computer Department
Urmia University
Urmia, Iran
st_g.taghizadehafshari@urmia.ac.ir

Mohammad Taghi Ahmadi
Institute of Power Engineering
Universiti Tanaga Nasional
Malaysia
mahmadi@uniten.edu.my

Amir Fathi
Electrical and Computer Department
Urmia University
Urmia, Iran
a.fathi@urmia.ac.ir

Abstract— **In this article, a sensor based on carbon nanoparticles has been designed and fabricated. Carbon nanoparticles have been synthesized by the Ac Arc Discharge device in suitable environmental conditions from the decomposition of butane gas. These nanoparticles are grown in different lengths between two metal electrodes. Considering this set as a field effect transistor in the role of an electrical sensor, its channel length changes according to the distance between the two electrodes. To check the correct functionality of the sensor, its sensitivity to different concentrations of salt-water has been checked and the resistance graph has been obtained according to the concentration for different Channel lengths.**

Keywords—nanoparticles, arc discharge method, sensor

I. Introduction

As a diagnostic platform, electrical sensors provide a measurable response to a physical phenomenon in the form of an electrical signal [1]. One of these types of sensors is electrical sensors based on carbon nanoparticles. These types of sensors have attracted a lot of attention due to their unique physical and chemical properties because of the size and type of nanoparticles used in them. Carbon nanoparticles have characteristics such as high electrical and thermal conductivity, biocompatibility, and high surface-to-volume ratio [2,3]. These nanoparticles include carbon nanotubes, graphene, graphene oxide, fullerene, and carbon-based

quantum dots. Fig.1 indicates the most popular structures by sp^2 hybridization [4,5].

Graphene is a two-dimensional nanoparticle in the form of a single-layer graphite sheet with a thin atomic thickness (0.345 nm). The carbon atoms are very tight together due to their hybridization and a form of hexagonal honeycomb structure [6,7]. Graphene has been theoretically discussed since 1947, but in 2004, two physicists from the University of Manchester obtained a layer with a thickness of one atom from a graphite sheet [8]. The characteristics of this nanoparticle include high surface area, high Young's modulus, and extremely high electrical conductivity[9]. The high electrical conductivity of graphene is due to the extremely high speed of electrons when passing through it. Also, graphene has a very high thermal conductivity even compared to carbon nanotubes. Graphene is harder than steel and diamond, but at the same time, it has high tensile strength [10].

Fullerene is a three-dimensional nanoparticle containing carbon molecules in the form of spheres or hollow ovals and has pentagonal and hexagonal faces. In 1985, its first sample was discovered as C_{60} which contained 60 carbon atoms [11,12]. These atoms are arranged in a spherical structure similar to a soccer ball. In this structure, due to existing hybridization, each carbon atom bonds with its three neighboring atoms. These bonds are single bonds in pentagons and double bonds in hexagons [13,14].

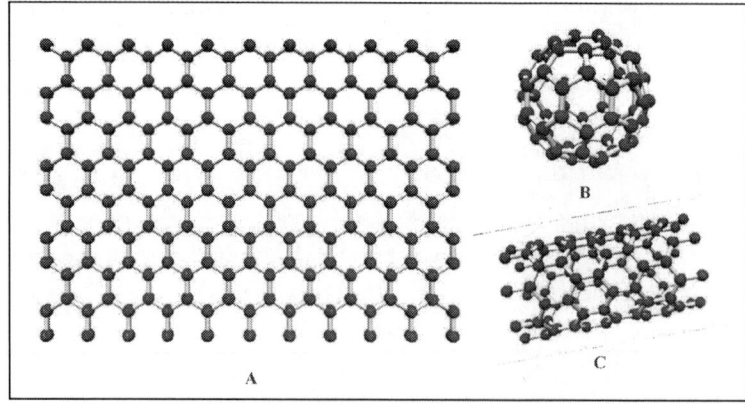

Fig. 1. A: Graphene sheet, B: Fullerene, C: Single-wall nanotube (SWCNT)

979-8-3503-6020-2/23 $31.00 © 2023 IEEE

Carbon nanotube as one-dimensional carbon nanoparticle is hollow cylinders and consist of one or more concentrically folded graphene layers with diameters of about 0.4 nanometers to tens of nanometers. One of these nanoparticles' important features is the high length-to-diameter ratio [5]. These nanoparticles are structurally divided into two categories based on the number of wall layers: single-wall nanotubes (SWCNT) and multi-wall nanotubes (MWCNT). Single-wall carbon nanotubes can include zigzag, armchairs, and chiral structures based on the roll-up direction. This is the fact that multi-wall carbon nanotubes include two states. In the first state, a graphene sheet rotates several times (Swiss), and in the next state, the graphene sheets are arranged as concentric cylinders (Russian doll) [15].

This article presents a fabrication method of an electrical sensor based on carbon nanoparticles. In the second part, the growth methods of nanoparticles are discussed. In the third part, the method of fabricating the sensor and in the fourth part of this article, the results of salt-water detection by this sensor are given.

II. NANOPARTICLES GROWTH METHODS:

In general, nanoparticles are synthesized by 2 top-down and bottom-up approaches [16]. In the top-down approach, the production of carbon nanoparticles occurs by breaking down larger carbon structures such as graphite, carbon black, and activated carbon. This approach is applicable through methods such as electric arc discharge, laser ablation, electrochemical oxidation, and ultrasonic methods. While in the bottom-up approach, small carbon organic precursors such as citrate, carbohydrates, and small alcohols are used for the synthesis of nanoparticles in larger and desired dimensions. In this method, the shape, dimensions, and characteristics of the grown nanoparticles are under control. Common methods of this approach include hydrothermal treatment, microwave synthesis, thermal decomposition, sol-gel, and chemical vapor deposition [17-21].

Among these methods, electric arc discharge, chemical vapor deposition, and laser ablation are more common. In the electric arc discharge method, a potential difference is applied between two graphite electrodes in the presence of inert gases, usually argon. The resulting arc discharge evaporates the graphite from the anode. Then this evaporated graphite is cooled and condensed to form of carbon

nanotubes in cathode [22]. The synthesis of nanoparticles with this method depends on things such as the type of power source, the shape and diameter of the electrodes, the gap between the electrodes, and the environment in which the electrodes are placed [23]. In the chemical vapor deposition method, carbon precursors such as hydrocarbons are used at high temperatures and in the presence of a catalyst. High temperature causes the decomposition of hydrocarbons, and after cooling, the synthesized nanoparticles will be available. This method is usually used for the mass production of carbon nanotubes. In this method, the morphology and structure of the synthesized carbon nanotubes can be controlled [24-25]. The laser ablation method irradiates a laser pulse to the solid target to create nanoparticles. For the synthesis of carbon nanoparticles in this way, the solid target is made of carbon. This process takes place in an aqueous or gaseous environment. Among the advantages of this method, it can be mentioned that it is an easy method and not using multi-step chemical methods, the processing time is not long, and it is compatible with the environment [26-27].

III. PROPOSED SENSOR CONSTRUCTION:

A. Proposed sensor construction:

According to Fig.2, the proposed sensor manufacturing system includes two metal electrodes inside a chamber, a distance meter with one-micrometer accuracy, and an Ac Arc Discharge system to apply the desired high voltage and a gaseous carbon source. The distance between the electrodes can be controlled. The proposed electric arc discharge device has two volumes to adjust the voltage range and frequency. Also, in this article, butane gas with the chemical formula C_4H_{10} is used as a source of gaseous carbon.

B. The growth mechanism of nanoparticles:

In general, by applying a voltage to two electrodes, the electric potential created between the electrodes causes the electrons to move from the positive pole to the negative pole and create an electric current. When ambient air acts as an electrical insulator between them, no electron flow occurs under normal conditions for low voltages. If the voltage is properly increased, the gas between the electrodes is ionized due to the discharge of the electric arc. In this condition, the medium between the electrodes,

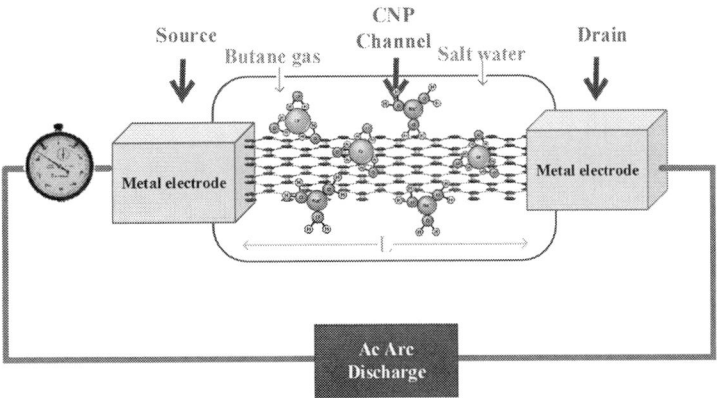

Fig. 2. General schematic of nanoparticle growth system

The 5th Iranian International Conference on Microelectronics (IICM2023)

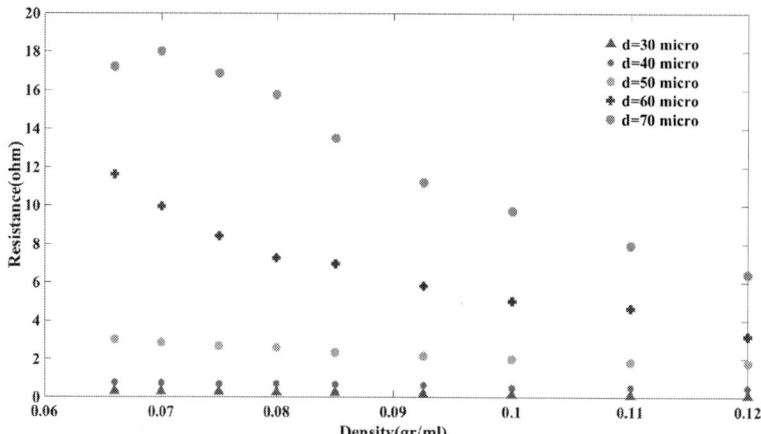

Fig. 3. Variations of resistance with respect to the changes in the Density

becomes conductive and allows charge carriers to pass through it. This phenomenon occurs very quickly and is usually accompanied by the discharge of its energy in the form of light.

By injecting butane gas between two electrodes, the electric field causes the decomposition of butane gas molecules and the production of carbon ions in this space. The resulting carbon ions also vibrate under the influence of a strong electric field between two electrodes and are deposited on one of the electrodes in pulse resonances, and this deposition growths to the other electrode in about a few seconds. As mentioned before, this method is a top-down approach. This set can be considered as a field effect transistor, where the electrodes are the drain and source and carbon nanoparticles are the channel of this transistor. After the completion of growth, this channel is always persistent and is responsible for the detection of different substances. Transistor manufacturing has been done with channel lengths of 30, 40, 50, 60, and 70 micrometers. In this channel, which is electrically neutral in general, due to the approach of charged salt-water molecules, an opposite charge is induced and changes in the electrical conductivity of the channel and changes in the threshold voltage and channel current. Finally, the current-voltage output parameter of this transistor changes. This shows the sensitivity of the fabricated sensor to this solution.

For the growth of nanoparticles at any given distance between two electrodes, there is a threshold voltage at which the electric arc discharge is established between the two electrodes. Threshold voltage depends on distance between the two electrodes, materials, and as environment conditions. The voltage required to start the growth of nanoparticles is slightly higher than the threshold voltage of the desired distance. Applying this voltage breaks butane and ionizes gas atoms. Carbon ions are deposited on one electrode and grow towards the other electrode. By applying a voltage between the two electrodes, this growth continues until the complete connection between the two electrodes. It is also considered constant in all frequency intervals.

IV. RESULTS

After the growth of nanoparticles with different lengths, different concentrations of salt-water have been injected on these nanoparticles to evaluate the functionality of the proposed sensor. Then the resistance between the two electrodes is measured and the results are shown in Fig.3. This experiment has been repeated for channel lengths of 30 to 70 µm and concentrations between 0.06 and 0.12 g/ml.

According to the results in Fig.3, increasing the concentration of salt-water has caused a decrease in the resistance of the sensor, which shows the sensitivity of the fabricated sensor to salt-water. When salt dissolves in water, its ions are separated to Na^+ and Cl^-. As a result, salt-water is considered a conductive solution due to the presence of these ions in it. By increasing the concentration of the salt water solution, these ions increase in the sensor channel and thus increase the conductivity of the sensor. In this case, according to the relation V=IR, the resistance of the sensor decreases.

It is crystal clear that the length of the channel, composed of carbon nanoparticles, has a significant effect on the behavior of the fabricated sensor. At low concentrations, the effect of channel length on the sensor resistance is significant, while at high concentrations, the sensor resistances at different distances are somewhat close to each other, and the effect of channel length decreases, and gradually with increasing concentration, the length effect decreases.

Also, the percentage of sensor resistance changes at different distances has been investigated.

According to Tables 1 and 2, at high concentrations of salt-water, 31.6% change in resistance has been achieved for the channel length of 60 micrometers and 18.8% for the channel length of 70 micrometers. While for low concentrations, the resistance changes are 14.2% for the channel length of 60 micrometers and 4.3% for the channel length of 70 micrometers. Therefore, it can be concluded that the fabricated sensor performs better in the longest channel lengths and in high concentrations.

In high and low concentrations of salt-water, the highest percentage of resistance-change occurs along the 60-micrometer channel. The Ac Arc Discharge device includes RLC circuits.

Channel length (μm)	R₁ (ohm)	R₂ (ohm)	Percentage of changes
30	0.31	0.3	3.2%
40	0.78	0.74	5.1%
50	3.01	2.85	5.3%
60	11.62	9.96	14.2%
70	18.01	17.22	4.3%

Table 1. Sensitivity comparison of fabricated sensor for different channel lengths- Low concentrations

Channel length (μm)	R₁ (ohm)	R₂ (ohm)	Percentage of change
30	0.1	0.098	2%
40	0.49	0.48	2%
50	1.83	1.76	3.8%
60	4.64	3.17	31.6%
70	7.91	6.42	18.8%

Table 2. Sensitivity comparison of the fabricated sensor for different channel lengths- High concentrations

It seems that in the length of the 60-micrometer channel, the frequency is matched to the frequency of the RLC circuit and the phenomenon of resonance has occurred. For this reason, the growth process in this length is stronger and has shown more resistance changes than other lengths of the channel.

V. CONCLUSION

This article describes the design and construction of a sensor based on carbon nanoparticles. This sensor is a field effect transistor and includes two electrodes as a source and drain and nanoparticles grown between the two electrodes as a permanent channel. The channel is grown in lengths of 30 to 70 micrometers by electric arc discharge method and contains carbon nanoparticles. To check the performance of this sensor, salt-water with concentrations of 0.06 to 0.12 was injected into channels of different lengths and the change in the resistance of the sensor was observed. In this way, in high concentrations, for every 0.01 change in salt-water concentration, 18.8% change in sensor resistance was observed in the longest channel length and 2% change in sensor resistance in the shortest channel length. At low concentrations, for every 0.005 change in salt-water concentration, 3.2% change in sensor resistance was observed in the shortest channel length and 4.3% change in resistance in the longest channel length.

REFERENCES

[1] Javaid, M., Haleem, A., Rab, S., Singh, R. P., & Suman, R. (2021). Sensors for daily life: A review. Sensors International, 2, 100121.

[2] Wang, F., & Hu, S. (2009). Electrochemical sensors based on metal and semiconductor nanoparticles. Microchimica Acta, 165, 1-22.

[3] Holmannova, D., Borsky, P., Svadlakova, T., Borska, L., & Fiala, Z. (2022). Reproductive and Developmental Nanotoxicity of Carbon Nanoparticles. Nanomaterials, 12(10), 1716.

[4] Kotia, A., Yadav, A., Rohit Raj, T., Gertrud Keischgens, M., Rathore, H., & Sarris, I. E. (2020). Carbon nanoparticles as sources for a cost-effective water purification method: A comprehensive review. Fluids, 5(4), 230.

[5] Lisik, K., & Krokosz, A. (2021). Application of carbon nanoparticles in oncology and regenerative medicine. International Journal of Molecular Sciences, 22(15), 8341.

[6] Zhang, T., Xue, Q., Zhang, S., & Dong, M. (2012). Theoretical approaches to graphene and graphene-based materials. Nano Today, 7(3), 180-200.

[7] Olabi, A. G., Abdelkareem, M. A., Wilberforce, T., & Sayed, E. T. (2021). Application of graphene in energy storage device–A review. Renewable and Sustainable Energy Reviews, 135, 110026.

[8] Anas, N. A. A., Fen, Y. W., Omar, N. A. S., Daniyal, W. M. E. M. M., Ramdzan, N. S. M., & Saleviter, S. (2019). Development of graphene quantum dots-based optical sensor for toxic metal ion detection. Sensors, 19(18), 3850.

[9] Holmannova, D., Borsky, P., Svadlakova, T., Borska, L., & Fiala, Z. (2022). Reproductive and Developmental Nanotoxicity of Carbon Nanoparticles. Nanomaterials, 12(10), 1716.

[10] Sukumaran, L. (2014). A study of graphene. Int. J. Educ. Manag. Eng, 4, 9-14.

[11] Isaacson, C. W., Kleber, M., & Field, J. A. (2009). Quantitative analysis of fullerene nanomaterials in environmental systems: a critical review. Environmental science & technology, 43(17), 6463-6474.

[12] 12. Kroto, H. W., Heath, J. R., O'Brien, S. C., Curl, R. F., & Smalley, R. E. (1985). C60: Buckminsterfullerene. nature, 318(6042), 162-163.

[13] Markovic, Z., & Trajkovic, V. (2008). Biomedical potential of the reactive oxygen species generation and quenching by fullerenes (C60). Biomaterials, 29(26), 3561-3573.

[14] Manawi, Y. M., Ihsanullah, Samara, A., Al-Ansari, T., & Atieh, M. A. (2018). A review of carbon nanomaterials' synthesis via the chemical vapor deposition (CVD) method. Materials, 11(5), 822.

[15] Sengupta, J., & Hussain, C. M. (2022). Decadal Journey of CNT-Based Analytical Biosensing Platforms in the Detection of Human Viruses. Nanomaterials, 12(23), 4132.

[16] Dhand, C., Dwivedi, N., Loh, X. J., Ying, A. N. J., Verma, N. K., Beuerman, R. W., ... & Ramakrishna, S. (2015). Methods and strategies for the synthesis of diverse nanoparticles and their applications: a comprehensive overview. Rsc Advances, 5(127), 105003-105037.

[17] Kokorina, A. A., Prikhozhdenko, E. S., Sukhorukov, G. B., Sapelkin, A. V., & Goryacheva, I. Y. (2017). Luminescent carbon nanoparticles: Synthesis, methods of investigation, applications. Russian Chemical Reviews, 86(11), 1157.

[18] Ealia, S. A. M., & Saravanakumar, M. P. (2017, November). A review on the classification, characterisation, synthesis of nanoparticles and their application. In IOP conference series: materials science and engineering (Vol. 263, No. 3, p. 032019). IOP Publishing.

[19] Kokorina, A. A., Prikhozhdenko, E. S., Sukhorukov, G. B., Sapelkin, A. V., & Goryacheva, I. Y. (2017). Luminescent carbon nanoparticles: Synthesis, methods of investigation, applications. Russian Chemical Reviews, 86(11), 1157.

[20] Lim, J. V., Bee, S. T., Tin Sin, L., Ratnam, C. T., & Abdul Hamid, Z. A. (2021). A review on the synthesis, properties, and utilities of functionalized carbon nanoparticles for polymer nanocomposites. Polymers, 13(20), 3547.

[21] Asadian, E., Ghalkhani, M., & Shahrokhian, S. (2019). Electrochemical sensing based on carbon nanoparticles: A review. Sensors and Actuators B: Chemical, 293, 183-209.

[22] Arora, N., & Sharma, N. N. (2014). Arc discharge synthesis of carbon nanotubes: Comprehensive review. Diamond and related materials, 50, 135-150.

[23] El-Khatib, A. M., Badawi, M. S., Ghatass, Z. F., Mohamed, M. M., & Elkhatib, M. (2018). Synthesize of silver nanoparticles by arc discharge method using two different rotational electrode shapes. Journal of Cluster Science, 29, 1169-1175.

[24] Manawi, Y. M., Ihsanullah, Samara, A., Al-Ansari, T., & Atieh, M. A. (2018). A review of carbon nanomaterials' synthesis via the chemical vapor deposition (CVD) method. Materials, 11(5), 822.

[25] Mohd Nurazzi, N., Asyraf, M. M., Khalina, A., Abdullah, N., Sabaruddin, F. A., Kamarudin, S. H., ... & Sapuan, S. M. (2021). Fabrication, functionalization, and application of carbon nanotube-reinforced polymer composite: An overview. Polymers, 13(7), 1047.

[26] Semaltianos, N. G. (2010). Nanoparticles by laser ablation. Critical reviews in solid state and materials sciences, 35(2), 105-124.

The 5th Iranian International Conference on Microelectronics (IICM2023)

[27] Chrzanowska, J., Hoffman, J., Małolepszy, A., Mazurkiewicz, M., Kowalewski, T. A., Szymanski, Z., & Stobinski, L. (2015). Synthesis of carbon nanotubes by the laser ablation method: Effect of laser wavelength. physica status solidi (b), 252(8), 1860-1867.

Synthesis of $TiNb_2O_7$ by mechanical alloying and subsequent heat treatment as an anode material for Li-ion batteries

Shiva Rashidi Kia
Faculty of Materials Science and Engineering,
K. N. Toosi University of Technology
Tehran, Iran
sh.rashidikia@gmail.com

Mehdi Khodaei
Faculty of Materials Science and Engineering,
Materonics and Materionics Research Group,
Advanced Materials and Nanotechnology Research Lab,
K. N. Toosi University of Technology
Tehran, Iran
khodaei@kntu.ac.ir

Abstract—**Lithium-ion batteries are widely used as rechargeable power sources for various devices. One of the main challenges in lithium-ion batteries is achieving high charging speeds. For this purpose, titanium-niobium-oxygen compounds are used as new anodes for lithium-ion batteries. Mechanochemical synthesis consisted of high energy ball milling of constituents followed by a heat treatment process is one of the solid-state methods that can be used to synthesize this material. In this work, the nanostructured $TiNb_2O_7$ has been synthesized by ball milling of TiO_2 and Nb_2O_5 raw materials and subsequent heat treatment. The structural, morphological and elemental characterizations have been done by XRD, SEM, and EDS analysis, respectively. The electrochemical properties of batteries were measured by rate performance and capacity retention tests. These tests show that the samples retain 65% of their initial capacity after 200 cycles, and batteries contain more than 85% of their initial capacity after 45 cycles.**

Keywords—*$TiNb_2O_7$, mechanochemical synthesis, ball milling, anode, Li-ion battery.*

I. INTRODUCTION

Lithium-ion batteries (LIB) were developed by Sony in 1991. Lithium-ion batteries offer several advantages over other battery types, including high energy density, excellent charge efficiency, safety and stability, low maintenance, long lifespan, high rate of discharge, and low self-discharge rate[1, 2].

For the first time, the Goodenough group in 2011 and 2014 and then the research and development center of Toshiba company in 2016 and 2017 presented TNO compounds as an alternative anode material to compound LTO anode materials to improve the properties of Li-ion batteries[3, 4]. $TiNb_2O_7$ is a promising alternative anode material with high charge/discharge rates, excellent cycling stability, and high energy density. The theoretical capacity of $TiNb_2O_7$ is equal to 387.6 mAh/g. But practically, anode material $TiNb_2O_7$ has weak ionic and electrical conductivity $TiNb_2O_7$ anodes for Li-ion batteries face challenges related to low electrical conductivity and limited lithium-ion diffusion. However, there are some techniques to improve electronic conductivity in Li-ion batteries and the overall performance of $TiNb_2O_7$ anodes such as surface nitridation, carbon coating, and nanostructuring[5].

The synthesis of $TiNb_2O_7$ (TNO) nanostructures can be achieved through various methods, including hydrothermal synthesis, solvothermal procedures, and mechanochemical synthesis.

B. Liu et al.[6] fabricated nano-sized $TiNb_2O_7$ by using an electrospinning technique. Electrospinning is a facile method to produce fibers with nano/microscale size; therefore this method can obtain shorter and less tortuous diffusion paths of lithium ions. V. Paulraj et al.[7] synthesized $TiNb_2O_7$ nanoparticles by co-precipitation technique. The discharge capacity of the $TiNb_2O_7$ nanoparticles was 143 mAh/g at 1C. K. Ise et al.[8] generated highly crystalline $TiNb_2O_7$ nanoparticles by a hydrothermal method and investigated the electrochemical properties of H- $TiNb_2O_7$ in the vast potential range of 0.6–3.0V. H. Aghamohammadi et al.[9] synthesized $TiNb_2O_7$ anode materials for Li-ion batteries by solvothermal method. They calcined the material at a temperature of 900 °C for two different hours (5 and 10 hours). The discharge capacity of their battery, which has been annealed for 10 hours, was 200 mAh/g at 1 C.

Mechanical alloying is a solid-state method that can be used to synthesize titanium niobium oxide (TNO) anodes for lithium-ion batteries. X. Wen et al.[10] synthesized V-doped $TiNb_2O_7$ nanoparticles by wet ball-milling method and investigated high electrochemical performance. The discharge capacity of their battery was 298.4 mAh/g at 1 C. T. Adhami et al.[11] prepared TNO compounds by solid-state method and annealed them under different atmospheres. They showed that different atmospheres in annealing treatment had no efficacy on the morphology of the synthesized compounds. While compounds annealed under an oxygen atmosphere showed a higher capacity than the argon atmosphere. G. Liu et al.[12] used the mechanical alloying method to make TNO/Ag anodes for Li-ion batteries. They made $TiNb_2O_7$ by ball-milling the materials for 6 hours, then calcined it at a temperature of 1100°C for 24 hours. Their battery's 2nd discharge capacity was 245 mAh/g at 1 C.

II. RESEARCH MATERIALS AND METHODS

A. Synthesis of TiNb₂O₇ nanoparticles

In this research, to synthesize TiNb₂O₇ nanoparticles by mechanochemical method, titanium dioxide, and niobium pentoxide have been used. In this method, 0.462 grams of titanium dioxide (≥98%, Neutron pharmachemical, Tehran, Iran), and 1.537 grams of niobium pentoxide (≥ 99%, Aldrich chemistry, Tehran, Iran) were measured and decanted into ball-mill cups. The weight ratio of powder to balls was considered 1 to 20. Two samples of the desired powder were milled for 20 hours at a speed of 400 rpm.

$$TiO_2 + Nb_2O_5 \rightarrow TiNb_2O_7 \tag{1}$$

Subsequently, to complete the reaction, a furnace was used for annealing the milled material under an air atmosphere. During this operation, the material was annealed at a temperature of 900°C for 2 hours in an air atmosphere. The sample was collected after being cooled to room temperature.

B. Characterizations

The formed phases of the samples were analyzed by X-ray (XRD, Rigaku Ultima IV, Japan) diffraction. The beam Cu-Kα with λ =1.541871 Å was used for the analysis. The crystal structure and morphology of the sample were analyzed by field emission scanning electron microscope (FE-SEM). Analysis Energy dispersive X-ray spectroscopy (EDS) was used to show the dispersion and weight percentage of the components of the samples.

C. Electrochemical measurement

Electrochemical investigations of the sample were carried out by making CR2032 half-cell coin batteries using lithium foil as an anode and covering TiNb₂O₇ particles on the copper foil as a cathode. The composition of the slurry used for coating on the copper foil included 80% TiNb₂O₇ particles, 10% carbon black, and 10% polyvinyl fluoride (PVDF). After mixing thoroughly and coating homogeneously the mentioned slurry on the copper foil, the hot rolling process was used for better adhesion of the coating to the substrate. The loading mass of active material was about 2.7~2.8 mg/cm².

The charge/discharge electrochemical tests were measured by a multichannel battery tester system (NEWARE Company). The rate performance tests were performed at different charge/discharge rates, and the capacity retention tests were investigated at a rate of 1C (387 mA/g) for 200 cycles.

III. RESULTS AND DISCUSSION

A. Structure and morphology

Figure 1. shows the XRD patterns of the synthesized samples with the Rietveld method (before and after calcination of TNO materials). Rietveld refinement of the XRD pattern shows the weight percentage of the TNO sample to be 99% (Figure 1 (b)). A comparison of the two graphs shows that before the heat treatment, the desired phase of TiNb₂O₇ is not formed, but by calcination of the material for 2 hours at a temperature of 900, the TiNb₂O₇ phase is formed.

Figure 2 shows the FE-SEM picture of the synthesized material. The morphology of the samples was analyzed by FE-SEM pictures. The image shows that the average size of the particles of TiNb₂O₇ is less than 1 micron.

Figure 3 shows the EDS analysis of the sample. The EDS map shows that Ti, Nb, and O elements are homogeneously distributed in the synthesized sample.

B. Electrochemical properties

The electrochemical performance results of Li-ion batteries are shown in Figures 4 and 5. Figure 4 shows the cycling performance of the TiNb₂O₇ anodes at 1C charge and discharge rates. The capacity of the prepared Li-ion battery at 1 C is about 245 mAh/g and after 200 cycles is about 140 mAh/g. This indicates that the battery retains more than 60% of its capacity after 200 cycles. Liu et al.[12] obtained the initial battery capacity of 245 mAh/g. These two measurements can be compared with each other. While Aghamohammadi et al.[9] measured the discharge capacity of their battery 200 mAh/g at 1 C.

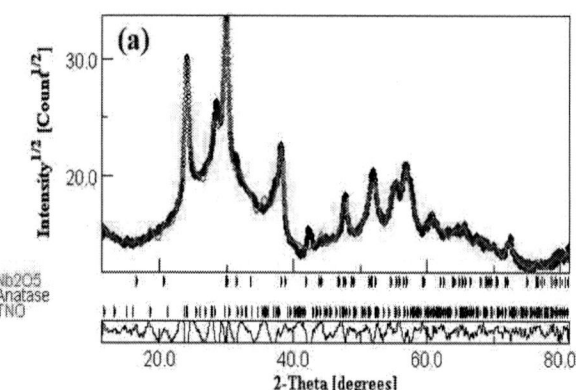

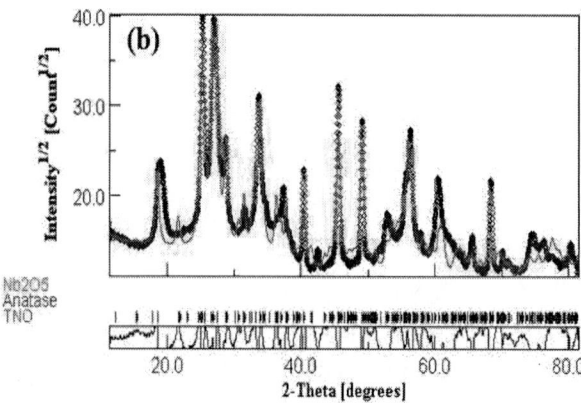

Fig. 1. The Rietveld refinement of XRD patterns for (a) before and (b) after calcination of the sample that has been milled for 20 hours.

The 5th Iranian International Conference on Microelectronics (IICM2023)

Fig. 2. The FE-SEM pictures of TiNb₂O₇.

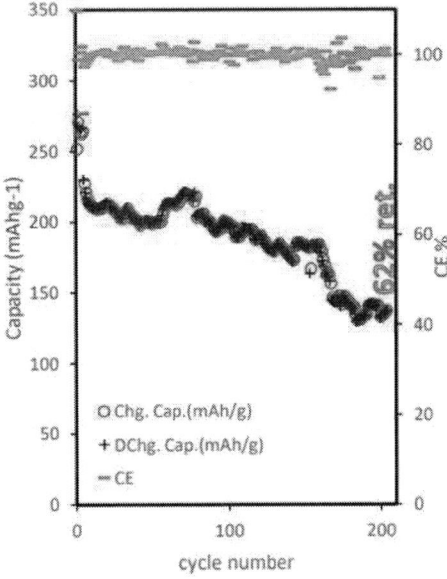

Fig. 4. Cycling performance of the TiNb₂O₇ battery at 1C charge and discharge rates.

Figure 5 shows the rate performance of the TiNb₂O₇ battery at various charge and discharge rates. The capacity of the battery has been measured 275, 250, 220, and 175 mAh/g respectively at 0.1, 0.5, 2, and 5C. The capacity of the desired battery has been measured 100 mAh/g at 20C. This value is two times the capacity that has been measured by Liu et al.[12] and also more than three times the capacity that has been measured by Aghamohammadi et al.[9] at 20C. Because the milling time in this research is longer so the size of the particles became smaller, also the short calcination time prevented the material from becoming agglomerate.

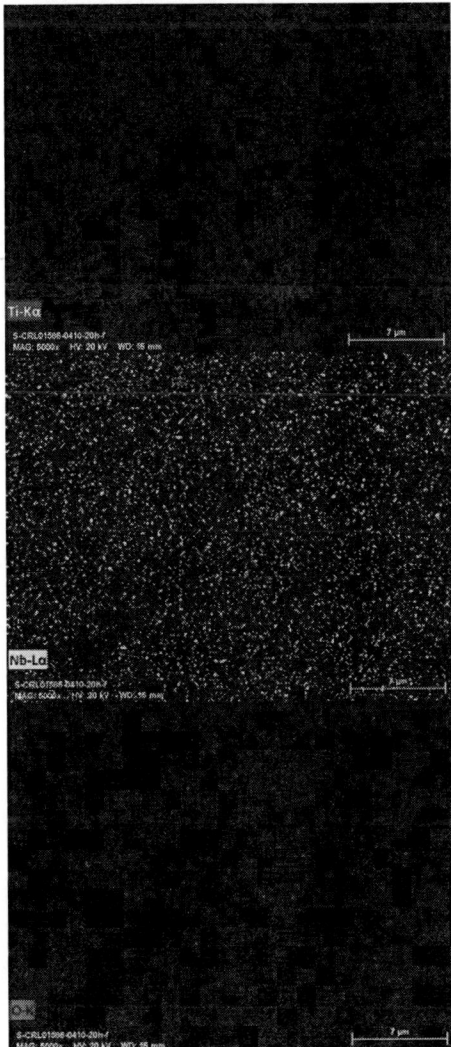

Fig. 3. EDS analysis of the TiNb₂O₇ material.

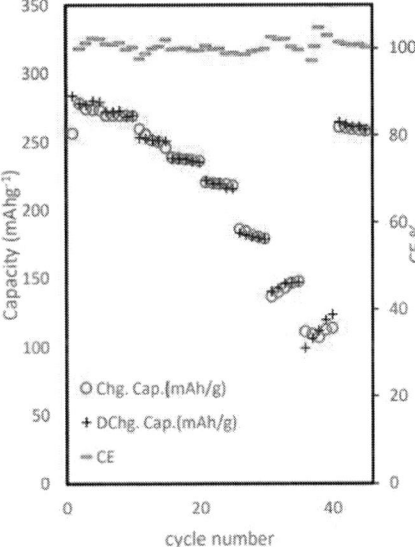

Fig. 5. Rate performance of the TiNb₂O₇ battery at various charge and discharge rates indicated.

The 5th Iranian International Conference on Microelectronics (IICM2023)

IV. CONCLUSION

In this study, $TiNb_2O_7$ was synthesized by mechanochemical method. The TNO-20hf sample was milled for 20 hours and then annealed at 900 ˚C for 2 hours in an air atmosphere. XRD patterns show that the desired TNO compounds were successfully prepared after heat treatment. The size of the particles is in the range of nanometers. In this study, the energy used for calcination is less than in other articles. Also, the shorter time and lower temperature for heat treatment have significantly increased the battery capacity.

REFERENCES

[1] M. Li, J. Lu, Z. Chen, and K. Amine, "30 years of lithium‐ion batteries," *Advanced Materials,* vol. 30, no. 33, p. 1800561, 2018.

[2] M. V. Reddy, A. Mauger, C. M. Julien, A. Paolella, and K. Zaghib, "Brief history of early lithium-battery development," *Materials,* vol. 13, no. 8, p. 1884, 2020.

[3] J.-T. Han and J. B. Goodenough, "3-V full cell performance of anode framework TiNb2O7/spinel LiNi0. 5Mn1. 5O4," *Chemistry of materials,* vol. 23, no. 15, pp. 3404-3407, 2011.

[4] J.-T. Han, Y.-H. Huang, and J. B. Goodenough, "New anode framework for rechargeable lithium batteries," *Chemistry of Materials,* vol. 23, no. 8, pp. 2027-2029, 2011.

[5] A. A. Pavlovskii, K. Pushnitsa, A. Kosenko, P. Novikov, and A. A. Popovich, "Organic Anode Materials for Lithium-Ion Batteries: Recent Progress and Challenges," *Materials,* vol. 16, no. 1, p. 177, 2022.

[6] B. Liu, J. Zhang, X. Wang, G. Chen, D. Chen, C. Zhou, and G. Shen, "Hierarchical three-dimensional ZnCo2O4 nanowire arrays/carbon cloth anodes for a novel class of high-performance flexible lithium-ion batteries," *Nano letters,* vol. 12, no. 6, pp. 3005-3011, 2012.

[7] V. Paulraj, T. Saha, K. Vediappan, and K. K. Bharathi, "Synthesis and electrochemical properties of TiNb2O7 nanoparticles as an anode material for lithium ion batteries," *Materials Letters,* vol. 304, p. 130681, 2021.

[8] K. Ise, S. Morimoto, Y. Harada, and N. Takami, "Large lithium storage in highly crystalline TiNb2O7 nanoparticles synthesized by a hydrothermal method as anodes for lithium-ion batteries," *Solid State Ionics,* vol. 320, pp. 7-15, 2018.

[9] H. Aghamohammadi and R. Eslami-Farsani, "Effects of calcination parameters on the purity, morphology, and electrochemical properties of the synthesized TiNb2O7 by the solvothermal method as anode materials for Li-ion batteries," *Journal of Electroanalytical Chemistry,* vol. 917, p. 116394, 2022.

[10] X. Wen, C. Ma, C. Du, J. Liu, X. Zhang, D. Qu, and Z. Tang, "Enhanced electrochemical properties of vanadium-doped titanium niobate as a new anode material for lithium-ion batteries," *Electrochimica Acta,* vol. 186, pp. 58-63, 2015.

[11] T. Adhami, R. Ebrahimi-Kahrizsangi, H. R. Bakhsheshi-Rad, S. Majidi, M. Ghorbanzadeh, and F. Berto, "Synthesis and electrochemical properties of TiNb2O7 and Ti2Nb10O29 anodes under various annealing atmospheres," *Metals,* vol. 11, no. 6, p. 983, 2021.

[12] G. Liu, X. Liu, Y. Zhao, X. Ji, and J. Guo, "Synthesis of Ag-coated TiNb2O7 composites with excellent electrochemical properties for lithium-ion battery," *Materials Letters,* vol. 197, pp. 38-40, 2017.

A Nonlinear, Low-Power, VCO-Based ADC for Neural Recording Applications

Reza Shokri[1], Yarallah Koolivand[2], Omid Shoaei[1], Orazio Aiello[3], Daniele Caviglia[3]

[1]Biomedical Integrated System Lab, University of Tehran, Tehran, Iran
[2]Khajeh Nasir Toosi University of Technology, Tehran, Iran
[3]DITEN, University of Genoa, Genoa, Italy

Abstract—Neural recording systems are essential for understanding the brain and developing treatments for neurological disorders. Analog-to-digital converter (ADC) is among the required building blocks of neural recording systems, as they convert the brain's electrical signals into digital data that can be processed and analyzed by processing units. In this paper, a new nonlinear ADC for spike sorting in biomedical applications has been introduced. The ADC is implemented with MOSFET varactors and voltage-controlled oscillators (VCO). By exploiting the nonlinear capacitance characteristics of MOSFET varactors, the ADC has a parabolic quantization function to suppress background noise in biomedical signals. Furthermore, nonlinear digitization gives an effective resolution that is almost 0.11 bit more than its physical number of bits. The resolution of neural signals can vary from 3.1 bits in the low amplitude range to 9.11 bits in the high amplitude range. The circuit was designed and simulated using a 180 nm CMOS process, taking up 0.102 mm² of silicon area. While operating at the sampling frequency of 25 kS/s and a supply voltage of 1 Volt, this ADC dissipates 62.4 μW.

Keywords—*Neural recording systems, VCO-based ADC, nonlinear quantization, parabolic function ADC*

I. INTRODUCTION

Neural recording systems have a significant role in a wide variety of neuroscientific and neurophysiological studies, as well as in many neuro-prosthetic applications. The front end of the system typically includes a low-noise amplifier, a filter, and an ADC, which is a critical part of the system. It is responsible for ensuring that the biological signals are acquired and processed accurately. There is no doubt that some design considerations like low power consumption and low area are vital for implantable chips. Especially in the recent high-density neural recording system, which records neural signals from a large number of electrodes, these considerations are even more important [1-4].

Fig. 1 shows an example of a neural signal recorded from the brain of a live rat. During the rest of the neurons, the recorded signal is quiet, and just the background noise (B-Noise) is sampled. While in the excitation phase, large spikes are seen called action potentials (APs) that play a critical role in extracting useful information from brain activities. In neuroscientific and neurophysiological studies, as well as in neuro-prosthetic applications, APs carry the information embedded in neural signals. While neural information is encoded in the time domain for single-neuron firings, the amplitude of APs is used to enhance the spatial resolution

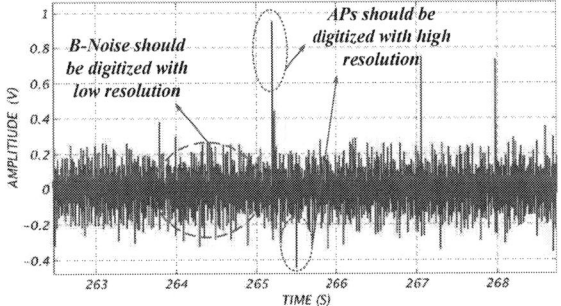

Fig. 1. Neural signal with depicting APs and background noise

and distinguish between APs from different neurons in high-density single-unit recording. Typically, linear ADCs are used in implantable neural recording systems. This means that the non-useful B-Noise is digitized with the same resolution as the useful APs. Since most of the time, the neurons are at rest, a large part of the outgoing bit-rate is wasted to carry the noise content present in the neural signal. Nonlinear ADCs are a promising technology for improving the performance of implantable neural recording systems. By reducing the amount of bit-rate that is wasted on noise, nonlinear ADCs can improve the signal-to-noise ratio (SNR) of the digitized neural signal, and they can also reduce the power consumption and area of the ADC. These nonlinear ADC can be beneficial for applications where the bandwidth is limited in the wireless telemetry block, such as implantable brain-machine interfaces with high-density microelectrodes. Moreover, although there have been significant advances in reducing the power consumption of analog-to-digital converters (ADCs), the power consumption of the digital backend processing still dominates in many portable always-on and multi-sensor systems. Reducing the amount of data streaming into the digital back end is another advantage of nonlinear ADCs [5-10].

In advanced CMOS processes, the supply voltage is typically scaled down to reduce power consumption. This can have a significant impact on the performance of traditional voltage-based ADCs, as the dynamic range and speed of these ADCs are inversely proportional to the supply voltage. As a result, traditional voltage-based ADCs are becoming difficult to design in advanced CMOS processes, and time-based ADCs are becoming the preferred choice for many applications in these processes [11].

This paper introduces a new nonlinear VCO-based ADC that is implemented with nonlinear characteristics of MOS varactors. The organization of this paper is as follows: A brief review of nonlinear neural recording ADCs and conventional VCO-based ADC is presented in section II. The proposed ADC is introduced in section III. The ADC simulation results are summarized in section IV.

II. BRIEF REVIEW

In [5], the 8-bit two-step SAR ADC is presented that uses a piece-wise-linearly-approximated exponential quantization function. The first step of the ADC operation is to determine the segment within which the input signal sample lies. This is done by comparing the input signal to a series of thresholds. The 3 MSBs of the output code are generated in this step. The second step of the ADC operation is to determine the remaining 4 LSBs of the output code, which is done by performing a linear in-segment SAR digitization process. Furthermore, one bit is allocated for the sign of the signal. The analog input voltage is compared with a threshold voltage that is temporarily set to the baseline voltage. The result of this comparison determines the half range in which the input voltage is located. In [6], a new mixed-signal interface that can be programmed to implement a wide range of non-linear transfer functions is presented. This allows the interface to be tailored to the specific needs of different applications. The interface consists of a standard 10-bit binary DAC, a dynamic comparator, and configurable digital logic. The digital logic controls the binary DAC in a closed-loop fashion, iteratively approximating the desired non-linear transfer function. In [8], a logarithmic DAC that uses a diode-connected MOS transistor to generate a logarithmic output voltage has been proposed. The transistor is biased in the subthreshold region, where its current is exponentially related to the gate-source voltage. The output voltage of the DAC is then proportional to the logarithm of the input current. The logarithmic DAC can be used to create an exponential ADC. All the previously presented nonlinear ADCs occupied a large area, which made them unsuitable for high-density neural recording systems. In addition to the large area, the ADC that is presented in the [5] also has large power consumption.

In [12-18] Voltage Controlled oscillator, VCO, has been used as the main core of ADC. The VCO basically converts the input signal to an output whose frequency is proportional to the average input voltage. As shown in Fig. 2-a, The VCO's output could act as a clock signal for a counter. The counter counts the number of clock edges it receives during a fixed sampling period. At the end of each sampling period, the counter output is digital code representing the input signal, called VCO-based ADC. The residual phase of the VCO output at the end of each sampling cycle becomes the initial phase of the next period. This results in first-order noise shaping for the VCO-based ADCs. There are two parameters that could be used for varying the frequency of the ring VCO: the current of VCO and the capacitor load of delay cells [11]. The ADC structure in [12-18] consists of a current mirror and a ring oscillator. The analog input voltage is converted to current feeding the current-starved inverters in the ring VCO affecting the output frequency of the oscillator. These works use differential architecture for enhancing the linearity of the ADC, and a subtractor to subtract the register contents related to positive and negative inputs, as shown in the Fig. 2-a. The subtractor occupies a large area in comparison to

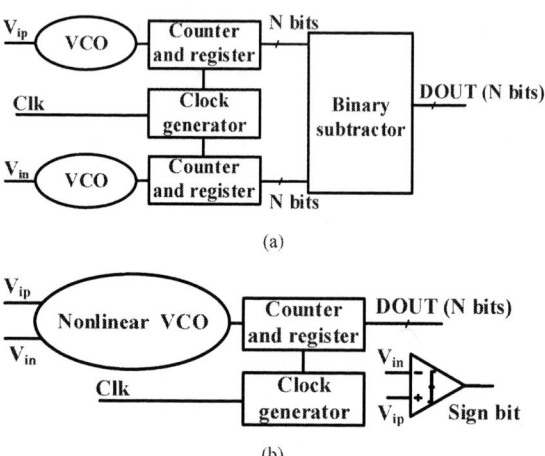

Fig. 2. VCO-based ADC architecture (a) conventional architecture, and (b) proposed nonlinear architecture.

all other subblocks in these ADC. For example, the 8-bit-subtractor, which consists of the 8 parallel 1-bit-subtractors is composed of 32 inverter gates and 64 Nand gates [12] and occupies a large area. In Section III, a new architecture for ADC based on VCO is proposed to reduce the area and power tailored for neural recording.

III. PROPOSED ARCHITECTURE

The proposed architecture for nonlinear ADC based on VCO is depicted in Fig. 2-b. In comparison to the conventional, in the proposed architecture, one of the VCO, counter/register, and the subtractor is removed. Furthermore, the differential input signal is just applied to the PMOS varactors, capacitor loads of the delay cells in VCO, serving the nonlinear output frequency of the VCO with respect to the input signal. Fig. 3 shows the capacitance of the two PMOS transistors versus the input signal. The gates of the transistors are biased to VDD/2 while the drain and source are connected together to the input. In case (a) the bulk is connected to the source and drain while in case (b) the bulk is connected to the supply voltage, VDD. In case (a) the bulk of the transistor experiences both accumulation and inversion at low and high input voltages respectively, while in case (b) the bulk is always in inversion mode, weak inversion at low input and strong inversion at high input signal [19-21]. Therefore, the capacitance behaviors of the two cases at the high input signal are the same while they are different at the low input signal. Since case (b) shows near-constant capacitance at a low input signal, it is considered for the proposed nonlinear ADC to better suppress the B-noise in this application. Fig. 4 shows the simplified VCO circuit employed as a core of nonlinear ADC. In the proposed VCO-based ADC, the input signal is connected to the drain and source terminals of the MOS varactors, while the gates are connected to the output nodes of the delay cells, biased to VDD/2 through the tail current source, and the bulk is connected to the highest voltage, VDD. As shown in Fig. 4, in each delay cell two pairs of PMOS varactors are used, one of them is connected to the positive input voltage, V_{ip}, while the other is connected to the negative ones, V_{in}, to keep symmetry around the y-axis. Furthermore, a dynamic comparator is used to detect the sign of the signal at the end of the conversion. Since the structure removes the B-noise,

979-8-3503-6020-2/23 $31.00 © 2023 IEEE

The 5th Iranian International Conference on Microelectronics (IICM2023)

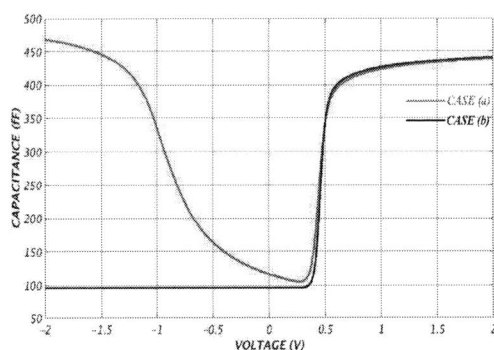

Fig. 3. Capacitance of two PMOS transistors, the bulk of which are connected to source and drain, case (a), or VDD, case (b)

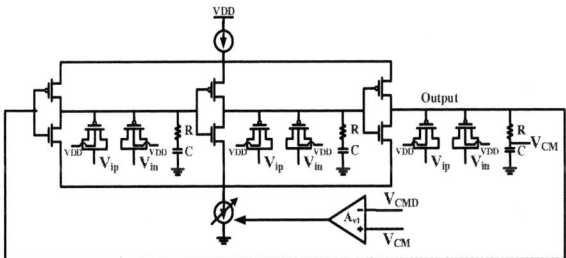

Fig. 4. The simplified schematic of nonlinear VCO used in nonlinear ADC.

the offset of the comparator has little effect on the performance of the ADC. In fact, the frequency variation of VCO around the zero input is negligible, so B-noise is represented with just one code. The extension of code width around the zero point can relax the design of the sign detection comparator against the mismatch. The architecture of the dynamic comparator is shown in Fig. 5. Due to the wide step around the zero point (±50mV), the mismatch and offset of the comparator is negligible, and, the sign detection comparator does not demand serious challenge in circuit design. It should be noted that as shown in Fig. 4, branches composed of a series high resistor and a capacitor are used at the output nodes of the delay cells of the VCO as low-pass filters to sense the DC voltages of the nodes serving as input to the difference amplifier, A_{v1}, to set the bias voltage of the high impedance nodes of VCO to VDD/2. MOS capacitors are sensitive to mismatches, process corners, and temperature variations. For implanted chips, the temperature variation is not a concern. To cope with a mismatch, the capacitor dimensions are chosen as large as possible. However, in VCO-based ADCs, the difference between the maximum frequency and the minimum frequency of the VCO must be greater than 2^N times the sampling frequency to achieve the desired resolution, that N is the number of ADC bits. Therefore, the size of these capacitors must be increased as much as possible, so that in addition to reducing mismatch, the difference between the maximum and minimum frequency of the VCO meets the ADC resolution as well. Despite the mismatch errors, the corner variation is not random and is the same for all the ADCs on a chip. So extra ADC is implemented beside others to calibrate the chip. The calibration ADC is turned on when the chip starts up. The ADC is exposed to a slow ramp signal and the output codes are transmitted to the external host, to make a lookup table for calibrating other life ADCs.

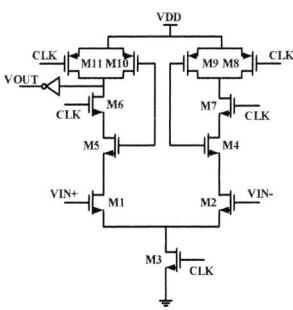

Fig. 5. Architecture of the conventional dynamic comparator

Although area and power are both the constraints of high-density neural recording systems, they are not critical for the host processor side. Therefore, it is possible to match the received code with the lookup table information and perform mathematical operations on the host side without major concern about the power and area. It should be noted that the ADC is susceptible to the common mode signal, and the preceding stage (amplifier/buffer) should provide a differential signal over VDD/2 as the common mode signal.

IV. SIMULATION RESULTS

The proposed architecture was designed using 180nm CMOS technology. Simulations of the architecture confirmed its validity. The power supply of the proposed nonlinear VCO-based ADC is 1V to reduce the power consumption as much as possible. The simulation results show that the output frequency of the VCO ranges from 15.73 MHz to 22.35 MHz when the differential input signal varies from -1.8V to 1.8V. This frequency range is sufficient to achieve the performance of a 9-bit 25 kS/s ADC. Fig. 6 shows the VCO output frequency for half of the input voltage. Shown in Fig. 7 the widths of the first code and the last codes are 50mV and 1.8mV respectively. The ADC resolution is 3.16 bits in the first code and 9.1 bits in the last code. This ADC can effectively eliminate background noise. The ADC power consumption is obtained from the simulation for the highest oscillation frequency ($V_{in} = V_{ip} = $ VDD/2) which is equal to 62.4 µW. Fig. 8 (a) and Fig. 8 (b) show the 25mV amplitude input sinusoidal signal at respectively 12 kHz and 12 MHz frequencies and their corresponding output frequency errors. Since the error frequency is below 25 kHz, no change in the output code will be experienced. To clarify the function of the proposed circuit a 0.9V amplitude differential signal is applied as the input, and the output frequency of the VCO and the sign bit are shown in Fig. 9. Table I summarizes the performance of the proposed recording system and some of the prior arts.

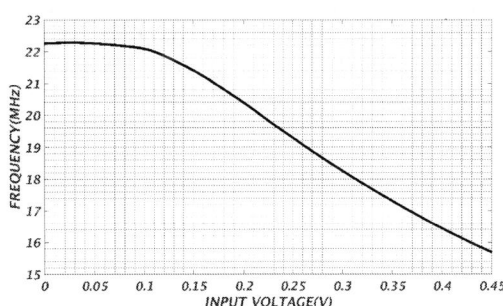

Fig. 6. Frequency of nonlinear VCO vs input voltage

979-8-3503-6020-2/23 $31.00 © 2023 IEEE   201

The 5th Iranian International Conference on Microelectronics (IICM2023)

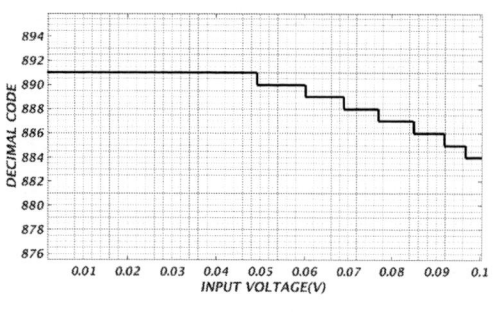

(a)

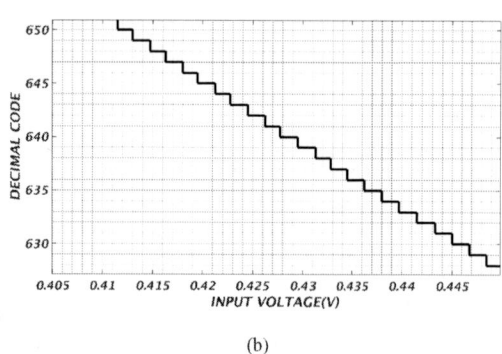

(b)

Fig. 7. Digital output codes of the proposed nonlinear ADC vs input voltage, (a) zoomed low input range, and (b) zoomed high input range.

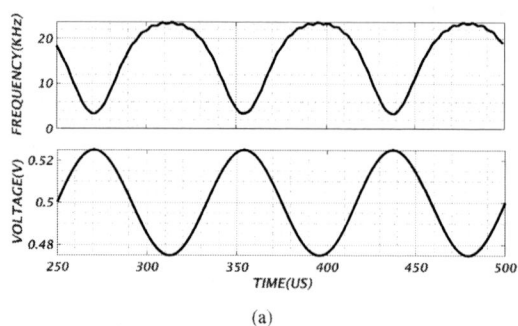

(a)

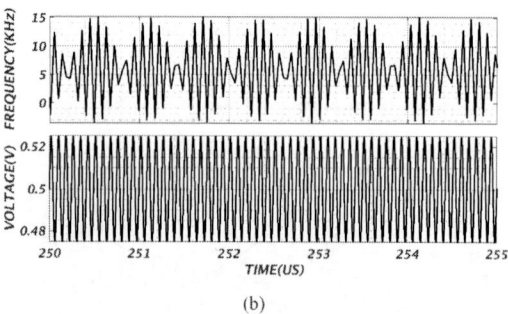

(b)

Fig. 8. 25mV amplitude input signal and the output frequency error when the input frequency is (a) 12 KHz, and (b) 12 MHz.

V. CONCLUSION

A nonlinear ADC is introduced in this paper for neural signal recording applications. Due to the nonlinear function

TABLE I.   RECORDING SYSTEM PERFORMANCE COMPARISION

	[12]	[5]	This Work
Process	0.18	0.18	0.18
ADC Type	Linear, VCO-based	Nonlinear, SAR	Nonlinear, VCO-based
Number of bits	10	8	9
Supply (V)	1	1.8	1
Sampling frequency (kS/s)	25	25	25
Input range (V)	0.16	1	1
Power consumption (µW)	20	87.2	62.4
Area (mm²)	0.027	0.036	0.102

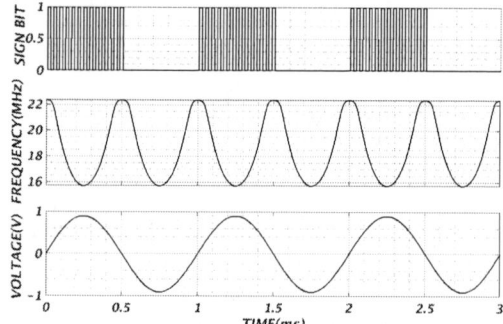

Fig. 9. VCO output frequency and the sign bit for 0.9V amplitude differential input signal

of this ADC, it can effectively eliminate the background noise in neural signals. Unlike conventional VCO-based ADCs, the input signal controls the capacitance of the PMOS varactors and subsequently the frequency of the VCO in nonlinear steps, making B- noise removal possible. Due to its low power consumption, this converter can be used in high-throughput recording systems. A relatively short period to complete the layout of these converters is one of the advantages of this design, as they are semi-synthesizable.

REFERENCES

[1] X. Tong and M. Ghovanloo, "Multichannel Wireless Neural Recording AFE Architectures: Analysis, Modeling, and Tradeoffs," IEEE Design & Test, vol. 33, no. 4, Aug. 2016.

[2] M. Sharma, H. J. Strathman, and R. C. Walker, "Verification of a Rapidly Multiplexed Circuit for Scalable Action Potential Recording," IEEE Transactions on Biomedical Circuits and Systems, vol. 13, no. 6, pp. 1655–1663, Dec. 2019.

[3] J. Chen, M. Tarkhan, H. Wu, F. Noshahr, J. Yang, and M. Sawan, "Recent Trends and Future Prospects of Neural Recording Circuits and Systems: A Tutorial Brief," IEEE Transactions on Circuits and Systems II: Express Briefs, vol. 69, no. 6, pp. 2654–2660, Jun. 2022.

[4] M. Reza Pazhouhandeh, H. Kassiri, A. Shoukry, I. Weisspapir, P. L. Carlen, and R. Genov, "Opamp-Less Sub-µW/Channel Δ-Modulated Neural- ADC With Super-GΩ Input Impedance," IEEE Journal of Solid-State Circuits , vol. 56, no. 5, pp. 1565–1575, May 2021.

[5] M. Judy, A. M. Sodagar, R. Lotfi, and M. Sawan, "Nonlinear Signal Specific ADC for Efficient Neural Recording in Brain-Machine Interfaces," IEEE Transactions on Biomedical Circuits and Systems, vol. 8, no. 3, pp. 371 381, Jun. 2014.

[6] K. Badami, J. Pablo, S. Lauwereins, and M. Verhelst, "Mixed-signal programmable non-linear interface for resource-efficient multi-sensor analytics," 2018 IEEE International Solid - State Circuits Conference - (ISSCC), Feb. 2018.

[7] L. Danial, K. Sharma, S. Dwivedi, and Shahar Kvatinsky, "Logarithmic Neural Network Data Converters using Memristors for

Biomedical Applications," 2019 IEEE Biomedical Circuits and Systems Conference (BioCAS), Oct. 2019.

[8] M. Jomehei, S. Sheikhaei, E. Hadizadeh Hafshejani, and S. Mirabbasi, "A Low-Power Logarithmic CMOS Digital-to-Analog Converter for Neural Signal Recording," IEEE Transactions on Circuits and Systems II: Express Briefs, vol. 69, no. 1, pp. 15–19, Jan. 2022.

[9] JC. Pena-Ramos, K. Badami, S. Lauwereins, and M. Verhelst, "A Fully Configurable Non-Linear Mixed-Signal Interface for Multi-Sensor Analytics," IEEE Journal of Solid-State Circuits, vol. 53, no. 11, pp. 3140–3149, Nov. 2018.

[10] Y. Sundarasaradula, T. G. Constandinou, and A. Thanachayanont, "A 6-bit, two-step, successive approximation logarithmic ADC for biomedical applications," 2016 IEEE International Conference on Electronics, Circuits and Systems, Dec. 2016.

[11] Behzad Razavi, Design of analog CMOS integrated circuits. New York, Ny: Mcgraw-Hill Education, 2017.

[12] X. Tong and J. Wang, "A 1 V 10 bit 25 kS/s VCO-based ADC for implantable neural recording," in 2017 IEEE Biomedical Circuits and Systems Conference (BioCAS), 2017, pp. 1–4

[13] P. Yeon, M. S. Bakir, and Maysam Ghovanloo, "Towards a 1.1 mm2 free-floating wireless implantable neural recording SoC," 2018 IEEE Custom Integrated Circuits Conference (CICC), Apr. 2018.

[14] T.-F. Wu and M. K. Chen, "A Noise-Shaped VCO-Based Nonuniform Sampling ADC With Phase-Domain Level Crossing," IEEE Journal of Solid-State Circuits, vol. 54, no. 3, pp. 623–635, Mar. 2019.

[15] S. Li et al., "A 0.025-mm2 0.8-V 78.5dB-SNDR VCO-Based Sensor Readout Circuit in a Hybrid PLL-ΔΣM Structure," 2019 IEEE Custom Integrated Circuits Conference (CICC), Apr. 2019.

[16] C. Pochet, J.-N. Huang, P. P. Mercier, and D. A. Hall, "A 174.7-dB FoM, 2nd-Order VCO-Based ExG-to-Digital Front-End Using a Multi-Phase Gated-Inverted-Ring Oscillator Quantizer," IEEE Transactions on Biomedical Circuits and Systems, vol. 15, no. 6, pp. 1283–1294, Dec. 2021.

[17] M. Bensenouci, M. A. Ali, Hammoudi Escid, Y. Savaria, and M. Sawan, "A VCO-Based Nonuniform Sampling ADC Using a Slope-Dependent Pulse Generator," 2020 32nd International Conference on Microelectronics (ICM), Dec. 2020.

[18] W. Jiang, V. Hokhikyan, H. Chandrakumar, V. Karkare, and D. Markovic, "A ±50-mV Linear-Input-Range VCO-Based Neural-Recording Front-End With Digital Nonlinearity Correction," IEEE Journal of Solid-State Circuits, vol. 52, no. 1, pp. 173–184, Jan. 2017.

[19] Chenming Hu, Modern semiconductor devices for integrated circuits. Upper Saddle River, N.J.: Prentice Hall, 2010.

[20] M. Li, C.-Y. Jiang, Y. Pan, H. Chen, and Y.-W. Hsu, "Using Inversion-mode MOS Varactors and 3-port Inductor in 0.18-μm CMOS Voltage Controlled Oscillator," 2019 International Symposium on Intelligent Signal Processing and Communication Systems (ISPACS), Dec. 2019.

[21] M. Margalef-Rovira et al., "Highly Tunable High-Q Inversion-Mode MOS Varactor in the 1–325-GHz Band," IEEE Transactions on Electron Devices, vol. 67, no. 6, pp. 2263–2269, Jun. 2020.

Design of 1-1-1 Cascaded Discrete-Time Delta-Sigma Modulator based on Tracking Quantizer

1st Mohsen Ghaemmaghami
dept. Electrical Engineering
University of Guilan
Persian Gulf Highway, Rasht 41996-13776, Iran
ghaemm.mohsen@gmail.com

2nd Shahbaz Reyhani
dept. Electrical Engineering
University of Guilan
Persian Gulf Highway, Rasht 41996-13776, Iran
shahbaz@guilan.ac.ir

Abstract— **In this article, a multi-bit discrete-time delta-sigma modulator based on tracking quantizer for use in telecommunication applications is presented. The proposed 4-bit quantizer uses a comparator, an internal digital-to-analog converter (DAC) and a digital control circuit to predict the integrator output within one clock pulse. In addition, the use of one comparator in the proposed quantizer reduces the power and area consumption of the proposed modulator. In order to study the performance of the proposed method, a 1-1-1 cascaded multi-bit discrete-time delta-sigma with 4-bit proposed quantizer is designed and simulated at the transistor level in 180 nm CMOS technology. The simulation results show a signal-to-noise and distortion ratio (SNDR) of 73.2 dB in a bandwidth of 500 kHz. The oversampling ratio (OSR), power supply voltage, power consumption, Walden's figure of merit (FoMw) and Schreier's figure of merit (FoMs) of the proposed modulator are 32, 1.8 V and 3.3 mW, 0.91 pJ/conv-step and 154.30 dB, respectively.**

Keywords—Analog to digital converter, digital to analog converter, cascaded discrete-time delta-sigma modulator, tracking quantizer

I. INTRODUCTION

In conventional multi-bit delta-sigma modulators, use of a flash quantizer with different voltage levels reduces OSR and quantization noise compared to single-bit delta-sigma modulators [1]. Also, the need for slew rate in the integrator of discrete-time modulators is greatly reduced [2]. On the other hand, if the number of flash quantizer bits increases by one unit, the number of comparators doubles. Therefore, the complexity, silicon area and power consumption of the flash quantizer increases exponentially with each additional bit [3].

The VCO quantizer structure has the same performance as a first-order delta-sigma modulator. Unfortunately, the VCO quantizer used in some delta-sigma modulators has a non-linear performance. Also, the high single-frequency noise components (tone) have limited the implementation of this type of modulators [5]-[7].

In addition, successive approximation register (SAR) ADCs can be used as a multi-bit quantizer in delta-sigma modulators [8],[9]. The SAR quantizer requires a long conversion time to complete its successive approximations

[10]. In other words, to predict the input signal by the SAR quantizer at a fixed bandwidth and OSR, more number of clock pulses are required than the flash quantizer. As a result, the high frequency blocks are needed and the power consumption increases. In the flash and tracking quantizer, the frequency of the clock pulse is equal to the sampling frequency, but for a 4-bit SAR quantizer, the minimum clock frequency must be 5 times the sampling frequency.

Also, tracking ADCs is used as a multi-bit quantizer in continuous-time delta-sigma modulator [11]. This type of tracking quantizer uses three comparators and reference voltages that are generated through a matrix switching circuit with 2^n resistors. The number of comparators in the tracking quantizer can be reduced by increasing the clock rate in the quantizer [12]. In this paper, a 1-1-1 cascaded multi-bit discrete-time delta-sigma modulator is presented, which uses a tracking quantizer with one comparator to predict the integrator output.

The structure of this article is as follows: in the second section, the structures of the conventional multi-bit delta-sigma modulator and the proposed modulator are introduced. In the third section, the circuit schematic of the proposed modulator and its main parts are presented along with their analysis results. In the fourth section, the simulation results of the proposed modulator are presented. Finally, the general conclusion is presented in the fifth section.

II. PROPOSED DELTA-SIGMA MODULATOR STRUCTURE

The cconventional multi-bit delta-sigma modulators usually use a flash or SAR quantizer to generate digital output related to the input signal. Fig. 1 shows a conventional delta-sigma modulator. Using the SAR quantizer, the conversion speed is related to the number of clock pulses that are required to predict the input signal. Therefore, the analog input bandwidth cannot be increased. Each sample of the input signal is compared with the DAC output and estimated using the successive approximation method.

The 5th Iranian International Conference on Microelectronics (IICM2023)

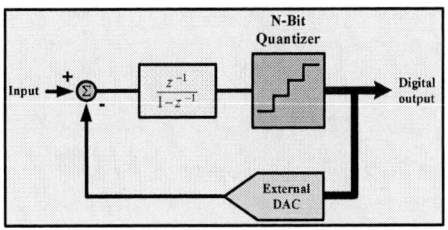

Fig. 1. The structure of a conventional multi-bit delta-sigma modulator.

The proposed delta-sigma modulator uses a tracking quantizer that eliminates the above limitation. The proposed scheme provides an approximate method that predicts the input signal in only one clock pulse. The structure of the proposed multi-bit delta-sigma modulator is shown in Fig. 2.

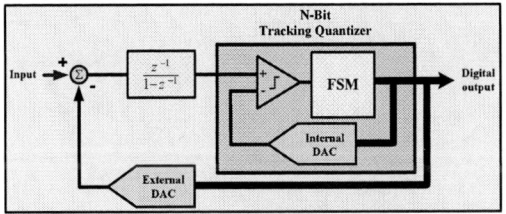

Fig. 2. The structure of a multi-bit delta-sigma modulator using the proposed tracking quantizer.

In the proposed modulator, the tracking quantizer consists of a comparator, a digital control circuit designed as a finite state machine (FSM), and an internal DAC. Also, a binary to thermometric decoder is used to convert 4-bit FSM outputs to 15-bit thermometric codes. Both DACs used in the modulator are of the unit-element type, which produce the required 16 voltage levels. In the proposed tracking quantizer, the input signal is continuously compared with the previous predicted value produced by the internal DAC. The basic problem of tracker quantizers is not following the input signal when its frequency is high, as shown in Fig. 3. Because of the proposed quantizer is used in a delta-sigma modulator, it is necessary to determine the minimum oversampling rate for the 4-bit quantizer. The full scale input voltage is considered to be 1.6V, so the difference between two consecutive voltage levels of the quantizer (produced by the internal DAC) will be 0.1V. In order to correctly estimate the input signal samples with the highest frequency (worst case), two consecutive input samples must be placed at two consecutive voltage levels and compared, otherwise the quantization error will increase and the efficiency of the modulator will decrease.

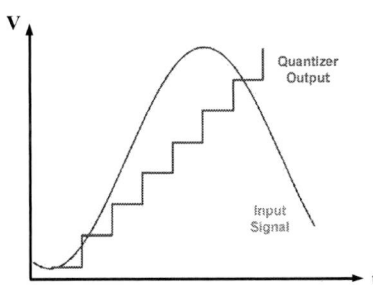

Fig. 3. Limitation of tracking quantizer for high slope input signal prediction.

To overcome this problem, the number of samples in one cycle is calculated for the signal with the maximum input frequency (input bandwidth). OSR is obtained from (1) [10]:

$$2^n < \frac{2}{\pi} OSR \qquad (1)$$

where n and OSR are the number of quantizer bits and the oversampling ratio of the modulator, respectively. This means that the maximum number of bits in the multi-bit modulator is limited by the oversampling ratio and the performance of the proposed quantizer improves with the increase of the oversampling ratio.

III. IMPLEMENTATION OF THE PROPOSED MODULATOR

The block diagram of the proposed 1-1-1 cascaded delta-sigma modulator is shown in Fig. 4. The modulator consists of three stages. Each stage includes an integrator, a 4-bit tracking quantizer, and a DAC. The output of each stage applies to the digital cancellation logic and the final output of the modulator is obtained.

Fig. 5 shows the non-overlapped clocks used in the modulator. The Φ_{1D} clock has a delay compared to Φ_1, and same goes for the other two clocks (Φ_{2D} and Φ_2).

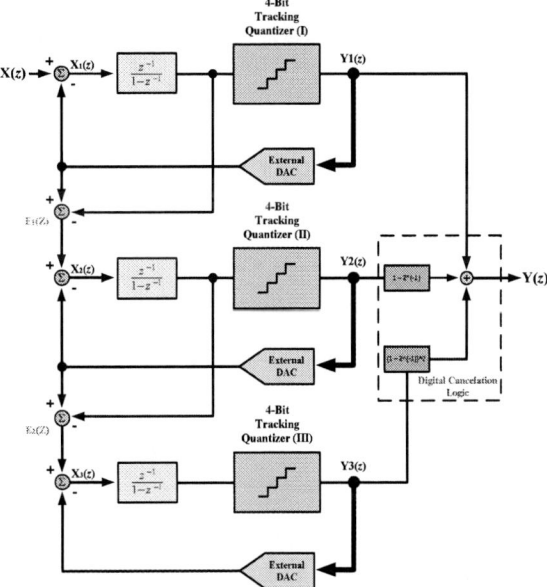

Fig. 4. Block diagram of the proposed third order delta-sigma modulator.

979-8-3503-6020-2/23 $31.00 © 2023 IEEE   205

The 5th Iranian International Conference on Microelectronics (IICM2023)

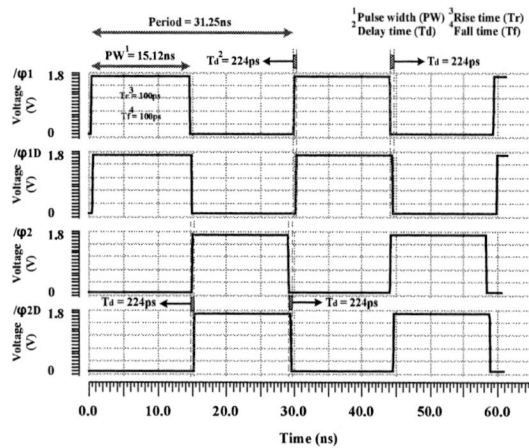

Fig. 5. Timing diagram of the proposed modulator.

The proposed third-order delta-sigma modulator was designed and implemented at the transistor level in a fully differential manner (Fig. 6).

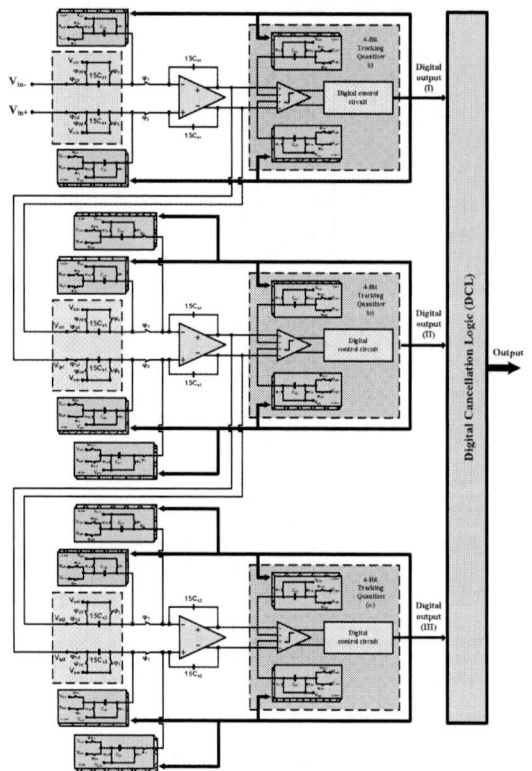

Fig. 6. the proposed delta-sigma modulator.

Each stage of the proposed modulator includes an integrator, a feedback DAC and a tracking quantizer. The proposed tracking quantizer is consisted of a fully differential comparator with a logic control circuit and a binary to thermometric decoder. In order to compare the output of the integrator with the previously predicted value, another switched-capacitor DAC is used.

In the following sections, the details of the important parts of the modulator are explained.

A. Switched-capacitor integrator and embedded DAC

The first stage of the designed modulator utilized a 16-level switched-capacitor DAC embedded with an integrator. In switched-capacitor DAC, 15 capacitors are used in each differential branch. Similar parts are used in the second and third stage of the modulator. In order to transfer the quantization error from current stage to the next stage, according to the transfer function of cascaded modulator, another switched-capacitor DAC has been embedded with the second and third stage integrators. The circuit of the switched-capacitor integrator and embedded DAC is shown in Fig. 7.

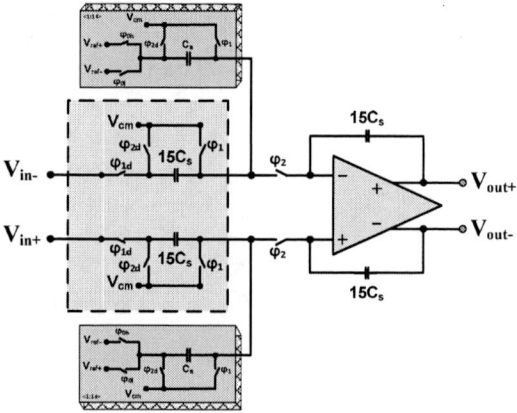

Fig. 7. Schematic of switched-capacitor integrator and embedded DAC.

B. Fully-differential operational amplifier

The fully-differential operational amplifier (op-amp) used in the integrator is shown in Fig. 8. The proposed amplifier is designed in two stages. In order to achieve high DC gain, a folded-cascode amplifier is used in the first stage. The second stage utilizes a common source amplifier to increase the output swing range. The cascade compensation method is used to increase bandwidth of the amplifier.

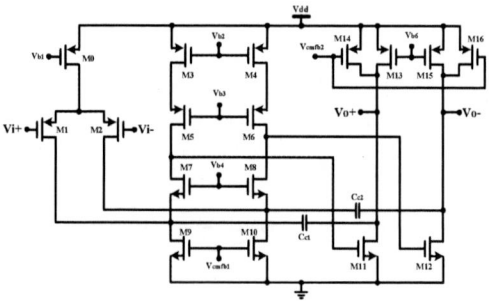

Fig. 8. Schematic of fully differential two-stage operational amplifier circuit.

TABLE I. Performances of the proposed op-amp.

Parameters	Value	Unit
Supply voltage	1.8	V
Bandwidth	213	MHz
Settling Time	3.2	ns
DC Gain	84.8	dB
Phase Margin	57	Deg.

TABLE II. Sizing of op-amp transistors.

Transistor	W(um)	L(um)
M0	70	0.36
M1-M2	32	0.18
M3-M4	7	0.36
M5-M6	5	0.18
M7-M8	6	0.18
M9-M10	12	0.36
M11-M12	9	0.36
M13-M14	5	0.36
M15-M16	3.5	0.36

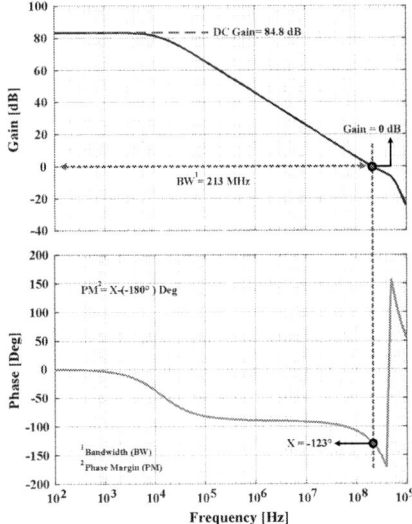

Fig. 9. Simulated Bode diagram of the proposed op-amp.

The proposed amplifier has a DC gain about of 85 dB, a unit gain bandwidth of 208 MHz, a phase margin of 58°, and a slew rate of 158 µv/s. The power consumption of the amplifier is 0.92 mW at a supply voltage of 1.8 V. The performance of the proposed op-amp is shown in Table I. Furthermore, the amplitude and phase responses of the designed amplifier are shown in Fig. 9. Table II shows the transistor sizes used in the op-amp.

C. Digital control circuit (DCC)

The DCC is used to predict the output voltage of the integrator. The output of the integrator is compared by the comparator with the previously predicted value and applies the value 0 or 1 to the DCC. According to the output of the comparator, the current value of the integrator output is one step higher or lower than its previous value. Then, the binary code is produced by the DCC and converted into thermometric code (using a decoder) and applied to DACs. The proposed DCC works as a FSM and consists of four JK flip-flops and a combinational circuit. The designed digital control circuit is shown in Fig. 10.

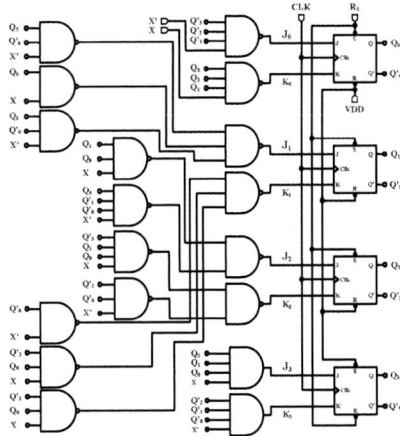

Fig. 10. Digital control circuit used in the proposed tracking quantizer.

D. Comparator

The main part of the tracking quantizer is comparator. Fig. 11 shows the source-coupled full-differential comparator circuit used in the proposed tracking quantizer.

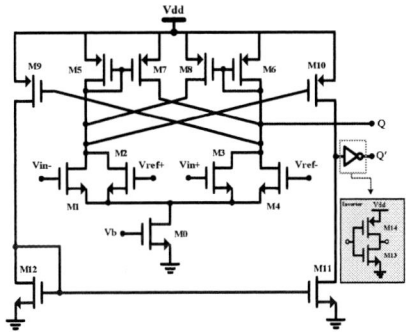

Fig. 11. Schematic of comparator circuit used in the proposed tracking quantizer.

TABLE III. Sizing of comparator transistors.

Transistor	W(um)	L(um)
M0	3.3	0.18
M1-M4	2.2	0.18
M5-M6	2.42	0.18
M7-M8	2.64	0.18
M9-M10	0.88	0.18
M11-M12	0.44	0.18
M13	0.44	0.18
M14	0.88	0.18

The propagation delay of the comparator is 550 ps and its power consumption at the supply voltage of 1.8 v is equal to 174 µW. The transistors size used in the comparator are shown in Table III.

IV. SIMULATION RESULTS

In this section, the performance of the proposed 1-1-1 cascaded delta-sigma modulator is evaluated using simulation at the transistor level in 180 nm CMOS technology. The oversampling ratio, sampling frequency and input signal bandwidth used in the simulation are 32, 32 MHz and 500 kHz, respectively. In the proposed modulator, the clock frequency of the digital part is equal to the sampling frequency. Thus, the need to use a high frequency clock is eliminated, when the input bandwidth is increased. The outputs of the proposed modulator are 4 bits. The input and output signals of the integrator are shown in Fig. 12. It can be seen that the output is following the input with rather small spikes. The power spectrum density (PSD) of the modulator output is shown in Fig. 13.

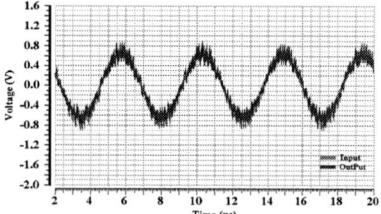

Fig. 12. The input and output signals of the integrator.

In order to compare the proposed modulator with other reported works, two types of figure of merits (FoM) are calculated [2]. The calculation formula for each FoM is mentioned in the footnote of Table VII. The calculated FoM values are equal to 0.88 pJ/conv-step and 155.42 dB, respectively.

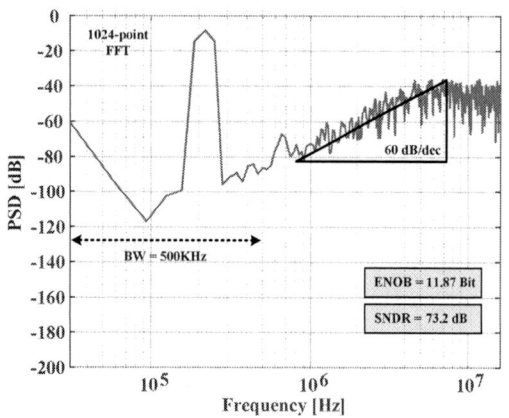

Fig. 13. The output PSD of the proposed modulator.

The performance of an ADC can be checked in the various process, voltage and temperature variations, which are known as PVT simulation. By examining the results obtained from PVT simulations, the relative stability of the proposed modulator can be concluded.

Fig. 14 shows the PSDs of the modulator output obtained from the various process variation simulations. The SNDR and ENOB values are listed separately in Table IV. According to the results obtained in the fast-slow (FS) corner compared to the typical-typical (TT) corner, the performance stability of the modulator is observed.

TABLE IV. Simulation results of process variations for the proposed modulator.

Parameter	TT	FF	SS	FS	SF
SNDR$_{dB}$	73.2	49.3	45.2	74.8	37.7
ENOB$_{Bit}$	11.87	7.90	7.21	12.13	5.97

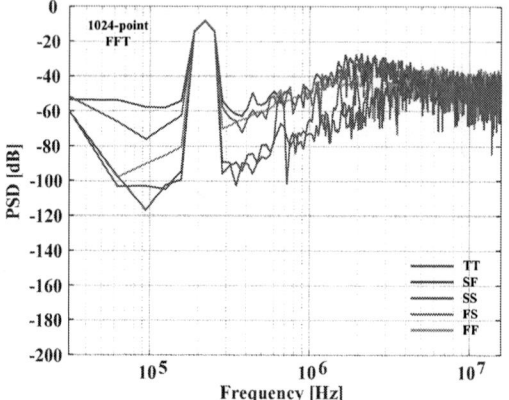

Fig. 14. The output PSDs for process variations.

Fig. 15 and Fig. 16 show the PSDs of the modulator output obtained from voltage and temperature variation simulations. Also, the results are listed in Tables V and VI, respectively. According to the results, the performance stability of the proposed modulator is relatively maintained and the percentage of changes in SNDR and ENOB values is low.

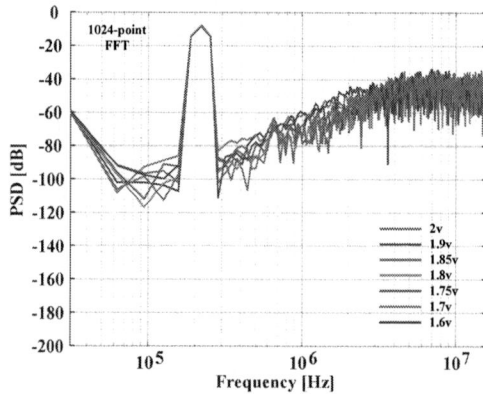

Fig. 15. The output PSDs for voltage variations.

TABLE V. Simulation results of voltage variations for the proposed modulator.

Parameter	1.6 VDD	1.7 VDD	1.75 VDD	1.8 VDD	1.85 VDD	1.9 VDD	2 VDD
SNDR$_{dB}$	67.5	62.2	73.64	73.2	71.7	70.00	69.00
ENOB$_{Bit}$	10.92	10.04	11.94	11.87	11.61	11.33	11.18

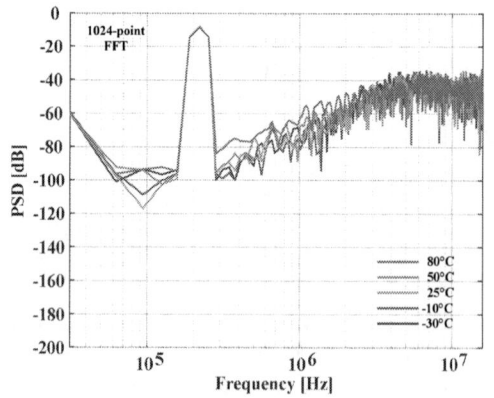

Fig. 16. The output PSDs for temperature variations.

TABLE VI. Simulation results of temperature variations for the proposed modulator.

Parameter	-30°C	-10°C	25°C	50°C	80°C
SNDR$_{dB}$	74.00	72.9	73.2	71.00	62.3
ENOB$_{Bit}$	12.01	11.81	11.87	11.50	10.06

Fig. 17 presents the variation of the SNDR versus the input signal amplitude. The designed modulator achieves a peak SNDR of 73.2 dB and a dynamic range of 80.24 dB.

The effect of transient noise on the proposed structure is shown in Fig. 18. The maximum and minimum SNDR values are 84.57 dB and 41.2 dB, respectively. Also, an average of 67.60 dB is obtained. The Monte Carlo analysis with 20 iterations was also calculated to evaluate the performance stability of the proposed modulator. Based on the obtained data in Fig. 19, the average SNDR is 71.64 dB.

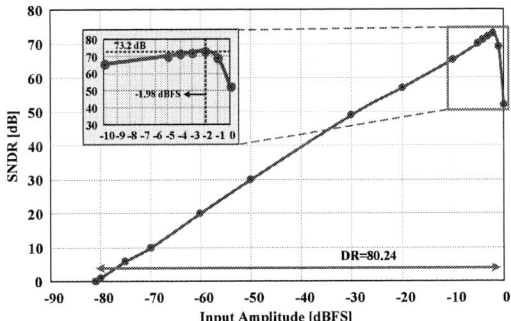

Fig. 17. Plot of SNDR versus input amplitude in the proposed modulator.

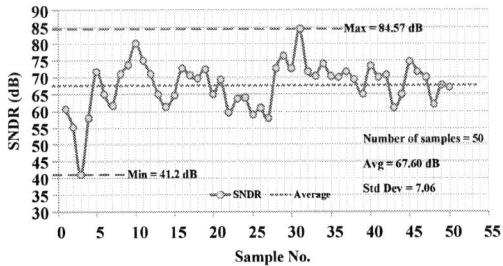

Fig. 18. Transient noise simulation results for the proposed modulator.

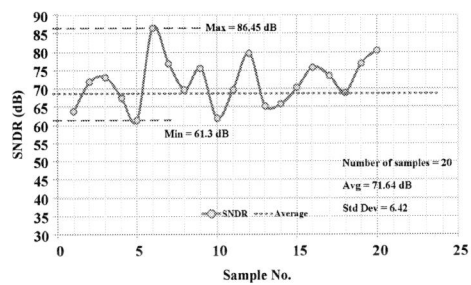

Fig. 19. Monte Carlo analysis results of the proposed modulator.

The details of power consumption for the various blocks in the proposed modulator are shown in Fig. 20. The largest share of the total power consumption is related to the op-amp used in the integrator. The proposed tracking quantizer has the lowest power consumption.

Table VII presents the simulation results of the proposed modulator and comparison with previous works. The improvement of some parameters compared to earlier designs

is noticed. As shown in Table VII, the proposed modulator provides optimal performance with low OSR and high bandwidth.

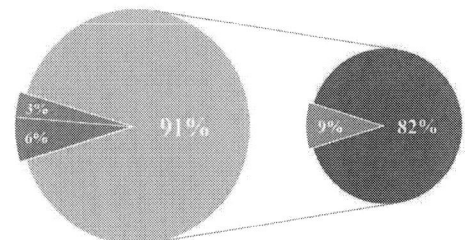

Fig. 20. Simulated power consumption of different sections of the proposed modulator.

Fig. 21 shows the layout of the proposed modulator. The area occupied by the proposed design is equal to 0.14mm².

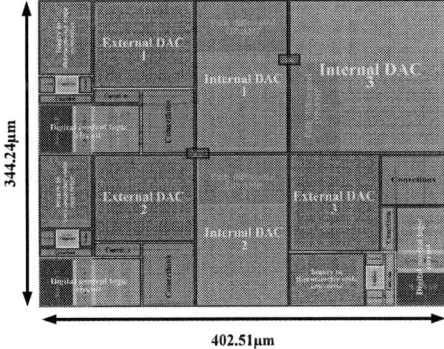

Fig. 21. Layout of the proposed 1-1-1 cascaded delta sigma modulator.

TABLE VII. Performance comparison of the proposed 1-1-1 cascaded delta-sigma modulator with previously reported modulators.

	This work	[13]	[14]	[15]	[16]	Unit
Technology	180	180	180	180	65	nm
Architecture	DT MASH	DT Loop	DT Loop	DT Loop	DT Loop	-
Shaping order	3rd	3rd	4th	2nd	2nd	-
Supply voltage	1.8	1.8	1.8	1.8	1.2	V
Area	0.14	0.22	N/A	N/A	0.21	mm²
BW	500	160	15	25	20	KHz
Fs	32	10.24	1	12.5	2.4	MHz
OSR	32	32	32	256	64	-
SNDR	73.2	87	93	87.05	91.3	dB
Power	3.3	2.52	8.28	3.41	0.04	mW
FoM$_w$^{I}	0.88	0.4	7.2	3.72	0.036	pJ/conv
FoM$_s$^{II}	155.42	165.5	155.8	155.7	177.9	dB

I. $FoM_{w(Walden)} = Power/(2 \times BW \times 2^{ENOB})$
II. $FoM_{s(Schreier)} = SNDR + 10\log_{10}(BW/Power)$

V. CONCLUSION

A cascaded 1-1-1 discrete-time delta-sigma modulator is presented that uses a tracking quantizer to perform 4-bit quantization in one sampling period. The proposed tracking quantizer uses a comparator, a DAC and a digital control circuit to predict the input signal. This feature reduces the occupied silicon area and the power consumption of the proposed delta-sigma modulator. The simulation of the proposed modulator has been done in 180 nm CMOS technology. The simulation results show that the proposed modulator achieves SNDR of 73.2 dB for the input bandwidth of 500 kHz, an OSR of 32 and power supply of 1.8 V. The power consumption and occupied area of the simulated modulator are equal to 3.3 mW and 0.14 mm², respectively.

REFERENCES

[1] R. Garvi and E. Prefasi, "A Novel Multi-Bit Sigma-Delta Modulator using an Integrating SAR Noise-Shaped Quantizer," in 2018 25th IEEE International Conference on Electronics, Circuits and Systems (ICECS), 2018: IEEE, pp. 809-812.

[2] S. Pavan, R. Schreier, and G. C. Temes, *Understanding delta-sigma data converters*. John Wiley & Sons, 2017.

[3] J. de la Rosa and S. D. Modulators, "Tutorial Overview, Design Guide, and State–of–the–Art Survey, Circuits and Systems I: Regular Papers," *IEEE Transactions on*, vol. 58, no. 1, 2011.

[4] J. De Maeyer, P. Rombouts, and L. Weyten, "Efficient Multibit Quantization in Continuous-Time Sigma-Delta Modulators," *IEEE Transactions on Circuits and Systems I: Regular Papers*, vol. 54, no. 4, pp. 757-767, 2007.

[5] H. Maghami, H. Mirzaie, P. Payandehnia, K. Mayaram, R. Zanbaghi, and T. Fiez, "A novel time-domain phase quantization noise extraction for a VCO-based quantizer," in 2018 IEEE 61st International Midwest Symposium on Circuits and Systems (MWSCAS), 2018: IEEE, pp. 145-148.

[6] B. Ferreira, M. Fernades, L. Oliveira, and J. Goes, "Impact of VCO non-linearities on VCO-based sigma-delta modulator ADCs," in 2018 International Young Engineers Forum (YEF-ECE), 2018: IEEE, pp. 97-102.

[7] M. Perrott, "A 12-bit 10-MHz bandwidth continuous-time ADC with a 5-bit 950-MS/s VCO-based quantizer," *IEEE J. Solid-State Circuits*, 2008.

[8] T. Mu and L. Liu, "An Improved Noise Canceling Sturdy 2-1 MASH Sigma-Delta Modulator with Multi-Bit SAR Quantizer," in 2023 China Semiconductor Technology International Conference (CSTIC), 2023: IEEE, pp. 1-3.

[9] L. Liu, D. Li, and Z. Wang, "A 0.6-V to 1-V audio $\Delta\Sigma$ modulator in 65 nm CMOS with 90.2 dB SNDR at 0.6-V," *VLSI Design*, vol. 2013, pp. 6-6, 2013.

[10] R. J. Baker, CMOS: *circuit design, layout, and simulation*. John Wiley & Sons, 2019.

[11] L. Dorrer, F. Kuttner, P. Greco, P. Torta, and T. Hartig, "A 3-mW 74-dB SNR 2-MHz continuous-time delta-sigma ADC with a tracking ADC quantizer in 0.13μm CMOS," *IEEE Journal of Solid-State Circuits*, vol. 40, no. 12, pp. 2416-2427, 2005.

[12] F. Colodro and A. Torralba, "Continuous-time sigma–delta modulator with a fast tracking quantizer and reduced number of comparators," *IEEE Transactions on Circuits and Systems I: Regular Papers*, vol. 57, no. 9, pp. 2413-2425, 2010.

[13] T.-G. Kim, K.-I. Cho, H.-J. Kim, J.-H. Boo, Y.-S. Kwak, and G.-C. Ahn, "A third-order DT delta-sigma modulator with noise-coupling technique," in *2020 International SoC Design Conference (ISOCC)*, 2020: IEEE, pp. 3-4.

[14] Y. Chen, Z. Wang, Y. Zhuang, and H. Tang, "Analysis and Design of Sigma-Delta ADCs for Automotive Control Systems," in 2021 IEEE 3rd International Conference on Circuits and Systems (ICCS), 2021: IEEE, pp. 235-241.

[15] M. Boncu, S. Pana, F. Draahlei, and G. Brezeanu, "A second order discrete-time ΔA analog to digital converter for audio applications," in 2022 International Semiconductor Conference (CAS), 2022: IEEE, pp. 209-212.

[16] L. Wang, S. Liu, Y. Zhang, L. Zhong, and Z. Zhu, "A 44-μ W, 91.3-dB SNDR DT $\Delta$$\Sigma$ Modulator With Second-Order Noise-Shaping SAR Quantizer," IEEE Transactions on Circuits and Systems I: Regular Papers, 2023.

The 5th Iranian International Conference on Microelectronics (IICM2023)

25 – 26 October 2023

Optimizing High Dynamic Range Current Measurement Circuit for IoT Applications

1st Yas Hosseini Tehrani
Department of Electrical Engineering
Sharif University of Technology
Tehran, Iran
yhtehrani@sharif.edu

2st Seyed Mojtaba Atarodi
Department of Electrical Engineering
Sharif University of Technology
Tehran, Iran
atarodi@sharif.edu

Abstract—**The power consumption of Internet-of-Things (IoT) devices is a major concern as it directly affects battery life. Therefore, measuring and monitoring the energy usage of IoT devices is vital. However, measuring the current consumption measurement of IoT devices is challenging as it must cover a wide range with low power consumption. In this paper, an improved design for the current measurement circuit is presented. The pass element which is placed in the current path to measure the current based on the charge and discharge of the capacitor is optimized to have reduced voltage drop, lower power consumption, decreased costs, and improved integration with IoT devices. The proposed circuit is simulated with LTspice and its proper operation is verified. The presented circuit is capable to be integrated into IoT devices, making monitoring and optimizing the power consumption of IoT networks feasible.**

Index Terms—**Current measurement, Internet of Things, Pass element, Integration**

I. INTRODUCTION

The Internet of Things (IoT) field plays a vital role in human life, in which its importance can not be overstated. The billions of physical devices worldwide are connected and exchanging data with the aim of IoT technology. However, IoT has its own set of challenges and limitations. One of the significant challenges is its power management as IoT devices rely on finite energy sources such as batteries. The significant growth rate in the number of IoT devices within networks makes power management more challenging. IoT devices include components such as processors, sensors, and communication radios. The various operation mode of IoT devices from sleep mode to fully activated mode leads to a significant dynamic range of current consumption from a few nAs to As within tens of μ seconds. Therefore, a high precision, fast response, and a high dynamic range current measurement system are required to estimate and enhance the battery lifespan of IoT devices. The current measurement system must be capable of being integrated with the IoT device node. In the recent study [1], the efficency of various power optimization methodologies for IoT networks was confirmed through an integrated power measurement circuit within each IoT device. The IoT devices perform real-time power measurements and transmit the obtained data to a central base station. The base station monitored and optimized the power consumption within the IoT network.

Selecting the proper current measurement method depends on the specific needs of the system in terms of cost, complexity, power dissipation, and dynamic range requirements. The existing current measurement techniques are primarily classified as direct and indirect approaches [2]–[5]. However, few methods meet the unique requirements of IoT applications. The performance comparison of the current measurement solutions is summarized in Table I. The authors have previously presented two current measurement circuits for IoT applications [6], [7]. In the first circuit, the authors introduced an accurate, real-time power measurement solution that features a dynamic range of 100 dB, specifically for IoT applications. The design of this circuit is based on a voltage regulating control loop by employing ultra-low offset amplifiers, high-precision Analog-to-Digital Converters (ADCs), and a reference voltage to achieve accurate and real-time monitoring of the current consumption of IoT devices [7]. In the second circuit, the authors presented a novel pass element designed to achieve higher accuracy and higher dynamic range in current measurements, making it more suitable for the diverse operational states of IoT devices. [6]. In this paper, we focus on improving the proposed current measurement circuit in our previous work [6]. Our aim is to develop a solution that addresses previous shortcomings and enhances its performance and optimizes the current measurement circuit for IoT applications.

Compared to the previous work, the originality and the main contributions of the work are given as follows: 1) a simpler circuit with a lower number of components is designed. 2) a new design is more feasible to be integrated within a System-on-Chip (SoC). 3) Significant improvements in performance have been achieved, including reduced power consumption, expanded dynamic range, improved precision, and reduced voltage drop.

This article is organized as follows. Section II provides the concept of the proposed current measurement solution in detail. Also, the block diagrams and design process of the system are presented. The simulation results are presented in Section III. Finally, the conclusions are given in Section IV.

979-8-3503-6020-2/23 $31.00 © 2023 IEEE

The 5th Iranian International Conference on Microelectronics (IICM2023)

TABLE I: Performance comparison of the current measurement solutions [6]

Method	Range	Accuracy	Bandwidth	Voltage drop	Power loss	DC capable	Size [mm^3]	Cost[1] [USD]
Shunt resistor	$mA - A$	0.1%–2%	$kHz - MHz$	$mV - V$	$mW - W$	Yes	> 25	< 0.5
Current transformer [8], [9], and [10]	$A - kA$	0.1%–1%	$kHz - MHz$	$\mu V - mV$	mW	No	> 500	> 0.5
Rogowski coil [11], [12], and [13]	$A - MA$	0.2%–5%	$kHz - MHz$	$\mu V - mV$	mW	No	> 1000	> 1
Hall effect [14] and [15]	$A - kA$	0.5%–5%	kHz	$\mu V - mV$	mW	Yes	> 1000	> 1
Current probe [16] and [17]	• $mA-kA$ • $\mu A-A$	1% – 3%	• $DC - MHz$ • $MHz - GHz$	μV	mW	Yes	> 300	> 1k
Fiber optic current sensor	$kA - MA$	0.1%–1%	$kHz - MHz$	$\cong 0$	W	Yes	> 10^6	> 1k
Authors previous work [7]	$1\mu A - 150mA$	0.32%	$5MHz$	$\cong 0$	< 0.1W	Yes	$\cong$ 900	$\cong$ 5
Authors previous work [6]	$3.3nA - 1.65A$	0.522%	$DC - 4kHz$	400mV constant	< 0.7W	Yes	$\cong$ 800	$\cong$ 4
Proposed method[2]	$3.3nA - 33A$	0.1%	$DC - 4kHz$	300mV constant	< 90mW	Yes	$\cong$ 100	$\cong$ 2

[1] In high volume production
[2] Simulated

II. PROPOSED CURRENT MEASUREMENT CIRCUIT

A. Review of the Previous Circuit

The measurement principle of the circuit is based on charging and discharging the measurement capacitor (C_m). C_m begins to charge with the current drawn from the Device Under Test (DUT), until it reaches the level of the positive voltage reference. At this stage, switch SW_{CNT} switch is activated and C_m discharges until its voltage drops down to the level of the negative voltage reference. The schematic of the current measurement circuit is shown in Fig. 1. In order to acheive a dynamic range about $174dB$, three stage of capacitor (C_{large}, C_{medium}, and C_{small}) are deployed. Due to the charging and discharging the measurement capacitor, the voltage ripple appears at the supply voltage of the load. The transistors $M_1 - M_3$ are used for voltage ripple compensation, with a voltage drop equals to V_d. According to (1), the current consumption of DUT is obtained by measuring the switching frequeny of SW_{CNT} switch (f_{sw}), with know values of the voltage reference (V_R) and C_m.

$$f_{sw} \cong \frac{1}{T_{sw}} = \frac{i_m}{2V_R C_m}, \tag{1}$$

B. Design Enhancements

To enhance the integrability, cost-effectiveness, dynamic range, and precision of the current measurement system, the following modifications are proposed.

1) Integrability: In the previous design, the addition of V_{R+} and V_{R-} voltages to the V_{C-} voltage required a separate circuit. Further, separate buffers and seperate operational amplifiers were employed for amplification. The resistors and npn transistors used for voltage addition give low input impedance characteristics to the comparator. Moreover, the supply voltage

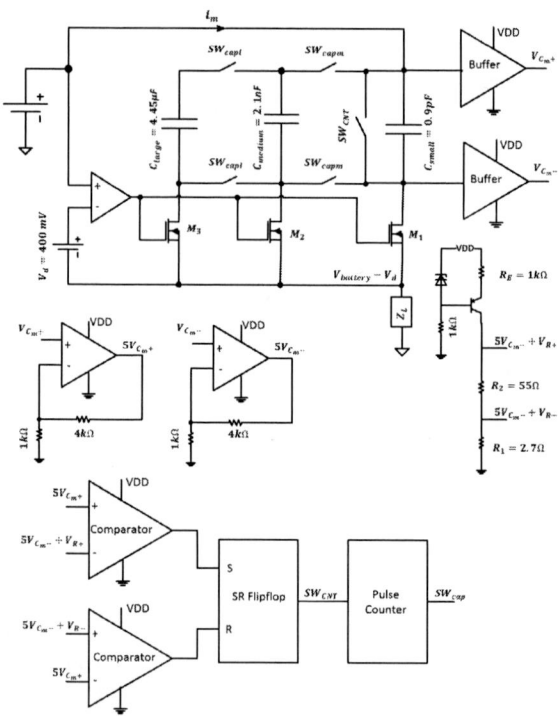

Fig. 1: The schematic of the current measurement circuit [6].

of $+15V/ -15V$ is relatively high. Therefore, the previous design is slow, dissipates too much power, is expensive, inefficient, and is not directly compatible with low voltage circuits. To address this issue, the mentioned are combined into a single circuit. The proposed circuit is shown in Fig. 2.

979-8-3503-6020-2/23 $31.00 © 2023 IEEE

The 5th Iranian International Conference on Microelectronics (IICM2023)

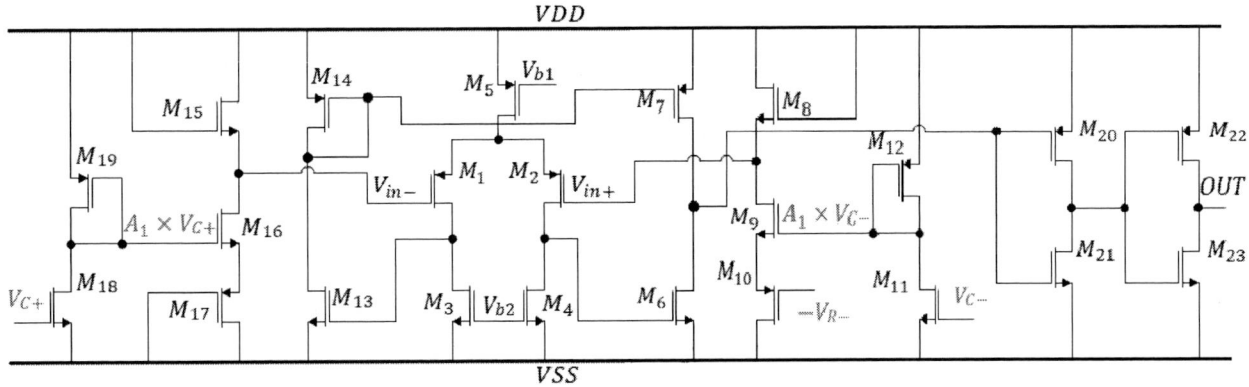

Fig. 2: The schematic of the proposed composite buffer-voltage adder-amplifier-comparator circuit.

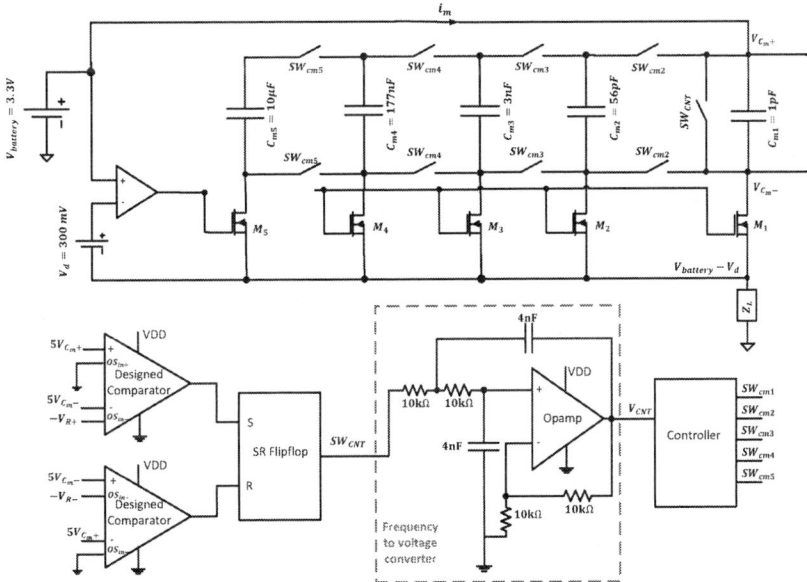

Fig. 3: The schematic of the simulated circuit.

The proposed circuit is a high-speed differential comparator with a controllable offset voltage and a pre-amplifier circuit. As shown in Fig. 2, the first pre-amplifier circuit (M_{18}, M_{19}, M_{11}, and M_{12} transistors) amplifies the input signals by the gain equals to $A_1 = -\frac{gm_{18}}{gm_{19}} = -\frac{gm_{11}}{gm_{12}}$. It should be noted that the transistor pairs (M_{17}, M_{10}), (M_9, M_{16}), (M_{18}, M_{11}), (M_{12}, M_{19}), (M_8, M_{15}), (M_6, M_{13}), (M_7, M_{14}), (M_1, M_2), and (M_3, M_4) are similar in design with the same size and characteristics. The predetermined offset voltages are given to the gates of M_{11} and M_{17} transistors. The M_9 and M_{10} (similarly, M_{16} and M_{17}) form a composite transistor (with K_{eq} and V_{Teq}) to provide high-input impedance for the offset and differential inputs, where $\frac{1}{\sqrt{K_{eq}}} = \frac{1}{\sqrt{K_9}} + \frac{1}{\sqrt{K_{10}}}$ and $V_{Teq} = V_{T9} + |V_{T10}|$. As the current flowing in M_8, M_9

and M_{10} are equal, we have:

$$V_{in+} = VDD - V_{T8} - \sqrt{\frac{K_{eq}}{K_8}}(-V_{R-} - A_1 V_{C-} - V_{Teq})$$
(2)

After performing a similar calculation, the equation for V_{in+} is obtained and the diffrential inputs to the comparator(gates of M_1 and M_2) is obtained by (3).

$$V_{in+} - V_{in-} = \sqrt{\frac{K_{eq}}{K_8}}(A_1(V_{C-} - V_{C+}) + V_{R-})$$
(3)

The $M_1 - M_7$, M_{13}, M_{14} transistors form a high-speed comparator. The two inverters ($M_{20} - M_{23}$) are connected in series to provide the digital output with the increase in gain and speed,. Therefore, as marked with red color in

Fig. 2, the proposed circuit performs a high input impedance direct comparison between V_{C-} and V_{C+} with the controllable offset.

2) Dynamic range and voltage drop: To enhance the dynamic range of our current measurement and mitigate the voltage drop of the pass element, the measurement capacitor with five stages is suggested. The proposed circuit is shown in Fig. 4 with $C_{m1} = 10pF$, $C_{m2} = 1.6nF$, $C_{m3} = 250nF$, $C_{m4} = 33\mu F$, $C_{m5} = 1000\mu F$. The value of the measurement capacitor is selected based on the current level and adjusted by controlling the SW_{cm2}, SW_{cm3}, SW_{cm4}, and SW_{cm5} switches. The states of the mentioned switches are determined by the switch control unit similar to the one presented in the authors' previous work.

The addition of an extra stage to the measurement capacitor unit allows the designer to select a lower voltage reference (V_R). This modification results in a reduced voltage ripple across the measurement capacitor. In the proposed circuit shown in Fig. 1, $M_1 - M_3$ transistors and V_d are employed to compensate for the voltage ripple. Therefore, by reducing the voltage ripple across the measurement capacitor, the lower V_d can be selected. Therefore, the voltage drop of the pass element is reduced.

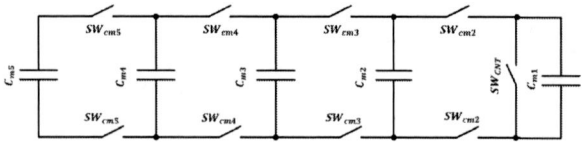

Fig. 4: The schematic of the proposed measurement capacitor circuit with five stages.

3) Percision: The measurement precision of the previous design relies on the clock frequency of the processor. The presented method counts pulses on the SW_{CNT} switch within specific time intervals. However, the error due to residual counts affects the measurement accuracy. The residual counts' term refers to the fraction of the pulses that may not be completely counted within the specified time frame. For example, if the counting is started at a random time point, the presented measurement method might be counting a pulse that had started before the count began. Therefore, to enhance the precision of the measurement circuit, a frequency-to-voltage converter is employed to transform the frequency of SW_{CNT} switch into a continuous voltage, providing a more accurate current measurement circuit. It is important to note that the precision of the measurements is dependent on the performance characteristics of the frequency-to-voltage converter. However, a complex frequency-to-voltage converter is not required, as a simple low-pass filter could be deployed.

III. SIMULATION RESULTS

To validate the modifications proposed in this paper, simulations are performed using both LTspice and Matlab, utilizing the 180 nm technology. The schematic of the simulated

circuit is shown in Fig. 3. The presented circuit in [6] is comprised of 11 distinct blocks: measurement capacitors, 2 buffers, 1 level circuit, 2 amplifiers, 2 comparators, an SR flip flop, a pulse counter, and a controller. The proposed circuit is optimized and consists of 6 blocks: measurement capacitors, 2 units of the composite buffer-voltage adder-amplifier-comparator circuit, an SR flip flop, a frequency-to-voltage converter, and a controller. Therefore, deploying the composite buffer-voltage adder-amplifier-comparator unit reduces the total number of blocks by nearly half, proving the efficiency of the proposed design. The simulated composite buffer-voltage adder-amplifier-comparator circuit has a gain of $72dB$ and unity gain of $1.6Ghz$ with the static current consumption of $2.23mA$. The V_{C_m} for load current is a range from $5\mu A$ to $15\mu A$ is shown in Fig .5. Therefore the circuit performs as expected.

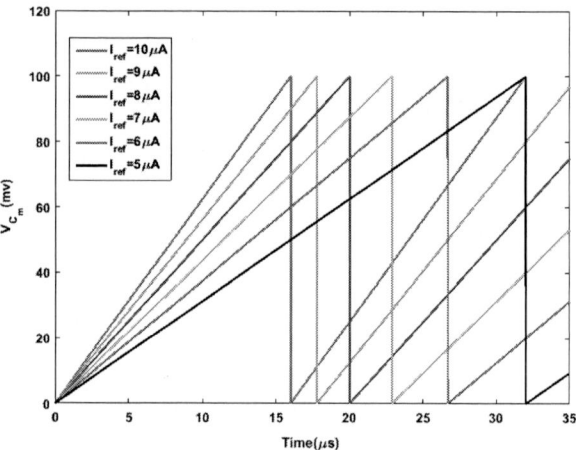

Fig. 5: The V_{C_m} for load current is a range from $5\mu A$ to $15\mu A$.

To examine the precision of the proposed circuit, a various sets of load values ranging from 0.1Ω to $10G\Omega$ are applied to the proposed circuit. The measurement error is obtained using (4).

$$\epsilon_i = \frac{I_{mes_i} - I_{ref_i}}{I_{ref_i}} \times 100, \qquad (4)$$

where I_{mes_i} is the measured current, and I_{ref_i} is the reference current level. The measurement error (ϵ_i) in terms of the reference current levels in dBA is shown in Fig. 6, where dBA is decibel relative to 1 Ampere.

As shown in Fig. 6, the measurement error of the first stage (C_{m1} is selected) is higher compared to the other stages due to the effects of parasitics and the leakage current from the switches and the designed comparators as their magnitudes are comparable to C_{m1} and the reference current level, respectively. The switching frequency of SW_{CNT} increases with a rise of the reference current level at each stage. Therefore, the propagation delays of the designed comparator, switches, and control unit decrease the measurement accuracy at the lower

979-8-3503-6020-2/23 $31.00 © 2023 IEEE

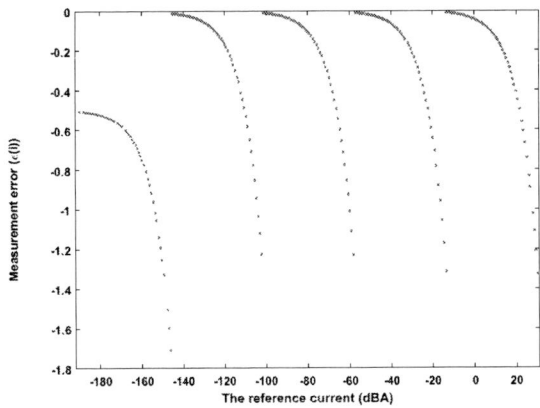

Fig. 6: The measurement error (ϵ_i) in terms of the reference current levels (dbA).

current level of each stage. It is worth mentioning that the propagation delay of the mentioned components is reduced in the proposed design, leading to an improved measurement error by approximately 2% and is reduced to 0.2% after performing the calibration. The performance comparison of the proposed method with other ones is presented in Table. I. The dynamic range is improved to $200dB$, from $174dB$ in the previous one, and its power consumption is reduced to $80mW$, from $700mW$ in the previous solution with the aim of composite buffer-voltage adder-amplifier-comparator uint. The power consumption of different blocks of the proposed circuit and the previous one is summarized in Table. II. It should be noted that the power consumption of the other blocks is not reported as their designs remain unchanged.

TABLE II: The power consumption comparison of the proposed current measurement solutions and the previous one [6].

Blocks	Poposed circuit	Previous work [6]
Buffers		$340mW$
Amplifiers	$14.8mW$	$60mW$
Level shifter		$12mW$
Comparators		$198mW$
Pulse counter / F-V converter	$< 5mW$	$< 5mW$

IV. Conclusion

The importance and necessity of a precise, real-time, and efficient current measurement circuit that can be integrated with IoT devices is crucial for IoT applications cannot be overstated. The present work has contributed to this ongoing field by proposing novel enhancements to the current measurement technique, with a focus on improving integrability, cost-effectiveness, dynamic range, and precision. The proposed circuit combines a buffer, comparator, amplifier, and voltage adder, forming an integrated block with simplified design. The addition of extra stages expands the dynamic range of the

current measurement with a reduced voltage dropout. Furthermore, a frequency-to-voltage converter is employed to improve the measurement accuracy. The proposed design is simulated using LTspice and Matlab. According to the obtained results, the proper operation and the mentioned improvement of the proposed circuit are verified, and it can provide a practical approach compatible with modern semiconductor manufacturing processes. Future research will focus on real-world validation and further optimization of the current measurement circuit for IoT applications.

References

[1] Y. H. Tehrani, A. Amini, and S. M. Atarodi, "A tree-structured LoRa network for energy efficiency," *IEEE Internet of Things Journal*, vol. 8, no. 7, pp. 6002–6011, 2020.

[2] M. Kinoshita, T. Inoue, K. Shimaoka, and K. Fujii, "Precise power measurement with a single-mode waveguide calorimeter in the 220–330 ghz frequency range," *IEEE Transactions on Instrumentation and Measurement*, vol. 67, no. 6, pp. 1451–1460, 2018.

[3] A. Pötsch, A. Berger, C. Leitner, and A. Springer, "A power measurement system for accurate energy profiling of embedded wireless systems," in *Proceedings of the 2014 IEEE Emerging Technology and Factory Automation (ETFA)*, pp. 1–4, IEEE, 2014.

[4] A. Al Mortuza, M. F. Pervez, M. K. Hossain, S. K. Sen, M. N. H. Mia, M. Basher, and M. S. Alam, "Pico-current measurement challenges and remedies: A review," *Univers. J. Eng. Sci*, vol. 5, no. 4, pp. 57–63, 2017.

[5] X. Jiang, P. Dutta, D. Culler, and I. Stoica, "Micro power meter for energy monitoring of wireless sensor networks at scale," in *2007 6th International Symposium on Information Processing in Sensor Networks*, pp. 186–195, IEEE, 2007.

[6] Y. H. Tehrani and S. M. Atarodi, "Design & implementation of high dynamic range current measurement system for iot applications," *IEEE Transactions on Instrumentation and Measurement*, vol. 71, pp. 1–9, 2022.

[7] Y. H. Tehrani and S. M. Atarodi, "Design & implementation of a high precision & high dynamic range power consumption measurement system for smart energy IoT applications," *Measurement*, vol. 146, pp. 458–466, 2019.

[8] N. McNeill, N. K. Gupta, and W. G. Armstrong, "Active current transformer circuits for low distortion sensing in switched mode power converters," *IEEE Transactions on Power Electronics*, vol. 19, no. 4, pp. 908–917, 2004.

[9] L. Dalessandro, N. Karrer, and J. W. Kolar, "High-performance planar isolated current sensor for power electronics applications," *IEEE Transactions on Power Electronics*, vol. 22, no. 5, pp. 1682–1692, 2007.

[10] L. Components, "Isolated current and voltage transducers," *CH24101*, 2004.

[11] W. Ray and C. Hewson, "High performance Rogowski current transducers," in *Conference Record of the 2000 IEEE Industry Applications Conference. Thirty-Fifth IAS Annual Meeting and World Conference on Industrial Applications of Electrical Energy (Cat. No. 00CH37129)*, vol. 5, pp. 3083–3090, IEEE, 2000.

[12] L. Zhao, J. Van Wyk, and W. Odendaal, "Planar embedded pick-up coil sensor for power electronic modules," in *Nineteenth Annual IEEE Applied Power Electronics Conference and Exposition, 2004. APEC'04.*, vol. 2, pp. 945–951, IEEE, 2004.

[13] J. Dupraz, A. Fanget, W. Grieshaber, and G. Montillet, "Rogowski coil: Exceptional current measurement tool for almost any application," in *2007 IEEE Power Engineering Society General Meeting*, pp. 1–8, IEEE, 2007.

[14] R. e. Popovic, Z. Randjelovic, and D. Manic, "Integrated Hall-effect magnetic sensors," *Sensors and Actuators A: Physical*, vol. 91, no. 1-2, pp. 46–50, 2001.

[15] C. Chien, *The Hall effect and its applications*. Springer Science & Business Media, 2013.

[16] Tektronix, "Tektronix current probe solutions." https://www.tek.com/en/products/accessories/current-probes. Accessed on 2021-12-12.

[17] Hioki, "Wide-band current probes, DC to 120 MHz." https://www.hioki.com/global/products/current-probes/wide-band. Accessed on 2021-12-12.

Few-layered phosphorene synthesis by CVD approach as an anode for sodium-ion battery

Pooya Dehghan
Energy Storage Laboratory, School of
Electrical and Computer engineering,
College of Engineering
University of Tehran
Tehran, Iran
pooyadehghan@ut.ac.ir

Soraya Hoornam
Energy Storage Laboratory, School of
Electrical and Computer engineering
College of Engineering
University of Tehran
Tehran, Iran
soraya.hoornam@gmail.com

Zeinab Sanaee*
Energy Storage Laboratory, School of
Electrical and Computer engineering,
College of Engineering
University of Tehran
Tehran,Iran
z.sanaee@ut.ac.ir

Shams Mohajerzadeh
Thin Film and Nanoelectronics Lab.,
School of Electrical and Computer
engineering, College of Engineering
University of Tehran
Tehran, Iran
mohajer@ut.ac.ir

Abstract— **Chemical vapor deposition (CVD) technique have been used to synthesis few-layered nanosheets of phosphorene. The nanosheets of phosphorene have been deposited on silicon substrate and transferred onto the stainless-steel substrate for electrochemical tests as an anode for sodium ion battery. The battery has been assembled in an Argon filled glovebox. The results show that the synthesized black phosphorene offers a steady special capacity of around 150 mAh/g.**

Keywords—Battery, sodium-ion, phosphorene, anode

I. INTRODUCTION

Ever growing demand for renewable and sustainable energy in the modern world leads to urgent need for high capacity and sufficient energy storage devices. Although the lithium-ion battery (LIB) dominates the market in portable electronics and recently in electric vehicles application, it is believed that LIBs are expensive to stationary, large scale energy storage systems due to the limited resource of lithium in earth crust. On the other hand, unlimited resource of sodium and resemblance in fundamental and operation principles of sodium-ion battery (SIB) to LIB, make SIB a promising alternative to LIB [1–6].

Among different anodic materials for SIBs, Phosphorus has attracted lots of attention in recent years, owing to its highest theorical capacity of 2596 mAh/g among all other potential materials [7,8].

Sodium storage in bulk phosphorus anode is dissatisfying owing to the a large volume change, poor electrical conductivity, and unstable solid electrolyte interface. Anode made up of nanostructure phosphorus is a prevalent solution to the mentioned drawbacks [7,9,10].

Phosphorus as a non-metallic element exists in three allotropes :red P, white P, black P. White P is highly reactive and toxic, therefor it is not appropriate for anode. Red P which is more stable, suffers from low conductivity (around 10-14 S/cm) and huge volume change (over 400%) during charge/discharge. Black P is a narrow band gap semiconductor (0.34 ev) with good electrical conductivity (around 300 Sm^{-1}) arris as promising candidate for energy storage application [11–13]. In this manuscript, black phosphorene has been synthesized and implemented as an anode material in the SIB.

II. EXPERIMENTAL DETAILS

A. red phosphorus deposition

Briefly, the red P was deposited on substrate and then the deposited layer transformed to black P.

To deposit the red P, the tablet of red P was made by pressing the powder of red P (purchased from Merck ≥ 97%, Hohenbrunn, Germany|) as a source of phosphor in CVD furnace (Figure 1a). The tablet and the silicon substrate were placed in the boat with 8cm distance. The boat was put in the CVD furnace. The chamber was vacuumed to 10 mtorr, then the temperature was increased to 510 °C. With 20 sccm flow of hydrogen gas the layer of red P was deposited on the substrate. The chamber was kept in that condition for 30 minutes and then cooled to the room temperature.

B. transforming red P into black P

The sample covered by red P was placed in PECVD furnace and was treated with H$_2$ plasma for 30 minutes in 310 °C and power density of 12 W. The flow of hydrogen gas was 20 sccm during the process, after the chamber was vacuumed to 1mtorr. This process transformed the red P to black P, and has increased the crystality of the deposited layer [14] .

The 5th Iranian International Conference on Microelectronics (IICM2023)

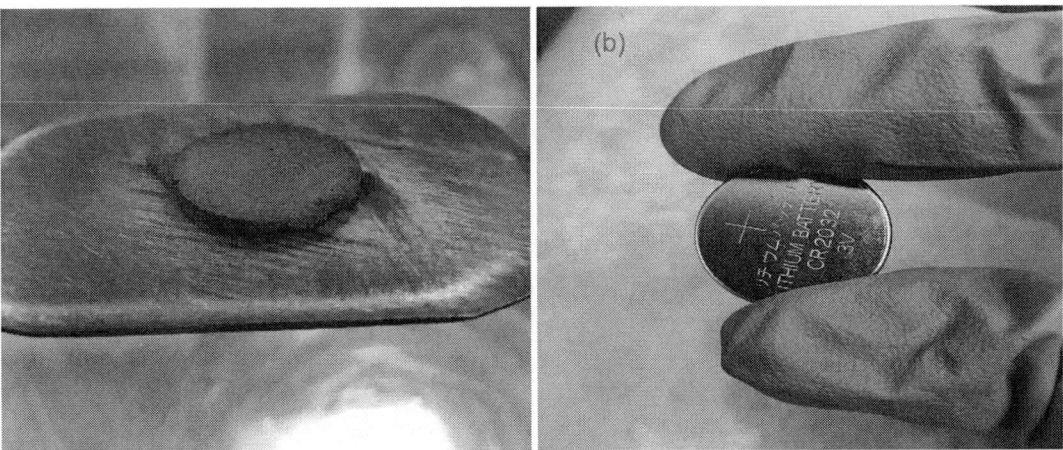

Figure 1 a) prepared tablet of red phosphor b) assembled coin cell

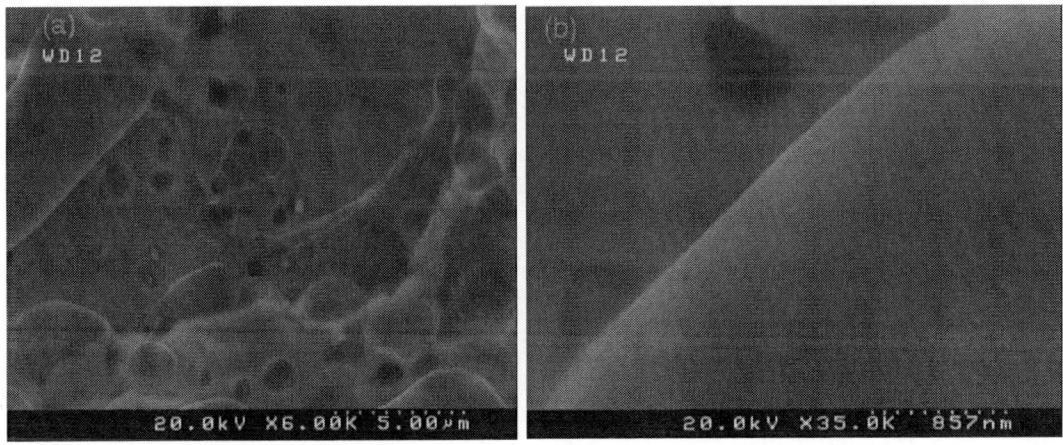

Figure 2 a,b) Scanning electron microscopy (SEM) images of the synthesised black P

C. Coin cell assembly for electrochemical test

The prepared black P nanosheets was dispersed in isopropyl alcohol (IPA) by use of probe sonicator and then transferred on stainless-steel substrate by drop casting method.

A Galvanometric charge/discharge test was performed to evaluate the electrochemical performance of the electrode. The metal Sodium electrode and prepared electrode were used to assemble the stainless-steel CR 2032 coin cell in an Ar-filled glovebox. Whatman glass-fiber membrane was implemented between the two electrodes as the separator. 1M $NaClo_4$ in 1:1 v/v ethylene carbonate/diethyl carbonate (EC/DEC) was implemented as the electrolyte.

III. RESULTS AND DISCUSSION

Figure 2 represent the scanning electron microscopy images of the prepared black P. Formation of thin layers of black phosphor is evident. As it is depicted in figure 3 the prepared electrode deliver the initial (2nd cycle) discharge capacity of 1546 mAh/g. The discharge capacity for 20 cycles of charge/discharge is depicted in figure 3. As can be seen in this figure, an initial capacity around 1550 mAh/g has been achieved. After the first formation cycles, the capacity gradually has reached a stable behavior, showing a steady specific capacity of around 150 mAh/g. This study shows the potential of synthesized black P as an anode of SIB, to offer high specific capacity. Further study can be done on improving the performance of this anodic material in the future, to have a better SIB.

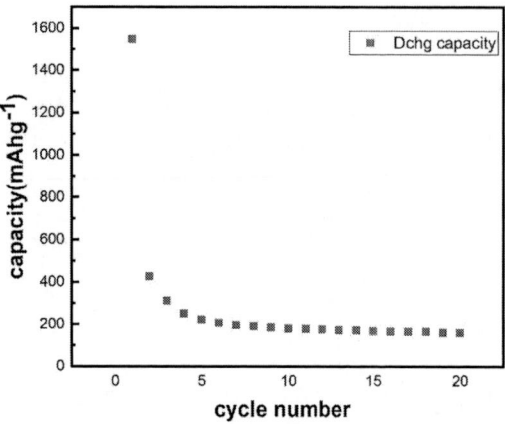

Figure 3 discharge capacity of black p electrode

CONCLUSION

In this paper, black P was synthesized using a CVD approach on a silicon substrate and transferred to stainless steel substrate. Then it has been implemented as an electrode material in Sodium ion battery. The battery was assembled in an Argon filled glove box with Na as an electrode and the synthesized black P as the other electrode, while using 1M $NaClo_4$ in 1:1 v/v ethylene carbonate/diethyl carbonate (EC/DEC) as the electrolyte. The result of galvanostatic charge discharge test revealed that the synthesized black P based electrode offers a steady special capacity of about 150 mAh/g.

REFERENCES

[1] K. Chayambuka, G. Mulder, D. L. Danilov, and P. H. L. Notten, " Sodium - ion battery materials and electrochemical properties reviewed," Adv Energy Mater, vol. 8, no. 16, p. 1800079, 2018.

[2] H. Tao, M. Zhou, R. Wang, K. Wang, S. Cheng, and K. Jiang, "TiS2 as an Advanced Conversion Electrode for Sodium-Ion Batteries with Ultra-High Capacity and Long-Cycle Life," Advanced Science, vol. 5, no. 11, Nov. 2018, doi: 10.1002/advs.201801021.

[3] T. Perveen, M. Siddiq, N. Shahzad, R. Ihsan, A. Ahmad, and M. I. Shahzad, "Prospects in anode materials for sodium ion batteries-A review," Renewable and Sustainable Energy Reviews, vol. 119, p. 109549, 2020.

[4] K. Kubota and S. Komaba, "practical issues and future perspective for Na-ion batteries," J Electrochem Soc, vol. 162, no. 14, p. A2538, 2015.

[5] T. L. Kulova and A. M. Skundin, "The use of phosphorus in sodium-ion batteries (a review)," Russian Journal of Electrochemistry, vol. 56, no. 1, pp. 1–17, 2020.

[6] L. Li, Y. Zheng, S. Zhang, J. Yang, Z. Shao, and Z. Guo, "Recent progress on sodium ion batteries: potential high-performance anodes," Energy Environ Sci, vol. 11, no. 9, pp. 2310–2340, 2018.

[7] T. Ramireddy et al., "Phosphorus–carbon nanocomposite anodes for lithium-ion and sodium-ion batteries," J Mater Chem A Mater, vol. 3, no. 10, pp. 5572–5584, 2015.

[8] M. Dahbi et al., "Black phosphorus as a high-capacity, high-capability negative electrode for sodium-ion batteries: investigation of the electrode/electrolyte interface," Chemistry of materials, vol. 28, no. 6, pp. 1625–1635, 2016.

[9] J. Sun et al., "A phosphorene–graphene hybrid material as a high-capacity anode for sodium-ion batteries," Nat Nanotechnol, vol. 10, no. 11, pp. 980–985, 2015.

[10] J. Sun et al., "Formation of stable phosphorus–carbon bond for enhanced performance in black phosphorus nanoparticle–graphite composite battery anodes," Nano Lett, vol. 14, no. 8, pp. 4573–4580, 2014.

[11] J. Ni, L. Li, and J. Lu, "Phosphorus: an anode of choice for sodium-ion batteries," ACS Energy Lett, vol. 3, no. 5, pp. 1137–1144, 2018.

[12] H. Liu et al., "Bridging covalently functionalized black phosphorus on graphene for high-performance sodium-ion battery," ACS Appl Mater Interfaces, vol. 9, no. 42, pp. 36849–36856, 2017.

[13] W. Yu et al., "Facile production of phosphorene nanoribbons towards application in lithium metal battery," Advanced Materials, vol. 33, no. 35, p. 2102083, 2021.

[14] Rajabali M, Esfandiari M, Rajabali S, Vakili - Tabatabaei M, Mohajerzadeh S, Mohajerzadeh S. High - Performance Phosphorene - Based Transistors Using a Novel Exfoliation - Free Direct Crystallization on Silicon Substrates. Adv Mater Interfaces 2020;7:2000774.

The 5th Iranian International Conference on Microelectronics (IICM2023)
25 – 26 October 2023

First Principles Study of Optical and Electrical Properties for Mixed-halide 2D $BA_2PbBr_{4-x}Cl_x$ (x=0, 2, and 4) as an Active Layer of Perovskite Light Emitting Diode

Samad Shokouhi
Faculty of Eelctrical and Computer
Engineering
Taribiat Modares University
Tehran, Iran
Samadben9125@gmail.com

Seyedeh Bita Saadatmand
Faculty of Eelctrical and Computer
Engineering
Taribiat Modares University
Tehran, Iran
bitasaadat73@yahoo.com

Vahid Ahmadi
Faculty of Eelctrical and Computer
Engineering
Taribiat Modares University
Tehran, Iran
v_ahmadi@modares.ac.ir

Abstract— **Two-dimensional (2D) perovskites with structure of Ruddlesden Popper (RP) type are extensively used in optoelectronic applications. This study proposes mixed-halide two-dimensional (2D) perovskites $BA_2PbBr_{4-x}Cl_x$ (x=0, 2, and 4) as a material in the active layer of perovskite light-emitting diodes (PeLED). The density functional theory (DFT) is used to examine the electrical and optical properties of $BA_2PbBr_{4-x}Cl_x$ which has not been reported yet. The results reveal that as the amount of x increases, the bandgap energy blueshifts, and the value of the refractive index decreases. In addition, we demonstrated the band-edge orbitals by analyzing the partial density of states (PDOS) and observed the effect of halide change on them. In the case of x=0 (BA_2PbBr_4), the thermodynamic stability is lower compared to the other two states: $BA_2PbBr_4Cl_2$(x=2) and BA_2PbCl_4 (x=4). All three materials are thermodynamically stable, with direct and large band gaps suitable for blue PeLED applications.**

Keywords—Density functional theory, 2d-Perovskite, Light emitting diode, Band structure, Thermodynamical stability

I. INTRODUCTION

In recent years, the use of perovskite semiconductor materials due to their interesting electrical, optical, and mechanical properties, has been greatly extended in several optoelectronic components such as LEDs, lasers, solar cells, and sensors [1-6]. A general chemical formula of Ruddlesden-Popper (RP) quasi-two-dimensional perovskite materials is $(L)_2(A)_{n-1}BX_{3n+1}$ (n is an integer), where L, A, B, and X represent spacer cation, small cation, metal cation, and anion, respectively. The quasi-2D perovskite materials for n=1 are two-dimensional while for n=∞ they are similar to three-dimensional (3D) perovskites. 2D perovskites exhibit superior environmental and thermodynamical stability compared to 3D perovskites. Generally, 2D perovskites have a structure similar to a multi quantum well (MQW), where quantum barrier is spacer cation(L), and quantum well is inorganic layer (BX_4) [7].

To achieve a white perovskite light-emitting diode, enhancing the efficiency of the blue PeLED is crucial. Two important strategies for achieving a blue PeLED with high efficiency are utilizing 2D perovskite and mixed-halide materials [8]. The material $BA_2PbBr_{4-x}Cl_x$ (x=0, 2, and 4) is experimentally used as one of the main candidates for fabricating blue PeLED [9].

In this work, using density functional theory (DFT), we examine the electrical and optical characteristics, and thermodynamic stability of $BA_2PbBr_{4-x}Cl_x$ (x=0, 2, and 4) structure to examine its potential use in blue PeLED applications. The results show that BA_2PbCl_4 is thermodynamically stable, and by changing the halide from bromine to chlorine, the bandgap value increases.

II. THEORY

DFT calculations are performed using the Kohn-Sham equation in the CASTEP module. The equation is defined as:

$$[T+U]\phi_i(r)=\varepsilon_i\phi_i(r)$$
$$T=-\nabla^2/2, \quad U=V_n(r)+V_H(r)+V_x(r)+V_c(r) \quad (1)$$

In this equation, the T represents the kinetic energy, while U is the potential energy. The term U consists of the nuclear potential ($V_n(r)$), Hartree potential ($V_H(r)$), exchange potential ($V_x(r)$), and correlation potential ($V_c(r)$). Also, the $\phi_i(r)$ and ε_i are Kohn-Sham wavefunctions, eigenvalues, respectively [10].

Using the eigenvalues of the Kohn-Sham equation, electronic properties such as the band structure, band gap, and partial density of states (PDOS) can be obtained. The imaginary part of complex dielectric function $(\varepsilon_2(\omega))$ can be calculated as [11]:

$$\varepsilon_2(\omega)=(2q^2\pi)/(V\varepsilon_0)\sum_{(k,v,c)}|<\psi_k^c|p.r|\psi_k^v>|^2\delta(E_k^c-E_k^v-\hbar\omega) \quad (2)$$

where $\hbar\omega$ is the photon frequency, q is the electronic charge, V is the volume of a unit cell, p is the polarization of the incident electric field, and ψ_k^c and ψ_k^v are the wave functions of electrons in conduction band and valence band at a certain k (wave vector), respectively. With Kramers-Kronig relations, the real part of the dielectric function

979-8-3503-6020-2/23 $31.00 © 2023 IEEE

($\varepsilon_1(\omega)$) can be obtained from its imaginary part ($\varepsilon_2(\omega)$). Therefore, the complex refractive index ($N=n+ik$) and other optical properties can be calculated with using of complex dielectric function.

To investigate the thermodynamical stability of 2D perovskites, the formation energy (FE) can be computed as [12]:

$$FE=E(L_2BX_4)-2E(LX)-E(BX_2) \quad (3)$$

where E is the total energy of ground state for components. The greater the absolute value of the FE, the more stable the material.

III. RESULTS AND DISCUSSION

The crystal data of $BA_2PbBr_{4-x}Cl_x$ ($x=0$, 2, and 4) at room temperature (RT) are presented in Table 1. All of the crystal structures exhibit an orthorhombic phase, as shown in Fig. 1 [8].

TABLE I. CRYSTAL DATA FOR THE PROPOSED 2D-PEROVSKITES

Materials of 2D-perovskite	Crystal structure	Lattice parameters($\mathring{A}$)
BA_2PbBr_4	Orthorhombic	a=8.33, b=8.21, c=27.55, $\alpha=\beta=\gamma=90°$
$BA_2PbBr_2Cl_2$	Orthorhombic	a=8.33, b=8.21, c=27.55, $\alpha=\beta=\gamma=90°$
BA_2PbCl_4	Orthorhombic	a=8.33, b=8.21, c=27.55, $\alpha=\beta=\gamma=90°$

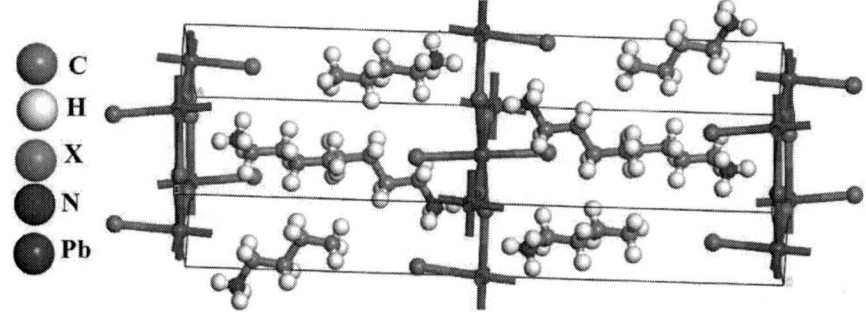

Fig. 1. Crystal structure of $BA_2PbBr_{4-x}Cl_x$ ($x=0$, 2, and 4).

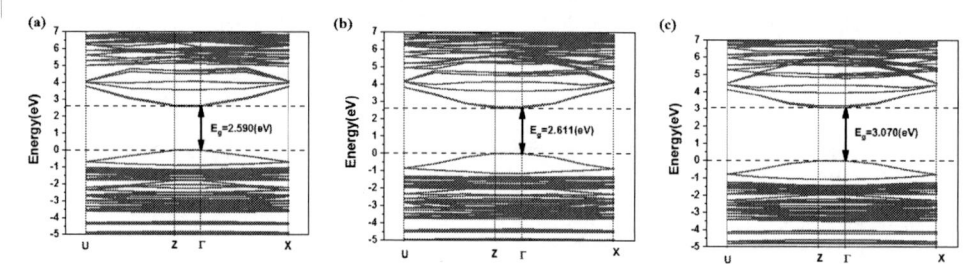

Fig. 2- Band structure for (a) BA_2PbBr_4, (b) $BA_2PbBr_2Cl_2$, and (c) BA_2PbCl_4

The band structure and bandgap energy (E_g) of $BA_2PbBr_{4-x}Cl_x$ ($x=0$, 2, and 4) are shown in Fig. 2. It is obvious that the bandgap energy increases as the halide changes from Br ($x=0$) to a mixed halide ($x=2$), and then to Cl ($x=4$). The E_g for BA_2PbBr_4, $BA_2PbBr_2Cl_2$, and BA_2PbCl_4, are 2.59 eV, 2.611 eV, and 3.07eV, respectively. All three materials have a direct bandgap at the Γ point. The bands near the valence band maximum (VBM) and conduction band minimum (CBM) in the Γ-Z region are flat and correspond to a barrier multiple quantum well, resulting in a large effective mass and restricted transport of carriers.

Fig. 3 show the PDOS of $BA_2PbBr_{4-x}Cl_x$ ($x=0$, 2, and 4). It can be concluded from these figures that the VBM as anti-bonding interaction between the s orbitals of Pb and the p orbitals of the halides (i.e, 6s-Pb with 4p-Br, or 3p-Cl), while the CBM arises from the weakly antibonding interaction between the 6p-Pb and the s orbitals of halide (X= Br, and Cl). Generally, in 2D perovskites, organic spacers do not directly affect the VBM and CBM but can indirectly influence the band gaps by inducing structural distortion.

The complex refractive index of $BA_2PbBr_{4-x}Cl_x$ ($x=0$, 2, and 4) as a function of frequency is displayed in Fig.4. The results show that in contrast to E_g, the refractive index at zero frequency (n_0) decreases with the change in halide from 1.90 in Br ($x=0$) to 1.830 in the mixed halide ($x=2$), and then to 1.752 in Cl ($x=4$). Therefore, the values of n_0 for BA_2PbBr_4, $BA_2PbBr_2Cl_2$, and BA_2PbCl_4, are 1.90, 1.830, and 1.752, respectively.

In Table 2, the thermodynamic stability of $BA_2PbBr_{4-x}Cl_x$ ($x=0$, 2, and 4) is investigated by computing the energy enthalpy of formation for the component materials. The results show the thermodynamical stability of $BA_2PbBr_2Cl_2$ is between BA_2PbBr_4 and BA_2PbCl_4. Also, BA_2PbBr_4 has a lower level of thermodynamic stability compared to $BA_2PbBr_2Cl_2$ and BA_2PbCl_4 due to its lower formation enthalpy energy.

TABLE 2. CRYSTAL DATA FOR THE PROPOSED 2D-PEROVSKITES

Materials	$E(L_2PbX_4)$ (eV)	$E(PbX_2)$ (eV)	$2*E(LX)$ (eV)	FE (eV)
BA_2PbBr_4	-10600.527	-4785.20	-5811.44	-3.88
$BA_2PbBr_2Cl_2$	-10771.207	-4785.608	-5981.49	-4.10
BA_2PbCl_4	-10942.509	-4956.052	-5982.26	-4.19

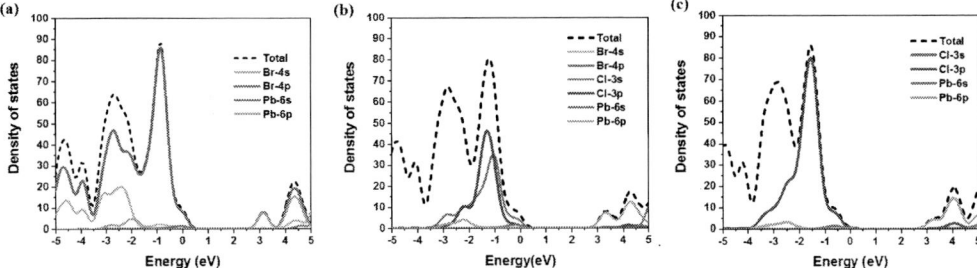

Fig. 3- PDOS curves for (a) BA_2PbBr_4, (b) $BA_2PbBr_2Cl_2$, and (c) BA_2PbCl_4

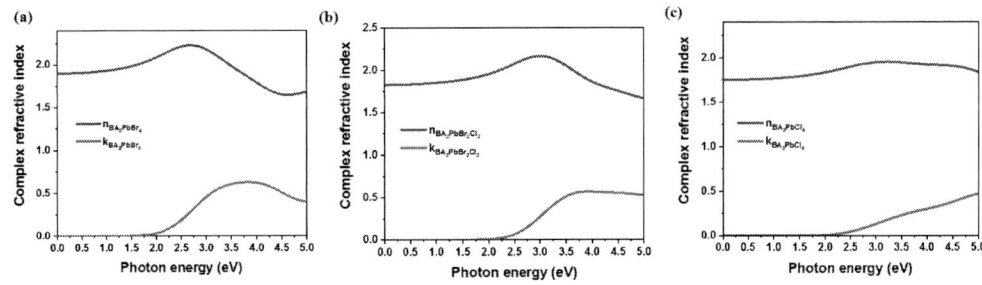

Fig. 4- Complex refractive index for (a) BA_2PbBr_4, (b) $BA_2PbBr_2Cl_2$, and (c) BA_2PbCl_4

These results reveal that the 2D perovskite-based mixed halide exhibits acceptable thermodynamical stability. In addition, it has a direct bandgap and a small refractive index, making it a suitable option for LED applications as an active layer.

IV. CONCLUSION

In summary, the electrical and optical properties of 2D-RP perovskites $BA_2PbBr_{4-x}Cl_x$ (x=0, 2, and 4) were investigated using DFT calculations. The results reveal that all three materials have a direct and tunable E_g. Increasing the amount of x in $BA_2PbBr_{4-x}Cl_x$ (x=0, 2, and 4) leads to a blueshift in E_g, and the value of the refractive index decreases. The PDOS plots indicate that the VBM is constructed by the antibonding interaction between the 6s-Pb and the 3p-Cl or 4p-Br orbitals, while the CBM dominates with 6p-Pb. All three materials are thermodynamically stable, as indicated by their negative formation energy. In conclusion, $BA_2PbBr_{4-x}Cl_x$ (x=0, 2, and 4) is an appropriate option for the emissive layer in PeLEDs.

[1] Veldhuis, S.A., Boix, P.P., Yantara, N., Li, M., Sum, T.C., Mathews, N. and Mhaisalkar, S.G., 2016. Perovskite materials for light-emitting diodes and lasers. *Advanced materials*, 28(32), pp.6804-6834.

[2] Yin, W.J., Yang, J.H., Kang, J., Yan, Y. and Wei, S.H., 2015. Halide perovskite materials for solar cells: a theoretical review. *Journal of Materials Chemistry A*, 3(17), pp.8926-8942.

[3] Chemerkouh, M.J.H.N., Saadatmand, S.B. and Hamidi, S.M., 2022. Ultra-high-sensitive biosensor based on SrTiO 3 and two-dimensional materials: ellipsometric concepts. *Optical Materials Express*, 12(7), pp.2609-2622.

[4] Saadatmand, S.B., Chemerkouh, M.J.H.N., Ahmadi, V. and Hamidi, S.M., 2023. Design and Analysis of Highly Sensitive Plasmonic Sensor Based on Two-Dimensional Inorganic Ti-MXene and SrTiO 3 Interlayer. *IEEE Sensors Journal*.

[5] Saadatmand, S.B., Shokouhi, S., Hamidi, S.M., Ahmadi, H. and Babaei, M., 2023. Plasmonic Heterostructure Biosensor based on Perovskite/Two dimensional Materials. *Optik*, p.171328.

[6] Saadatmand, S.B., Shokouhi, S., Ahmadi, V. and Hamidi, S.M., 2023. Design and analysis of a flexible Ruddlesden–Popper 2D perovskite metastructure based on symmetry-protected THz-bound states in the continuum. *Scientific Reports*, 13(1), p.22411.

[7] Zhang, L., Sun, C., He, T., Jiang, Y., Wei, J., Huang, Y. and Yuan, M., 2021. High-performance quasi-2D perovskite light-emitting diodes: from materials to devices. *Light: Science & Applications*, 10(1), p.61.

[8] Xing, J., Zhao, Y., Askerka, M., Quan, L.N., Gong, X., Zhao, W., Zhao, J., Tan, H., Long, G., Gao, L. and Yang, Z., 2018. Color-stable highly luminescent sky-blue perovskite light-emitting diodes. *Nature Communications*, 9(1), p.3541.

[9] Zhou, G., Li, M., Zhao, J., Molokeev, M.S. and Xia, Z., 2019. Single-Component White-Light Emission in 2D Hybrid Perovskites with Hybridized Halogen Atoms. *Advanced Optical Materials*, 7(24), p.1901335.

[10] Segall, M.D., Lindan, P.J., Probert, M.A., Pickard, C.J., Hasnip, P.J., Clark, S.J. and Payne, M.C., 2002. First-principles simulation: ideas, illustrations and the CASTEP code. *Journal of Physics: Condensed Matter*, 14(11), p.2717.

[11] Palik, E.D. ed., 1998. Handbook of Optical Constants of Solids (Vol. 3). Academic Press.

[12] Yang, Y., Gao, F., Gao, S. and Wei, S.H., 2018. Origin of the stability of two-dimensional perovskites: a first-principles study. *Journal of Materials Chemistry A*, 6(30), pp.14949-14955.

Light Trapping in InAsSb-based barrier Photodetectors for Enhanced Mid-Wave Infrared Bio-Medical Sensing: A Study on Jurkat Biomarker Detection

Maryam Shaveisi
Aerospace Research Institute (Ministry of Science, Research and Technology)
Tehran, Iran
M_Shaveisi@ari.ac.ir

Peiman Aliparast
Aerospace Research Institute (Ministry of Science, Research and Technology)
Tehran, Iran
Aliparast@ari.ac.ir

Sha Shiong Ng
Institute of Nano Optoelectronics Research and Technology (INOR), Universiti Sains Malaysia
Penang, Malaysia
shashiong@usm.my

Abstract— **This work presents the performance results of InAsSb-based n-cantact/barrier/n-absorber mid-wave infrared detectors, employing a Ge/BaF₂ 2-stacks mirror reflector with exceptional reflectivity exceeding ~90%. The focus of the investigation is on evaluating their effectiveness for Jurkat biomarker detection. The analysis includes a comprehensive examination of the dark current characteristics exhibited by this proposed device. Significantly, the findings reveal a remarkably low dark current level of approximately 2.032×10^{-5} A/cm² at 300 K, indicating its suitability for high-temperature operation. Additionally, we assess the spectral response of the device and observe peak current, responsivity and quantum efficiency values of 1.19 A/W and 0.37 respectively, when subjected to an incident light density of 1 W/cm². These metrics demonstrate impressive levels even under lower incident power densities, reaching values as high as 502.74 A/W for peak current responsivity and 155.85 for quantum efficiency when exposed to an incident power density as low as 10^{-3} W/cm². These significant findings validate the importance and effectiveness of utilizing light trapping methods based on Ge/BaF₂ mirror reflector in enhancing detector performance within this context.**

Keywords—InAsSb, nBn, Photodetector, Light trapping, Bio-medical sensing, Mid-wave infrared.

I. INTRODUCTION

In a wide range of applications, infrared photodetectors have gained recognition as valuable tools for detecting and measuring infrared light. These detectors possess unique capabilities that make them well-suited for analyzing the mid-wave infrared (MWIR) range, which encompasses wavelengths ranging from 3 to 5 µm [1]. Among the various areas of application, MWIR technology show immense promise in the detection of blood biomarkers, particularly when it comes to analyzing Jurkat cells [2]. Jurkat cells, propelled from human leukemia and belonging to the T lymphocyte lineage, have found wide-ranging utility as a model system in biomedical research and clinical

investigations [2, 3]. By discerning specific biomarkers associated with Jurkat cells, such as surface antigens or intracellular molecules, valuable insights into immune responses, disease progression, and treatment efficacy can be gained. However, the conventional methods employed to detect Jurkat biomarkers often involve arduous and time-consuming techniques like flow cytometry or immune-histochemistry. The utilization of MWIR detectors in detecting Jurkat biomarkers in blood samples presents numerous advantages [4]. Firstly, the MWIR range is appropriate for examining the mid-infrared region's molecular vibrations and absorption bands of biomolecules. This unique capability facilitates accurate identification and quantification of Jurkat biomarkers by utilizing their distinct spectral signatures. Moreover, MWIR photodetectors provide the ability to detect Jurkat biomarkers in blood samples without the need for labeling, and in real-time [4, 5]. This eliminates the necessity for complex sample preparation procedures and decreases the overall analysis time. Furthermore, it enables the continuous monitoring of Jurkat biomarkers, facilitating dynamic studies and providing valuable information on disease advancement and the effectiveness of treatments. Additionally, incorporating MWIR photodetection in the analysis of blood biomarkers offers the advantage of non-invasiveness. The ability of MWIR light to penetrate biological tissues allows for in vivo measurements without the need for invasive procedures [4]. This feature is particularly beneficial when monitoring Jurkat biomarkers in patients, as it offers a less invasive and more patient friendly approach. In order to maximize the potential of MWIR photodetectors for detecting Jurkat biomarkers in blood, ongoing research is focused on the advancement of specialized sensing platforms and methodologies. This entails the exploration of cutting-edge detector technologies, including barrier photodetector or XBn devices, known for their outstanding sensitivity and selectivity within the MWIR range [6-9]. Through the integration of these detectors and bio-sensing components, it becomes feasible to develop compact and portable systems suitable for point-of-care testing and remote monitoring applications. Barrier architecture infrared photodetectors are part of an advanced class of photodetectors that operate in the infrared spectrum, particularly in the medium wave infrared (MWIR) and long wave infrared (LWIR) regions. The mentioned photodetectors are meticulously designed using various semiconductor materials, including InAsSb, HgCdTe, InAs/GaSb, and InAs/InAsSb, known for their exceptional performance characteristics in detecting infrared radiation [8, 10-12]. XBn infrared photodetectors offer an added notable benefit in their capacity to operate at elevated temperatures [13]. These devices demonstrate efficient performance at or in the vicinity of room temperature, obviating the need for costly and cumbersome cryogenic cooling systems, the nBn infrared imaging devices are highly suitable for a wide range of applications. By employing specific materials in n-contact/barrier/n-absorber photodetectors, an impressive capability of band gap engineering is unlocked. This empowers precise customization of absorption characteristics across the entire infrared spectrum. As a result, XBn photodetectors excel in efficiently detecting and distinguishing specific infrared wavelengths, making them highly suitable for a wide array of applications that encompass multispectral and hyperspectral imaging. The progress and refinement of XBn infrared photodetectors have been driven by advancements in epitaxial growth techniques, material design, and device engineering. Continuous research endeavors are focused on enhancing the performance parameters of XBn devices, encompassing responsivity, dark current, and noise equivalent power [14-16]. These improvements aim to optimize the sensitivity and overall performance of XBn photodetectors. In this research paper, we present the application of InAsSb-based nBn infrared photodetectors that utilize a light trapping technique (Germanium (Ge) and barium fluoride (BaF$_2$) stacks) for precise, real-time, and non-invasive detection of Jurkat biomarkers in blood samples. The integration of these photodetectors has immense potential to drive advancements in biomedical research, diagnostics, and personalized medicine. It creates new avenues for gaining comprehensive understanding of immune responses, disease progression, and treatment efficacy, thereby offering valuable insights for medical applications.

II. DESCRIPTION OF DEVICE AND SIMULATION METHOD

Figure 1 illustrates the detailed schematic structure of the nBn (n-contact/barrier/n-absorber) barrier photodetector, which has undergone meticulous calibration based on a previously fabricated devices mentioned in reference [6, 7]. In contrast to conventional XBn structures, the proposed design incorporates specific epilayers that serve distinct functions. These epi-layers consist of multiple layers, including a highly doped n-type top contact layer (InAs$_{1-x}$Sb$_x$, x_{Sb}=0.11) with a thickness of 100 nm, an n-type absorber layer (InAs$_{1-x}$Sb$_x$, x_{Sb}=0.11) with a thickness of 3000 nm, an un-doped barrier layer utilizing the AlSb material system with a thickness of 200 nm, and an n-type GaSb bottom contact layer with a thickness of 100 nm. The precise arrangement of these layers on a substrate with the same bottom contact layer material ensures both the structural integrity and compatibility of the photodetector.

For more comprehensive specifications and detailed information regarding the material layers, please refer to Table I. At the upper part of the suggested structure, germanium (Ge) is designated as the material possessing a high refractive index (HRI), while barium fluoride (BaF$_2$) is chosen as the material with a low refractive index (LRI) to construct reflector Ge/ BaF$_2$ 2-stacks with λ/4-thickness for using as a mirror reflector section. More comprehensive information about design of this particular configuration can be found in the discussion presented in reference [14]. To gain a more profound understanding of the photodetector's performance characteristics, an extensive theoretical analysis is conducted. This analysis involves the utilization of numerical solutions for Poisson's equation and the electron-hole current continuity equations, which are solved using Silvaco device software. The software employs a sophisticated drift-diffusion model that takes into account the electrical and optical properties of semiconductors. By considering the influences of the Shockley-Read-Hall, radiative, and Auger mechanisms, the software enables accurate estimation and prediction of the device's performance [15, 16]. In this study, the refractive index (n) and extinction coefficient (k) values of Ge and BaF$_2$ were used from references [14, 17]. This information is highly valuable for further evaluation, optimization, and refinement of the photodetector's design.

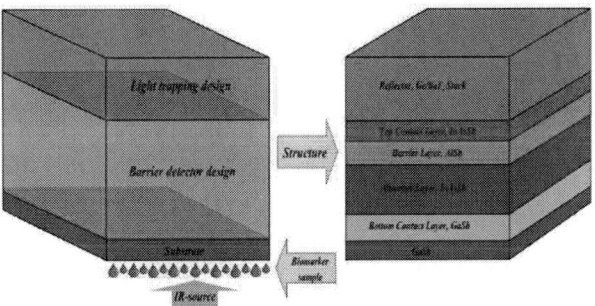

Fig. 1. A schematic representation of the proposed optical device.

TABLE I. THE MAIN PROPERTIES AND SPECIFICATIONS OF THE PROPOSED OPTICAL DEVICE

Region	Material	Doping density	Depth
HRI-layer	Ge	-	220 nm
LRI-layer	BaF$_2$	-	700 nm
Top contact	InAs$_{1-x}$Sb$_x$ (x_{Sb}=0.11)	n-2×10^{16} cm^{-3}	100 nm
Barrier	AlSb	Non-doped	200 nm
Absorber	InAs$_{1-x}$Sb$_x$ (x_{Sb}=0.11)	n-2×10^{16} cm^{-3}	3000 nm
Bottom contact	GaSb	n-5×10^{17} cm^{-3}	100 nm
Substrate	GaSb	n-5×10^{17} cm^{-3}	-

III. SIMULATION AND ANALYSIS

Figure 2(a,b) illustrates the simulated energy band structure of the proposed nBn device at both zero bias and a bias voltage of -500 mV. The depicted band structure corresponds to an n-type absorber layer made of InAs$_{1-x}$Sb$_x$, where the Sb composition is x_{Sb}=0.11, and it has a doping density of 2×10^{16} cm^{-3}. Additionally, there is an un-doped barrier layer in the structure. In the energy band diagram, it is crucial to focus on two key objectives: increasing the barrier of the conduction band and decreasing the barrier of the valence band. The presence of a conduction band barrier between the absorber layer and the barrier layer acts as an obstacle, blocking the movement of electrons from the absorber layer to the barrier layer. This barrier serves the purpose of decreasing the dark current. Furthermore, it has been observed that a barrier has emerged in the valence band, which obstructs the trapping of optically generated holes within the absorber layer. Therefore, in this specific design, our goal is to establish a conduction band barrier with the highest energy and a valence band barrier with the lowest energy. To accomplish this objective, we strive for optimal conditions where the conduction band barrier is approximately 1.263 eV and the valence band barrier is around 79 meV under zero bias conditions. By increasing the bias voltage, the reduction in the valence band barrier occurs. However, this simultaneous creation of a depletion region within the absorber region contributes to an increase in dark current. Based on Fig. 2(c), it is evident that the recombination rate following the Shockley-Read-Hall (SRH) model demonstrates an increase with higher applied voltages. This can be attributed to the augmented injection of carriers and intensified recombination of electrons and holes under conditions of elevated applied bias. Consequently, this heightened SRH recombination leads to a noticeable reduction in electron density within the absorber region. As a result, there are negative implications for overall device performance and efficiency since it hinders the effective collection and utilization of generated electrons for desired functionalities such as current generation or signal detection. Therefore, it becomes imperative to minimize SRH recombination by optimizing voltage conditions in order to maintain higher electron densities and enhance device performance pertaining to charge transport and functionality. The findings illustrated in Fig. 2(d) reveal a significant observation, wherein an increase in the applied bias voltage corresponds to an elevation of the electric field within the barrier region. This unintended consequence carries substantial implications for device performance due to its correlation with dark current generation. The intensified electric field observed in the barrier region can be attributed to enhanced carrier injection and improved carrier transport under conditions of elevated bias. This heightened electric field facilitates electron tunneling or surmounting energy barriers that would typically impede their mobility. Consequently, this phenomenon leads to the occurrence of dark current, which represents an undesired component independent of external illumination. The presence of dark current exerts adverse effects on device performance by introducing additional noise, diminishing signal-to-noise ratio, and compromising overall sensitivity. Moreover, it restricts dynamic range capabilities and undermines image quality in imaging devices such as photodetectors or cameras that rely on precise detection of light signals. To decrease this undesired effect while enhancing device performance, various strategies can be implemented, including optimizing material properties or engineering barrier structures. These measures aim to minimize the generation of dark current and improve overall device efficiency.

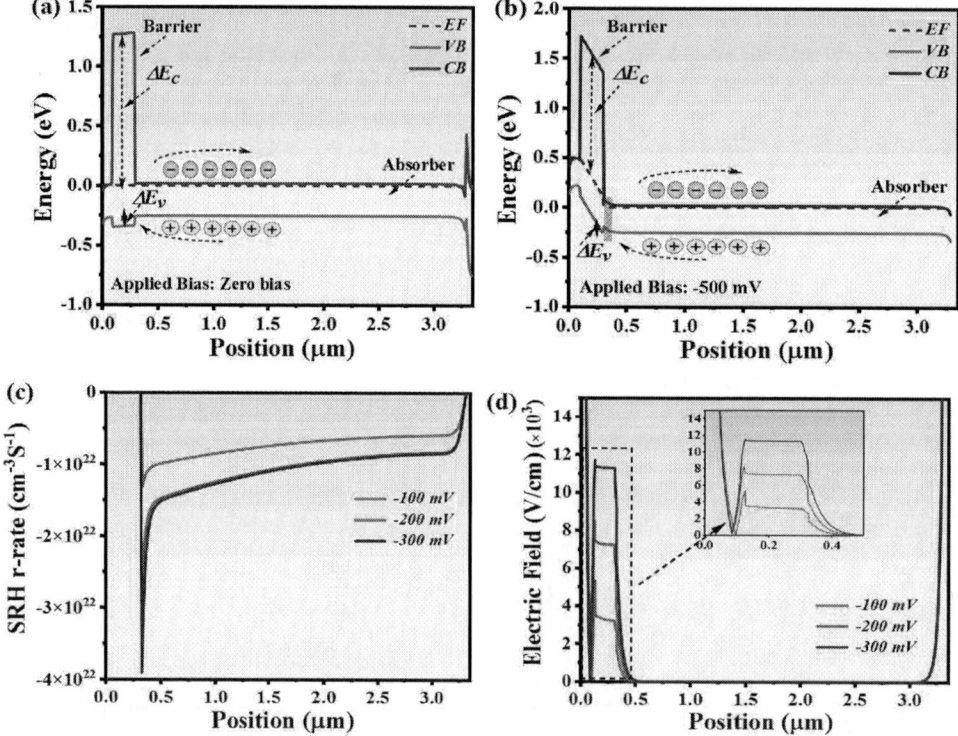

Fig. 2. The simulated energy band arrangement of the proposed infrared nBn structure is shown in a) a zero bias condition and b) at a bias of -500 mV, and the variation of c) the SRH recombination rate and d) the electric field magnitude as a function of position under various applied biases.

The 5th Iranian International Conference on Microelectronics (IICM2023)

In Fig. 3(a), the dark current (J_{dark}) characteristic curve of a photodetector with dimensions of 15×15 µm² is presented, showing the relationship between voltage and dark current at temperatures of 200 K and 300 K. The purpose of this analysis is to examine how the dark current behaves under different bias conditions and temperatures. Specifically, when the device operates at a temperature of 300 K with an applied bias voltage of -200 mV, the computed dark current density is an impressive 2.032×10⁻⁵ A/cm². This finding indicates that the device performs exceptionally well in suppressing undesired leakage currents under these specific operating conditions. Achieving such low levels of dark current is essential for maintaining a high signal-to-noise ratio and minimizing noise interference in photodetection applications. It also reflects positively on the overall quality and efficiency of the proposed photodetector design. According to available literature, it is clear that diffusion mechanisms primarily dominated the dark current in nBn photodetectors at lower bias voltages [6, 18, 19]. In our investigation, we conduct

an optical analysis where the proposed device is exposed to infrared radiation with a wavelength of 4 µm and a light intensity of 1 W/cm², the illumination is directed from the bottom side of the device. Based on experimental findings reported in [14], it has been observed that the Ge/BaF₂ 2-stacks, when utilized as a mirror reflector on top of the device, exhibit a reflectance exceeding 90% across the entire wavelength range of the mid-wave infrared (MWIR) band. Figure 3(b) presents the spectral current responsivity and quantum efficiency of the photodetector at a temperature of 300 K. Impressively, this photodetector demonstrates peak optical response in detecting Jurkat biomarkers within blood samples. When a bias voltage of -200 mV is applied, the device exhibits a responsivity of 1.19 A/W and quantum efficiency (QE) value of 0.37. These findings indicate that this specific device configuration achieves high sensitivity in capturing signals associated with Jurkat biomarkers. As a result, it is well-suited for targeted detection applications in blood sample analysis.

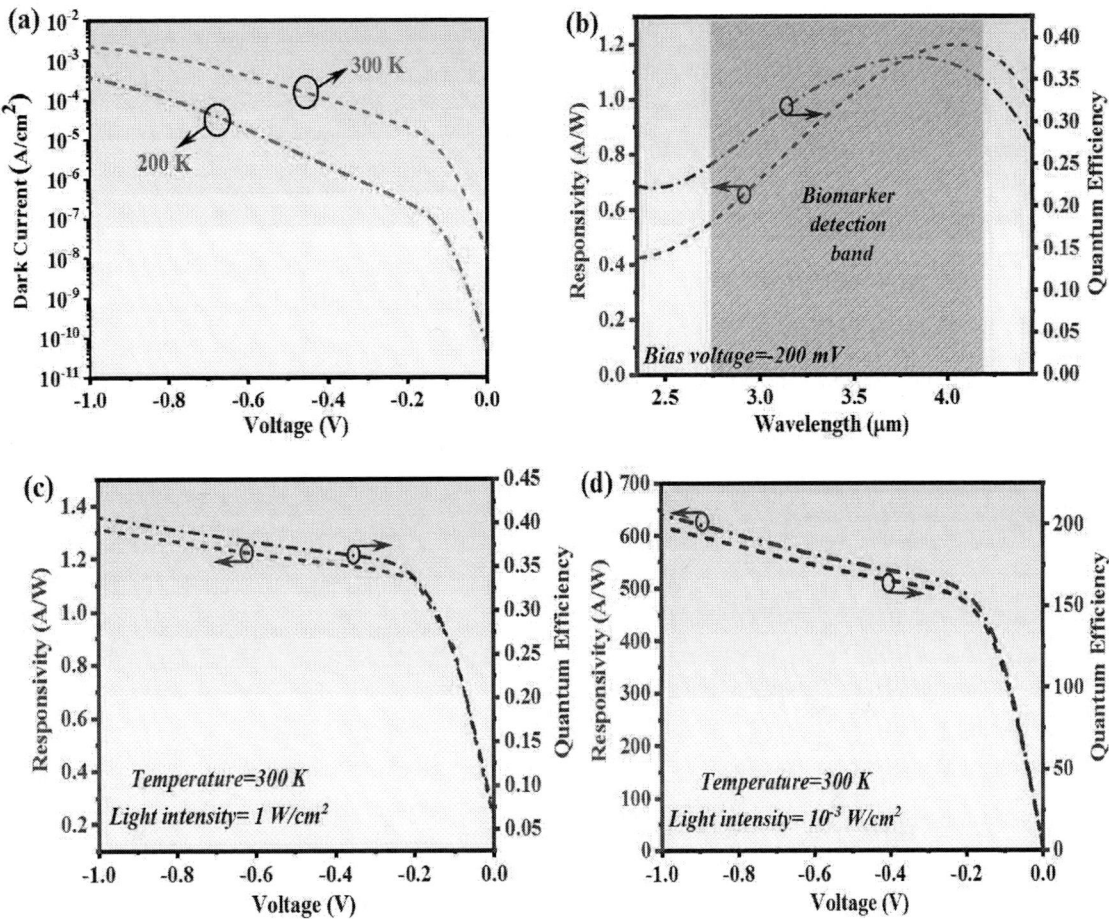

Fig. 3. a) The computed dark current density at different temperature conditions b) the spectral response, including current responsivity and quantum efficiency (QE), c) and d) the current responsivity and QE plotted across a range of applied biases, from zero to -1 V, under different light intensities of the proposed optical device at 300 K.

Figure 3(c,d) displays the current responsivity (R_λ) and quantum efficiency (QE) for biases ranging from 0 to -1 V (V_b), considering light intensities of both 1 W/cm² and 10^{-3} W/cm². The results indicate that incorporating a mirror reflector consisting of Ge/BaF₂ 2-stacks yields a noticeable enhancement in optical response, particularly at lower light intensities. By utilizing the current responsivity and dark current density, incorporating Johnson and shot noise [18], the specific detectivity (D*) and noise equivalent power (NEP) can be estimated. In this case, under a constant irradiance of 1 W/cm² and a bias voltage of -200 mV, the proposed device exhibits outstanding performance. The peak detectivity reaches an impressive value of 2.1×10^{10} cmHz$^{1/2}$/W, while the NEP measures at 5.21×10^{-11} WHz$^{1/2}$ (refer to Fig. 4). Table II presents a comprehensive performance comparison of barrier photodetectors that have been specifically designed for the MWIR region.

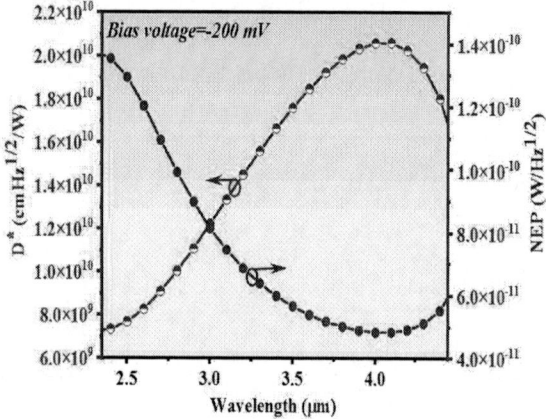

Fig. 4. The calculated spectral D* and NEP under -200 MV at 300 K.

TABLE II. COMPARISON OF INASSB-NBN DEVICE (1 W/CM² AND 4 μM) WITH RECENTLY MID-WAVE PHOTODETECTORS

Ref.	V_b (mV)	J_{dark} (A/cm²)	R_λ (A/W)	D* (cmHz$^{1/2}$/W)
Ref. [20]	0	0.2	0.3	1.5×10^9
Ref. [21]	-500	0.6	0.1	3×10^8
Ref. [22]	-50	3.4×10^{-4}	0.6	10^{10}
This work	-200	2.032×10^{-5}	1.19	2.1×10^{10}

CONCLUSION

In this study, we present the performance outcomes of InAsSb-based nBn MWIR detectors in detecting Jurkat biomarkers, employing a Ge/BaF₂ 2-stacks mirror reflector. The investigation encompasses an analysis of the dark current characteristics, which revealed that this device achieved a low dark current density of approximately 2.032×10^{-5} A/cm² at 300 K. This characteristic renders it suitable for high-temperature operations. Furthermore, the spectral response analysis demonstrated that the proposed device exhibited peak current responsivity and QE values of 1.19 A/W and 0.37, respectively under an illumination intensity of 1 W/cm². These metrics improved to 502.74 A/W and 155.85 respectively, when subjected to an illumination intensity of 10^{-3} W/cm². These findings underscore the significance of implementing light trapping

techniques based on the Ge/BaF₂ mirror reflector to enhance detector performance. By utilizing such methods, efficient photon absorption is achieved along with higher current responsivity and QE values. Our results highlight the promising potential of InAsSb-based nBn MWIR detectors integrated with a Ge/BaF₂ mirror reflector design in sensitive detection applications involving Jurkat biomarkers or similar targets.

REFERENCES

1. Rogalski, A., *Infrared detectors: an overview.* Infrared physics & technology, 2002. **43**(3-5): p. 187-210.
2. Titus, J., et al., *Early detection of cell activation events by means of attenuated total reflection Fourier transform infrared spectroscopy.* Applied Physics Letters, 2014. **104**(24).
3. Mihály, J., et al., *Characterization of extracellular vesicles by IR spectroscopy: fast and simple classification based on amide and CH stretching vibrations.* Biochimica et Biophysica Acta (BBA)-Biomembranes, 2017. **1859**(3): p. 459-466.
4. Miller, L., G. Smith, and G. Carr, *Synchrotron-based biological microspectroscopy: from the mid-infrared through the far-infrared regimes.* Journal of Biological Physics, 2003. **29**(2-3): p. 219-230.
5. Mostaço-Guidolin, L.B. and L. Bachmann, *Application of FTIR spectroscopy for identification of blood and leukemia biomarkers: A review over the past 15 years.* Applied Spectroscopy Reviews, 2011. **46**(5): p. 388-404.
6. Klipstein, P. *"XBn" barrier photodetectors for high sensitivity and high operating temperature infrared sensors.* in *Infrared Technology and Applications XXXIV.* 2008. SPIE.
7. Klipstein, P., et al. *MWIR InAsSb XBn detectors for high operating temperatures.* in *Infrared Technology and Applications XXXVI.* 2010. SPIE.
8. Klipstein, P.C., *Perspective on III–V barrier detectors.* Applied Physics Letters, 2022. **120**(6).
9. Shaveisi, M. and P. Aliparast, *Mid-wave infrared optical receiver based on an InAsSb-nBn photodetector using the barrier doping engineering technique for low-power satellite optical wireless communication.* Applied Optics, 2023. **62**(10): p. 2675-2683.
10. Zavala-Moran, U., et al. *Antimonide-based Superlattice Infrared Barrier Photodetectors.* in *PHOTOPTICS.* 2020.
11. Shaveisi, M. and P. Aliparast. *Dark Current Evaluation in HgCdTe-based nBn Infrared Detectors.* in *2021 Iranian International Conference on Microelectronics (IICM).* 2021. IEEE.
12. Arounassalame, V., et al. *Electro-optical characterizations to study minority carrier transport in Ga-free InAs/InAsSb T2SL XBn midwave infrared photodetector.* in *Electro-Optical and Infrared Systems: Technology and Applications XVIII and Electro-Optical Remote Sensing XV.* 2021. SPIE.
13. Rogalski, A. and P. Martyniuk, *Mid-Wavelength Infrared n B n for HOT Detectors.* Journal of electronic materials, 2014. **43**(8): p. 2963-2969.
14. Gill, G.S., et al., *Ge/BaF2 thin-films for surface micromachined mid-wave and long-wave infrared reflectors.* Journal of Optical Microsystems, 2022. **2**(1): p. 011002-011002.
15. Shaveisi, M. and P. Aliparast, *Design and modeling of high-performance mid-wave infrared InAsSb-based nBn photodetector using barrier band engineering approaches.* Frontiers of Optoelectronics, 2023. **16**(1): p. 5.
16. Shaveisi, M. and P. Aliparast. *Performance Analysis of Mid-Wave Optical Receiver based on Barrier Upside-down nBn Photodetectors for Free Space Optical Wireless Communication Systems.* in *2022 Iranian International Conference on Microelectronics (IICM).* 2022. IEEE.
17. Querry, M.R., *Optical constants of minerals and other materials from the millimeter to the ultraviolet.* 1998. Chemical Research, Development & Engineering Center, US Army Armament ....
18. Shaveisi, M. and P. Aliparast, *Design of a high-sensitivity extended mid-wave infrared InAsSb-based nBn photodetector by utilizing barrier band engineering technique: an outstanding device for biosensing applications.* Optical and Quantum Electronics, 2023. **55**(10): p. 848.
19. Shaveisi, M. and P. Aliparast. *High Performance MWIR InAsSb-based nBn Photodetectors.* in *2022 Iranian International Conference on Microelectronics (IICM).* 2022. IEEE.
20. Lotfi, H., et al., *High-frequency operation of a mid-infrared interband cascade system at room temperature.* Applied Physics Letters, 2016. **108**(20): p. 201101.
21. Huang, J., et al., *High speed mid-wave infrared uni-traveling carrier photodetector.* IEEE Journal of Quantum Electronics, 2020. **56**(4): p. 1-7.
22. Korkmaz, M., et al., *Performance evaluation of InAs/GaSb superlattice photodetector grown on GaAs substrate using AlSb interfacial misfit array.* Semiconductor Science and Technology, 2018. **33**(3): p. 035002.

Possible Teleportation of Quantum States using Squeezed Sources and Photonic Integrated Circuits

Mobin Motaharifar
Electrical Engineering Department
Photonics Research lab.(PRL)
Amirkabir University of Technology
Tehran, Iran
Email: mobinmotahari@gmail.com

Hassan Kaatuzian
Electrical Engineering Department
Photonics Research lab.(PRL)
Amirkabir University of Technology
Tehran, Iran
Email: hsnkato@aut.ac.ir

Mahmood Hasani
Electrical Engineering Department
Photonics Research lab.(PRL)
Amirkabir University of Technology
Tehran, Iran
Email: mahmoodhasaniph@gmail.com

Abstract—Emergent quantum technologies are opening up new horizons for researchers of different disciplines. One such promising technique is the teleportation of a quantum state. Theoretical physics and telecommunication are different fields for which quantum teleportation can be fruitful. Additionally, squeezed state of light is introduced in this paper as a tool for measurement accuracy improvement. Moreover, with the help of integrated Mach-Zehnder photonic cells and squeezed light source, we propose the use of a quantum photonic computation method (continuous variable quantum states) for the sake of quantum teleportation. We will further try to implement a quantum key distribution (QKD) algorithm based on this method and validate its feasibility as well.

Keywords—quantum information processing, squeezed light, Mach-Zender modulator, teleportation, quantum key distribution

I. INTRODUCTION

In the context of quantum computation, squeezed light[1] can be employed to create squeezed qubits. A squeezed qubit is a quantum state that exhibits reduced noise in one of its observables, typically the amplitude or phase. By squeezing the noise in one of these properties, we can enhance the precision and sensitivity of measurements and operations performed on the qubit without violating the uncertainty principle.

Squeezed light qubits have several advantages in quantum computation. Firstly, they can improve the accuracy of quantum gates and measurements, leading to more reliable and robust quantum operations. Secondly, they can enhance the resilience of qubits against certain types of noise and decoherence, making them more suitable for long-term storage and manipulation of quantum information. Lastly, squeezed light qubits can enable the implementation of novel quantum algorithms and protocols that take advantage of the unique properties of squeezed states.

Moreover, it is well known that the actual breakthrough in digital computation was the development of new methods to increase the density of transistors on a specific

integrated surface area. This improves both the speed and the power efficiency of processors. While bulk photonics is very familiar around the world, photonic integrated circuits (PICs) can bring similar advantages as the ICs did to the field of electronics. Aside from their vast potential in classical technologies like telecommunications and computation, PICs are attracting growing attention for their role in quantum technologies. Using the squeezed states of light in quantum photonic integrated circuits (qPICs) could be very useful in the three following domains [2]:

A. Quantum communication

It follows two goals of developing quantum key distribution (QKD) systems and distributed quantum computation via a quantum internet. Many large-scale QKD networks are already built and tested in different parts of the world, like Switzerland [3], Japan [4], and the United States of America [5]. However, the concept of miniaturizing the apparatus is a new one and is still under development. Overall, squeezed light sources are anticipated to play a role in quantum networks for both QKD and distributed quantum computation purposes.

B. Quantum computation and simulation

With the help of the integrated generation of squeezed states of light, researchers were able to vastly improve the sampling rate in Gaussian boson sampling experiments and circumvent the probabilistic nature of the parametric downconversion source [6]. The demonstration of quantum supremacy with the help of squeezed light sources [7, 8] shows a promising future for this particular state of light in the field of quantum computation. The squeezed state also helps reduce certain types of quantum noise in qPICs [9-11].

C. Quantum sensing and metrology

This field seems to be the most fruitful domain in the near future, and the role of squeezed light is of pivotal importance in its success. The key difference between squeezed and coherent light is the improvement in the accuracy of one quadrature at the cost of less accuracy in another quadrature. Such characteristic makes measurements with unprecedented accuracy possible. LIGO is improving its capabilities to detect gravitational waves by the implementation of squeezed light techniques [12, 13].

In Section II, the concept of squeezed light will be reviewed, and with the help of that, our suggested approach to continuous variables quantum teleportation will be introduced in Section III alongside further discussion of obstacles in Section IV. Simulation results of this approach are presented in Section V. We will also have a conclusion in Section VI.

II. SQUEEZED STATE OF LIGHT

Squeezed state of light is a quantum state of light that can be interpreted by its different variants. Similar to any quantized oscillator, the oscillation of the field strength is characterized by two quantities that do not commute with each other. In other words, the standard deviation ranges for both position and momentum (as two parameters that don't commute with each other) are equal to ½; thus, the quadrature uncertainty forms a circular shape for a coherent light beam. The minimum value of the uncertainty product between position and momentum in coherent-state light is ¼ when both the wave functions Ψ(x) and φ(p) are Gaussian in nature. The Ψ(x) function is related to exp[-(x-α).(x-α)], and φ(p) is related to exp[-(p-β).(p-β)], where α and β represent the mean values of position (x) and momentum (p) respectively. In coherent light, the probability of photon counting follows a Poisson distribution, unlike in electromagnetic optics. Coherent light in quantum optics is not deterministic. When dealing with a vacuum state (zero photons) or a single photon, there is an uncertainty in time measurement that is not suitable for the precise measurements required in different applications of integrated photonics[14].

To address this, it becomes necessary to utilize a nonclassical squeezed-state light source, which allows for reducing the uncertainty of one of the quadrature components below ½. This squeezing of the uncertainty circle of the electric field phasor transforms it into an ellipse. This asymmetry enables measuring the field only at times when the uncertainty is minimal, resulting in reduced noise compared to a coherent state. By heterodyning the squeezed field with coherent light, it becomes suitable for the precise measurements needed in quantum technologies. Photon-counting probability in squeezed light follows a sub-Poisson distribution. With the assistance of nonlinear optics, parametric oscillation, and twin-beam light, it becomes possible to generate the required squeezed light as the primary source to be used for Mach-Zehnder Interferometer (MZI) cells [15].

In the case of electromagnetic waves, these quantities are the amplitude quadrature (X) and the phase quadrature (Y) of the electric field. Squeezed states exhibit a reduced uncertainty in either the strength of the electric field or its phase. These two situations are depicted in Fig. 1. Squeezed state operator can be written as:

$$\hat{S}(\xi) = \exp[(\xi \hat{a}^2 - \xi^* \hat{a}^{\dagger 2})/2] \quad (1)$$

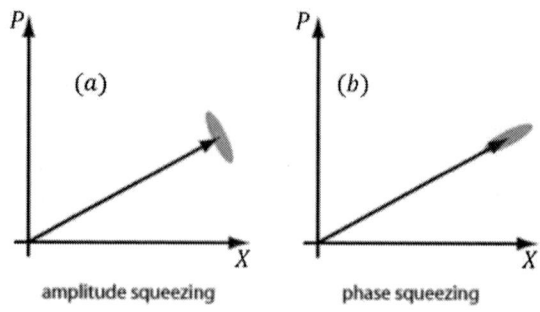

Fig 1. (a) is the amplitude-squeezed state of light and (b) is the phase-squeezed state of light

Where here $\hat{S}(\xi)$ is one mode squeezed light operator and $\hat{a}$ and $\hat{a}^\dagger$ are annihilation and creation operators and ξ is a squeezing parameter. While the two-mode squeezed light operator is:

$$\hat{S}_2(\xi) = \exp\left[\left(-\xi \hat{a}\hat{b} + \xi^* \hat{a}^\dagger \hat{b}^\dagger\right)\right] \quad (2)$$

Where parameters are the same as the last equation except for the $\hat{b}$ (and $\hat{b}^\dagger$) operator which is the annihilation (and creation) operator for the second mode.

III. CONTINUOUS VARIABLES QUANTUM TELEPORTATION

A. Quantum Teleportation

The transfer of an undisclosed quantum state between physically distant qubits or qumodes is made possible through quantum teleportation. This process is dependent on a combination of classical communication and quantum entanglement and is an important protocol in the field of quantum information. It has diverse applications, including in quantum communication and distributed information processing for quantum computation.

In general, the fundamental principle behind all quantum teleportation circuits remains the same. There are two observers, Alice and Bob, who are located at a distance from each other. They share a maximally entangled quantum state, which can be one of the four Bell states in discrete variables or a maximally entangled state for a fixed level of energy in continuous variables. Additionally, they have access to a classical communication channel.

To transport an unknown state from Alice to Bob, Alice performs a joint measurement on her half of the entangled state and the unknown state. This measurement is done by projecting onto the Bell basis. Alice then transmits the measurement results to Bob. Using this information, Bob can transform his half of the entangled state into an accurate replica of the original unknown state. This transformation involves performing a conditional phase flip for qubits or displacement for qumodes.

The original concept of quantum teleportation was initially developed for discrete-variable quantum communication involving qubits. However, its adaptability extends to continuous-variable qumodes when they are spatially separated. In Fig. 2, we can observe the circuit representation specifically designed for the purpose of continuous-variable quantum teleportation.

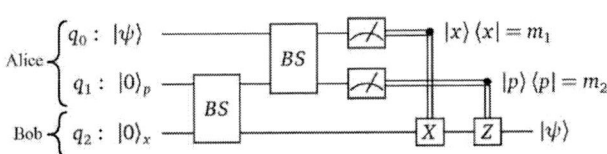

Fig. 2. The implementation of quantum teleportation on a continuous quantum variable circuit. [16]

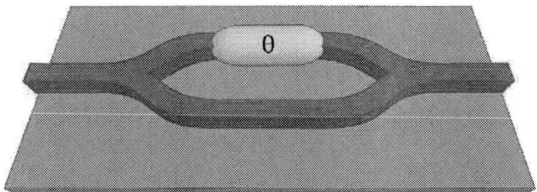

Fig. 3. A modified integrated unit cell inspired by the Mach-Zehnder Cell.

B. Integration of the elements

As indicated in [17], any unitary transformation on single qubits can be represented as a sequence of three rotations, one around $\hat{y}$ axis and two around $\hat{z}$ axis. Inspired by the mentioned fact, a Mach-Zehnder integrated cell is suitable for the realization of an arbitrary single-qubit unitary transformation [18]. A modification to the Mach-Zehnder interferometer cell can be applied so that it would fit the structure shown in Fig. 2. The new integrated cell is depicted in Fig. 3.

One can simply redesign elements of the Fig. 2 structure using two of these newly modified cells, two homodyne detectors, and two directional couplers. The resulting circuit will be like Fig. 4. The two directional couplers are needed to make entanglements between Alice's and Bob's states. They act the same as beamsplitters in Fig. 2. Such a circuit will also need two homodyne detectors to make measurements of the average value of arbitrary quadrature components. Note that in such configuration, the quantum states are not created using single photons, but rather continuous variables quantum states that are able to be squeezed.

The process of continuous variables quantum teleportation in Fig. 4 can be described as follows:

1) Initially, the qumodes q_1 and q_2 are prepared as infinitely squeezed vacuum states in momentum and position space, respectively. These states can be represented as:

$$|0\rangle_x \sim \lim_{z \to \infty} S(z)|0\rangle \qquad (3)$$

$$|0\rangle_p \sim \lim_{z \to -\infty} S(z)|0\rangle = \frac{1}{\sqrt{\pi}} \int_{-\infty}^{\infty} |x\rangle dx \qquad (4)$$

2) Next, q_1 and q_2 are maximally entangled using a 50-50 beamsplitter, represented as:

$$BS(\pi/4, 0)\left(|0\rangle_p \otimes |0\rangle_x\right) \qquad (5)$$

These qumodes are then spatially separated, with q_1 held by Alice and q_2 held by Bob. They are connected through classical communication channels c_0 and c_1.

3) In order to teleport her unknown state $|\psi\rangle$ to Bob, Alice needs to perform a projective measurement of her entire system using the maximally entangled basis states. This is achieved by entangling $|\psi\rangle$ and $|q_1\rangle$ through another 50-50 beamsplitter. Alice then performs two homodyne measurements, one in the x quadrature and the other in the p quadrature.

4) The measurement results are transmitted to Bob, who uses them to perform a position displacement (conditional on the x measurement) and a momentum displacement (conditional on the p measurement). These displacements allow Bob to accurately recover the transmitted state $|\psi\rangle$.

IV. FEASIBILITY AND CHALLENGES

Although the use of squeezed light provides unprecedented accuracy, it doesn't come without trade-offs. Applying squeezed state of light takes some critical considerations, from the difficulties regarding generating these states [19], to designing and fabricating compatible Photonic Integrated Circuits (PICs). In this part, we will review three of these complications and efforts made to overcome them:

A. Suggested methods to generate squeezed state of light

1) Suppressing pump fluctuation in semiconductor laser:

This comprehensive method takes into account both the thermal fluctuations in the flow of minority carriers and the quantum-mechanical generation-recombination noise they produce [20].

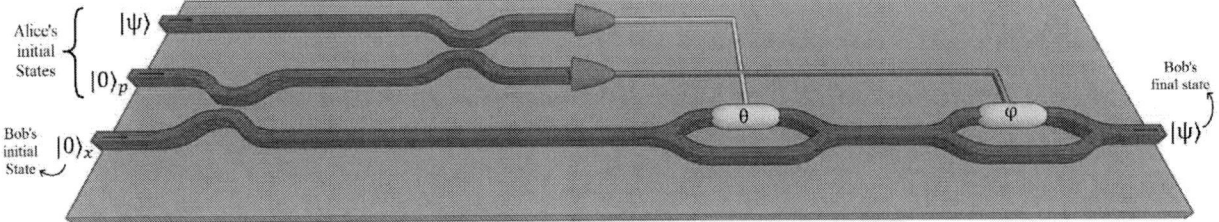

Fig. 4. A 3D model of a PIC design to realize quantum teleportation on a continuous quantum variable circuit. Input states enter the circuit from the left and at the end (right side) Bob will have Alice's initial state of $|\psi\rangle$. Two blue cones represent homodyne detectors to make measurements of the average value of arbitrary quadrature components. The result of these measurements is used as controls for X and Z Pauli gates by controlling the respective phase shifters.

To better understand this phenomenon, it is crucial to consider the injection-current-pumped semiconductor laser. In this laser, the carrier-injection process is regulated by a similar effect. The rate at which carriers are injected, known as the junction current, is determined by the forward-biased voltage applied across a p-n junction. This voltage opposes the built-in potential within a depletion layer, thereby allowing the carrier-diffusion process to dominate over the reverse-directed drift process.

When the junction current surpasses its average rate, the junction voltage decreases due to the increased voltage across the series resistance, denoted as Rs. Consequently, more minority carriers (in this case, electrons as depicted in Fig. 5) return to the n-type bulk layer due to the prevailing built-in field within the depletion region. Conversely, when the injection current falls below its average rate, the junction voltage increases, enabling more electrons to diffuse into the active layer.

It is worth noting that this modulation effect on the junction voltage, induced by the presence of the series resistance, plays a crucial role in regulating the process of minority carrier injection into the active layer. This understanding of the injection process and its regulation is vital for comprehending the behavior of semiconductor lasers.

The scheme depicted in Fig. 5 has a notable characteristic: it can achieve a high quantum efficiency. In this scheme, a large portion of the electrons injected into the active layer can be converted into coherent photons emitted from the laser cavity. This characteristic helps prevent the dilution of the amplitude squeezing (or sub-Poissonian features) of the emitted photons due to random deletion (partition) noise.

2) Nonlinear optics:

Nonlinear optics is a field of study that explores the interaction between light and materials with nonlinear properties. One interesting application of nonlinear optics is the generation of squeezed states of light.

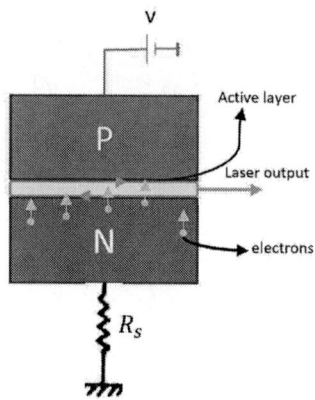

Fig. 5. A semiconductor laser pumped by a high-impedance source.

While suppressing pump fluctuation in semiconductor lasers is one way of squeezing light, another method, that is more common, typically involves passing a laser beam through a nonlinear crystal or medium. The nonlinear properties of the material cause the light to undergo a nonlinear interaction, resulting in the generation of squeezed states. This method has shown more promising results in the experiments than the former method (Fig. 6).

There are different techniques used in nonlinear optics to generate squeezed states, such as parametric down-conversion and four-wave mixing. These techniques involve manipulating the properties of the input laser beam, such as its intensity or frequency, to induce the desired nonlinear interaction and generate the squeezed state.

3) Integrated optical ring resonator:

Aside from the many different purposes that ring resonators can serve, several materials utilized in the fabrication of ring resonator circuits exhibit nonlinear responses to intense light. This nonlinearity enables frequency modulation processes like four-wave mixing and

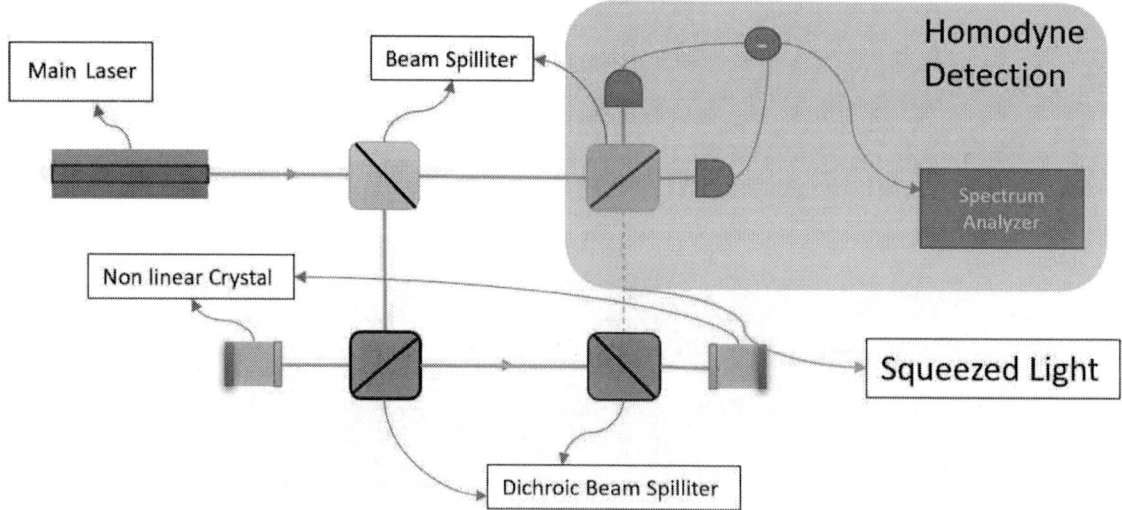

Fig. 6. A schematic drawing and analysis in this study to re-introduce a method for generating squeezed light. Infrared laser light is doubled in frequency to green laser light (SHG) and is then turned into squeezed infrared laser light using an optical parametric amplifier (OPA). The properties of the squeezed light experiment have already been measured with a homodyne detector [21].

spontaneous parametric down-conversion, which generate pairs of photons [22]. This can be further used for squeezing light beams in photonic integrated circuits which will offer tremendous advantages for many different fields ranging from telecommunication to classic and quantum computation.

B. Proper integration for squeezed light:

Optimizing PICs for the use of squeezed state of light is another challenge, for most PICs are designed and dedicated to coherent light or single photons in order to operate. One of the main barriers in this way is the integration of squeezed light source with other parts of the PIC. However, there are some efforts to overcome such problems, one of which is an experimental chip design by Xanadu [23]. This device is considered to be a programmable nanophotonic chip that can perform quantum circuitry tasks by utilizing squeezed beam of light.

According to Fig. 7(a), the central component of Xanadu's programmable device is a photonic chip with dimensions of 10 mm × 4 mm. This chip has the capability to generate squeezed light in up to eight optical modes. For the chip to achieve this, it is initialized into four distinct two-mode squeezed vacuum states, which are independent of each other. The squeezing occurs between specific bichromatic mode pairs, and each pair occupies one of the four spatially separated waveguide modes. To manipulate these spatial modes effectively, we employ an interferometer that employs a network of beam splitters and phase shifters. This interferometer allows for the implementation of a user-programmable gate sequence, which corresponds to an SU(4) transformation (where SU(n) represents the particular unitary group of degree n) applied to the spatial modes. By utilizing this setup, our photonic chip synthesizes an eight-mode Gaussian state. Subsequently, this state is measured on the Fock basis using eight independent photon-number-resolving detectors. For a visual representation of the machine's functioning, refer to the equivalent quantum circuit diagram in Fig. 7(b).

C. Photon loss:

Photon loss is a significant phenomenon in quantum optics that can have various effects on quantum systems. When photons are lost or absorbed by the environment, it can lead to a degradation of the quantum state, such as

Decoherence and State Dephasing, Reduced Fidelity, etc., but this protocol exhibits robustness [24], effectively functioning even in the face of photon loss, non-ideal sources, and imperfect detection. Scaling the protocol to accommodate large numbers of photons provides compelling evidence of its resilience, underscoring its suitability for practical applications. This scalability represents a significantly simpler task compared to constructing a universal quantum computer [25].

V. SIMULATION RESULTS

To effectively simulate the squeezed states and the evolution of their quantum states through the proposed algorithm, we used Strawberry Fields Library in Python [16]. This library enables the simulation of such photonic apparatus for quantum applications.

Let's assume we want to teleport a coherent function as depicted in Fig. 8. The entangled states of Bob and Alice should be a combination of states as depicted in Fig. 9. They are considered entangled to each other as a result of the beam splitter transformation being applied to the initial states. by doing the third and fourth steps on the states, Bob will achieve the same state as the Fig. 8.

VI. CONCLUSION

Squeezed states of light have various applications in quantum information processing, quantum communication, and precision measurements. They can be used to enhance the sensitivity of certain measurements, improve the performance of quantum communication protocols, and enable the implementation of quantum information processing tasks.

We demonstrated the possible teleportation of quantum states for the purpose of realization of quantum key distribution (QKD). However, the combination of squeezed light and integrated photonics promises new horizons for different types of quantum technology [26]. With the help of ever-developing simulation software and libraries, we hope to be able to better stimulate the integration features of the analyzed structure (for example, investigation of balanced homodyne detectors in this circuit or the photon loss of the structure) or even more complex circuits for future quantum applications.

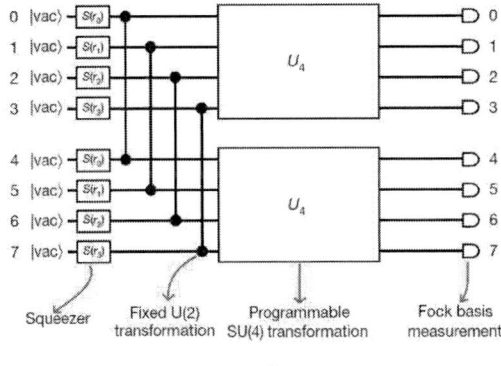

a

b

Fig. 7. (a) A reintroduction of the chip, derived from a micro-graph of the physical device, is depicted. (b)The quantum circuit diagram provided offers an equivalent representation of the photonic hardware, highlighting its operational characteristics [23].

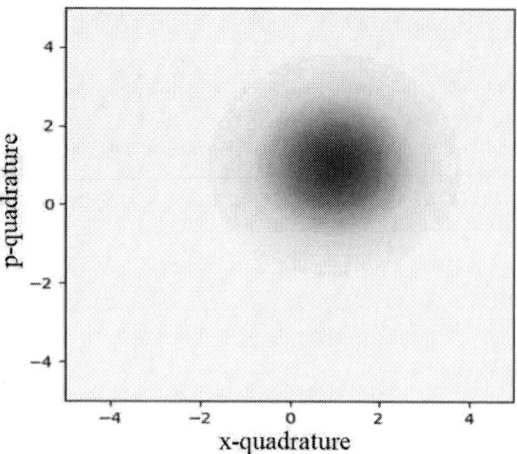

Fig. 8. Simulation results of quadrature diagram of the coherent state, which will be teleported by Alice to Bob.

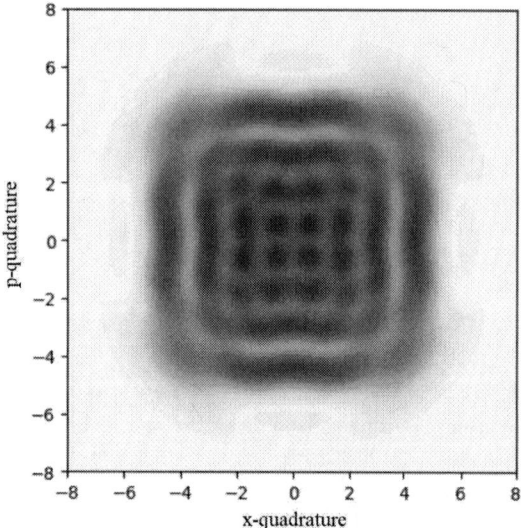

Fig. 9. Simulation results of quadrature diagram of Bob and Alice entangled states after going through beamsplitters.

REFERENCES

[1] Walls, Daniel F, "Squeezed states of light", Nature, volume 306, P.141–146, 1983, Nature Publishing Group UK London.

[2] E. Pelucchi *et al.*, "The potential and global outlook of integrated photonics for quantum technologies," *Nature Reviews Physics*, vol. 4, no. 3, pp. 194–208, Dec. 2021, doi: 10.1038/s42254-021-00398-z.

[3] D. Stucki *et al.*, "Long-term performance of the SwissQuantum quantum key distribution network in a field environment," *New Journal of Physics*, vol. 13, no. 12, p. 123001, Dec. 2011, doi: 10.1088/1367-2630/13/12/123001.

[4] M. Sasaki *et al.*, "Field test of quantum key distribution in the Tokyo QKD Network," *Optics Express*, vol. 19, no. 11, p. 10387, May 2011, doi: 10.1364/oe.19.010387.

[5] R. J. Hughes, J. E. Nordholt, K. McCabe, R. Newell, C. G. Peterson, and R. D. Somma, "Network-Centric Quantum Communications with Application to Critical Infrastructure Protection," *arXiv (Cornell University)*, May 2013, doi: 10.48550/arxiv.1305.0305.

[6] C. S. Hamilton, R. Kruse, L. Sansoni, S. Barkhofen, C. Silberhorn, and I. Jex, "Gaussian Boson Sampling," *Physical Review Letters*, vol. 119, no. 17, Oct. 2017, doi: 10.1103/physrevlett.119.170501.

[7] H.-S. Zhong *et al.*, "Quantum computational advantage using photons," *Science*, vol. 370, no. 6523, pp. 1460–1463, Dec. 2020, doi: 10.1126/science.abe8770.

[8] H.-S. Zhong *et al.*, "Phase-Programmable gaussian boson sampling using stimulated squeezed light," *Physical Review Letters*, vol. 127, no. 18, Oct. 2021, doi: 10.1103/physrevlett.127.180502.

[9] R. Schnabel, "Squeezed states of light and their applications in laser interferometers," *Physics Reports*, vol. 684, pp. 1–51, Apr. 2017, doi: 10.1016/j.physrep.2017.04.001.

[10] H. Vahlbruch *et al.*, "Observation of Squeezed Light with 10-dB Quantum-Noise Reduction," *Physical Review Letters*, vol. 100, no. 3, Jan. 2008, doi: 10.1103/physrevlett.100.033602.

[11] H. Vahlbruch, M. Mehmet, K. Danzmann, and R. Schnabel, "Detection of 15 dB Squeezed States of Light and their Application for the Absolute Calibration of Photoelectric Quantum Efficiency," *Physical Review Letters*, vol. 117, no. 11, Sep. 2016, doi: 10.1103/physrevlett.117.110801.

[12] S. Barzanjeh, S. Guha, C. Weedbrook, D. Vitali, J. H. Shapiro, and S. Pirandola, "Microwave quantum illumination," *Physical Review Letters*, vol. 114, no. 8, Feb. 2015, doi: 10.1103/physrevlett.114.080503.

[13] H. Wang *et al.*, "Observation of Intensity Squeezing in Resonance Fluorescence from a Solid-State Device," *Physical Review Letters*, vol. 125, no. 15, Oct. 2020, doi: 10.1103/physrevlett.125.153601.

[14] H. Kaatuzian, Photonics, vol.1, 6th printing, Amirkabir University Press, 2020.

[15] B. E. A. Saleh, *Fundamentals of Photonics*, 2 ed. John Wiley & Sons, Inc., 2007.

[16] "Basic tutorial: quantum teleportation — Strawberry Fields." https://strawberryfields.ai/photonics/demos/run_teleportation.html

[17] M. A. Nielsen and I. L. Chuang, *Quantum Computation and Quantum Information*. Cambridge University Press, 2010.

[18] M. Motaharifar, H. Kaatuzian, "Mach-Zehnder Interferometer Cell for Realization of Quantum Computer; A Feasibility Study," accepted and presented at *31st International Conference on Electrical Engineering (ICEE 2023)* Tehran, Iran.

[19] M. Hasani, H. Kaatuzian, M. Motaharifar, "Monte Carlo simulation for the generation of squeezed state of light," accepted in *Frontiers in Optics + Laser Science 2023 (FIO, LS)*, Tacoma, Washington, United States of America.

[20] Y. Yamamoto, S. Machida, and O. Nilsson, "Amplitude squeezing in a pump-noise-suppressed laser oscillator," *Physical Review*, vol. 34, no. 5, pp. 4025–4042, Nov. 1986, doi: 10.1103/physreva.34.4025.

[21] H. Vahlbruch, M. Mehmet, K. Danzmann, and R. Schnabel, "Detection of 15 dB Squeezed States of Light and their Application for the Absolute Calibration of Photoelectric Quantum Efficiency," *Physical Review Letters*, vol. 117, no. 11, Sep. 2016, doi: 10.1103/physrevlett.117.110801.

[22] E. Engin *et al.*, "Photon pair generation in a silicon micro-ring resonator with reverse bias enhancement," *Optics Express*, vol. 21, no. 23, p. 27826, Nov. 2013, doi: 10.1364/oe.21.027826.

[23] J. M. Arrazola *et al.*, "Quantum circuits with many photons on a programmable nanophotonic chip," *Nature*, vol. 591, no. 7848, pp. 54–60, Mar. 2021, doi: 10.1038/s41586-021-03202-1.

[24] M. A. Broome *et al.*, "Photonic boson sampling in a tunable circuit," *Science*, vol. 339, no. 6121, pp. 794–798, Dec. 2012, doi: 10.1126/science.1231440.

[25] A. Molina and J. Watrous, "Revisiting the simulation of quantum Turing machines by quantum circuits," *Proceedings of the Royal Society A: Mathematical, Physical and Engineering Sciences*, vol. 475, no. 2226, p. 20180767, Jun. 2019, doi: 10.1098/rspa.2018.0767.

[26] H. Kaatuzian, "Quantum Supremacy Versus IoT Conspiracy in Smart Cities," *IEEE Xplore*, Sep. 01, 2020. https://ieeexplore.ieee.org/document/9250203 (accessed Aug. 04, 2023).

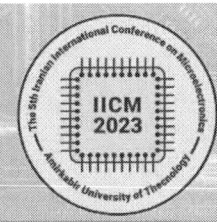

The Fifth Iranian International Conference on Microelectronics

iicm2023.aut.ac.ir

Amirkabir University of Technology (Tehran Polytechnic)

KEYNOTE SPEAKERS

Title: Neural and Tissue Electrical Stimulation and Acquisition Electronics

Prof. Omid Shoaei
Professor of Electrical and Computer Engineering,
University of Tehran, Iran

Abstract:

Implantable neural electrical stimulators can be used to treat a variety of neurological disorders and/or restore some body functions such as DBS (Deep Brain Stimulation) for Parkinson disease, SCS (Spinal Cord Stimulation) for chronic pain, Cochlear Implants for stimulating cochlear nerves for inner hearing loss, Epiretinal prosthesis for treating retinal degenerative diseases, etc. Also, in cardiac pacemakers and ICD's an electrical impulse can depolarize cardiac tissue near the pacing electrode, which then propagates through the heart to restore a normal cardiac rhythm. Neural signal recording and bio-impedance spectroscopy are essential for efficient closed-loop treatment. In the meantime, concurrent stimulation and recording remain a challenge for high dynamic range AFE requirement as well as the overall system power constraints. The power efficiency and safety of the electrical stimulators are uncompromisable. Also, the characteristics of the stimulator such as the voltage compliance, and current/voltage resolution are among the design challenges defining the ASIC technology node. Some charge balancer circuits and systems particularly for multipolar stimulators to ensure the electrical stimulation safe for both the tissue and the electrode are presented. Multi-channel concurrent stimulation and generating different waveforms to increase the efficacy are also discussed. Different types of neural stimulation circuitries are introduced, and as an example, an energy efficient multichannel adiabatic switching-based stimulator with high driving current capability (up to 10 mA) is presented in more details. Also, some techniques for Neural signal recording and bio-impedance spectroscopy are presented.

Biography:

Omid Shoaei (M'96) received the B.Sc. and M.Sc. degrees from the University of Tehran, Iran, in 1986 and 1989, respectively, and the Ph.D. degree from Carleton University, Ottawa, ON, Canada, in 1996, all in electrical engineering. In 1995, he was with Philsar Electronics, Inc., Ottawa, working on the design of a bandpass delta-sigma data converter. From December 1995 to February 2000, he was a Member of Technical Staff with Bell Labs, Lucent Technologies, Allentown, PA, USA, where he was involved in the design of mixed analog/digital integrated circuits for LAN and Fast Ethernet systems. From February 2000 to March 2003, he was with the Design Center, Valence Semiconductor, Inc., Dubai, United Arab Emirates, as the Director of the Mixed-Signal Group, where he has been working on pipelined and delta-sigma analog-to-digital converters. From January 2008 to February 2012, he was Qualcomm, San Diego, CA, USA, where he was the chip lead and a supervisor of a team of about 20 designers for two codec development projects for smart phone, and tablet OEMs. Since January 2014, he has been the Principal Investigator of the Deep Brain Stimulator (DBS) Project supported by the CSTC. He is currently working on the development of a new IC generation for DBS with the University of Tehran. He has also been an Associate Professor with the Department of Electrical and Computer Engineering, University of Tehran, since 1999. He has received three U.S. patents, and has authored or coauthored more than 180 international and national journal and conference publications. His research interests include biomedical integrated circuits and systems, analog-to-digital converters, precision analog/mixed-signal circuits and systems, and automotive electronics.

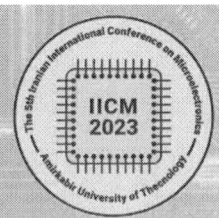

The Fifth Iranian International Conference on Microelectronics

iicm2023.aut.ac.ir

Amirkabir University of Technology
(Tehran Polytechnic)

KEYNOTE SPEAKERS

Title: Neuromorphic Architecture-based Neuromodulation for the Diagnosis and Treatment of Brain Disorders

Prof. Mohamad Sawan, FIEEE, FRSC, FCAE
Chair Professor, School of Engineering, Westlake University,
Hangzhou, China
Emeritus Professor, Polytechnique, University of Montreal,
Canada

Abstract:

Closed-Loop Systems intended for efficient diagnosis and treatment of neurodegenerative diseases are targets to mimic brain regular operation. Consequently, neuromorphic based learning techniques are the heart parts of emerging control units to be embedded in proposed Neuro modulation Systems. This talk covers the implementation of wearable and implantable medical devices based on custom system-on chip (SoC) integrated platforms. The latter are intended for the diagnosis, treatment, and prediction of health conditions. These devices include signal processing methods, design and tests of SoCs and system assembly of bio-electronic closed-loop systems for brain interfaces. These methods deal with multidimensional design challenges such as efficient power management, very low-power and high-data rate wireless communication methods, and reliable systems. In these neuro modulation applications, priority could be given to non-invasive approaches, however for some healthcare dysfunctions, wearable systems can not apply, implantable devices should be used. Also, optoelectronic methods are becoming the winning approaches to build proposed advanced closed-loop systems for both non invasive Nano imaging, and transcranial stimulation. Case studies include several applications such as epilepsy, vision, addictions, and early and fast virus detection.

Biography:

Mohamad Sawan is Chair Professor in Westlake University, Hangzhou, China, and Emeritus Professor in Polytechnique Montreal, Canada. He is founder and director of the Center of Excellence in Biomedical Research on Advanced Integrated onchips Neurotechnologies (CenBRAIN Neurotech) in Westlake University, Hangzhou, China. Also, he is founder of the Polystim Neurotech Laboratory in Polytechnique Montréal. He received the Ph.D. degree from University of Sherbrooke, Canada. Dr. Sawan research activities are bridging micro/nano electronics with biomedical engineering to introduce smart medical devices dedicated to improving the quality of human life. He is cofounder and was Editor-in-Chief of the IEEE Transactions on Biomedical Circuits and Systems (2016-2019). He hosted the 2016 IEEE International Symposium on Circuits and Systems, and the 2020 IEEE International Medicine, Biology and Engineering Conference (EMBC). He was a Canada Research Chair in Smart Medical Devices (2001-2015), and was leading the Microsystems Strategic Alliance of Quebec, Canada (1999-2018). Dr. Sawan published more than 1000 peer reviewed papers and many books and patents. Among the numerous received honors, Dr. Sawan received the Chinese National Friendship Award, The Lebanese's President Medal of Merit, the Shanghai International Collaboration Award, the Queen Elizabeth II Golden Jubilee Medal. Dr. Sawan is Fellow of the Royal Society of Canada, Fellow of the Canadian Academy of Engineering, Fellow of the IEEE, and "Officer" of the National Order of Quebec.

979-8-3503-6020-2/23 $31.00 © 2023 IEEE

The Fifth Iranian International Conference on Microelectronics

iicm2023.aut.ac.ir

Amirkabir University of Technology
(Tehran Polytechnic)

INDEX TO AUTHORS

A

Abbasi, Seyed Peyman
Abrishamifar, Adib
Ahangar Darband, Maryam
Ahangari, Zahra
Ahmadi, Mohammad
Ahmadi, Mohammad Taghi
Ahmadi, Vahid
Aiello, Orazio
Akrami, Ferdos
Alibakhshi, Mahmood
Alijani, Mahdi
Aliparast, Peiman
Armin, Amin
Atarodi, Seyed Mojtaba

B

Babasafari, Maryam
Babazadeh, Farshad
Baleghi, Yasser
Barahimi, Behdad

C

Caviglia, Daniele
Chegini, Marzieh

D

Darbari, Sara
Dehghan, Nima
Dehghan, Pooya
Dortaj, Hannaneh

E

Ebrahim Kafoori, Kian
Ebrahimi, Emad
Ehsanian, Mahdi

F

Fathi, Amir

G

Ghaemmaghami, Mohsen
Ghasemi, Mehrdad

Ghouchani, Arman
Gozalpour, Farshad

H

Hamedi, Samaneh
Hamedvasighi, Fatemeh
Hasani, Mahmood
Hashemi Bani, Soroush
Hassanzadeh, Alireza
Heydari, Hamid Reza
Hodaei, Arash
Hoornam, Soraya
Hosseini Tehrani, Yas
Hosseiny, Adib

J

Jafari Touchaei, Behnam
Jahanirad, Hadi
Javanbakht, Pardis
Javanmardi, Mohammadmahdi

K

Kaatuzian, Hassan
Kamarei, Mahmoud
Kashef, Seyed Sadra
Kashi, Amir
Khodaei, Mehdi
Koolivand, Yarallah

M

Mahmoodpour, Mohammad-Ali
Mashoufi, Behboud
Masoumi, Nasser
Matloub, Samiye
Menbari, Ahmad
MirAlvandi, Reza
Mohajerzadeh, Shams
Mohammadian, Sajjad
Mojarad, Mortaza
Monajati, Mehrnaz
Moradzadeh Rezaei, Yeganeh
Moravvej-Farshi, Mohammad Kazem
Mostafavi, Mohammad
Motaharifar, Mobin

INDEX TO AUTHORS

N

Najafiaghdam, Esmaeil
Namaki, Pouya
Nemati, HojjatAllah
Nemati, Reza
Nematian, Hamed
NezhadAhmadi, Mohammad-Reza
Ng, Sha Shiong
Nikzad, Askandar
Noroolahi, Amir
Norooz Oliaei, Mahdi

P

Paknazar, Pegah
Pakravan, Elaheh
Pazira, Milad

R

Rahimi, Hemin
Rahimpour, Hamid
Rahmani, Adibeh
Rashidi Kia, Shiva
Reyhani, Shahbaz
Rezaei, Ali

S

Saadatmand, Seyedeh Bita
Salighe, Erfan
Sanaee, Zeinab
Shafiei, Alireza
Shakiba, Maryam
Shalchian, Majid
Sharifkhani, Mohammad
Shaveisi, Maryam
Shoaei, Omid
Shokouhi, Samad
Shokri, Reza
Soleimanpour, Farzaneh

T

Taghizadeh Afshari, Golsa

Y

Yargholi, Mostafa
Yavari, Mohammad

Z

Ziraksaz, Fazel

The Fifth Iranian International Conference on Microelectronics

iicm2023.aut.ac.ir

Amirkabir University of Technology
(Tehran Polytechnic)

SCIENTIFIC COMMITTEE

Name	Affiliation
Abdolali Abdipour	Amirkabir University of Technology
Mohammad Abdolahad	University of Tehran
Mohammad Mahdi Ahmadi	Amirkabir University of Technology
Vahid Ahmadi	Tarbiat Modares University
Peiman Aliparast	University of Pavia
Amirreza Alizadeh	University of California Santa Barbara
Gholamreza Ardeshir	Babol Noshirvani University of Technology
Mohammad Reza Ashraf	Shahrood University of Technology
Rasoul Dehghani	Isfahan University of Technology
Mehdi Ehsanian	K. N. Toosi University of Technology
Mehdi Fardmanesh	Sharif University of Technology
Ebrahim Farshidi	Shahid Chamran University
Javad Forounchi	Tabriz University
Asghar Gholami	Isfahan University of Technology
Mohammad Gholami	University of Mazandaran
Mohsen Hayati	Razi University
Seyed Ebrahim Hosseini	Ferdowsi University of Mashhad
Amir Jahanshahi	Amirkabir University of Technology
Saeideh Kabirpour	Yazd University
Hossein Kassiri	York University
Hassan Kaatouzian	Amirkabir University of Technology
Fabian Khateb	Brno University of Technology- Czech Republic
Yaralah Koolivand	K. N. Toosi University of Technology
Mohammad Hossein Maghami	shahir rajaee Teacher Training University
Hamed Majedi	University of Waterloo
Negin Manavizadeh	K. N. Toosi University of Technology
Samieh Matloub	University of Tabriz
Masoud Meghdadi	Shahid Beheshti University
Hossein Miar Naimi	Babol Noshirvani University of Technology
Ali Mirvakili	Yazd University
Mohammad Hossein Moaiyeri	Shahid Beheshti University
Kambiz Moez	University of Alberta

979-8-3503-6020-2/23 $31.00 © 2023 IEEE

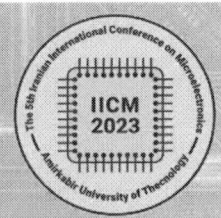

The Fifth Iranian International Conference on Microelectronics

iicm2023.aut.ac.ir

Amirkabir University of Technology (Tehran Polytechnic)

SCIENTIFIC COMMITTEE

Mohsen Moezzi .... Amirkabir University of Technology

Morteza Mojarad .... Urmia University

Mohammad Kazem Moravvej-Farshi .... Tarbiat Modares University

Abdolreza Nabavi .... Tarbiat Modares University

Esmaeil Najafiaghdam .... Sahand University of Technology

Amir Nikpaik .... Tarbiat Modares University

Mehdi Saberi .... Ferdowsi University of Mashhad

Saeed Saeedi .... Tarbiat Modares University

Morteza Saheb Zamani .... Amirkabir University of Technology

Mohsen Saneei .... Shahid Bahonar University of Kerman

Majid Shalchian .... Amirkabir University of Technology

Hossein Shamsi .... K. N. Toosi University of Technology

Mostafa Shaterian .... Shahid Bahonar University of Kerman

Samad Sheikhaei .... University of Tehran

Akram Sheikhi .... Lorestan University

Mohammad Taherzadeh-Sani .... Ferdowsi University of Mashhad

Sirous Toofan .... University of Tabriz

Mostafa Yargholi .... University of Zanjan

Javad Yavand Hasani .... Iran University of Science and Technology

Mohammad Yavari .... Amirkabir University of Technology

Mohammad Hasan Yavari .... Shahed University

Ashkan Zandi .... Georgia Institute of Technology

Mohammad Hossein Zarifi .... The University of British Columbia

979-8-3503-6020-2/23 $31.00 © 2023 IEEE   238

The Fifth Iranian International Conference on Microelectronics

iicm2023.aut.ac.ir

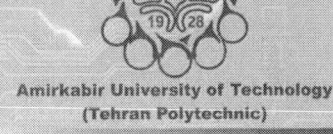

Amirkabir University of Technology
(Tehran Polytechnic)

ORGANIZING COMMITTEE

Prof. Ahad Tavakoli .... General Chair

Prof. Mohammad Yavari .... Chair

Prof. Hassan Kaatuzian .... Technical Program Chair

Prof. Majid Shalchian .... Industrial Relations

Prof. Samad Sheikhaei .... International Relations

Prof. Mohsen Moezzi .... Workshop and Tutorials

Prof. Amir Jahanshahi .... Information Technology

Prof. Sirous Toofan .... Communications and Information

Mr. Abbas Khalili .... Finance Chair

Mr. Amir Kashi .... Publications Chair

Mr. Farshad Gozalpour .... Executive Secretariat

Mr. Ali Sajadi .... Executive Secretariat

Mr. Rasoul Fathipour .... Executive Secretariat

Mr. Soroush Hashemi Bani .... Executive Secretariat

Ms. Mahboubeh Liaghatirad .... Executive Secretariat

979-8-3503-6020-2/23 $31.00 © 2023 IEEE

IEEE
445 Hoes Lane
Piscataway, NJ 08854-4141

ISBN 979-8-3503-6020-2